S0-ATH-549

Studies and reports in hydrology 19

Recent titles in this series

15. Mathematical models in hydrology: Proceedings of the Warsaw Symposium, July 1971 / Les modèles mathématiques en hydrologie : Actes du colloque de Varsovie, juillet 1971. Vol. 1-3. *Co-edition IAHS-Unesco-WMO / Coédition AISH-Unesco-OMM.*
16. Design of water resources projects with inadequate data: Proceedings of the Madrid Symposium, June 1973 / Élaboration des projets d'utilisation des resources en eau sans données suffisantes : Actes du colloque de Madrid, juin 1973. Vol. 1-3. *Co-edition Unesco-WMO-IAHS / Coédition Unesco-OMM-AISH.*
17. Methods for water balance computations. An international guide for research and practice. *Published by the Unesco Press.*
18. Hydrological effects of urbanization. Report of the Sub-group on the Effects of Urbanization on the Hydrological Environment. *Published by the Unesco Press.*
19. Hydrology of marsh-ridden areas. Proceedings of the Minsk Symposium, June 1972.

For details of the complete series please see the list printed at the end of this work.

Hydrology of marsh-ridden areas

Proceedings of the Minsk Symposium
June 1972

A contribution to the International Hydrological Decade

Avec résumés en français

The Unesco Press
IAHS

Paris 1975

Published jointly by
The Unesco Press
Place de Fontenoy, 75700 Paris
and
the International Association of Hydrological Sciences (President: J.-A. Rodier),
19, rue Eugène-Carrière, 75018 Paris

Printed by Offset Aubin, Poitiers

ISBN 92-3-101264-9

Printed in France

PREFACE

This work is the nineteenth volume to appear in the "Studies and Reports in Hydrology" series, publication of which was begun by Unesco, along with that of the series "Technical Papers in Hydrology", when the International Hydrological Decade was launched.

The International Hydrological Decade, which has just come to an end, in 1974, was launched in 1965 by the General Conference of Unesco at its thirteenth session. Its purpose was to advance knowledge of scientific hydrology by promoting international co-operation and by training specialists and technicians. At a time when the demand for water is constantly increasing as a result of the rise in population and of developments in industry and agriculture, all countries are endeavouring to make a more accurate assessment of their water resources and to use them more rationally. The IHD has been a valuable means to this end.

In 1974 National Committees for the Decade had been formed in one hundred and eleven of Unesco's one hundred and thirty-five Member States to carry out national activities and contribute to regional and international activities within the programme of the Decade. The implementation of this programme was supervised by a Co-ordinating Council, composed of thirty Member States selected by the General Conference of Unesco, which studied proposals concerning the programme, recommended the adoption of projects of interest to all or a large number of countries, assisted in the development of national and regional projects and co-ordinated international co-operation. The promotion of collaboration in developing hydrological research techniques, diffusing hydrological data and organizing hydrological networks was a major feature of the programme of the IHD, which encompassed all aspects of hydrological studies and research. Hydrological investigations were encouraged at the national, regional and international level, to strengthen and improve the use of natural resources in view of both local and global needs. The programme enabled countries well advanced in hydrological research to exchange information and developing countries to benefit from such exchanges in order to elaborate their own research projects and plan their own hydrological networks, taking advantage of the most recent development in scientific hydrology.

The purpose of the series "Studies and Reports in Hydrology" is to set forth the data collected and the main results of hydrological studies undertaken within the framework of the Decade, as well as to provide information on the hydrological research techniques used. The proceedings of symposia dealing with this subject are also included. It is hoped that these volumes will furnish material of both practical and theoretical interest to hydrologists and governments and meet the needs of technicians and scientists concerned with problems of water in all countries.

But a great deal remains to be done in this field, in particular with regard to the study of the water cycle and the scientific methods used in the world to assess water resources, and also the evaluation of the influence of man's activities on the water cycle in relation to environmental problems as a whole.

For this reason the General Conference of Unesco has decided to launch, in 1975, a long-term intergovernmental programme, to be known as the International Hydrological Programme, which will be focused on the scientific and educational aspects of hydrology and in particular on the study of the problems mentioned above.

Unesco will therefore continue to publish "Studies and Reports in Hydrology", which will remain an important means of collecting and disseminating the accumulated experience of hydrologists throughout the world.

C O N T E N T S

INTRODUCTION

An International Symposium on the Hydrology of Marsh-Ridden Areas was held in Minsk, USSR from June 17 to 21, 1972. The Symposium was organised by the State of the Byelorussian Soviet Socialist Republic and sponsored by UNESCO and IAHS under the IHD Programme.

The aim of the Symposium was the scientific exchange between experts from different countries working on the hydrology of marsh-ridden areas of the temperate zone and the further development of international scientific and engineering co-operation.

Scientists and experts of the Byelorussian Soviet Socialist Republic, the Federal Republic of Germany, Finland, the German Democratic Republic, Hungary, Ireland, the Netherlands, Norway, Poland, Romania, the Ukrainian Soviet Socialist Republic and the United States of America, discussed the following problems of the hydrology of marsh-ridden areas -

Methods and results of water budget calculations (analytical procedures, simulation, etc.), their use in the design of drainage and irrigation systems in the temperate zone.

Modifications of regime in the temperate zone by modern reclamation measures and forecasting the effect upon hydrometeorological conditions of the environment.

Because of reclamation measures and transformation of natural conditions in countries of the temperate zone, the study of the hydrology of marsh-ridden areas has become extremely urgent.

In view of the difference between research work and practical needs it was suggested that the present Symposium be held within the framework of the IHD Programme.

Forty - nine papers were presented at the Symposium, including the opening lecture "Reclamation in the Soviet Union and some aspects of the hydrology of marsh-ridden areas" by B. G. Shtepa, Assistant Minister of Reclamation and Water Economy of the USSR. This volume contains the complete texts of all the papers presented at the Symposium and discussions.

LAND RECLAMATION IN THE SOVIET UNION AND SOME ASPECTS OF HYDROLOGY OF MARSH-RIDDEN AREAS

B. G. Shtepa

Assistant Minister,
Ministry of Reclamation and Water Management of the USSR,
Moscow, USSR.

The hydrology of marsh-ridden areas and land reclamation is directly related to the problems of food supply for the increasing world population. Using achievements in science and technology, improvements in land reclamation, its improvement and increased fertility, can achieve both an increase in the yield and the quantity of crops; this optimistic viewpoint is widely held in the USSR.

Agriculture in the Soviet Union is practised under a wide range of natural conditions. Some areas endure droughts while others suffer with an excess of moisture. But the most effective and lasting means of enhancing the fertility of the soils is efficient reclamation. This brings into agricultural production such lands which are either of low efficiency or which cannot be used at all; reclamation increases the effect of fertilizers, permits the mechanization and intensification of agriculture; stabilizes the agricultural production process, diminishes the negative effect of natural factors and permits increases in the yields of cereals, forage and other crops, irrespective of weather conditions.

Of vital importance is the role of improved lands in the production of fodder for cattle. Since the 24th Congress of the Communist Party of the Soviet Union (CPSU), much attention has been given to the establishment of cultivated pastures on drained marshlands. About 2.5 million ha of irrigated pastures will be put to use in the years 1972-1975.

The total area of irrigated and drained lands in this country is 12 and 10 million ha, respectively; the area of irrigated pastures increased by 33 million ha during the last five year period alone to 186 million ha in 1970.

Reclamation and irrigation projects are being executed in almost every part of the country, but in recent years special attention has been paid to the development of irrigation in the regions of marketable grain production (such as the Volga region, the North Caucasus, and South Ukraine). Drainage of marshes and marsh-ridden lands in the Baltic Republics, Polessie, western parts of the RSFSR and the Ukraine, in the humid subtropical zone of Georgia as well as in the Far East, is well under way.

This, however, is only the beginning of a new stage in the development of land reclamation. Speaking in June 1970 and stressing the importance of land improvement, L. I. Brezhnev said: "The future of our agriculture lies in the extensive development of land reclamation".

Under the present Five-Year Plan 27.8 million roubles have been allocated to land reclamation projects. This is more than twice the amount which was allotted by the 8th Five-Year Plan.

State investments alone allotted under the 1971-1975 Five-Year Plan will meet the cost of commissioning 3.2 million ha of irrigated lands, (or twice as much as during the period 1966-1970) and 5 million ha of drained land which represents an increase of 34%.

Moreover, it is planned to irrigate 41.2 million ha of pasture and to cultivate 9.5 million ha, as compared with 33.4 and 6.3 million ha, respectively, during the 8th Five-Year Plan.

When bringing land reclamation engineering projects into effect paramount importance is attached to scientific and technological progress. Perfected reclamation systems that meet up-to-date technical requirements are being built on a large scale.

Whereas previous practice was to design reclamation works, comprising mainly open drainage or irrigation ditches, now only main canals and large distributors will be open. Approximately 90% of the whole irrigation network and 75% of the drainage network will be equipped with subsurface pipes. These will provide better conditions for the mechanization of agriculture, favour weed control, sharply decrease the evaporation and seepage losses, and substantially reduce labour expenditure involved in such systems.

In the Soviet Union the extension of land reclamation is also planned for a period of 15 years (1971-1985) involving an increase in improved land area of 48-50 million ha. Ambitious plans exist involving the transfer of part of the discharge of northern Siberian rivers towards the Caspian and Aral Seas. Possibilities of land reclamation are enormous in the Soviet Union since marsh-ridden areas cover 212 million ha, of which 68 million ha are true bogs (peat depths of over 30 cm). The share of the Euopean part of the USSR of the total area is about 30 million ha, concentrated mainly in the non-chernozem zone of the country and including 3 million ha in Byelorussia.

Because of the vast area of the USSR and sharp differences in physiographic conditions of the various regions, the hydrological characteristics of marshes are most diverse. In the European part of the USSR the period of excess moisture occurs in the spring, whereas in the Far Eastern region it occurs during the monsoon rains, July-August, and in the marshes of the Kolkhida Depression in the summer.

The depth of frozen soil varies widely: from 30 cm in the west to 80 cm in Siberia, increasing to permafrost in the northeast of the Far East region. Besides marshes which are saturated with fresh water, there are a great number of marshes in both west and east Siberia which show a tendency to salinisation after drainage.

The crops grown on drained lands in the European part of the country, suffer from water deficiency in mid and late summer, whilst in west Siberia the deficiency occurs at the beginning of the vegetation period.

These specific and diverse conditions must be taken into consideration when planning, designing and operating land improvement measures. At present the total area of drained arable lands in the USSR amounts to 10.2 million ha including 3.7 million ha in the Baltic Republics, 3.3 million ha in the BSSR and the Ukr.SSR, and 3.2 million ha in the RSFSR.

Progress in the design of drainage systems has led to a wider use of subsurface drainage pipes, involving the use of new materials, such as plastics (in place of ceramic pipes) and the complete mechanization of installation techniques.

Operational methods have been developed for a combined drainage and irrigation control system. This permits the drainage of excess water during wet periods and subsequent irrigation of the fields in the dry season.

Polders, combined with pump drainage, are numerous in Lithuania, Latvia, Byelorussia and the RSFSR (in Kaliningrad, Kostroma, Moscow and some other regions).

Polessie is an example of large-scale land reclamation works in this country. It is a vast flat plain of 132.000 km^2, wherein marshes occupy 4.54 million ha, or 35% of the total area. It is planned to drain 1.1 million ha of the marshes in Byelorussian Polessie and 4 million ha in Polessie as a whole in the period 1971-1975.

Drainage schemes give high economic returns. After drainage, the marsh-ridden lands in Byelorussian Polessie, in the western regions of the Ukraine and in the Baltic Republics, produce acceptable crop yields: cereals from 3.5 to 4.5 metric tons per ha; perennial grasses from 7 to 10 metric tons per ha; cabbage from 50 to 100 metric tons per ha; potatoes from 30 to 40 metric tons per ha.

Drainage reclamation programmes of marsh-ridden lands have a favourable effect on the economy of collective and state farms.

In the USSR Soviet Lithuania is the region in which land reclamation is progressing fastest. The area of lands is over 1 million ha, or 37% of the total arable land, which represents 22% of the potential area. In the last five years almost half a million ha were cleared of stones and stumps, and as many ha were limed. During the same five year period the crop yields of cereals, grown on drained fields, increased twofold.

Large-scale land reclamation makes high demands on science. The most urgent drainage problems now being investigated are the study of the fundamental laws of movement of moisture, nutrient materials and heat transfer in the drained soils. These may control biological processes and can therefore be used in agricultural programming.

The solutions to these problems depend upon the understanding, by experimentation, of the soil moisture balance in the unsaturated zone as well as on the movement of surface water and the fluctuation of groundwater levels. The variation of soil temperature and salt content of the land before and after reclamation must also be studied. The application of fundamental theoretical sciences, above all physics and mathematics is also necessary.

The further development of reclamtion methods and the construction of controlled drainage and irrigation networks require that the optimum conditions for crop development be defined. It is also necessary to prepare elaborate cropping patterns for the optimum use of drained areas.

Protection of the environment, whilst at the same time making efficient use of natural resources, is one of the most important problems of present times. The intervention of human activity upon the natural equilibrium must not have undesirable effects. We must hand down to future generations a better planned and a more beautiful land, and, at the same time, we must preserve the natural environment: even marshes have a charm of their own.

The major subjects which require to be investigated are the influence of drainage on water resources, streamflow and groundwater reservoirs, etc., on the climatic conditions, natural landscape, on the flora and fauna of the marshes and of the adjoining areas. In recent years numerous papers have been published in which the undesirable consequences of marsh reclamation are reported. In particular attention is drawn to the shallowing of lakes and rivers, drying up of production wells, decrease in productivity of hunting and fishing areas, depletion of peat bogs, aesthetic damage to landscape, loss of ecological stability, etc.

All this should be taken into consideration when planning land improvement programmes and constructing reclamation systems for each specific region. Byelorussian Polessie, the Luban Depression in Latvia, the Meshchera lowland in the centre of the RSFSR are examples of comprehensive solutions of the problem. Of great significance in the solution of these problems is the role of hydrologists who should clearly work in close co-operation with specialists in land reclamation.

The hydrology of marsh-ridden areas and land reclamation hydrology are specialized fields, each having specific features of their own.

First of all, it should be noted that marsh reclamation is a complex problem. Besides the requirements of agriculture, the demands of industrial and urban development, forestry, pisciculture and recreation should be considered.

Improved land and marshes are distributed in various climatic zones which makes it necessary to investigate these conditions in depth, starting with the source of the water supply. The hydrological regime of marshes is dynamic and depends upon changes in weather conditions: the main hydrophysical properties of marshy soils and their hydrological parameters are stochastic.

The solution of the above mentioned problems will make it possible to forecast more accurately the effect of land reclamation upon the water regime and also to improve upon the accuracy of drainage system designs and their reliability under given probability criteria.

All this work requires long-term hydrological observations. The Soviet Union has achieved much in this field and fairly complete hydrological data is produced from networks which include 6,500 permanent observation stations and the same number of special observation stations.

In the last 10 or 15 years hydrological investigations of the water and temperature regime of marsh-ridden areas have been regularly carried out at specialized stations set up within the framework of the Hydrometeorological Service. Since 1965 the State Institute for Hydrology has been proceeding with the field exploration of marshes and marsh-ridden areas in the regions of west Siberia. At present extensive work is being done with a view to the automation of observations and processing of data. The collected data is processed regularly and published in papers and year books on hydrology.

Soviet experimental hydrology has been developing successfully. Attention should be drawn in particular to the achievements of the Valdai Hydrological Laboratory; the Laboratory is well-known both in this country and abroad, for its contribution to the advancement of field hydrology. Much research work on the marshes of Siberia and the Far East is being done by the Novosibirsk and Khabarovsk branches of the USSR Academy of Sciences.

Soviet hydrology has followed a long path in its development from random observations and analysis of the simplest components of the hydrological cycle to the conduct of extensive water balance and water management investigations aimed at solving complicated problems of land reclamation. The most important investigations in the field of the hydrology of land reclamation are co-ordinated by the State Committee for Science and Technology of the Council of Ministers of the USSR.

The most important investigations of the hydrology of marshy areas, and of the movement of groundwater are carried out by the USSR Ministry of Geology, which has hydrogeological stations in all the principal zones of the country. This work is supervised by the All-Union State Research Institute of Geology. This Institute has completed the geological and hydrogeological mapping of marshes and the classification of marshland into geological and reclamation regions. The Research Institute for Geochemistry and Geophysics of the Academy of Sciences of the BSSR and the Byelorussian Geological Research Institute have carried out a series of hydrological investigations of marshy areas, estimating potential groundwater availability etc.

Investigations of peat deposits, their hydrological and temperature regimes are being made by the Dokuchaev Soil Institute (Moscow), Moscow State University, the All-Union Research Institute for Peat Industry (Leningrad), and by a number of departments of high schools as well as other scientific research organizations.

The theoretical justification of land reclamation techniques, including the theory of movement of water through the unsaturated zone, has received considerable attention at the specialized laboratory of the Moscow Hydrology and Reclamation Institute under the supervision of Academician S. F. Averyanov.

The various components of the water balance equation, which in turn influence the availability of water resources, are closely linked to meteorological events. A significant amount of data in this sphere is received from weather satellites. This data is widely used for meteorological and hydrological forecasting. Satellites will probably be adopted for land reclamation, water management and hydrological research work in the near future.

Scientific research and design institutions concerned with water resources have elaborated methods of hydrological investigation, prepared engineering designs and defined hydrological standards for water-logged soils, drainage rates and supplementary irrigation of marshes, etc.

A great deal has been done in organising operational hydrometry in typical reclamation systems; hydrometric stations and observation wells have been installed and equipped; observations of soil moisture content are made. As a result it has become possible to specify hydrological parameters more accurately and soil moisture to be controlled in order to meet crop requirements.

Great attention is paid to engineering and economic justification for adoption of selected probability levels of hydrological parameters (water discharge, evaporation, etc.).

Institutes of the Ministry of Reclamation and Water Economy of the USSR are carrying out diverse scientific work in the fields of hydrological research, hydrology of marshes and, in particular, on the influence of drainage on soil moisture. At present, the Institutes of the Ministry of Reclamation and Water Economy includes 60 design and research units, which employ 70 000 people.

The Byelorussian Research Institute of Reclamation and Water Economy has carried out research on drainage and cultivation of peaty soils, on the operation of drainage systems, and on hydrology of reclamation of marshes.

The Byelorussian Research Institute of Reclamation and Water Management has accumulated much experience in drainage and development of peaty soils, evolved methods of flood protection and the application of plastic pipe systems for marsh drainage.

As a result of long-term research (from 1957 to 1971) field data has been obtained on evaporation and flow from marshes as well as the influence of different types of utilization of peaty soils upon these parameters. Determination of the flood regime on drained lands has been studied.

The Central Research Institute on Multi-Purpose Use of Water Resources (Minsk) investigates the technical merit of plans for the comprehensive utilization of water resources, examines the effect upon the environment of water management measures, elaborates methods for the optimum use of water resources, studies the hydrological properties of groundwater reservoirs and any changes in their behaviour due to the influence of land reclamation. This Institute pays great attention to the investigation of water balance problems, which are included in the programme of the International Hydrological Decade in the Soviet Union.

The All-Union Research Institute for Water Engineering and Reclamation elaborates methods and procedures for drainage, determines drainage standards of peaty soils and the volume of supplementary irrigation. Extensive work is done on studying changes of such hydrological and hydrogeological characteristics as vertical permeability, soil moisture capacity and many others; on changes in streamflow and groundwater movement as a result of land reclamation; on developing methods of effective utilization of peaty soils and on the technology of reclamation.

The Lithuanian Research Institute of Water Engineering and Reclamation studies the problems of the drainage of water-logged mineral soils and their subsequent irrigation. The Institute also does research into the pump drainage of lands liable to floods. (polder-type drainage). Investigations have also been made into the water balance in periodically wetted mineral soils, drained with earthenware pipe networks.

The Latvian Research Institute of Water Engineering and Reclamation is the leading scientific research institution for the study of problems concerning the application of plastic and other new materials in reclamation engineering. It has carried out much research into the effectiveness of different types of marsh drainage in the Baltic region.

The Ukrainian Research Institute for Water Engineering and Reclamation has carried out extensive research into the subsurface irrigation of peaty soils and hydrological substantiation of marsh reclamation in the forest and steppe zones.

Regional problems of marsh reclamation, including the problems of the water balance of peat marshes, are being studied by the North Research Institute of Water Engineering and Reclamation, the Far East Research Institute of Water Engineering and Reclamation, the Georgian Research Institute of Water Engineering and Reclamation and by other Institutes.

Soviet specialists in land reclamation and water management take an active part in international scientific and technological collaboration in the spheres of land reclamation, hydrology and hydrogeology. This co-operation is both bilateral and multilateral. In the joint research work of specialists from the USSR and the GDR on the problems of reclamation and marsh hydrology is an example of bilateral co-operation. They have developed methods of simulation of groundwater movement, etc. The joint Soviet-Finnish research group is engaged in urgent problems of the design and putting into practice of reclamation measures. Investigations of reclamation, regulation and utilization of water resources are included in the co-operation programmes, which have been negotiated between the USSR and the United Kingdom, the USSR and Italy and a number of other countries.

Multilateral co-operation is usually sponsored by various international organizations: CMEA, UNESCO, UNECE (United nations) or such international scientific and technological organizations as the International Commission of Irrigation and Drainage, International Association of Scientific Hydrology, International Association of Hydraulic Research, etc.

Thus, for example, the problems of reclamation in the CMEA countries are solved by the Standing Working Group for Irrigation and Reclamation of Agricultural Land.

The Council of Mutal Economic Assistance is involved with improving the methods for best usage of water resources; the development of the fundamentals for highly efficient reclamation systems; elaboration of mathematical simulation and calculation of groundwater regime and drainage systems using computers; perfection of operation methods and automation of drainage systems; application of polymers in reclamation engineering and protection of drains against sedimentation and establishing methods of liming sour soils, etc.

The Committee on Water Problems of the ECE of the United Nations has prepared and is implementing a long-term programme of co-operation on comprehensive utilization of water resources. The USSR is participating in this work.

Urgent problems of reclamation and water economy are also discussed at the ICID Congresses, at which Soviet specialists participate. This organization has held a number of symposia, including sprinkler irrigation and soil erosion as well as on other subjects.

UNESCO is sponsoring a wide programme of work under the International Hydrological Decade, which aims at a global study of the Earth's hydrology. The IHD programme calls for hydrological research in compliance with a pattern common to all countries during the period 1965-1975. It also promotes the organization of conferences of various working groups, international symposia and sessions.

One of the benefits of this type of co-operation is that it helps to reduce the time necessary for solving problems and promotes a free exchange of ideas among scientists of different countries.

The problem of the hydrology of marsh-ridden areas, to which the present symposium is devoted, is an important subject in the IHD programme and it is of great interest to many countries. Many aspects of the hydrological, hydrogeological, climatic and other fields of research of marsh-ridden areas will be discussed in detail at the Symposium. The analysis of modern achievements and an exchange of experience in this field of science will prove helpful, and at the same time the Symposium will stimulate a further development of hydrological science in the world and will be a useful contribution to the implementation of the IHD programme.

TOPIC I

METHODS AND RESULTS OF INVESTIGATIONS OF THE HYDROLOGY, CLIMATE, HYDROGEOLOGY, SOILS AND VEGETATION IN MARSH-RIDDEN AREAS OF THE TEMPERATE ZONE

RESULTS OF LONG TERM INVESTIGATIONS INTO THE WATER BALANCE OF TIDAL AREAS IN THE MARSHY PLAINS OF THE RIVER EIDER IN THE NORTH COASTAL REGION OF FRG

U. Schendel

Institute of Water Economy and Reclamation, Christian-Albrechts University, Kiel, FRG

SUMMARY: A water balance has been drawn up of the 952 km^2 catchment of the tidal river Eider in the coastal region of Schleswig-Holstein bordering the North Sea. Nearly one half of the lower plains of the catchment is covered with grass on peat soils, the rest being cropped land on areas of greater elevation with podzolic soils. The lower plains are slightly above or even below sea-level.

Rainfall was registered at twelve sites.

Discharge was measured by means of digital event recorders in pumping stations in six smaller catchments; all the other components of the water balance equation were measured additionally in two small experimental basins specially selected for this particular purpose within the total catchment. The data of actual evapotranspiration obtained in one of the two small experimental basins were extrapolated to the total catchment.

Under the influence of back-water from the North Sea, and with nearly no slope, the average groundwater level in those areas of the catchment where there are peat soils is about 250 mm below the surface. Mole drainage lowered the average groundwater level to 600 mm below surface without giving rise to any harmful subsidence of the area to date. The area had, however, been drained by ditches a long time ago. The average soil moisture content was also considerably reduced by mole drainage.

Principal water balance components (annually, 1957-1967) are: Rainfall, 853 mm; potential evapotranspiration, 570 mm; actual evapotranspiration, 458 mm (520 mm in the lower peat soil plains and 422 mm on high-level cultivated area of the same catchment; runoff, 428 mm (366 mm in the lower peat soil and 464 mm in the higher situated cropped area). Actual evapotranspiration in the summer period is more than three times that of the winter period. Runoff in the summer amounts to only one half of that in winter.

Due to lower runoff and higher rates of evapotranspiration it is concluded that, from the point of view of the hydrological budget, the water balance situation of cultivated peat soil in the region is more favourable than that of the cropped areas in the higher region of the same catchment. Average discharge per unit area is 14 and peak flow 149 litres per second per km^2.

Results of statistical analysis characterizing the correlations between rainfall and runoff and between runoff and pumping costs are given.

The data obtained by long term measurements are suitable for extrapolation to other regions having similar conditions by means of mathematical models.

RESULTATS DE L'ETUDE PLURIANNUELLE DU BILAN HYDRIQUE DES PLAINES LITTORALES SOUMISES A L'INFLUENCE DE LA MAREE MONTANTE DES REGIONS MARECAGEUSES DE LA RIVIERE EIDER EN REPUBLIQUE FEDERALE D'ALLEMAGNE

RESUME : L'auteur étudie le bilan hydrique du bassin versant (952 km^2) de l'Eider, rivière à marée dans la région littorale du Schleswig-Holstein, en bordure de la mer du Nord. La partie basse du territoire du bassin versant est pour près de la moitié couverte d'herbe sur sol tourbeux ; le reste, zone haute cultivée, est constitué de sols podzoliques. Les basses contrées sont situées au niveau de la mer et même au-dessous de ce niveau.

Les précipitations atmosphériques ont été mesurées en douze points. Le débit a été enregistré à l'aide d'appareils de contrôle installés aux stations de pompage de six petites prises d'eau ; les autres paramètres de l'équation du bilan d'eau ont été mesurés dans deux petits bassins expérimentaux choisis spécialement à cet effet dans les limites du bassin considéré. Les données obtenues dans ces deux bassins pour l'évapotranspiration ont été extrapolées à tout le bassin versant.

Influencé par le remous de la mer du Nord et ayant une pente hydraulique à peu près nulle, le niveau moyen des eaux souterraines de cette région, où prédominent des sols tourbeux absolument non décomposés, se trouve à 250 mm au-dessous de la surface. Le drainage en "taupe" a rabaissé le niveau moyen des eaux souterraines à 60 mm de la surface sans provoquer jusqu'à présent un affaissement dangereux de ce territoire. Cependant, il y a longtemps déjà ce territoire avait été drainé par des rigoles. Le degré moyen d'humidité du sol a été considérablement réduit par le drainage en "taupe".

Les données du bilan hydrique (années 1957-1967) sont les suivantes : précipitations atmosphériques : 853 mm ; évapotranspiration potentielle : 570 mm ; évapotranspiration réelle : 458 mm (520 mm sur les tourbières basses et 422 mm sur les hauts terrains cultivés de ce bassin versant) ; écoulement : 428 mm (366 mm pour les tourbières basses et 464 mm pour les hauts terrains cultivés). L'évapotranspiration réelle en été est trois fois plus forte qu'en hiver. L'écoulement d'été n'est que la moitié de celui d'hiver. Cette réduction de l'écoulement et l'accroissement de l'évapotranspiration amènent à conclure que, du point de vue du budget hydrologique, les conditions où se trouvent les terrains tourbeux en matière de bilan hydrique sont plus favorables que celles des terres cultivées plus haut. Le débit moyen de l'eau est de 14 l/s/km^2 et le débit maximal de la crue est de 149 l/s/km^2.

L'auteur donne ensuite les résultats des calculs sur la relation entre précipitations et écoulement et aussi sur la relation entre l'écoulement et le coût du pompage.

Les résultats ainsi obtenus grâce à des mesures sur de longues périodes peuvent être extrapolés à d'autres régions, dont les conditions sont analogues, au moyen de modèles mathématiques.

Introduction and characteristics of the region

In 1957 discharge measurements were started in six individual catchments of the large tidal and mostly grass-covered plains of the river Eider in Schleswig-Holstein by means of digital event recorders. The area of the six small catchments is 708 km^2, that of the total catchment 952 km^2.

Due to the problem of continuous measurement of runoff in tidal streams with little or no slope, or even backslope, and of determining the groundwater contribution to evapotranspiration in such areas where the groundwater table fluctuates from close to the soil surface to one metre below, very little information on the water balance situation of these areas was available until now.

Large parts of the dike-protected lower plains are at, or even below, sea level. 42% of the total area has peat soils with high

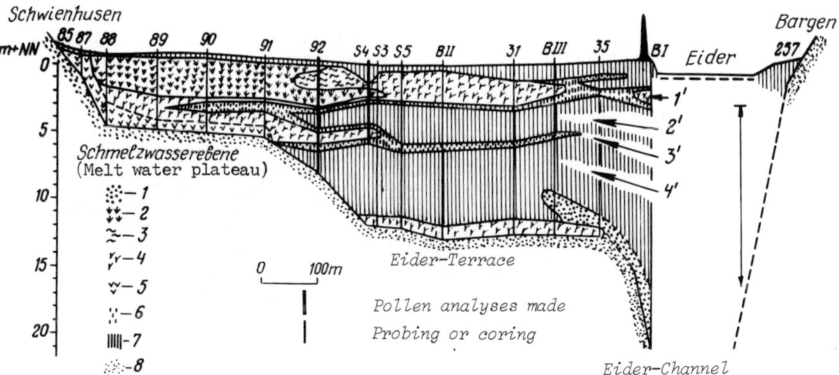

Fig. 1 Geological cross section of the Eider-plain between Schwienhusen and Bargen

1. Pleistocene subsoil
2. Sphagnum moss and calluna peat
3. Brown moss and menyanthes peat
4. Rush, sedge and fern peat
5. Swamp, woodland peat
6. Clayey peat
7. Clay
8. Holozaner sand

groundwater levels, the rest has mineral soils of the podzolic type. The tide cannot flow up the river because a dam has been built across the river bed in the north-west of the catchment. This does not mean, however, that the influence of tidal movement is eliminated. During high flood-tide in the North Sea, the gates of the dam are closed causing the water to pond up and creating high water tables. The meadows covering the peat soil plains are protected by dikes built centuries ago along the river bank. The inadequate drainage and occasional flooding caused by peak tidal flows combined with certain climatic features gave rise to the large peat covered areas in this region. The geological formation of the lower plains in the Eider-catchment is characterised by a base of diluvium and overlain by alternate layers of organic materials of different origins and marine sediments. The process of the successive formation of organic materials and mineral sediments of a marine origin (mostly silty clay) is derived from the transgression - and regression movements of the North-Sea which lasted centuries. Figure 1 shows a typical geological cross-section of this region.

Alternating layers of organic material and marine sediments occur for 500 m on each side of the river, while in the remaining part of the plain there is only peat. The depth of the holocenic horizon is about 10 metres (from 1 metre above, to 11

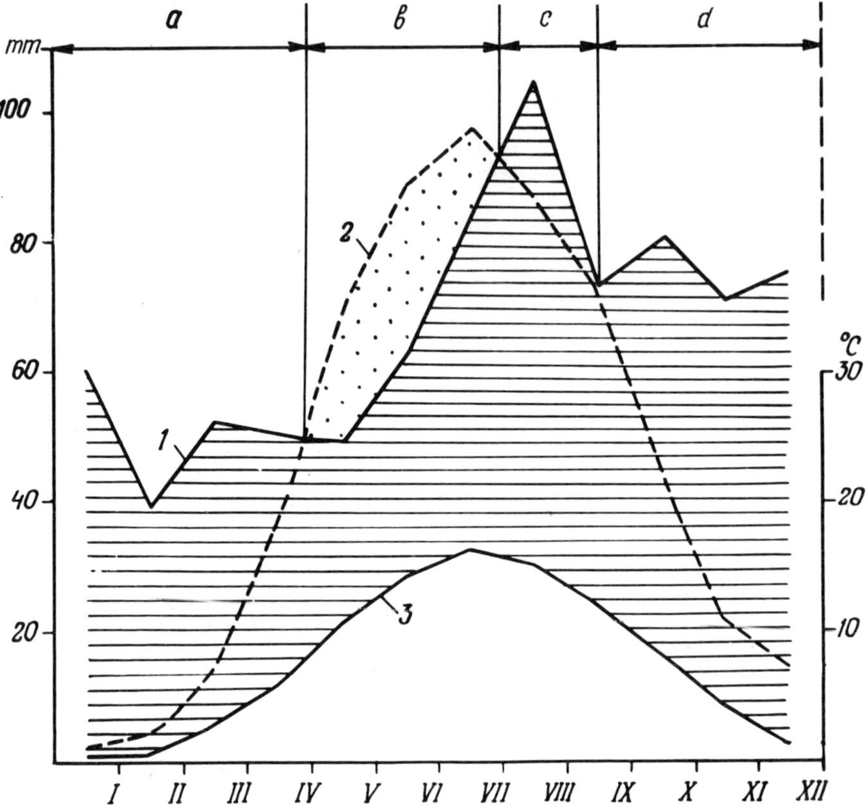

Fig. 2. Climatic diagram of the Eider region (long-term averages): a, period of water surplus; b, period of water recharge; d, period of water surplus. I - rainfall; 2, pot. evaporation; 3, temperature. Rainfall 812 mm, temperature 7.7 °C, relative humidity 84%, potential evaporation 535 mm, water deficit 61 mm (May - July).

metres below sea-level) for a distance of 500 metres from the river, and about 6 metres (from 1 metre above to 5 metres below sea-level) in the remaining part adjacent to the 'Geest' (diluvial formation of the 'Saale-Riss' glacial period).

The peat materials are generally decomposed, but the degree of decomposition varies. The percentage of organic material in the holocenic horizon ranges between 30 and 90%, pH between 3.5 and 5.5 and hydraulic conductivity between 6.10^{-2} cm/s and 1.10^{-5} cm/s.

Due to a low elevation above sea-level (some parts are even below sea-level), there is no natural slope and therefore very little natural drainage. All the excess water has to be removed by pumping. 32 pumping stations with a total installed capacity of 90 cm/s have been built along the river since 1948. The ditches carrying the excess water to the pumping stations are constructed to discharge 120 l/s km^2 which corresponds to the long-term average peak flows in the region. This drainage system in combination with the protection afforded by the dikes on both sides of the river provides the necessary security against flooding when exceptional tides occur in the North Sea.

The Eider-region has a climate characterized by Figure 2. It belongs to the oceanic cool-temperature zone, the principal mean annual characteristics are: rainfall 812 mm temperature 7.7°C, relative humidity 84%, potential evaporation 535 mm. As the diagram shows, the period of water surplus is of greater duration than that of water deficit (the last mentioned being 61 mm in the long-term average for the period May-July).

Measurements

Rainfall: Rainfall was registered at 12 sites in the catchment area (Standard Hellmann raingauge, 1 metre above the ground).

Discharge: Digital-event discharge recorders were installed in pumping stations along the river. The discharge recorders record the difference in level between the main ditch for which water is being pumped, and the river. The recorded difference is automatically integrated to give the total volume of discharged water in cumecs. The corresponding discharge period is registered on a special time scale enclosed in the measuring system. Discharge was registered over a period of 11 years in each of the six individual catchments separately, of which the smallest has an area of 15 km^2 the largest 267 km^2.

Evapotranspiration: Evapotranspiration was measured in a small experimental catchment (6.9 km^2) within the same watershed by means of specially constructed groundwater-lysimeters (Schendel, 1970). It was also determined in the experimental catchment by summating the other components of the water-balance equation. As the experimental catchment does not differ from the overall catchment to any great extent, evapotranspiration values from the relatively small experimental catchment were extrapolated to the whole area.

Groundwater and soil moisture: The groundwater and soil moisture situation was also regularly observed in the above mentioned experimental catchment and in one other experimental catchment. Soil moisture samples were taken once a month, the level of the groundwater table was measured every week.

Results

Rainfall

The average annual rainfall during the period of the investigation (1957-1957) amounted to 853 mm and ranged from 495 mm (1959) to 1030 mm (1960).

Groundwater and soil moisture

The complete data will be published elsewhere (Schendel, *in press*) Only a small section of the data is presented in Figure 3, showing the effect of mole-drainage on soil moisture and depth of groundwater table.

The depth of the mold drain is about 1000 mm at the point of discharge into the ditches. During the period of observation from June 1968 to November 1969 the presence of the mole drains resulted in a lowering of the average groundwater table from 250 mm to 600 mm below ground level. It should be mentioned however that the area had already been drained by ditches a long time ago. Soil moisture measurements carried out from December 1968 to October 1969 confirmed the favourable effect of mole drainage. Although subsidence of this fairly undecomposed peat soil cannot be entirely excluded, no unfavourable effects in this respect have been observed so far. This applied also to the other parts of the area, where mole drainage was carried out ten years ago on a larger scale, and can partly be attributed to the capillary rise of the water from the groundwater table which is about 500 to 800 mm in undecomposed peat soils. From these results it would seem that a lowering of the groundwater-level from 200 to 300 mm below ground level to 500 to 700 mm by mole drainage can be recommended for these areas.

Evapotranspiration

Annual potential evaporation (calculated from air temperature and relative air humidity) amounts to 570 mm (for the period 1957-1967).

Annual actual evapotranspiration in the total catchment area with 42% grass covered peat soil is 458 mm, of which 23% occurs in winter (November-April) and 77% in summer (May-October).

The data on the water balance of the total catchment and of the groundwater-lysimeter measurements (the last mentioned representing the situation in the grass covered peat soil areas) enabled a separate determination to be made of actual evapotranspiration of the higher podzolic areas and of the grass covered peat soil areas. In the peat soil areas, with groundwater levels fluctuating between the soil surface and one metre below, annual evapotranspiration amounts to 520 mm of which 231 mm (44%) is due to water loss from the groundwater table and 289 mm from soil moisture. However, it must be kept in mind that the groundwater contribution to evapotranspiration cannot be separated from the soil moisture component, on account of the capillary rise of water from the groundwater table into the soil moisture zone. The data can therefore only give an indication of the soil moisture and groundwater contributions to evapotranspiration.

Actual evapotranspiration from the higher cropped areas (Geest) with podzolic soils amounts to 422 mm per annum. The higher evapotranspiration (nearly 100 mm) in the peat soil areas is due to the higher groundwater table (between almost the soil surface and one metre below the surface in the peat soil plains as compared to one to two metres below the surface in the higher cropped areas). In addition, the grass cover consumes water over a longer period of time than the crop vegetation which also accounts for the higher

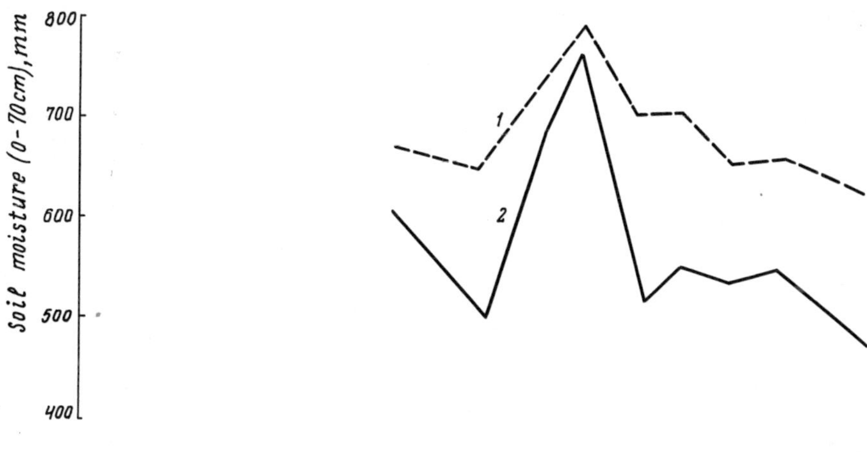

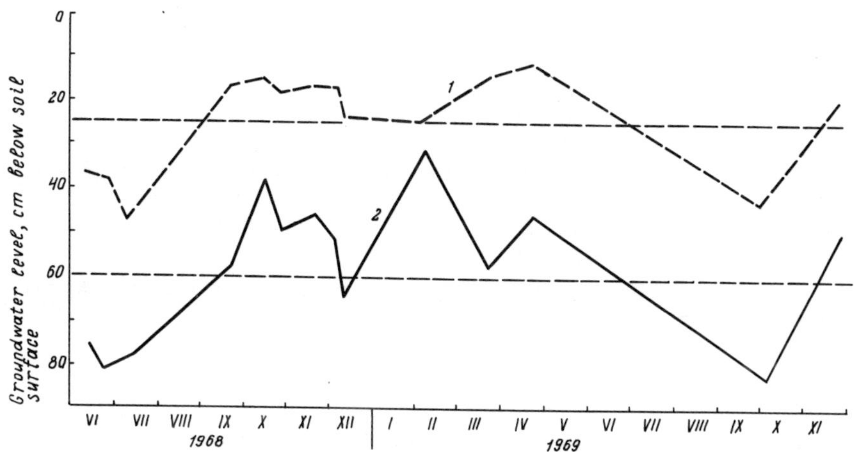

Figure 3. Effect of mole-drainage on the fluctuation of soil moisture (0-70 cm) and groundwater movement in a fairly undecomposed peat-soil of the 'Dellstedter Moor' 1 - without mole-drainage, 2 - mole-drainage.

values of evapotranspiration of the grass cover.

Runoff

Runoff from the total catchment amounts to 428 mm with variations between 213 (driest year) and 680 mm (wettest year). Of the total runoff, 464 mm originated from the higher cropped areas and 366 mm from the lower peat soil plains. This means that the greater amount of evapotranspiration of the grass-covered peat soil areas reduced runoff as compared with the higher cropped areas of the same catchment. During the eleven year period of the investigation, the sum of the runoff and evapotranspiration exceeded rainfall by 33 mm per year. This volume must have been contributed by neighbouring catchment areas.

The runoff-to-rainfall ratio is 0.53 which means that in comparison with other rivers in low lying regions of the moderate-cool climatic zone the tidal river Eider and its soil areas has to be placed in the category of rivers having a moderate to high runoff coefficient.

Specific and total discharges were as follow:-
Long-term annual mean discharge , 14 l/s km^2 (13 cumecs, 469 cusecs), long-term average of annual high flows, 46 l/s km^2 (45 cumecs, 1595 cusecs), highest recorded peak flow, 149 l/s km^2 (142 cumecs, 5013 cusecs). Frequency of high flows, 7% of all high flows occur in the range between 100 and 150 l/s km^2, and involve a danger of inundation, 21% of all high flows occur in the range between 60 and 100 l/s km^2.

Pumping

An analysis of runoff with respect to the running costs of pumping has been made. The annual running costs of pumping per hectare in the peat soil area are approximately 11 Deutsche Mark (at 1972 prices). The pumping costs for discharging 14 l/s km^2, which represents the long-term average discharge and which corresponds to approximately 1 mm per day, are 0.03 Deutsche Mark per ha day or 0.90 Deutsche Mark per ha month, respectively. The linear regression equation characterizing the relation between specific discharge, x, (per unit area of the catchment in litre per sec and km^2) and the monthly pumping costs in DM per ha, y, is of the form:

$y = 0.3909 + 0.0394\ x$, the correlation coefficient being 0.96 and the value of r^2 is 0.92

Stochastic analysis

Regressions of rainfall against runoff and rainfall against actual evapotranspiration have been calculated for winter and summer periods. There are close correlations between rainfall, x, and runoff, y, both for the winter and the summer period.

Winter: $y = -122.70 + 1.14\ x$. Correlation coefficient $r = 0.97$, $r^2 = 0.94$;

Summer: $y = -92.87 + 0.49\,x$. Correlation coefficient $r = 0.90$, $r^2 = 0.81$;

Year: $y = -241.95 + 0.80\,x$. Correlation coefficient $r = 0.88$, $r^2 = 0.78$.

Water balance of the complete catchment

Table 1 summarises the principal results of the water balance in the Eider catchment.

Table 1. Water balance of the Eider catchment (Binneneider) 1957-1967

	Winter period Nov.-April	Summer period May-October	Total or mean
1. Rainfall (mm)	372	481	853
2. Potential evapotranspiration (mm)	114	456	570
3. Actual evapotranspiration (mm)			
a. peat soil area	119	401	520
b. higher situated cropped are	97	325	422
c. total catchment	105	353	458
4. Runoff (mm)			
a. peat soil area	245	121	366
rainfall to runoff ratio	1:0.66	1:0.24	1:0.43
b. higher situated cropped	311	153	464
area rainfall to runoff ratio	1:0.84	1:0.34	1:0.54
c. total catchment	285	143	428
rainfall to runoff ratio	1:0.77	1:0.30	1:0.53
d. discharge per unit area $l/s.km^2$	19	9	14
e. peak flow per unit area $l/s.km^2$	147 (Dec.+Jan.)	149 (Sept.)	

As a result of water-inflow from neighbouring catchment areas, runoff and actual evapotranspiration exceed rainfall by 18 mm during the winter period and by 15 mm during the summer period (total water inflow 33 mm).

Actual evapotranspiration is considerably higher from the grass covered peat soil area than from the higher cropped area, and run-off is reduced (see also the ratio of rainfall and runoff in the peat soil area).

Thus it follows that the water balance of the cultivated peat soil areas in the moist coastal region of the FRG can be considered to be more favourable than that of the higher cropped areas of the same catchment which show a larger runoff. This is in accordance with the findings of Baden and Eggelsmann (1964) and Eggelsmann (1964) in the Königsmoor of North-West FRG which is situated in a similar climatic zone.

Actual evapotranspiration in the summer period is more than three times that of the winter period in the peat soil area as well as in the higher cropped areas, whereas runoff in summer is reduced to almost one half of the winter value.

The highest recorded flow in the Eider catchment is 149 l/s km^2. The drainage ditches in the river plains however only have a capacity of 120 l/s km^2. This is a compromise between the cost of possible damages which may occur and the cost of the drainage system.

EXTRAPOLATION

The data presented here are based on measurements of all the components of the water balance equation in relatively small experimental basins. They are fairly representative of large areas of peat soils in the north and north-west coastal region of the FRG. When extrapolating these data, only slight alterations will have to be made to allow for variations of the meteorological and hydrological parameters, thus simplifying the application of mathematical models to other regions with similar climatic, morphological and physiographical conditions.

References

Baden, W.,&Eggelsmann, R. 1964. Der Wasserkreislauf eines nordwestdeutschen Hochmoores. Schriftenreihe des Kuratoriums für Kulturbauwesen. H.12, Verlag Wasser und Boden, Hamburg.

Baumann, H.,&Schendel, U. 1968. Untersuchungen über Schleswig-Holsteins J. of Hydrology, 6, 373-384.

Eggelsmann, R. 1964. Die Verdunstung der Hochmoore und deren hydrographischer Einfluss, DGM, 8, H.6, 138-147.

Lange W.,& Menke, B. 1967. Beiträge zur frühpos glazialen erd- und vegetationsgeschichtlichen Entwicklung im Eidergebiet, insbesondere zur Flussgeschichte und zur Genese des sogenannten Basistorfes, Meyniana, 17, 29-44.

Petersen, M. 1967. Der Eiderdamm Hundeknöll-Vollerwiek als Folge kunstlicher Eingriffe in den Wasserhaushalt eines Tideflusses. Das Unternehmen Landentwicklung, Programm Nord, Eiderraum, Materialsammlung der Agrarsozialen Gesellschaft, Nr. 62, 158-173.

Schendel, U. 1970. A newly developed groundwater lysimeter for measuring evapotranspiration from different groundwater levels in a small catchment area of the North German coastal region. IASH-

UNESCO-Symposium, Publication No. 96, Wellington (New Zealand), 1970.

Schendel, U. In press. Uber den Wasserhaushalt im norddeutschen Flachland erlautert am Beispiel Schleswig-Holsteins, Teil III: Der Wasserhaushalt in Teilniederschlagsgebieten der tidebeeinflussten Binneneider. Besondere Mitteilungen zum Deutschen Gewasserkundlichen Jahrbuch, Koblenz, Nr. 34 (in press).

Witt, K. H. 1967. Die Wasserwirtschaft im Gebiet des Eiderver bandes. Das Unternehmen Landentwicklung, Programm Nord, Eiderraum, Material-sammlung der Agrarsozialen Gesellschaft, Nr. 62, 158-173.

DISCUSSION

V. N. Litvin (USSR) - 1. Is a constant or variable (as at the surrounding area) water level maintained in the lysimeter?
2. In your paper you have stated that at the area considered, drainage works have been made for a long time over the area of 1000 km^2. How much do hydraulic works cost?

U. Schendel (FRG) - 1. The level in the lysimeter is generally constant and usually depends on the moisture content of the surrounding soil. 2. I cannot give you exact figures, but can tell you that hydraulic works cost very much.

Prof. L. Wartena (The Netherlands) - 1. In your paper some data are given of potential evaporation, in what way is this calculated?
2. Do you believe the actual evaporation depends on the soil moisture content?

U. Schendel (FRG) - 1. Potential evaporation is calculated from the relation of the temperature to the relative humidity.
2. I think no correlation exists for the two values.

STUDY OF WATER BALANCE OF STRATIFIED AQUIFERS BASED ON THE ANALYSIS OF THEIR REGIME DURING HYDROLOGICAL INVESTIGATIONS ON MARSHY LANDS AND ADJACENT AREAS

P. A. Kiselev

All-Union Research Institute of Hydrogeology and Geological Engineering, Moscow, USSR

SUMMARY: This paper presents the main points arising from a study of balance in stratified formations of the Polessie lowland. The water balance equation and recommendations for the location of observations wells in stratified formations with reference to the problems of the water balance study are given. The water balance of the marshy lands of the Upper Yaselda river basin is derived with a view to obtaining the order of magnitude of the recharge by infiltration of confined groundwater on marshy lands of Polessie.

The results obtained may be considered as the justification for observing the groundwater table and its equilibrium in order to forecast the behaviour of the aquifers in connection with land reclamation projects.

ETUDE DU BILAN D'EAU DES NAPPES STRATIFIEES A PARTIR DE L'ANALYSE DE LEUR REGIME AU COURS DE RECHERCHES HYDROGEOLOGIQUES SUR DES TERRES MARECAGEUSES ET DANS LEUR VOISINAGE

RESUME : Cette étude expose les principes qui président à l'étude du bilan hydrique des nappes stratifiées en prenant pour exemple les conditions hydrogéologiques de la plaine de Polésie. L'auteur propose des équations du bilan et formule des recommandations quant à l'implantation des forages d'observations dans les nappes stratifiées à partir des problèmes que pose l'étude du bilan. Le calcul du bilan d'eau dans le périmètre marécageux à l'amont du bassin de Jasiolda est destiné à donner un ordre de grandeur de l'alimentation par infiltration des nappes captives dans les zones marécageuses de Polésie.

Les résultats obtenus peuvent être considérés comme une base scientifique pour l'étude du régime de la nappe phréatique en relation avec la mise en valeur des terres.

A study of the behaviour of the groundwater table in marshy and adjacent lands is of great scientific and practical significance for the solution of different problems associated with land reclamation, water control, predicting the resulting unconfined groundwater trends, water supply, etc.

When studying groundwater resources, stratified aquifers must not be overlooked. The deposition of glacial sediments may be the cause of a stratified structure of aquifers in marshy and adjacent areas.

The marshes of Byelorussia, typical of such formations are described below. Here a stratified structure of the Quaternary period is the principal feature of the aquifer. It is characterised by alternate layers of more or less permeable deposits resulting from the interbedding of moraine, lacustrine, alluvial sandy loams and loams with older alluvial and fluvioglacial sands. The presence of peat bogs overlying sandy and sandy loam-loamy deposits should also be noted.

This type of formation is typical of the greater part of the marshy lands under development in the south of Byelorussia, namely, the Polessie lowland, and covers part of the Ukrainian territory as well. Here the superficial Quaternary deposits have a thickness of about 20-75 m and sometimes more than 100 m. The area under reclamation has four moraines separated by intermorainic formations composed of grey and yellow-brown loams and sandy loams with a variable sand content. The formation also contains boulders, gravels and pebbles. The two upper moraines which are known as Moscow and Dnieper are the most extensive ones; two lower moraines are isolated individual areas. Submorainic, intermorainic and supramorainic deposits are composed predominantly of fluvioglacial sands with local isolated areas of interglacial, fluvial and lacustrine-swampy formations with peat interbeds enclosed. The older alluvial sand deposits comprising the first and second terraces above the flood-plain of the Pripyat river are widely developed over the Polessie Lowland. Loess like rocks are encountered locally in patches. Recent sandy-clay and peat-swamp formations of small depth are confined to shallow depressions and river valleys. The material of the Quaternary period overlies the Tertiary, Upper Cretaceous and locally older Mesozoic, Paleozoic, Proterozoic and Archean formations.

The stratified Quaternary formation of Polessie includes hydraulically connected confined and unconfined groundwater, which, in general saturate fluvioglacial sands, sandy formations of the river terraces and peat bogs. Below, in the Tertiary, Cretaceous, Jurassic and local older deposits, occurs the bulk of the confined fresh groundwater. This is the zone of active water exchange and it plays an obvious part in swamping the area. The fresh groundwater storage is partly recharged by mineralized water from deeper aquifers which ascends along the zone of tectonic dislocations and through permeable rocks. The author observed unconfined highly mineralized groundwater in the swamp areas of Polessie.

The groundwater regime of inter-connected aquifers of the zone of active water movement in Polessie is characterized by the regularity of the behaviour of the water level. During the hydrological year these are two main rises of the piezometric level of the free water table and of the aquifer of the Quaternary, Tertiary and older deposits. These peaks are also observed in the water levels of rivers, lakes and swamps. The two peaks occur in spring and early winter. Sometimes in summer when precipitation takes place, a third but smaller rise of water level is observed. Annual amplitudes of unconfined groundwater level fluctuations vary in general between 0.5 and 2 m, increasing near rivers to 2 - 4 m. The unconfined groundwater depth ranges between 1 m and 4 m, decreasing to nothing in marshy areas and increasing to

8-10 m and more in the interfluve areas and slopes of the Polessie lowland.

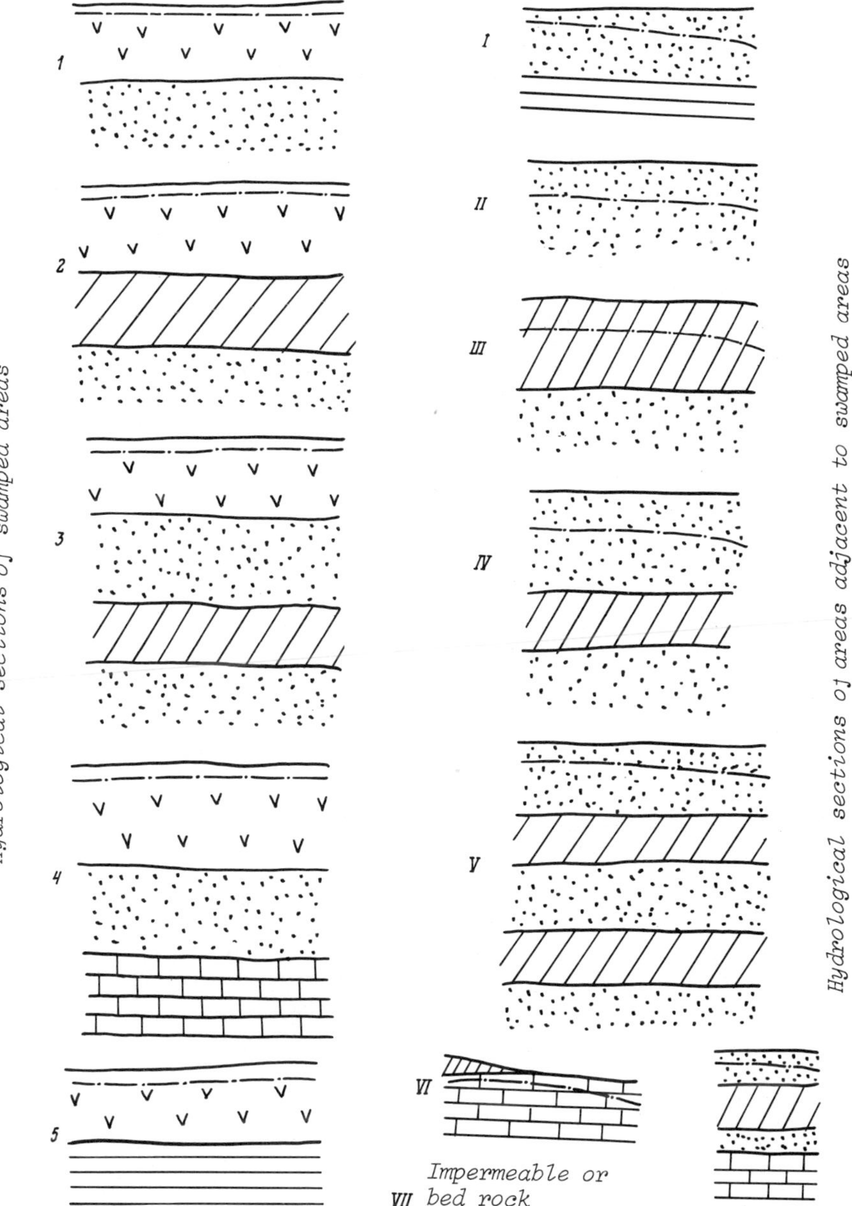

Fig 4. Schematic well logs for computations of groundwater balance in stratified formations corresponding to the major hydrogeological cross-sections through swamp and adjacent lands in Polessie.

The average annual precipitation is 500 to 600 mm and the average annual temperature is between 6 to 7°C.

Rivers, swamps and lakes play an important role in the groundwater regime of Polessie. The main drainage network is the Pripyat river and its numerous tributaries. The largest swamp areas known as the Pinsk swamps, are located in the region of the city of Pinsk: These swamps are in the basins of the Pripyat river tributaries: Big lakes cover an area in the western part of Polessie. In the upper Pripyat river, in the region where cretaceous rocks occur near the surface are located the lakes associated with Karsting (Svityazskoe, Pulemetskoe, Lyutsemer, etc).

In the study of behaviour of groundwater in stratified aquifers on the basis of the water levels observed in marshy and adjacent lands, a description of hydrogeological conditions is of great importance. From this point of view, it is necessary to distinguish between marshy and adjacent lands located in the area of groundwater recharge and runoff. Downward flow of groundwater is observed in the area of recharge. Swamps may serve as sources of groundwater recharge and such swamps are found on the northern and southern slopes of the Polessie lowland. Upward flow of groundwater is observed in the area of discharge. Here swamps are recharged from below by confined groundwater. Such swamps are located on the flood plain terrace or the first terrace above the

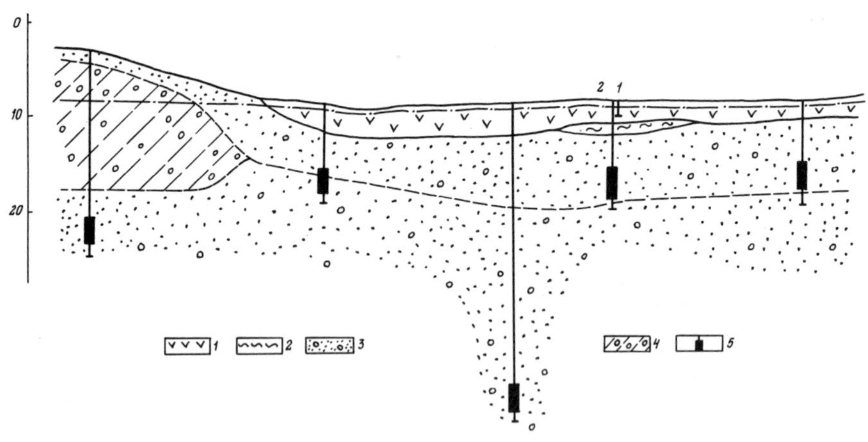

Fig 5. Hydrogeological sections of the swamp area in Polessie, the upper Yaselda region.

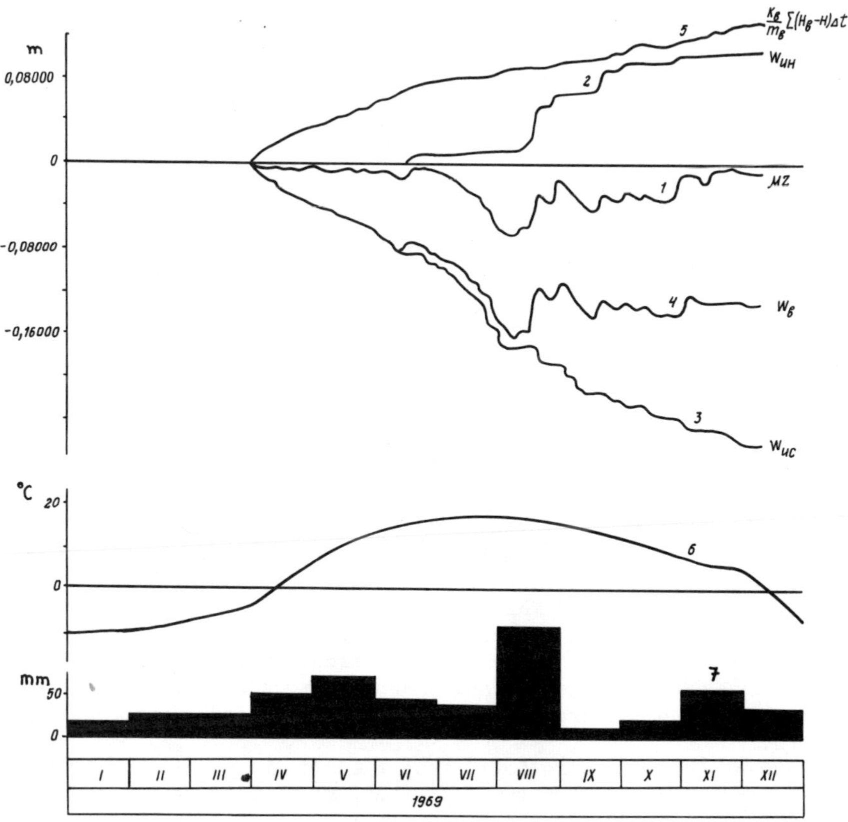

Fig 6. Plot of the unconfined groundwater balance dynamics of the swampy area in Polessie, the upper Yaselda region:
1. change in unconfined groundwater storage;
2. total infiltration recharge;
3. total evaporation from the water table;
4. difference between infiltration recharge and evaporation from the water table;
5. recharge from confined aquifer;
6. monthly average temperatures;
7. precipitation.

flood plain of the Pripyat river. In the tributaries they are found in the lower parts of the second terrace above the flood plain, or where the bottom moraine horizon drops and the swamps are recharged by the confined groundwater occurring in inter-morainic, submorainic, Tertiary and older deposits. In the south-western part of Polessie, in the upper Pripyat, the Cretaceous waters play an active part in swamp recharge. The swamps located in the intermediate zone between the zones of groundwater recharge and discharge may be supplied by confined groundwater and may also serve as sources of groundwater recharge. Examples of such swamps occur on the moraine deposits of the Moscow glaciation to the north of the Pripyat river, between the Sluch and Dnieper rivers.

The recharge of the unconfined groundwater of marshy and adjacent lands is ensured by precipitation over the whole Polessie lowland area. The recharge of swamps from surface runoff occurs where the geomorphological conditions and the upper rocks are favourable. On the low terraces of the Pripyat and its tributaries the swamps are periodically recharged by floods.

For the purpose of the study of groundwater behaviour on the basis of water level data on marshy and adjacent lands, all the swamps located in the Polessie lowland may be divided into two groups: a first group of swamps receiving water from a confined aquifer or recharging the aquifer, and a second group of swamps underlain by mainly impermeable rocks with no vertical connection with the underlying groundwater. In both groups of swamps lateral recharge is generally received from unconfined groundwater and is absent only in rare cases. For the more detailed descriptions of hydrogeological conditions that are necessary for hydrogeological calculations, the typical major hydrogeological cross-sections through marshy and adjacent lands are described. These are shown in Table 2 and Figure 4 which show one-, two- and three-layered aquifers on marshy and adjacent lands in the upper part of the section of importance for land reclamation. Different variations in the hydrogeological sections shown are accounted for by interruptions in the occurrence of low permeability and impermeable layers.

Swamp deposits also form stratified aquifers. In some cases two-layered (water-saturated peat-silt, peat-silty sand) and one-layered aquifers (water-saturated peat) are formed in the deposits, in others, three- four- and multi-layered aquifers are formed due to interbedding of peats, silts, silty sands, sandy loams and loams. The peat is of variable silt content and degree of decomposition; its thickness varies from 1 to 5-10 m or more.
The equation for the groundwater balance of a multilayered aquifer obtained by the author is applicable to the conditions of the Polessie area discussed above. It has the form:

$$\mu z = Q + W \qquad (1)$$

where

- μ is the saturation deficiency of rocks in the zone of fluctuations of the water level of the unconfined groundwater table;

z is the change in unconfined groundwater level at the moment under consideration; with respect to the initial position

Q is the lateral outflow of a multilayered aquifer underlaid by an aquiclude (particularly in the case of one or two water-bearing layers, etc) at the same moment of time;

W is the actual infiltration recharge of unconfined groundwater from above and evaporation from the surface at the same moment of time;

μz is the change in unconfined groundwater storage; a positive sign shows recharge and a negative sign indicates loss. The values of μz, Q and W are expressed in terms of depth of water per unit area.

Equation (1) is valid for both one-and two-dimensional flows. The difference is in different expressions for Q. In the former case Q is computed from observation of wells located along a flow line, in the latter case from observations of a network of wells, usually a square.

When unconfined groundwater occurs in low permeability rocks underlain by a confined aquifer (two-layered aquifer) or when they are separated by a low permeability layer (three-layered aquifer), ie for almost all the types of hydrogeological section shown in Table 2 and Figure 4 (except cases 5, I, VII), equation (1) may be used in the form:

$$\mu z = Q + W - \sum_{1}^{n} \frac{K_B (H_B - H) \Delta t}{M_B}, \qquad (2)$$

where:

K_B and M_B are the permeability coefficient and thickness of a low permeability layer;

H_B is the unconfined groundwater level;

H is the piezometric level of a confined aquifer;

n is the number of terms in the sum and equal to the number of time intervals taken Δt (Δt_1, Δt_2, ..., Δt_n). Q in equation (2) expresses the lateral outflow of unconfined groundwater. With a negative sign the sum from 1 to n in equation (2) is the value of the unconfined groundwater recharge by confined groundwater from below, and with a positive sign, it is the value of the losses of unconfined ground water to the lower aquifers.

To compute the water balance by equation (2) it is necessary to have data on the water levels in four wells, three of which are located along the unconfined groundwater flow and the fourth located near the middle well in a confined aquifer. For two-dimensional flow it is necessary to have a set of five wells in the unconfined groundwater and a further well, located near the central one, in the confined aquifer.

If the value of the lateral outflow of the unconfined groundwater is negligible, $Q \approx 0$ may be assumed. In this case equations (1) and (2) take for form:

$$\mu z = W \tag{3}$$

and

$$\mu z = W - \sum_{1}^{n} \frac{K_B (H_B - H) \Delta t}{M_B} \tag{4}$$

To compute the water balance by equation (4) two wells located in the unconfined and the confined groundwater horizon, are sufficient and for equation (3) one well in the unconfined groundwater is sufficient.

The value of W in equations (1)-(4) is equal to the difference $W_{in} - W_{ev}$, where W_{in} and W_{ev} are the total values of infiltration recharge of unconfined groundwater and evaporation from the surface for a given time.

For four- and five-layered aquifers (cases 3, V in Table 2 and Figure 4) three pairs of specially designed observation wells are sunk into the unconfined groundwater and the uppermost confined aquifer in the direction of groundwater flow; the mean point of each of the paired wells is determined by a means of a third well which penetrates the second confined aquifer; in all, seven wells are required. For two-dimensional flow nine wells are required. Thus in each additional aquifer under investigation three additional wells are required for unidimensional flow and four additional wells for two-dimensional flow.

The installation of multi-staged wells on marshy and adjacent lands in Polessie is vital in the study of groundwater behaviour and has made possible the determination of vertical water exchange between the aquifers, the evaluation of the role of infiltration and confined recharge in the unconfined groundwater balance of swamps and evaporation from their surface as well as the investigation of the groundwater flow cycle.

The order of magnitude of swamp recharge by infiltration and confined groundwater in the Polessie lowland, can vary according to the location of the area. This is illustrated by the results obtained for the groundwater balance for 1969 of one of the marshy areas of the upper Yaselda region.

On the surface of the marshy area there is a well-decomposed peat layer 2.20 m thick containing unconfined groundwater. The peat layer is partially underlain by a layer of water-saturated silt about 0.7 m thick (Fig. 5). The level of the groundwater table in the peat is situated at 0.02 to 0.65 m below the surface according to season. These swamp deposits are underlain by fluvioglacial sands, the thickness of which is not known, containing confined groundwater. Within the area adjacent to the marsh, dry sands are encountered at the top and above the flood plain underlain by morainic loams and sandy loams of the Moscow Glaciation which overlie the fluvioglacial sands containing confined groundwater. Here unconfined groundwater is known as non-artesian groundwater of the upper layer, whereas confined groundwater is water of the underlying layers whether or not an impermeable stratum occurs between them (Kiselev 1971). The given section can be treated either as a three-layered formation (peat, silt, sand) or as a two-layered formation where the peat and silt form a low permeability layer, of mean permeability coefficient K_B, overlying sand. From the hydrodynamic point of view this section is typical of almost all the hydrogeological sections of Polessie presented in Table 2 and Figure 4 (except cases 5, I, VII).

In order to estimate the unconfined groundwater equilibrium a two-layered structure was assumed and water level measurements from two wells were used: these were wells 1 and 2 in Figure 5; the filter of well 1 was located in peat, that of well 2 in the underlying fluvioglacial sands. The water level in well 1 was lower than that in well 2 by 0.05-0.15 m or less; in rare instances their levels were the same. The mean coefficient of permeability K_B of the low permeability layer (peat-silt) was determined using the water level observations in two-stages wells 1 and 2. The value of the saturation deficiency μ in the zone of unconfined groundwater level fluctuations in peats was equal to 0.12. The average thickness of the low permeability layer (peat-silt) m_b was 2.02 m, and the value of K_B was 0.0165 m/day.

The unconfined groundwater balance was computed by equation (4) inasmuch as Q was negligible and was not taken into account for the period from April to December; in December, January, February and March, the water level in well 1 was not measured because of ice. The results are given in Table 3 and in Figure 6.

As can be seen from Table 3 the inflow to the unconfined aquifer for the period April to December amounted to: 125.1 mm of infiltration recharge and 131.7 mm of recharge from confined groundwater giving a total of 256.8 mm. The outflow component of the water balance equation was evaporation which amounted to 262.8 mm, a value which exceeds the inflow component by 6 mm and indicates losses of unconfined groundwater of this magnitude.

Precipitation in this period totalled 458.9 mm, 125.1 mm of which (27.3%) was unconfined groundwater recharge.

The above data on the unconfined groundwater budget of this marshy area are representative of the individual items of the unconfined groundwater budget of the marshy areas of the Polessie lowland and can be used when solving problems associated with the control of its water resources and its economic development.

In conclusion it should be noted that the results obtained for stratified aquifers on marshy and adjacent lands of the Polessie lowlands as well as the suggested balance relations and the computational procedure of the balance under these conditions constitute a scientific basis for locating observation wells in Polessie taking into account the technique of constructing observation wells in stratified formations.

Table 2. Typical Hydrogeological Cross-Sections through Marshy and Adjacent Lands in Polessie (with or without flow of water between the confined aquifer and the swamp)

Hydrogeological sections of a layer of swamp deposits underlying marshy lands	Hydrogeological sections of lands adjacent to marshes with or with no lateral groundwater inflow	
1. Confined aquifer in sands of great thickness (usually unknown)	with lateral inflow.	1. Unconfined groundwater flow in sands or sandy loam-loamy deposits overlaying loams and clays assumed to be impermeable
2. Aquifer in low permeability rocks (loams, sandy loams) underlain by a confined aquifer in sands	" – "	II. Unconfined groundwater flow in sands of great thickness (usually unknown)
		III. Unconfined groundwater flow in low permeability loamy-sandy loam rocks underlain by a confined aquifer in sands
3. Confined aquifer in sands separated by a low permeability layer from the underlaying confined aquifer in sands	with lateral inflow.	IV. Unconfined groundwater flow in sands separated by a low permeability layer (sandy loams, loams) from the underlying confined aquifer in sands
4. Marly-cretaceous aquifers usually separated by Quaternary sands and sandy-loam-loamy rocks from swamp deposits	" – "	V. Unconfined groundwater flow in sands overlying low permeability rocks underlain by a confined aquifer

		separated from the underlying confined aquifer by a low permeability or impermeable layer.
		VI. Unconfined or confined groundwater flow in marly-cretaceous stratum and overlying sands and sandy-loamy rocks
5. Clays and morainic loams assumed to be impermeable	No lateral inflow	VII. Dry and practically impermeable rocks (clays and morainic loams)

Table 3. Unconfined Groundwater Budget in the Upper Yaselda Region of Polessie, in 1969

Quarter (of the year)	Date	Change in unconfined groundwater storage	recharge of unconfined groundwater coming from confined aquifer	Total infiltration recharge of unconfined groundwater	Total evaporation from the water table	Difference between total infiltration recharge of unconfined groundwater and evaporation from water table
II	4.IV.1969	O	O	O	O	O
	I.VII.1969	- 4.8	78.3	8.3	- 91.9	- 83.1
III	I.X.1969	-33.6	104.4	89.7	-227.7	-138.0
IV	7.XII.1969	- 6.0	131.7	125.1	-262.8	-137.7

References

Kiselev, P. A. 1971. Izuchenie vodnogo balansa sloistykh vodonosnykh gorizontov na osnove analiza ikh rezhima pri gidrogeologicheskhikh issledovaniyakh dlya melioratsii zemel. (Study of water balance of stratified aquifers and analysis of the regime in hydrogeological investigations connected with land reclamation). In: Tsikl formirovaniya podzemnogo stoka i gidrogeologicheskoe kartirovanie. Moscow.

DISCUSSION

G. Kovacs (Hungary) - Can you give us some assistance with similar works initiated in Hungary, namely, measurment of groundwater losses by evaporation? In your paper you have reported losses of about 267 mm. How did you measure these?

A. P. Kiselev (USSR) - We calculated the groundwater losses by evaporation.

G. Kovacs (Hungary) - Do you mean the evaporation was not measured but calculated when the level of the groundwater table fell?

A. P. Kiselev (USSR) - The evaporation equation involves a number of quantities such as evaporation from the groundwater table, groundwater depth, etc. In the present case the periods considered were those during which the groundwater level was lower.

A. I. Ivitsky (Byelorussian SSR) - If you only calculated evaporation from changes of groundwater level, how did you estimate the fall due to outflow along rivers and canals?

A. P. Kiselev (USSR) - The outflow effects were included in the equation. In the present case it was insignificant.

P. I. Zakrzhevsky (USSR) - What was the value of groundwater flow for the season which was neglected in your calculations and can you also specify, if possible, in what soil layer it occurs.

A. P. Kiselev (USSR) - This value is 2 or 3 mm for the season.

V. M. Zubets (USSR) - Does the slope change due to subsidence or does it remain the same?

A. P. Kiselev (USSR) - The slope and subsidence depend primarily on peat layer thickness. We worked with the peat whose slope does not change.

INVESTIGATION OF GROUNDWATER REGULATION IN PEAT SOILS IN THE GDR

A. Scholz, G. Wertz

The Pulinenaue Meadow, Pasture and Mire Institute.
The GDR Academy of Agricultural Sciences.
Land Reclamation and Plant-Growing Department,
Rostock University, Rostock.

SUMMARY: Systems of two-way groundwater regulation are required to ensure high and stable crop yields on sand, sand-underlined turf and thick peat soils. The transition to industrialized methods in agriculture is highly conducive to the replacement of open drainage with closed drainage. To this end, novel-type plastic drain pipes of a bigger diameter are being used in the GDR both in sand and in peat soils, in the latter case the pipes being used in combination with mole drains. Depending on how extensively the grasslands are utilized, groundwater regulation is carried out with due regard to water saving with the maximum utilization of the soil water-holding capacity and differential water provision.

RECHERCHE SUR LA REGULATION DES EAUX SOUTERRAINES DANS LES TOURBIERES DE LA REPUBLIQUE DEMOCRATIQUE ALLEMANDE

RESUME : Des systèmes à double action pour la régulation des eaux souterraines sont nécessaires pour assurer des récoltes abondantes et stables sur des sols sablonneux, des terres tourbeuses situées sur une couche de sable ou des tourbières épaisses. Le passage à des méthodes industrialisées en agriculture conduit au remplacement du drainage à découvert par le drainage fermé. Dans ce but, des conduites de drainage en plastique de type nouveau, de plus gros diamètre, sont utilisées dans la République démocratique allemande à la fois dans les sols sablonneux et tourbeux ; dans le dernier cas les conduites étant utilisées en combinaison avec des drains en coulée de taupe. Suivant la manière plus ou moins intensive dont les herbages sont exploités, la régulation des eaux souterraines est entreprise en prenant soin d'économiser l'eau, en utilisant au maximum la capacité de rétention d'eau du sol et en prévoyant des approvisionnements en eau différenciés.

There are about 480 thousand hectares of peat-soil land under farming in the German Democratic Republic, 470 and 10 thousand hectares falling on lowland and upland moor areas, respectively. This represents 7.5% of the total cultivated area in the GDR. In addition, there are some 20 or 30 thousand hectares of peaty soils and about 60 to 80 thousand hectares of swampy and marsh soils under forest. In contrast to the Soviet Union, where peat soils are often sown with agricultural crops (Jangol, 1969, Zubec, 1971), those in the German Democratic Republic are used exclusively as meadows and pastures, and represent one-third of the whole area of grassland.

Approximately 50 per cent of the peat soils of the GDR need reclamation, the success of which depends, first of all, on the location and surveying of peat soils areas. This task, which started in 1953 is now well under way. The classification and mapping of peat soils is based on the soil types differentiated by Wojahn and Illner, (1962). The data concerning the optimum and permissible range of fluctuations in the groundwater level in individual soil types are defined on the basis of experimental yields over a period of many years (Wadbrink 1969, Gebhardt 1971). The upper limit of the permissible fluctuation range is characteristic of the winter level of groundwater whereas the lower limit is the groundwater level which is acceptable to plants for a short period of drought without any significant harm being caused. With regard to meadows and pastures on peat soils, Okruszko (1969), Jangol (1969) and Zubec (1971) furnish similar evidence of the desirability of preventing excessive drops in the level of groundwater and maintenance of water consumption at an economic rate. The data on the requirements for additional water were based on the results of water flow measurements that were made over a period of two years under conditions of backwater in valley sand for an area of 930 hectares (Scholz, 1972). In addition, Gebhardt (1971) made use of the measurements of moisture content in a deep, lowland moor soil. In practice, the crop yields that have been obtained are only 25% below the yields from experimental plots. It should be noted that differential distribution of water results in water economy. The regulation of groundwater is most effective in areas where the ground level does not deviate more than ± 20 cm from the mean plane of the relief. Such lands, which are considered optimum for groundwater regulation, should be given priority for water supply to ensure their intensive utilization. With the perfection of drainage, the role of the soil, as compared with that of relief, decreases.

Shallow lowland peat soils lying on sand comprise 40% of the boggy lands and play an important role in moor reclamation and agricultural exploitation. Regulation of groundwater in such soils commenced as early as the first half of the XVIIIth century (Scholz, 1969). Yet it was only the XXth century that saw genuine development of such lands, where reclamation involved deepening of canals and ditches, exclusion of winter floods and construction of surge water retaining installations and dams. As was the case with pure sandy soil, regulation of groundwater was undertaken here by means of open ditches only. Closed drainage was, in the main, used to reclaim heavy-textured soils and deep peat soils.

Deep lowland peat soils with a peat depth of 120 cm and more represent about 40% of boggy lands and areas where the depth of peat is less, account for another 20% of these lands. These soils generally require drainage. Peat soils which are a minimum of one-metre deep, yield crops of 5.0 to 7.0 tonnes of dry substance per hectare without being provided with additional water. They require supplementary irrigation, when fertilized with nitrogen (from 100 to 200 kg per hectare) or in dry-weather years, in order to increase yields and obtain more reliable harvests.

Considerable success in groundwater regulation has been achieved in recent years as a result of the development of plastic drainage pipes (see Figure 7). The changes in farming demand better water management and new techniques and the use of plastic materials has greatly favoured the development of drainage technology.

Table 4 shows a trend towards the adoption of pipes of greater diameter. In gently sloping lowlands such a tendency seems essential when it is proposed to reduce water use over large areas. The employment of small-diameter drain pipes involves the creation of an artificial slope in the drainage system, deepening of collector ditches and a greater flow of water in the ditches whereas, large-diameter drain pipes do not require any slope.

Special attention should therefore be given in groundwater regulation to the recently developed two-piece drain pipes comprising two longitudinal halves. This design, as shown in Figure 8, combines the scientific achievements and practical experience gained over the past 13 years in the employment of plastic drains in the GDR. Another advantage of large-diameter pipes is that their walls are thin. Longitudinal half pipes are much easier to store or to transport than complete pipes. Moreover, the latest investigations (Wertz, 1967, Dahms, 1968, Kisperth, 1969, Hilliger, 1970; Scholz and Hilliger, 1972) have shown that it is also easier to perforate or slot the half pipes. The following combinations are possible:

The table shows that a pipe consisting of two longitudinal halves can be used in sand and peaty soils; no additional filter materials are required.

Flexible pipes comprising two longitudinal halves are laid with the "Meliomat-Universal" trenchless pipelayer. With this machine the rate of pipelaying is 240 metres per hour. In the case of the greater drain-to-drain distances (from 50 to 150 m), it is thus possible to achieve 20 to 30 hectares per shift.

Agriculture is making severe demands on groundwater regulation in lowlands with sand and peaty soils having high groundwater levels. Agriculture in the GDR is embarking upon a new stage of development characterized by further intensification of production and gradual transition to industrial organization and management of production. The new systems of groundwater regulation meet requirements among which the most important are the following.

1. Fields of 8 tonnes and over of dry matter per hectare under regular production conditions.
2. Creation of 50-hectare plots and over, free of ditches, on at least 70% of the reclaimed area with a ditch network density below 25 metres per hectare.
3. Maintenance of adequate soil moisture in dry periods, accessibility and possibility of use of pasture in excess moisture periods.

It has been established by investigation of the regions typical of the Republic that approximately two-thirds of the peat soil hay fields and pastures meet the above-mentioned requirements and are suitable for industrial harvesting. Other methods of harvesting are, of course, applicable under unfavourable conditions such as high relief, excessive moisture due to spring water, complex strati-

Table 4. The most important types of plastic drainage pipes (of high-density polyvinylchloride) tested under field conditions in the GDR.

Pipe		Diameter mm	Wall thickness mm	Water-inlet area. cm^2/m	Application
Plastic drain pipe (Greifswald design)	1959	36	0.8	about 10	drainage
Plastic-film drain pipe (Weimar design)	1962	36	0.4	about 10	drainage
Perforated drain pipe	1965	50 75	1.3 1.8	about 10 about 10	drainage
Corrugated plastic drain pipe (chequered)	1967	90 60 75 110	0.5 to 0.8	12+) to 20	drainage and groundwater regulation
Pipe of two longitudinal halves	1971	145 170	1.0	160 to 326	preferably groundwater regulation

Table 5. Utilization of longitudinal half-pipes

Version	Application	Water-Inlet Opening			Water-Inlet Area	
		round mm	slots width mm	slots length mm	cm^2/m	Percentage of pipe surface
1.	peaty soil	3-5	-	-	330	5.2
2a.	sandy soil	3-5 (at bottom)	0.5-0.6 (on top)	12	180	2.8
2b.	sandy soil	3-5 (at bottom)	0.8-1.0 (on top)	12	235	3.7
3a.	sandy soil	-	0.5-0.6	12	160	2.5
3b.	sandy soil	-	0.8-1.0	12	270	4.2

+) with 6 rows of slotted openings

graphic structure and small or narrow plots.

The groundwater regulation systems with no slope drains have been in experimental use since 1971.

The main aspects of slopeless drainage are as follow:

1. The flow in ditches and drains is always ensured by the natural slope. Artificial slope is permissible up to 0.3% only. At present the employment of such systems is limited to land slopes of 0.5%.
2. The drain depth must be in the range of 1 to 1.2 metres.
3. The drain-to-drain distance is between 50 and 150 metres, when large-diameter plastic drain pipes are used, and depends on the permeability of the ground and other hydrogeological factors.
4. The daily drainage runoff in the conditions pertaining in the GDR ranges from 5 to 10 mm. It is planned to provide for a daily water supply of 2 to 5 mm, depending on the intensity of use of the system.
5. The distance between ditches is fixed by doubling the maximum length of a drain pipe. The longitudinal effect is determined by the energy gradient along the drain. The energy gradient should be calculated with due regard to the drain pipe parameters and inflow to the drain by applying the usual hydraulic design procedure. In practice the required slope varies between 0.2% and 3.3%.
6. The precision of drain-pipe laying is a decisive factor. The difference between altitude marks must not exceed 5 cm in the direction of flow.
7. The connection of each end of the plastic drain pipe to the open ditch makes for ease of maintenance. At the same time this allows for the raising of the level in one ditch to flush the drain pipes.
8. Two-way water-elevation installations can help considerably to enhance the operational reliability of groundwater regulating systems.

In peat soils having a layer of turf at least 1.5 metre deep and a vertical permeability of less than one metre per day, it is possible to employ slopeless drains with an outlet into ditches at each end in combination with mole drains.

The developments described above and applied to groundwater regulation in peat soils in the GDR are the result of constant and collective scientific investigation. They are aimed at creating the optimum conditions for advanced and highly-intensive farming. In principle, groundwater regulation should be linked with water saving measures which make full use of winter precipitation and the ability of soils to store and supply the plants with adequate quantities of water. Efficient groundwater management exerts a smoothing effect on the hydrological regime of a given area. This is felt especially in the periods of excess when water is retained in controlled storage and subsequently released to supply those areas which need water.

References

Baden, W. und Eggelsmann, R. 1971. Maulwurfdränung im Moor. Zeitschr. für Kulturtechnik 2 S.146-166.

Gebhardt, E. 1971. Untersuchungen zum Wasser- und Lufthaushalt tiefgrundiger Niedermoorstandorte mit Aussagen uber die Ziele der Wasserregulierung. Forschungsbericht.

Dahms, P. Ein Beitrag uber die Funktion der Dranrahrein-trittsoffnungen und deren Wechselwirkungen mit dem dranrohrnahen Bodenkorper. Dissertation, Rostock.

Hilliger, D. 1970. Der Einflub von Rohrparametern und des drännahen Raumes auf die Aufnahmeleistung der Dranrohre und den Dränabstand. Dissertation, Rostock.

Jangol, A. M. 1969. O normach osusenija pri dwustoronnem regulirowanii wodnowo rezima torfjanych pocv Ukrainy. Sposoby i normy osusenija.

Kirchner, H. 1972. Ein Beitrag zur Klassifizierung von Niedermoorstandorten in den 3 Nordbezirken der DDR unter für die Melioration bedeutungsvollen hydrologischen Gesichtspunkten. Dissertation, Rostock.

Kisperth, V. 1969. Ein Beitrag zur Dranabstandsbestimmung. Dissertation, Rostock.

Okruszko, H. 1969. Richtungen und Grundsatze für die Bewirtschaftung von Moorgebieten. PWR und L.

Scholz, A. 1969. Grundwasserregulierung im Havelländischen Luch. Zeitschr. für Landeskul tur 10, S. 297-311.

Scholz, A. und Hilliger, D, 1972. Nichtfilternde Durchtrittsöffnungen in basaler Anordung fur Drane zur zweiseitigen Wasser-regulierung in Grundwasserniederungen mit Sand im Dranbereich. Meliorationsinformationen im Druck

Scholz, A. 1972. G rundwasserregulierung auf Niedermoor und Grundwassersand - ein effektives Mittel zur Sicherung und Steigerung der Pflanzenerträge. Broschüre des VEB Ingenieurbüro fur Meliorationen, Berlin im Druck.

Scholz, A. 1967. Maulwurfdränung in Moorböden mit pressenden und schneidenden Werkzeugen. Zeitschr. für Landeskultur 8. S. 29-44.

Waydbrink, W. 1969. Rohe und Sicherheit der Erträge des Grunlandes in verschiedenen Niedermoorgebieten. Zeitschr. für Landeskul tur 10, S. 279-295.

Wertz, G. 1967. Ein Beitrag zur Bemessung von Dränanlagen unter besonderer Berücksichtigung der Erhaltung und Steigerung der Funktionstuchtigkeit. Habilschrift, Universität Rostock.

Wertz, G. Dahms, P. Kisperth, V. 1969. Aufgaben der Entwässerung und Bemessung der Anlagen im Rahmen der Wasserhaushaltsregulierung. Wiss. Zeitschr. der Universität Rostock, 18.

Wojahw, E. und Illner, K. 1962. Die Standorttypen der Niedermoore als Grundlage der Meliorationsplanung und -projektierung. Wasserwirtschaft, Wassertechnik 12, S. 139-165.

Zubec, V. M. 1971. Schemata fur die zweiseitige Wasserregulierung auf Moorböden (für die Bedingungen Belorublands) Gidrotechnika i melioracija, H. 8.

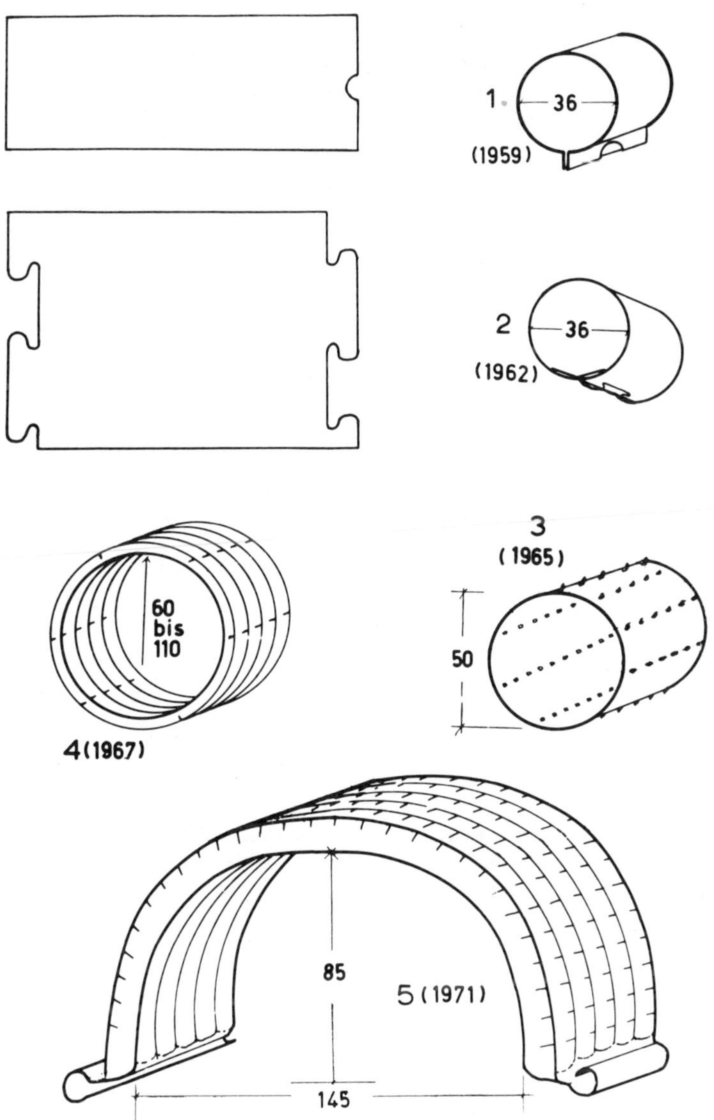

Fig 7. The main stages in the development of plastic drain pipes in the GDR

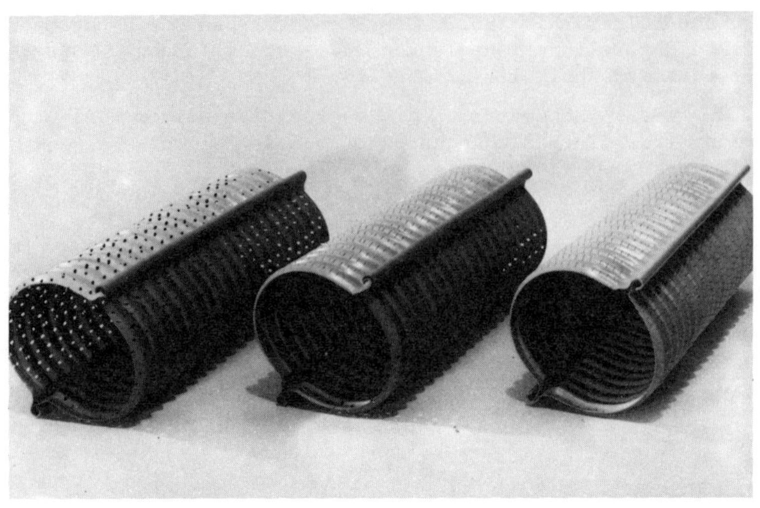

Fig 8. Photograph of recently developed drainage pipes.

DISCUSSION

G. Kienitz (Hungary) - How are the half-tubes connected?

A. Scholtz (GDR) - Half-tubes, 4.5 m long, are locked automatically.

W. H. Van der Molen - 1. I have some experience in the use of plastic drain pipes and have found that the pipes become clogged by decaying peat. Why don't you use drains of a large diameter? 2. Have you observed peat subsidence due to drainage?

A. Scholtz (GDR) - 1. First concerning the diameter. At first we made slits in the pipes but these were clogged by decayed peat. Now we make circular holes, 4 mm in diameter and the drains operate adequately. We use a combination of plastic pipes and soil drains. Plastic drains are buried at a depth of 1 m; soil drains at 8 cm. 2. As to the second question, subsidence occurs here as elsewhere.

HYDROLOGICAL RESEARCH ON A PEAT-BOG IN THE UPPER SUPRASL BASIN

Z. Mikulski and E. Lesniak

Warsaw University, Warsaw, Poland

SUMMARY: Hydrological research on the Michalowo-Imszar peat-bot (in the upper Suprasl basin, North-Eastern Poland) was carried out in 1958-1963. The peat-bog covers almost 60 km^2, of which the major part is at the level of the groundwater table and has a peat layer thickness of about 2.5 m. The investigation lasted five years. It began one year before the drainage works were installed, which lasted two years, after which there was a twelve-month period of farm development which was concluded a year after the drainage was completed. The drainage caused a lowering of the groundwater level of over 30 cm; the variations in groundwater level showed a distinct correlation with stage in the river which crosses the peat-bog. A 15% decrease of evaporation shows that this parameter is closely dependent on the hydrological regime of the peat-bog and on the level of its agricultural development. In summer and autumn, runoff from the peat-bog increased by about 20%, a clear indication of the effect of the drainage undertaken. Changes in the value of groundwater storage in the water balance equation showed positive differences, as a rule, in autumn and winter, whilst negative differences were generally observed in summer. The runoff increase was essentially attributable to a lowering of the groundwater table by the drainage network involving intensive water extraction from the peat-bog and the surrounding catchment area.

RECHERCHES HYDROLOGIQUES PORTANT SUR UNE TOURBIERE DU BASSIN DE LA HAUTE SUPRASL

RESUME : De 1958 à 1963 furent menées des recherches hydrologiques portant sur la tourbière de Mikhlowo-Imchar (bassin de la Haute Suprasl, dans le nord-est de la Pologne), qui couvre 60 km^2. Une grande partie de cette tourbière est occupée par un marais tourbeux qui se trouve au niveau de la nappe phréatique et dont les couches de tourbe ont une profondeur de 2,5 mètres. Les recherches ont duré cinq ans : elles ont commençé une année avant les travaux de drainage et continué pendant deux années de mise en valeur agricole après la fin de ces travaux. Le drainage a entrainé un abaissement de plus de 30 cm du niveau des eaux souterraines ; les variations de ce niveau se sont révélées être en relation directe avec le niveau des eaux de la rivière coulant à travers le marais. L'évaporation a diminué de 15 %, ce qui montre à quel point elle est tributaire du régime hydraulique du marais et de la mise en valeur du terrain. Pendant l'été et l'automne, l'écoulement a augmenté de 20 %, conséquence directe du drainage réalisé. Les modifications intervenues dans la rétention des eaux souterraines dans l'équation du bilan d'eau sont caractérisées par des différences positives pendant les mois d'automne et d'hiver, les différences négatives prédominant en été. L'accroissement de l'écoulement était surtout dû à l'abaissement du niveau de l'eau souterraine, car le drainage a provoqué une intense extraction d'eau du massif marécageux et du bassin environnant.

There are approximately 24 000 peat-bogs in Poland, covering an area of nearly 15 000 km^2 which is about 5% of the total area of the country (Maksimov, 1965); these are mostly lowmoor, rheophilic peat-bogs, i.e. fed mainly by groundwater (Figure 9). The total volume of peat soil is estimated to be about 26 km^3; this volume is capable of storing a large quantity of water. Assuming that the porosity is 0.95 (Ivanov, 1953), the volume of water accumulated in peats is about 25 km^3 or 13% of the total annual precipitation in Poland. Only that part of the aquifer which is situated between the maximum and minimum levels of the ground water table participates in the hydrological cycle. Under present conditions these levels range from 0.40 to 1.40 m below the surface of the ground (Stolevska, 1962). Consequently the active water volume is about 1.35 km^3 which is barely 6% of the total water capacity of the peat-bogs in Poland. The largest peat-bog areas of Poland are situated in the north-eastern region (Bialystok province). They are mostly remnants of ancient lakes and cover an area of about 2320 km^2, amounting to more than 3.6 km^3 of peat.

The lack of knowledge of the hydrologic regime of peat-bog areas encouraged the Polish Hydrological Service to embark upon a programme of research. It was decided to study the peat-bog of Michalowo-Imszar which is in the basin of the upper Suprasl. This bog covered almost 60 km^2 and is one of the largest in Poland. It can be divided into three basic types of which the low bog occupies the largest part and the remaining small area is divided between the high and transitional types. The upper part of the Suprasl catchment area, covering about 50 km^2, is outside the bog area. Downstream of where the Suprasl outfalls from the bog the land is about one-third marshland and the remainder, slightly higher and partly afforested. Together they have a catchment area of 180 km^2. The whole area is covered with quaternary deposits, which belong to a moraine of the Mid-Polish glaciation. The thickness of the peat layer attains 7 m with an average value of 2.5 m. Below the peat there is a clay layer, 2-3 m thick, and this is underlain by fine-grained sand and silt (down to about 21 m) which outcrops in some places near the watershed. The permeability of the surface formations is high.

The peat-bog is a remnant of a lacustrine system, as evidenced by the presence of gyttja, lacustrine muds and marls (Marksimov, 1965). It was formed as a result of the growth of rush associations. The Imszar peat-bog, which received precipitation water only, took the form of a high ombrophilous peat-bog. On the other hand Mikhalowo peat-bog which was continuously supplied with the Suprasl river water, retained a low moor form. The topmost layer of the Mikhalowo consists of sedge peats, covering reed peats, with some xylem, about 30% decayed. There are two small lakes on the bog: Gorbacz (46 ha) and Wielki (10 ha), drained by tributaries of the Suprasl (Figure 10).

The peat remained in its natural state until 1938. Drainage carried out in 1938-1939 resulted in a lowering of the groundwater level thus affording possibilities for meadow farming. But during the 2nd World War the drainage installations were destroyed, the surface ditches became over grown and the peat-bog reverted to its natural state. Its use, at the beginning of the investigation, was as follows: 36% meadows and grassland, 19% forests and shrubs, 45% rushes and other waste lands. The hay crop from meadows and grassland produced 1.5 t/ha.

The installation of the observational network was started during the summer of 1958, with the siting of standard raingauges on the catchment and two river gauges on the Suprasl (one upstream where the river flows into the peat-bog, and the other where it flows out). Both the river gauges were used for systematic discharge control. Along the axis of the peat-bog 20 wells were built to observe groundwater behaviour, two of these being located on either bank of the peat-bog in mineral soil. It can be seen that the well network satisfactorily covered the needs of the bog. The most important observation station was installed in the middle of the peat-bog and equipped with a set of instruments to investigate the water balance. Observations of air temperature and humidity at the 0.5, 2.0 and 5.0 m levels as well as peat temperatures in the vegetation zone were undertaken at the station. In summer, the evaporation gradient (at 0.5 and 2 m) was measured with the Piche evaporimeter. An observational well was built and equipped with a Valday-type limnigraph and a raingauge and recording rain-gauge were installed. A field chemical laboratory was also put into service to investigate water and peat (Bortkiewicz 1959) (Figure 11).

After one year, in the autumn of 1959, new drainage works were started which covered 84% of the whole peat-bog area. The following reclamation works were built (Lesniak 1964):

(a) river training works of the Suprasl and its tributaries in the bog area, involving a total length of 40 km

(b) restoration of the existing network of drainage ditches, almost 40 km in total length

(c) building about 25 km of new ditches

The density of the drainage network, not including the Suprasl and its tributaries, was of the order of 1.2 km/km^2 with an average distance between ditches of about 800 m. The new drainage works were built to lower the groundwater level as happened in the previous reclamation.

The reclamation works took approximately two years, and were completed in the autumn of 1961. Then followed a year of post-reclamation farm development (up to the autumn of 1962). During this time much of the former waste land was cultivated and transformed into grassland. The lack of irrigation, and therefore the impossibility of controlling soil moisture, dictated the adoption of extensive farming. The following pattern of land use was achieved: 60% meadows and grassland, 19% forests and shrubs, 21% rushes and partial waste land. The hay yield increased to 4 t/ha for unstocked meadows and 80 t/ha for stocked ones: in 1963 it was an average yield of 55-60 t/ha. These are estimated values only (Lesniak, 1964).

The results of reclamation works and farm development distinctly affected the water balance of the peat-bog. The five-year overall period (September 1958 - August 1963) which was taken for water balance calculations can be divided into the following characteristic periods (Lesniak, 1964).

September 1958 - August 1959 - period before drainage
September 1959 - August 1961 - period of drainage works
September 1961 - August 1962 - period of farm development
September 1962 - August 1963 -

The above time-table made it necessary to express the water balance for hydrological years lasting from September until the following August. Special attention was paid to those hydrological parameters which changed considerably during the overall balance period, namely, evaporation from the basin, soil moisture retention and runoff. As could be expected, most evident changes took place in the behaviour of the groundwater levels. These were lowered considerably, especially during the period when the drainage works were constructed (Figure 12). The level of the groundwater table was found to have a fundamental influence upon evaporation and runoff from the basin.

The changes in groundwater levels were observed during 5 years at the bog station. The first year of observations (1958/59) established the characteristic behaviour of the peat-bog under natural conditions, the groundwater level corresponding on average with the surface of the ground. Maximum levels occurred in winter and during the snowmelt flood when the water level was from 10 to 20 cm above ground; the minimum levels were observed in summer, and sometimes in autumn. As a result of two years of drainage works the groundwater table dropped by over 30 cm on average. Similar levels were observed during the two following years (1962-1963), with notable oscillations; the spring maximum accompanied by an emergence of water on the soil surface was especially marked. The poor correlation between the groundwater level and precipitation is noticable. The correlation between the river stage and groundwater level was, in contrast, quite apparent as can be seen in Figure 12.

Observations of the range of groundwater levels covering almost the whole peat-bog area, made it possible to attempt to calculate the groundwater storage by the so-called finite difference method of Kamienski (Orsztynowicz 1963). Calculations made for the periods May 1959 to April 1960 and May 1960 to April 1961 (time-limited by the necessary peat analyses), only showed the situation resulting from current drainage works. The period selected for the calculations coincided with the period of the lowering of the groundwater table. The decrease in storage, calculated for the first year (May 1959 - April 1960), illustrates the influence of drainage works, with resulting flow of groundwater to the river. Subsequently, as the hydrological regime of the peat-bog became stabilized (May 1960 - April 1961), the river supplied the peat-bog with water, and the intensity of infiltration, determined by the retention-to-precipitation ratio, decreased with a corresponding increase in surface runoff via the drainage ditches (Table 6). Because of a lack of sufficient data which restricted the application of Kamienski's method, the results obtained are only approximate.

It was initially intended to calculate evaporation from the

basin using the heat-balance method, considered to be the most exact. The attempts that were made did not, unfortunately, give the expected results as it was difficult to obtain values of all the parameters involved. For this reason use was made of Konstantinov's method (Konstantinov 1963, 1968), modified for Polish conditions (Debaki 1960). The results for the drained peat-bog showed that the evaporation decreased by about 15% as compared with the evaporation during the pre-reclamation period. In 1959 the annual evaporation was 526 mm; in 1960-1963 it ranged from 420 to 460 mm, the average being 444 mm. Comparing seasonal distributions, the decrease of evaporation was especially evident in summer (May to August). Only in autumn (September-October) were no distinct changes observed (Figure 13). No obvious change in air temperature and precipitation was noted.

An analysis of the annual distribution of evaporation shows a significant dependence on the hydrological regime of the peat-bog as well as on the state of farm development. The high evaporation rate in 1959 seems to be a result of the high groundwater level. Part of the decrease of evaporation in 1960-1961 was caused by the lowering of the groundwater level during a time when the original vegetation still existed. The considerable decrease of evaporation in 1962, especially in summer, resulted primarily from the start of post-reclamation farm development, ie from the destruction of the original flora before new species could develop fully. The high value of evaporation in the autumn was a result of unusually high air temperatures in September (11.7°C) and October (8.7°C). Finally, in the last year (1963) there was a marked increase in evaporation during the vegetation period as a result of an increase in meadow crops. This confirms the general opinion expressed by Shebeko (Shebeko, 1965) on the basis of research carried out in Byelorussia. Bavina and Malankina (Bavina & Malankina, 1969) also came to a general conclusion that a change of vegetation cover is the main factor which changes evaporation from a drained peat-bog.

The peat-bog runoff was calculated from the difference between the river profiles downstream and upstream of the peat-bog. The results show a 20% increase in runoff, as compared to the pre-reclamation period. This represents more than the decrease in evaporation from the basin. The runoff coefficient (runoff-to-precipitation ratio) also showed a tendency to increase, being on an average about 17%. There were increases in the runoff, in summer (May to August) and in autumn (September-October). Bulavko in his monograph notes the conspicuous increase of the runoff due to a decrease in evaporation during the first years after drainage.

By analysising the individual hydrological components, the water balance of the peat-bog for a five-year period (1959-1963) was compiled. This used the method of selected wells developed by Débski (Débski 1963). Observations from two wells on each side of the peat-bog were used to calculate the states of storage. Total storage was included in the water-balance definition as a sum of groundwater storage and surface retention. It was assumed that in summer only groundwater storage occurs, whilst in winter the total storage was calculated from the water-balance equation (Lesniak, 1964). The structure of the water balance is clearly illustrated by the changes in storage. A positive storage difference can be found, as a rule, in autumn and in winter (October to March). In 1962 this positive difference was observed until summer which is exceptional. In summer (May to July) negative differences are normal, which may continue until autumn (Table 7, Figure 14).

The analysis has revealed no evidence to permit the attribution of the increase in runoff to precipitation. This increase was found to be due essentially to a decrease of storage. As a result of reclamation, water was drained from the peat-bog and the surrounding catchment are. This can be illustrated by the situation during summer of 1963, when towards the end of the balance period quite a high runoff was accompanied by a high evaporation value, almost twice as high as the precipitation of that period. Thus it can be concluded that drainage of a low peat-bog surrounded by small areas of high permeability, can disturb the balance, not only of the peat-bog itself but also of the whole surrounding catchment area. This shows the need for investigations over the whole catchment area since only in this way is it possible to understand the changes which may follow reclamation works.

References

Bavina, L. G., Malankina, Y. P. 1969. Vliyaniye ossusheniya na elementy girometeorologicheskogo rezhima poimennogo nizinnogo bolota (na primere Luninskogo bolotnogo massiva). (Effect of drainage on the components of the hydrometeorological regime of a low-moor illustrated by the Luninskiy bog). Trudy GGI, issue 177. Leningrad.

Bortkiewicz, A. 1969. Stacja begienne PIHM w Grodku woj. bialostockie. (Swamp station of SIHM in Grodek/Bialystok province). Cosp. wodna-Biul. PIHM, No. 4.

Bulavko, A. G. 1971. Vodny balans rechnikh vodosborov. Osnovnye zakonomernosti, metody rascheta i problemy preobrazovaniya. (Water balance of river catchments. Basic relationships, methods of computation and problems of transformation). Leningrad, Gidrometeozdat.

Debaki, K. 1960. Szczegolowy bilans wodny rzeki Wieprze w Kosminie jako przyklad rozwiazania rownan bilansu metode studzien wybranych (Detailed water balance of Wieprz river in Kosmin as an example of the solution of balance equations by the method of selected wells). Rocz. Nauk roln., No. F-74-3.

Débski. 1963. Ptzystosowanie nomogramu Konstantinova do obliczeñ parowania terenowego w Polsce. (Use of the Konstantinov monograph for the computation of evaporation from watersheds in Poland). Prace i Studia KIGW PAN. No. 6.

Ivanov, K. E. 1953. Gidrologiya bolot. (Hydrology of swamps). Leningrad, Gidrometeoizdat.

Konstantinov, A. P. 1963, 1968. Ispareniye v prirode. (Evaporation in nature). Leningrad, Gidrometeoizdat.

Lesniak, E. 1964. Wplyw melioracji na odplyw na przykladzie zlewni gornej Suprasli (Effect of drainage on runoff as exemplified by a drainage basin in the Upper Suprasl). (Doctor's thesis). Warsaw,

Institute of Technology, Faculty of Sanitary and Water Engineering, Warsaw.

Maksimov, A. 1965. Torf i ego uzytkowanie w rolnictwie. Wyd.II (Peat and its use in agriculture). 2 ed. Warszawa, PWRIL.

Orsztynowicz, J. 1963. Proba ustalenia retencji gruntowej na torfowisku Michalowo-Imszer metoda koncowych ròznic Kamienskiego. (An attempt to evaluate groundwater storage in the Michlowo-Imszer peat-bog by the Kamienski method of finite differences). Biul. PIHM, No. 7 also summary in: Cosp. wodna - Biul. PIHM, No. 3.

Stolerska, A. 1962. Pròba ustalenia pojemnosci wodnej zloz torfowych na terenach Polski. (An attempt to evaluate the water capacity of peat deposits in Poland). Prace i Studia KLIGW, No. III, Warsawa.

Shebeko, V. F. 1965. Ispareniye s bolot i balans pochvennoi vlagi. (Evaporation from swamps and soil moisture balance). Minsk, "Urozhai".

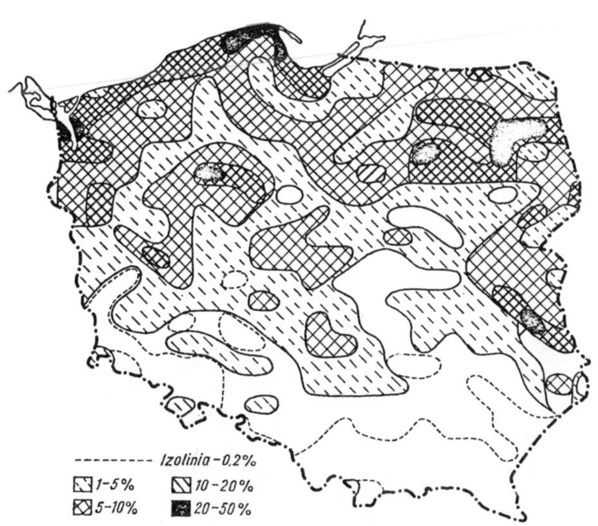

Fig. 9. Map of peatbogs in Poland (after Maksimov, 1965)

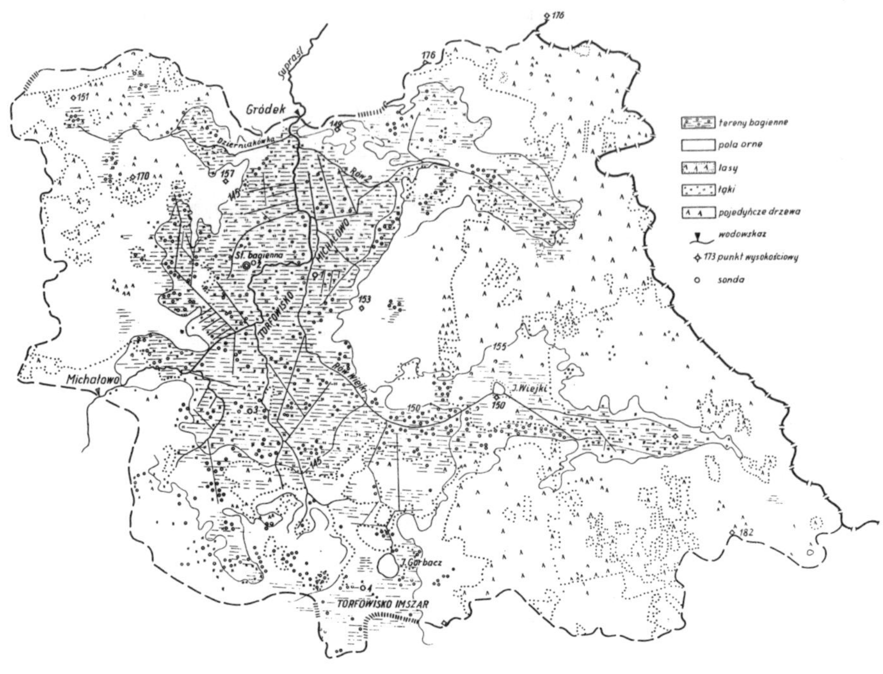

Fig 10. Map of the Michalowo-Imszar peatbog.

Fig 11. General view of the Swamp Station at Grodak (phot. by Z. Mikulski)

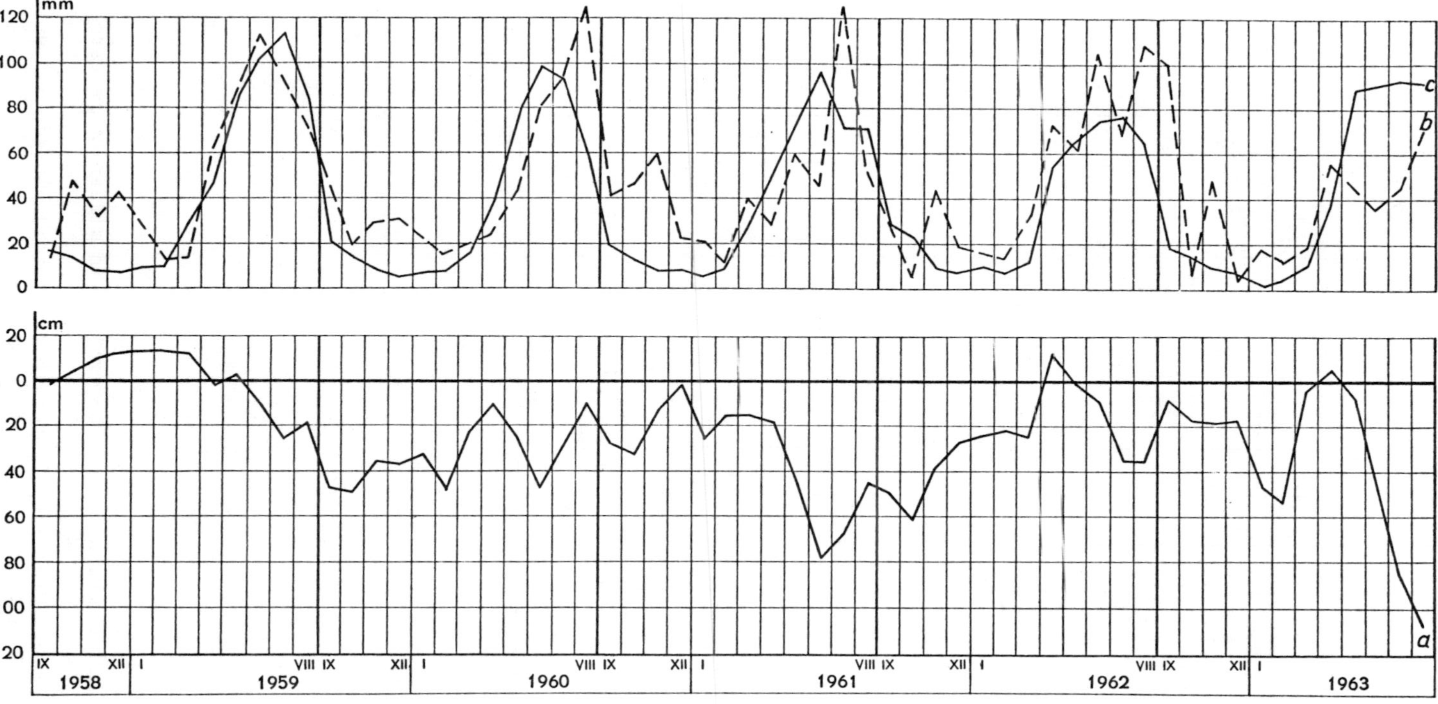

Fig. 12. Distribution of ground-water levels (a), precipitation (b), and evaporation (c).

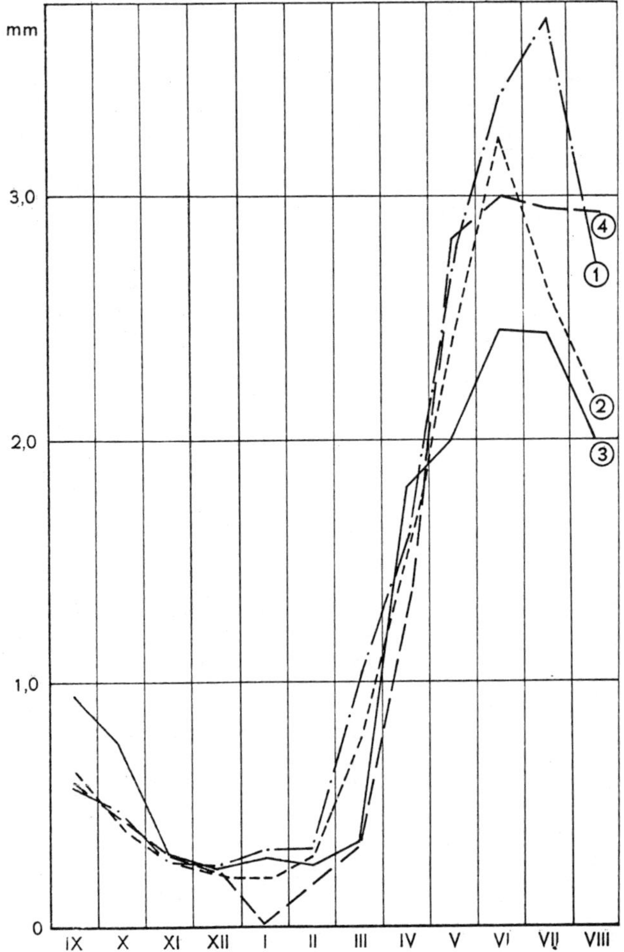

Fig 13. Annual distribution of evaporation in different research periods.
1. before drainage, 2. period of drainage works, 3. period of farm implements, 4. after drainage.

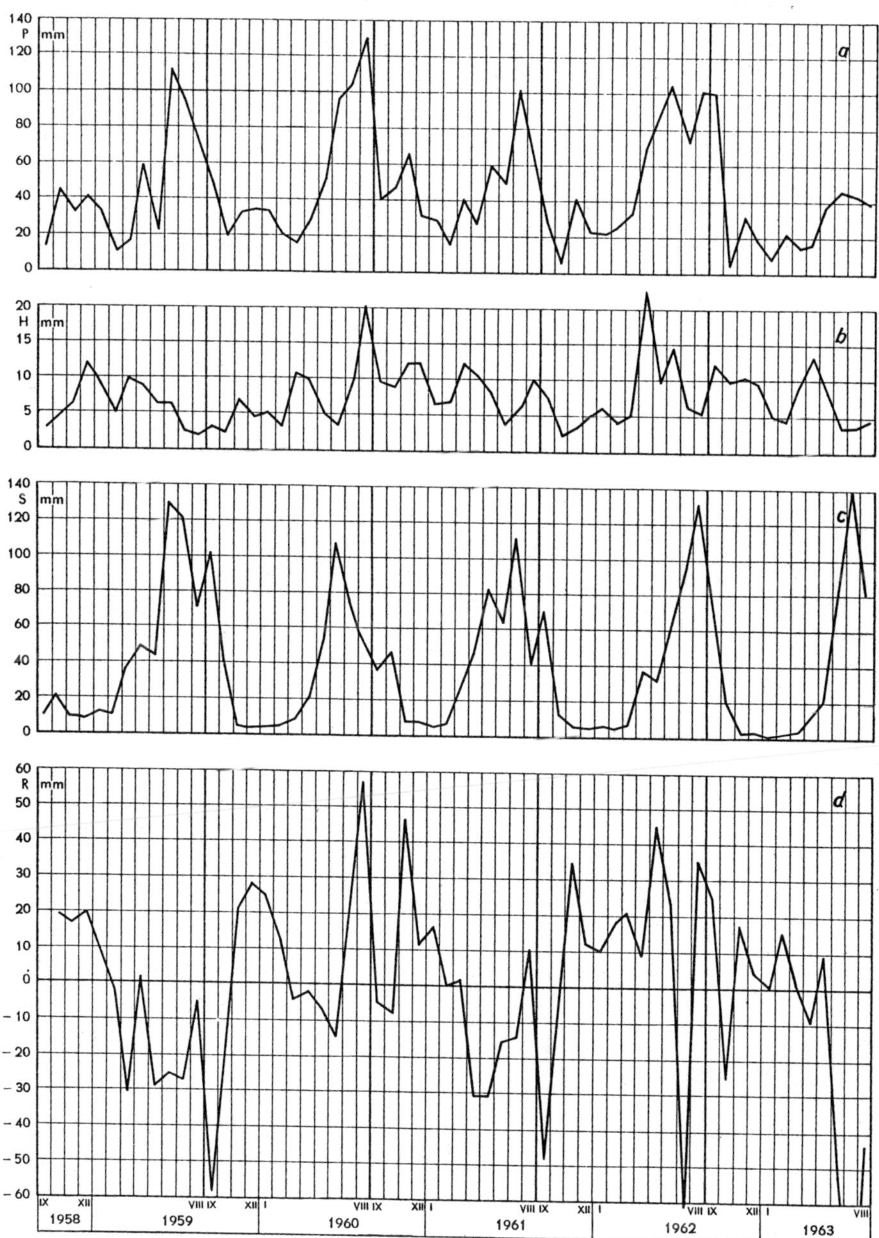

Fig. 14. Distribution of water-balance elements: precipitation (a), runoff (b), runoff deficit (balance losses) (c), retention differences (d).

Table 6. Computation of Ground Retention by Kamienski's method of finite differences (Orsztynowicz, 1963)

Period	Storage increase	Groundwater storage W	Precipitation P	Ratio W P	Evaporation from groundwater table -W	Groundwater inflow R	Groundwater outflow -R
	mm	mm	mm	%	mm	mm	mm
May 1959 - April 1960	-9.0	138.0	483	28.5	151	4.20	0.05
May 1960 - April 1961	-18.9	86.4	643	13.4	102	0.27	3.67

Table 7. Water Balance of the Michalowo-Imszar Peat-bog (1959-1963

Year	input		Output			Storage		
							R=R - Z	
	mm	mm	mm	mm	mm	mm	+	-
1958/59	162.6	543.8	706.4	76.5	523.4	106.5	32.1	88.2
1959/60	106.5	612.9	719.4	85.0	473.0	161.4	135.1	80.2
1960/61	161.4	567.9	729.3	106.7	480.6	142.0	44.7	64.1
1961/62	142.0	613.0	755.0	91.6	466.5	196.9	113.3	58.4
1962/63	196.9	376.2	573.1	92.9	440.1	40.1	29.8	186.6

INTRODUCTION

Adequate drainage is essential for the cultivation of peat soils in the temperate zone, but such drainage causes physical changes which never end. The macropore space and permeability decrease whilst bulk density and compaction increase. This paper gives information on this topic of importance from the agricultural engineering point of view.

The principal water balance data of the north-western regions of the FRG (Table 8) shows the necessity of drainage.

Table 8. Principal water balance data for high-bog grassland (1951/58)

Period	Rainfall	Run-off	Evaporation	Storage
Winter	364	176	93	+95
Summer	378	75	398	-95
Year	742	251	491	0

Subsidence of peat surface resulting from drainage

Figure 15 shows how the solid matter content, macropore space, subsidence and permeability are related for the complete drainage period. This has been found to hold in all drained peat areas of Western Europe, Canada, Scandinavia and Yugoslavia. The magnitude of the alterations change with the kind of peat and degree of moisture content as with the intensity and duration of drainage. The external indication of the physical effects of peat drainage is the surface subsidence which can be seen around piles which have been driven into the mineral subsoil. In the past two decades predictions have been made about the anticipated subsidence in sixty moorlands. Where it was possible to verify these predictions by measurements after drainage a good correlation was obtained.

In order to predict subsidence the apparent density of the peat was measured and Haleakorpi's empirical formula for peat subsidence was applied (Segeberg 1960). This states that

$$S = a\ (0.080\ T + 0.066)$$

where: S is the subsidence, a is factor of relative peat density, and T is the thickness of the peat layer (m).

PHYSICAL EFFECTS OF DRAINAGE IN PEAT SOILS OF THE TEMPERATE ZONE AND THEIR FORECASTING

R Egglesmann

Peat Soil Research Institute
Bremen, FRG.

SUMMARY: Information is given on the influence of drainage. The utilization of peat soils is impossible without drainage, but their macropore space and their permeability decrease after drainage, and the micropore space, bulk density and consolidation increase. Data is given to estimate drain depths with regard to subsidence. Lastly a diagram is presented showing drain spacing in the fen and highbog in the north-west of the Federal Republic of Germany relative to the degree of moisture content and depending on the duration and intensity of pre-draining.

EFFETS DU DRAINAGE SUR LES CARACTERISTIQUES PHYSIQUES DES TOURBIERES DE LA ZONE TEMPEREE. LEUR PREVISION.

RESUME : Cette étude nous donne des informations sur les effets du drainage. Il est impossible d'utiliser les tourbières sans drainage préalable. Le drainage entraîne une réduction de l'espace occupé par les macropores et une diminution de la perméabilité, en même temps qu'un accroissement de l'espace occupé par les micropores et une augmentation de la densité et de l'affaissement. Quelques données sont fournies permettant de calculer la profondeur des drains en rapport avec l'affaissement du sol. Enfin un diagramme montre l'espacement des drains dans les marais hauts et bas de la région nord-ouest de la République fédérale d'Allemagne, en rapport avec le degré d'humidification, avec la durée du drainage et avec l'intensité de celui-ci.

Drain depth before and after draining

All peat layers consolidate after draining. Several measurements made after installing drains have shown that even peat layers below drains are subject to consolidation as shown in Figure 16. This figure shows diagramatically the subsidence of a peat surface and of a drain, and quantifies the amount of consolidation for different thicknesses of peat layers in centimetres and as a percentage of the original layer. It can be seen that the thicker the peat layer, the smaller the compaction. The compaction curve is cumulative.

Table 9 shows the figure to be added to the depths of ditches or subsoil drains to allow for subsidence for various qualities of peat expressed as a percentage of the desired depth (Segeberg 1960).

Table 9. Quantity to be added (%) to desired depths of drains according to relative peat density

Relative peat density	Desired depth of drain (m)					
	0.8	1.0	1.2	1.4	1.6	1.8
	% to be added					
Dense	10	12	14	16	18	20
Rather dense	15	17	20	23	25	28
Rather loose	21	26	30	34	38	42
Loose	31	38	45	51	58	65
Nearly floating	no drainage by pipe					

Peat permeability and drainage

The correlation shown in fig. 17 shows that as moisture content increases the influence of the subfossil plant residuals on the permeability of the peat decreases. This conclusion was reached after statistical analysis of more than 2000 measurements of permeability and moisture content. Phragmites peat has the highest permeability followed by Carex peat. Sphagnum peat has the lowest permeability.

The duration and intensity of drainage have an important influence on permeability. After draining the permeability decreases very quickly at first but later the decrease becomes asymtotic to a minimum value (Figure 18).

When a moorland is drained it is only the water contained within the macropore space of the peat that is discharged. When this free water has been removed, the macropore spaces of peat are transformed into micropore spaces. This results in a smaller permeability and higher bulk density as well as consolidation of the peat soil.

Drain spacing before and after drainage of peat soils

A great number of formulae exist for calculating drain spacing according to soil permeability. Mufller (1967) has compiled 61 formulae from world-wide sources. There are several nomograms by which drain spacing may be determined according to the permeability of the soil.

With the help of Hooghudt's formula (Van Beers 1965) the drain spacing was calculated for fen and high-bog under the conditions prevailing in the north-west of the FRG but instead of permeability (according to Figure17) moisture content was plotted against drain spacing (Figure 19). This diagram is suitable for moorlands with a peat thickness of 2 m. The wider drain spacings are suitable for moorland with only slightly predrained peats, the closer drain spacings are required for intensive and long-term drained moorland. The diagram can be used for subdrains using all kinds of pipe and mole drains.

Mole-drains in peat soils

Twenty years ago a draining machine was developed which was designed especially for high-bogs in the north-western regions of the FRG. It has since been used to cut more than 30 000 km of mole-drains in nearly 25 000 ha of peat in the FRG, Ireland and Switzerland, and has been effective in peat areas used for meadows, forests and in the peat industry.

This drain milling machine produces a subsoil mole-drain channel with a rectangular cross-section of 20 cm x 15 cm (height x width) with a slope at a depth of between 0.8 m and 1.6 m. Two milling wheels run in opposite directions and bring the milled peat material to the drain centre. The scrape chain running in the sward brings this material to the soil surface.

In peat soils mole-drains are generally single drains which discharge into open ditches. The cost for mole-drains made by a drain milling machine amounts to 0.30 DM/m (1972 price).

The service life of the unlined mole-drains usually depends on the relative peat density, expressed in terms of volume percentage of solid matter content (Table 10).

Table 10. Relationship between relative peat density, volume percentage of solid matter content and unlined mole-drain service life

Relative peat density	Solid matter content (Vol. -%)	Service life of unlined mole-drain (years)
Dense	12	8
Rather dense	12.0 - 7.5	8 - 5
Rather loose	7.4 - 5.0	5 - 3
Loose	4.9 - 3.0	3 - 1
Nearly floating	3	1

Although contemporary experience suggests the effective service life of drains to be as shown in Table 3 there are some hand-dug drains in the Frisian high-bogs which were constructed four decades ago and which are still effective. Peat density is obviously a vital factor in drain performance and life but good maintenance and the requirement that the drain shall discharge above the water level in the main ditch are practical points of good husbandry which cannot be overestimated.

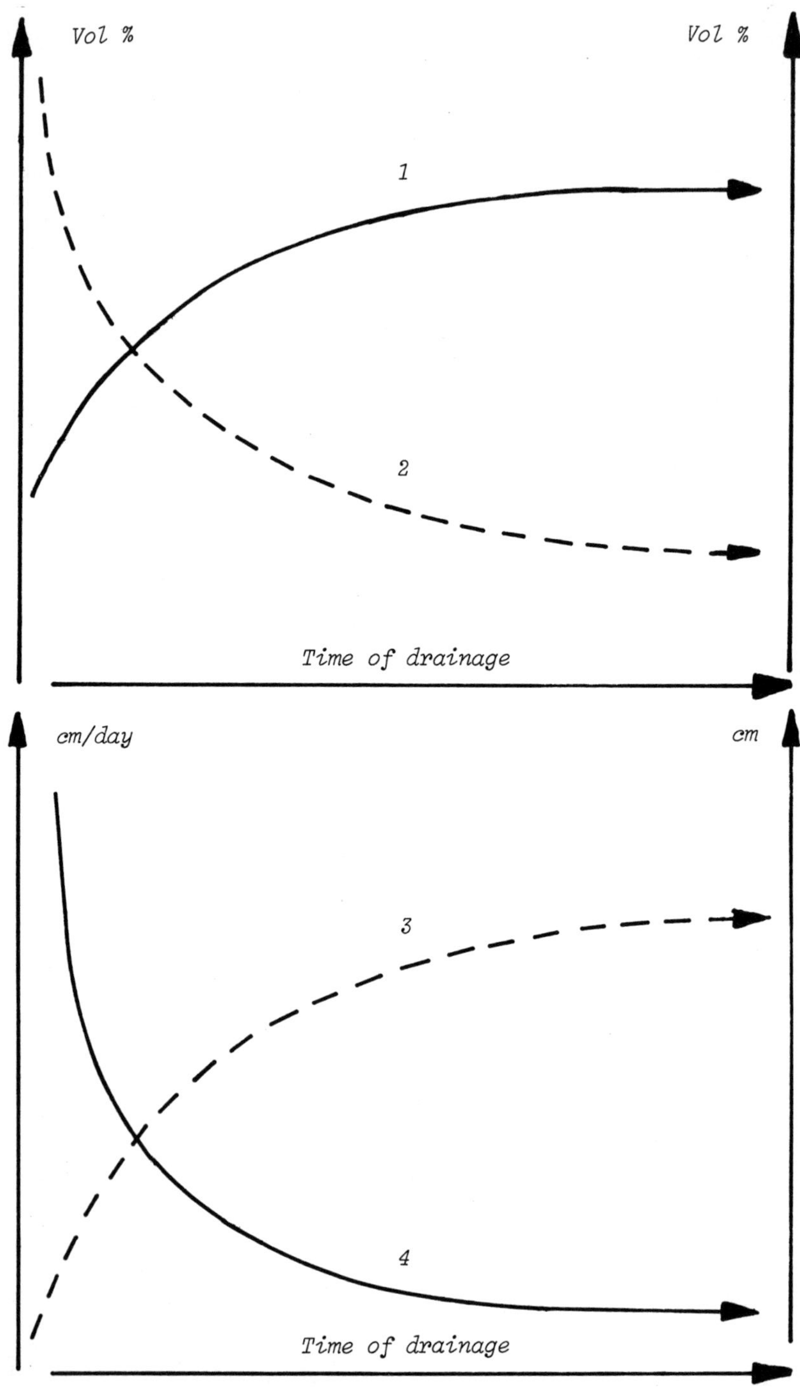

Fig 15. Influence of peat drainage on soil structure.
1. Bulk density; 2. Drainage pore space;
3. Subsidence; 4. Permeability.

References

Van Beers, W. F. J. 1965. Some nomograms for the calculation of drain spacings. Wageningen, Veenman & Zonen.

Segeberg, H. 1960. Moorsackung durch Grundwasserabsenkung und deren Virausberechnung mit Hilfe empirischer Formeln. Zeitschrift für Kulturtechnik, 1: 144-161.

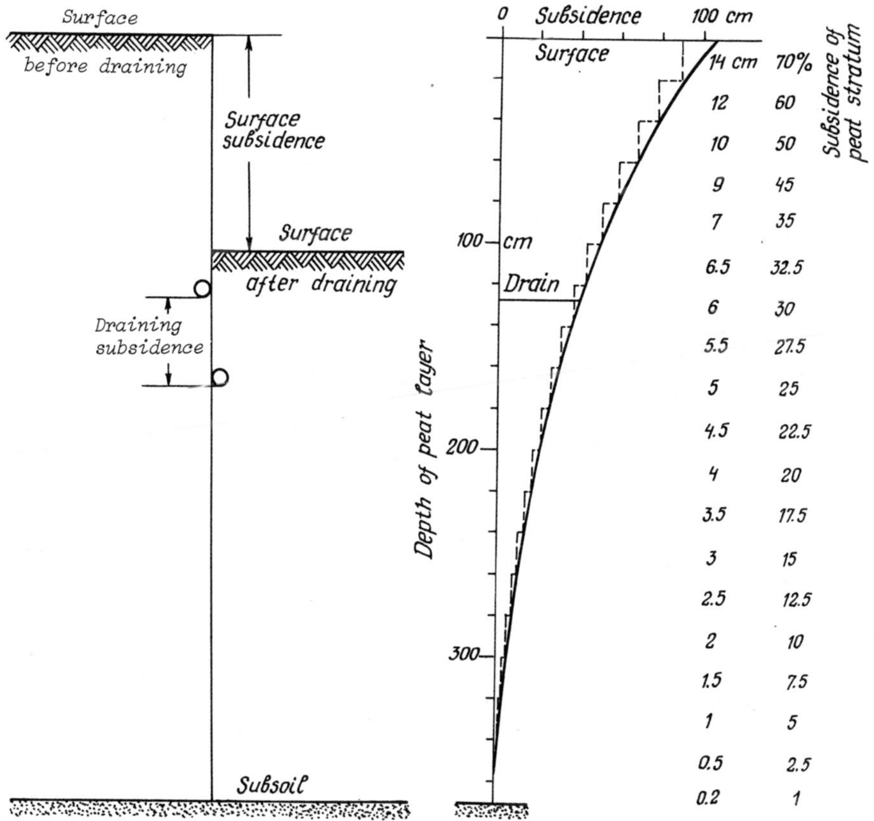

Fig 16. Subsidence in different peat layers.

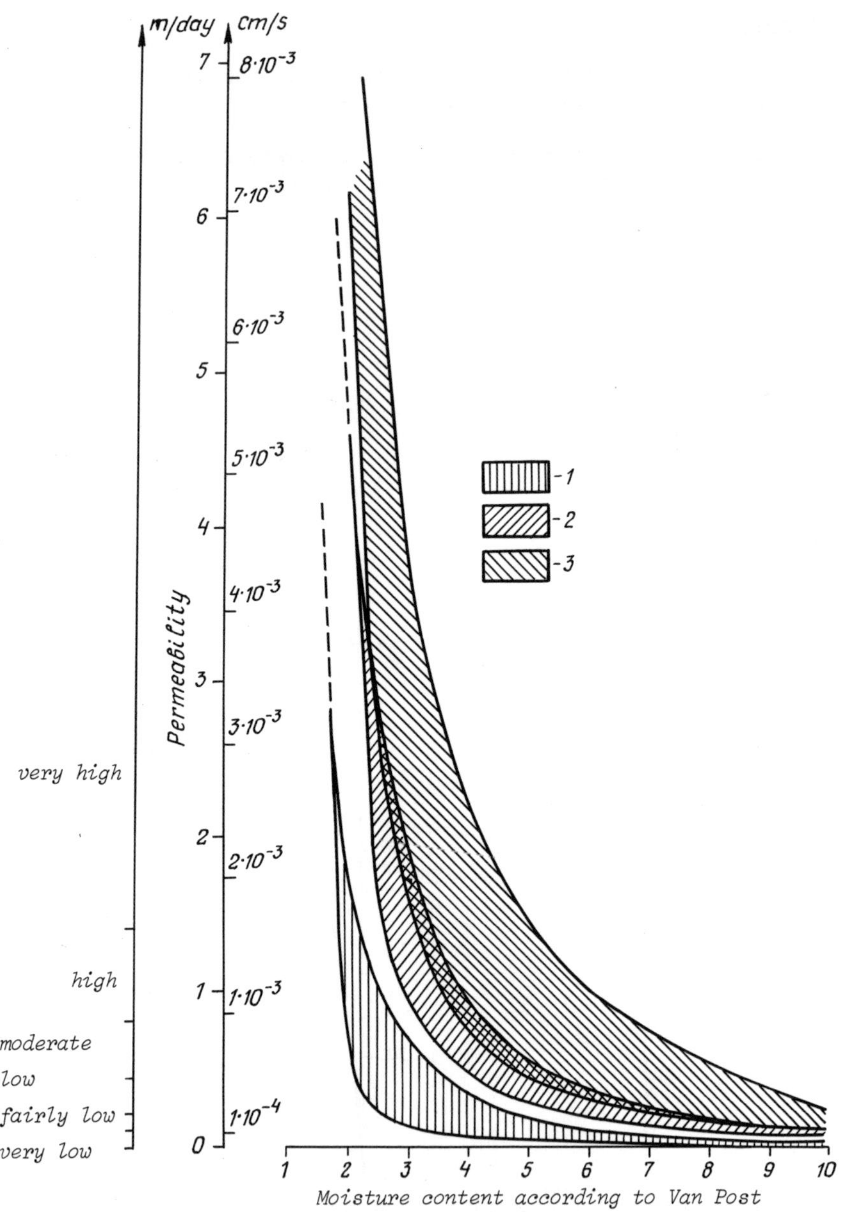

Fig 17. Relationship between permeability and soil moisture for average peats.
1. Sphagnum peat; 2. Bryales-sedge-peat;
3. Phragmites and Carex peat.

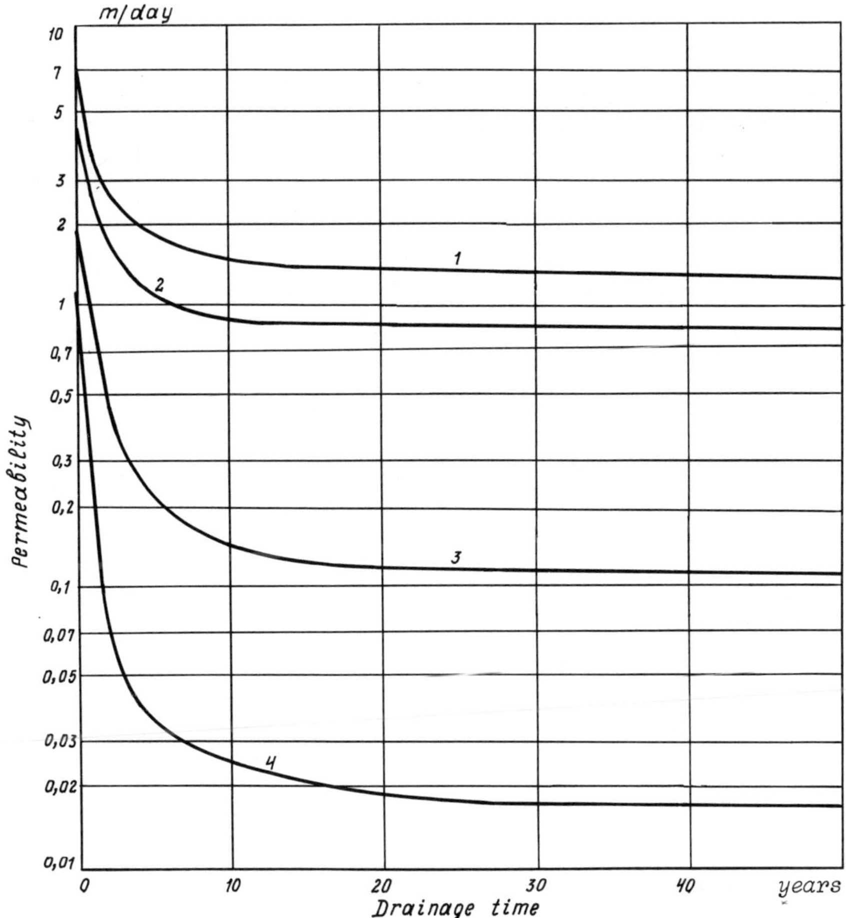

Fig 18. Decrease with time of peat permeability due to drainage.
1. Woody peat; 2. Phragmites peat;
3. Sphagnum Peat; 4. Mud.

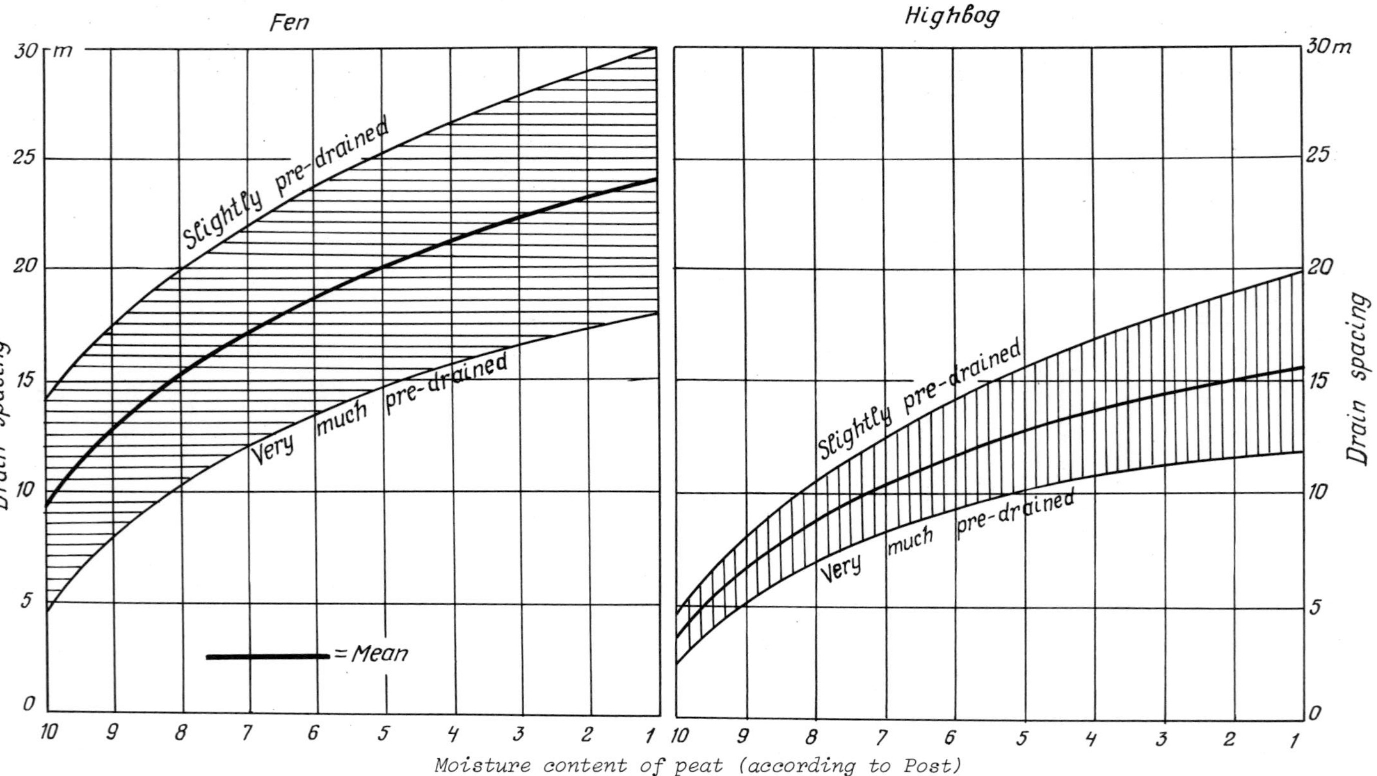

Fig 19. Subdrain spacing for fen and highbog with peat layer of 2 m with respect to moisture content.

HYDROPHYSICAL INVESTIGATIONS OF BOGS IN THE USSR

V. V. Romanov, K. K. Pavlova, I. L. Kalyuzhny, P. K. Vorobiev

State Hydrological Institute, Leningrad, USSR.

SUMMARY: The paper presents the results of investigations into the physical properties of the layer of active moisture and heat exchange in bogs (active layer of swamps) and the processes occurring in this layer.

The measured and predicted values of the hydraulic properties of the layer are compared in respect of inter-cell capillary moisture capacity, water yield, permeability and capillary properties. Thermal properties of the layer (heat capacity, thermal conductivity, thermal diffusivity) are also presented and the interrelation of thermal and hydraulic properties is considered.

The results of studies of evaporation including interception evaporation are given. The relations describing this process and the results of investigations of frozen swamps are presented.

RECHERCHES HYDROPHYSIQUES SUR DES MARAIS EN URSS

RESUME : Cette étude présente les résultats de recherches sur les propriétés physiques de la couche active du marais ou s'effectue l'échange d'humidité et de chaleur, ainsi que sur les phénomènes qui se déroulent dans cette couche.

L'étude compare les résultats des mesures avec les prévisions effectuées sur les propriétés hydrauliques de cette couche : pouvoir de rétention intercellulaire et capillaire, apport en eau, propriétés de filtration et propriétés capillaires ; les propriétés thermiques (conductivité thermique, capacité calorique, diffusion de la température) sont également étudiées, comme aussi la corrélation entre propriétés thermiques et propriétés hydrauliques.

Sont présentés également les résultats des recherches sur l'évaporation (y compris l'évaporation de l'eau interceptée) ainsi que sur le processus de ce phénomène, de même que les résultats des recherches sur des marais gelés.

The following topics are characteristic of the development of hydrophysical studies of bogs in the USSR:

a. The composition and hydraulic properties of the active layer of a peat deposit and the relevant hydrodynamic processes under molecular and gravitational forces.
b. The thermal properties and processes of heat exchange with the atmosphere; the thermal balance of bogs.
c. The phase transformation of water (bog freezing and melting), physical properties of the active layer when frozen, evaporation and condensation (vapour transfer within the active layer).

Since the forest zone of the USSR is mainly covered by convex high (sphagnum) bogs, particular attention was paid to these bogs from the outset.

The active layer of peat deposits is characterized on high bogs, by a considerable variability of properties in both vertical and horizontal directions. Vertical changes are related to changes in the compression rate and partial decay of the vegetative matter. Horizontal changes may be explained by the distribution of micro-relief features of different height and variations in vegetal cover over the area. Therefore, the study of the physical properties of the active layer within each of the bog micro-landscapes should be performed individually for each different micro-relief element. Thus attention should be paid to the study of the surface shape of the bog micro-landscape (area distribution curves) covered by micro-relief features of different height (Vobobiev, 1966).

The density of dry vegetal material in the active layer does not exceed 0.02-0.08 g/cm^3. The porosity of this layer is high.

The external surfaces of plant stems in sphagnum moss usually does not exceed 0.01-0.1 cm^2/cm^3, and the amount of water absorbed onto this surface is small. The high water absorptivity of this layer may be attributed to substances within the cells of plants. There are two types of pores in the active layer which contain capillary and gravitational water. These pores are inside the cells of living or dead, but not decayed, plants. The amount of inter-cellular water depends on the plant species and the rate of compression and decay of the sample; the amount of capillary moisture decreases with increase in height above the groundwater level, generally following a hyperbolic law. More accurate data on capillary moisture distribution may be obtained using a capillarimeter.

If the capillary moisture distribution curve is known, it is possible to calculate the pore size distribution curve. The latter may be used to obtain the hydraulic conductivity as a function of the degree of wetting and its limit, namely the coefficient of permeability (in a two-phase system). It is also possible to calculate capillary outflow, this being the amount of water raised by capillary forces per unit of time at some given height, and the volume of water yield.

The coefficient of permeability in the active layer tends to decrease rapidly with depth from the order of 100 cm/sec at 1 to 2 cm depth, to the order of .001 cm/sec at the lower boundary of the active layer. The parameters for these curves were obtained for different micro-relief features in different bog micro-landscapes and a method of calculating infiltration outflow as a function of the bog water level was developed, (Ivanov, 1967).

The curves $\xi_z = f(Z)$ of water yield coefficients versus the bog water level obtained experimentally for different micro-relief features, progressively change their form, depending on the height of a micro-relief feature relative to the predicted micro-landscape surface (PMS)*. The curves obtained for micro-relief features of the same height, relative to the PMS, practically coincide. Therefore, a single calculation of the water yield cannot describe the water yield of the micro-landscape as a whole; samples for these cal-

---

*On non-drained bogs this change depends on the depth of water level.

culations should be selected for all micro-relief heights. The general water yield curve of the micro-landscape is plotted on the basis of partial water yield curves and data on the proportion of micro-landscape area covered by specific micro-relief features, (Vorobiev 1967).

Thermal properties of the active layer (heat capacity, thermal conductivity and thermal diffusivity) were also studied for high bogs in melted and frozen states (Romanov & Rozhanskaya, 1948; Romanov, 1961) poorly drained and undrained moors (Shebeko, 1957; Pavtova, 1969).

It was shown that thermal conductivity of the active layer depends primarily on the volumetric moisture content of the layer, its temperature, density and pore structure. Heat is transferred not only by molecular heat conduction, but also by moving liquid water and water vapour. The contribution of the separate heat transfer modes depends on the size of pores and the degree of wetting. With the decrease of temperature below zero some water changes into the solid phase; the nature of the phase transformation depends on the degree of dispersion, compression, porosity and mineralization of the water in the active layer.

In sphagnum moss, phase transformations occur at temperatures close to zero; in more compact sphagnum peat they take place at a temperature of 0° to -3°C. In fen peats the amplitude of intensive phase transformations is extended to -10° to -12°C (Pavlova, 1969).

Internal evaporation of a peat deposit represents a small (2-10%) portion of the overall evaporation; however, its role in the process of heat transfer is significant. It is possible in some cases to calculate (on the basis of experimental determination of heat flux) the amount of vapour moving both towards the bog surface and towards the water surface, the dependence of the intensity of the above process on the temperature gradient and the amount of liquid water transferred (Belotsetkooskaya, *et al*, 1969.

The temperature regime and heat balance of high bogs were studied by I. I. Katyuzhnaya and I. L. Kalyuzhny (1909), as well as by other researchers.

Much attention has been paid to elucidating the mechanisms of frosts in drained bogs and to measures for their control (Goltsberg, 1955; Goncharova *et al*., 1957, Pavlova, 1969).

Evaporation and heat losses due to evaporation constitute the most important expenditure components of both the water and heat balances of bogs. For this reason much attention has been paid to these processes at bog stations of the network of the Hydrometeorological Service of the USSR.

Despite the great amount of research work devoted to evaporation from bogs (Gerasimov, 1925; Dubahk 1936, Romanov, 1953; Shebeko,   ), much remains to be explained: only one aspect of the process has been investigated, namely, that of water losses by evaporation, whilst energy income and heat losses for phase transformation of water have been neglected.

Different evaporimeter techniques and methods were applied by the Hydrometeorological Service (turbulent diffusion method, heat balance method) for the study of evaporation from bogs.

Investigations of evaporation from high bogs in combination with heat exchange studies show that evaporation is directly proportional to net radiation. The coefficient of proportionality or "specific evaporation" (in mm/cm^2 kg cal^{-1}) depends on the phase and stage of plant development, and within each phase it depends on the degree of wetting (Romanov, 1955;). Having found the value of specific evaporation and the seasonal variations thereof it was then possible for the normal mean annual evaporation values from bogs to be calculated using net radiation data for the bog surface. These values were used to compile an evaporation map for high bogs and low moors of the European territory of the USSR (Romanov, 1962).

More recently, specific evaporation values and the evaporation rate in different regions of the USSR have been evaluated more precisely on the basis of the heat balance method and observed data from GBI-B-1000 and some other types of evaporimeters installed at the bog stations.

Knowledge of the dependence of evaporation upon meteorological parameters, depth of bog water table, runoff and water yield (also water level) has made possible, by means of a simultaneous solution of the equations of evaporation, runoff and water yield, the development of a mathematical model of the system. This model describes input and output of moisture and calculates the water balance and groundwater level hydrograph in sphagnum low-bush-micro-landscapes for any given meteorological conditions. By introducing appropriate correction coefficients, the evaporation rates can be forecast in sphagnum-lowbush and other high-bog micro-landscapes (Romanov, 1960).

Computations have shown that evaporation in May and June, when moisture stored in the spring is still available, depends mainly on net radiation and only slightly on precipitation. On the contrary, in July and August, evaporation depends mainly upon the occurrence of precipitation.

The methods used for the measurement of evaporation have certain advantages but also disadvantages; for example a heat balance method which allows the main relationships of evaporation to be determined in a comparatively short time, gives average evaporation rates over fairly large areas. In most cases this method may be regarded as perfectly adequate, but at the same time it is unsuitable for the study of evaporation from different micro-relief features (ridges, pools, etc). In such cases it is advisable to use specific evaporimeters (GGI-B-1000 and GGI-B-1000M). Micro-landscapes with pronounced micro-relief are characterized by considerable variation in evaporation over the area. It is therefore necessary to install 5 to 7 GGI-B-1000 evaporimeters on typical micro-relief features and to maintain an appropriate bog water level in order to determine mean evaporation (Tyuremnov, 1928).

Bog freezing has a major influence upon spring runoff. The initiation of peat extraction and of agricultural work on drained bogs depends on the time of melting of the frozen layer.

In the 1930s, the frozen layer of bogs was considered to be qualitatively homogeneous and capable of being characterised by a single value, namely the total depth (Dubahk, 1936; Pechkurov & Kaplan, 1937). It is that the properties of the frozen layer, especially of high bogs vary within wide limits. Depending on the density and ice content of the frozen layer, the thermal conducti-

---

* On undrained bogs it depends on the depth of the water table.

vity can vary by as much as 16 times (from 14 x 10^{-5} cal/cm sec deg) and the fracture resistance by more than 80 times, etc. (Romanov & Rozhanskaya, 1948).

For this reason, a two-term classification was developed for the visual characteristics of the frozen layer and this classification is used on all bog stations of the Hydrometeorological service of the USSR.

The rate of freezing of the positive micro-relief features (tussocks, ridges) is much greater at first than that of the negative elements (depressions, pools). The properties of the frozen layer of these micro-relief features differ sharply (Romanov, 1961).

Because of the very slight wetting of the active layer of the positive micro-relief features, the frozen layer of these is loose (crumbles in the hand) and has a low thermal conductivity and heat capacity. Only when the depth of freezing reaches the highly moistened layers does a very dense layer (specific gravity 1.0) form at the bottom of the frozen layer with high fracture resistance and high thermal conductivity. On negative micro-relief features even the upper layers are characterized by high density, fracture resistance and thermal conductivity.

Calculations of the freezing depth in bogs (Goltsberg, 1955; Deryugin *et al*; 1969) are in good agreement with the observed values only when time-dependent thermal conductivity and moisture content values of the frozen layer rather than average values are introduced into the formulae.

Since the thermal conductivity of the active layer in the liquid and frozen states depends on the moisture content, which in turn depends on the height above the bog water level, all the parameters, required for the calculations may be determined on the basis of the data on the relative level of the water in the bog.

Table 11. Conditions for confining bed formation in drained bogs

Type of bog	Density g/cm³	Initial critical moisture content at the critical temperature					
		0°	- 1°	- 3°	- 5°	-10°	-20°
Drained low moor, sedge-hypnum peat P_{tot} = 0.887	0.180	0.81 / 0.91	0.80 / 0.90	0.77 / 0.87	0.75 / 0.85	0.70 / 0.79	0.65 / 0.73
Drained low moor, saline carbonate peat P_{tot} = 0.856	0.230	0.80 / 0.94	0.77 / 0.90	0.73 / 0.85	0.71 / 0.83	0.66 / 0.77	0.60 / 0.70
Drained high bog, sphagnum peat P_{tot} = 0.93	0.110	0.86 / 0.92	0.85 / 0.91	0.835 / 0.90	0.82 / 0.88	0.79 / 0.85	0.74 / 0.80

Note: The upper figure gives the initial critical moisture content in cm³/cm³, the lower figure indicates the filling of pores at a given moisture content.

All available data on swamp freezing have been assembled and reviewed by S. A. Chechkin (1970). Who also developed a method for calculating the freezing depth using variable coefficients.

In regions where deep freezing of bogs occurs, the operation of drainage in spring greatly depends on the depth of freezing and the physical properties of the frozen layer. When the initial critical moisture content and temperature (see Table 1) at which the melt water penetrates the soil to fill vacant pores and freeze, are known, it is possible to calculate the depth of the confined bed in which further infiltration ceases with the result that seepage water does not enter the drains.

References

Belotserkovskaya, O. A.; Largin, I. F.; Romanov, V. V. 1969. Issledovanie poverkhnostnogo i vnutrizalezhnogo ispareniya na verkhovykh bolotakh (Surface and internal evaporation from high bogs). Trudy GGI, issue 177, Leningrad.

Vorobiev, P. K. 1969. Opredelenie vodootdachi torfyanoi zalezhi estestvennykh bolot (Determination of water yield from natural peat deposit). Trudy GGI, issue 177, Leningrad.

Vorobiev, P. K. 1966. Provierka rezultatov experimentalnykh issledovanii vodno-fizicheskihk svoistv torfa k poverhknosty bolot. (Application of experimental data, on hydrophysical properties of peat, to swamp surfaces) Trudy GGI, issue 177, Leningrad.

Gerasimov, D. A. 1925. Iz rezultatov statsionarnykh issledovanii na verkhvom bolote. (Some results of stationary research on a high bog). Torfyanoe Delo, No. 6.

Goltsberg, L. A. 1955. Zamorozki na osushennykh bolotakh (Frosts on drained bogs). Trudy GGI, issue 49 (III), Leningrad.

Goncharova, V. I.; Zavyalova, I. N.; Petrova, I. A.; Romanov, V. V.; Ryvkina, V. B. 1957. Nekotorye voprosy gidrologii, gidrofiziki bolot. (Some aspects of swamp hydrophysics). Trudy GGI, issue 60, Leningrad.

Deryugin, A. G.; Ivanov, K. E.; Kuznetsov, E. N.; Novikov, S. M.; Romanov, V. V. 1969. Promorazhivanie bolot dlyaustroistva letnihk dorog v bolotno-taezhnoi zone Sibiri. (Bog freezing for construction of summer roads in the swamp-taiga zone of West Siberia). Trudy GGI, issue 157, Leningrad.

Dubahk, A. D. 1936. Ocherky po gidrologii bolot. (Papers on swamp hydrology). Moscow, Izd. GGI.

Ivanov, K. E. 1957. Osnovy gidrologii bolot lesnoi zony. (Elementary hydrology of swamps in the forest zone). Leningrad, Gidrometeoizdat.

Pavlova, K. K. 1969. Teplovye svoistva deyatelnogo sloya bolot (Thermal properties of the Active layer in swamps). Trudy GGI, issue 177, Leningrad.

Pechkurov, A. F.; Kaplan, M. A. 1937. Opredelenie glubing promerzaniya i ottaivaniya bolot. (Determination of freezing and melting depths of swamps). In: Osvoenie zabolochennykh zemel. (Collected Papers). Moscow - Leningrad, Izd. VASKhNIL.

Romanov, V. V.; Rozhanskaya, O. D. 1948. Opyt issledovaniya fizicheskikh svoistv promerzshego sloya bolot. (Generalized experience of investigation of physical properties of swamps). Trudy GGI, issue 7(61), Leningrad.

Romanov, V. V. 1961. Gidrofizika bolot. (Hydrophysics of swamps). Leningrad, Gidrometeoizdat.

Romanov, V. V. 1953. Issledovanie ispareniya so sfagnovykh bolot (Evaporation from sphagum bogs) Trudy GGI, issue 39(93), Leningrad.

Romanov, V. V. 1962. Isparenie s bolot evropeiskoi territorii SSSR (Evaporation from swamps in the European part of the USSR). Leningrad, Gidrometeoizdat.

Romanov, V. V. 1960. Izmerenie vodnogo balansa bolot v zasushlivye i vlaznye gody. (Water-balance variations in dry and wet years). Trudy GGI, issue 89, Leningrad.

Tyuremnov, S. N. 1928. Mikroklimaticheskie nablyudeniya na verkhovom bolote. (Observation of the micro-climate of high bogs). Trudy nauchno-issledovatelskogo torfyanogo instituta, issue 1. Moscow.

Shebeko, V. I. 1957. Teplovoi rezhim torfyanykh pochv. (Thermal balance of peat soils). Trudy Konferentsii po osvoeniyu bolotnykj i zabolochennykh pochv. Minsk, Izd. Akad. nauk BSSR.

Shebeko, V. I. Vliyanie osvoeniya bolot na rezhim ispareniya. The effect of cultivation on the rate of evaporation from swamps). Trudy 3 Vsesoyuznogo gidrologicheskogo s'ezda, vol. 3, Sect. gidrofiziki. Leningrad., Gidrometeoizdat.

Chechkin, I. A. 1970. Vodno-teplovoi rezhim neosushennykh bolot i ego raschet. (Water and thermal balance of swamps and their calculation). Leningrad, Gidrometeoizdat.

DISCUSSION

L. Wartena (The Netherlands) - Your say these values cannot be neglected when dealing with the water table? When can they be neglected? Please give an example, if possible.

I. L. Kalyuzhny (USSR) - These values were obtained from special formulae in works on moisture bedding. We found that vapour can migrate towards the groundwater table, but this quantity is infinitesimal in the bog balance and can be neglected in calculations.

L. Wartena (The Netherlands) - Do your formulae include the thermal conductivity?

I. L. Kalyuzhny (USSR) - The quantity of water migrating to the groundwater table is small. Its importance in the thermal balance of swamps is small, and it is not included in the formulae.

MOISTURE ACCUMULATION IN DRAINED PEATLANDS

K. P. Lundin

Byelorussian Research Institute for Reclamation and Water Economy, Minsk, BSSR

SUMMARY: In this paper the accumulation of moisture in peat soil subjected to drainage is examined. It is also shown that bog drainage leads to an increase of surface runoff.

From the relation between the moisture potential and water content, the solution for equilibrium water storage in the unsaturated zone and groundwater level is obtained.

As a result of experiments carried out on different soils it has been possible to determine the water yield and field capacity of the upper layer of 0.5 m for various drainage rates.

It is known that on account of the rather large pore space of the shallow peats which are cultivated in the Polessie region, the summer precipitation is retained in the form of soil moisture within the saturated zone and hence no rise in the groundwater level occurs.

L'ACCUMULATION D'EAU DANS LES TOURBIERES SOUMISES AU DRAINAGE

RESUME : L'auteur expose la façon dont le drainage provoque l'accumulation de l'eau dans les tourbières. Il analyse l'effet du drainage des marais sur l'accroissement de l'écoulement de surface.

En partant de la corrélation entre l'humidité potentielle et la teneur en eau, il calcule la valeur du stockage d'eau en équilibre dans la zone d'aération au niveau des eaux souterraines.

L'exploration de différents sols a permis de déterminer quel est l'apport en eau et quel est le stockage d'eau de la couche supérieure (0,50 m de hauteur) pour divers taux de drainage.

La plaque de Polésie étant cultivée et ses dépôts tourbeux étant de faible épaisseur, il en résulte que son sol est partiellement libre d'humidité, de sorte qu'en été les précipitations atmosphériques restent en suspension dans la zone d'aération sans provoquer la montée de la nappe souterraine.

In the present paper two aspects of the occurrence of moisture in peat soils are considered. The first aspect concerns the initial stage of bog drainage during preparation for agricultural use and the second aspect concerns the use of the soil for agriculture.

Total water accumulation in a 1 m layer of undrained low-land bog amounts to 930 - 940 mm solid material forming 6 - 7% of the volume. As the groundwater level falls to 1 metre below the surface of the soil, the volume of solid material in the top layer increases from 6 - 7% to 10 - 12% as a result of water yield and peat subsidence. The thickness of the peat deposit is reduced from 100 - 150 cm to 50 - 100 cm, the soil moisture storage being as much as 430 - 800 mm.

Thus, initial bog drainage removes a secular layer of water storage of 500 - 600 mm or 5 000 - 6 000 m^3/ha. This, naturally, causes an increase in the runoff for the period. If the total area of watershed is 1 000 km^2, of which 50% is bog, the volume of water removed by initial drainage will be (5 000 to 6 000) x 50 000 = (250 to 300). 10^6 m^3. Calculations show that for a drain system with spacing of 200 - 300 m and depth of 1.8 - 2.0 m the groundwater level of a low bog falls for 1 - 2 years. The stabilization period of a peat deposit is about 3 years. During this period the average increase of runoff due to bog drainage is 3 m^3/sec for the 1 000 km^2 watershed. These values are somewhat tentative. However they illustrate the considerable water yield which results from bog drainage and show that the hydrological regime of rivers is affected.

Observations show (Lundin, 1964) that the properties of peat such as density and porosity, once these have become stabilized after drainage, are not liable to change with seasonal variations in the groundwater level. Soil moisture storage however, is subject to seasonal variations depending on weather conditions and water consumption by plants. In spring, when the soil is fully saturated, water storage is near maximum (350 - 400 mm in the 0 to 0.50 m layer) and in summer it is near minimum (150 - 175 mm).

Let us now examine the characteristics of soil moisture within the upper 0.5 m layer of a drained field which depend on the level of the groundwater table. Here a layer of this thickness has been chosen as representative of the (seasonal) average thickness of the rooting layer. The moisture storage is assumed to be in a state of equilibrium when the groundwater table is at a certain level and there is no evaporation, infiltration of precipitation or water consumption by plants. An idea of the nature of the moisture distribution and character of the moisture reservoir in the unsaturated zone of the soil layer may be obtained from the relation between the moisture potential and moisture content. It shows an equilibrium distribution of moisture within the soil profile after the draining off of free water and moisture re-distribution is completed.

The moisture potential at a point above the groundwater table in the unsaturated zone is proportional to the suction pressure.

In the capillary zone ($h < h_k$) the moisture potential (Φ) distribution with height above the groundwater table in a state of equilibrium is expressed by:

$$\Phi = -gh$$

(gf.em/g)

where

h is the height above the groundwater table (measured in cm);
g is the acceleration of gravity (cm sec^{-2}).

The value of the moisture potential at a point, expressed in terms of depression in centimetres of a water column, is equal to the height of the point (cm) above the free water surface, when in a state of equilibrium with the moisture equally distributed throughout the soil profile (Korchunov *et al*, 1960). At a point whose level is equal to that of the groundwater table, $h = 0$, and the potential Φ is equal to zero. At a point where $h = 50$ cm, $\Phi = -50$ (ie the potential energy gained by a mass of 1 gm when lowered by 50 cm).

Water storage in the layered soil under consideration depends on the distance above the water table and may be determined with the help of the relationship between moisture potential Φ and moisture content. This relationship is well described by the empirical equation obtained by Korchunov's method (1961):-

$$\Phi = \Phi_0 e^{-bv} \tag{2}$$

Figure 20 shows a family of curves for typical soils.

Experimental values of Φ and v plotted on semi-logarithmic co-ordinates fall on a line, the slope of which gives the value of b. This is the rate of moisture content decrease with increase in the absolute value of the potential.

Φ_0 is a conventional quantity since it is obtained by extrapolation of the relationship (2) with linear exponent up to $v = 0$, and at a low moisture content the absolute value of the potential increases at a greater rate than in the range of weakly bound water.

It is known that the lowering of the groundwater table results in reduction of moisture content within the upper soil layers. With no evaporation or infiltration of precipitation, the water suction continues until equilibrium is reached. At this moment the value of potential Φ at any point is equal to the height of the point above the groundwater level (cm).

The water stored in a given soil layer may be found from the relation $\Phi(v)$ and is the area between the curve $\Phi(v)$ and the axis Φ. The area may be evaluated by integration of equation (2) or geometrically.

From (2) it follows that

$$v = \frac{1}{b} (\ln\Phi_0 - \ln\Phi). \tag{3}$$

If we take a section between Φ_1 and Φ_2 on the curve $\Phi(v)$, the water stored in the corresponding soil layer will be

$$W = \int_{\Phi_1}^{\Phi_2} v d\Phi = \frac{1}{b} \ln \Phi_0 \int_{\Phi_1}^{\Phi_2} \ln \Phi d\Phi.$$

Upon integration we obtain the following expression

$$W = \frac{1}{b} \left| (\Phi_2 - \Phi_1) \ln\Phi_0 - \Phi_2(\ln\Phi_2 - 1) + \Phi_1(\ln\Phi_1 - 1) \right|. \tag{4}$$

Since $\Phi = h$ in the capillary zone at equilibrium and expressing the storage in mm, we obtain for the layer considered between h_1 and h_2 the following equation:

$$W_{mm} = \frac{10}{b}(h_2 - h_1)\ln\Phi_0 - h_2(\ln h_2 - 1) + h_1(\ln h_1 - 1)\Big| . \quad (5)$$

Based on the analysis of the relationship between moisture potential and moisture content for different soils of the Polessie region, the values for equilibrium water storage were determined for varying depths of the groundwater level.

The resulting estimates are given in Table 11 and plotted in Figure 21.

The mean water yield for the upper layer ($h_2 - h_1 = 500$ mm) was found to be

$$\bar{\delta} = v_t - \frac{1}{h_2 - h_1} \int_{h_1}^{h_2} v dh. \quad (6)$$

The data shows Table 11a) that due to the high water yield the arable layer and sandy soils respond most to drainage. In the sub-arable layers, especially the transition layers between peat and sand under drainage, water storage is less subject to change. Thus comparing water storage at drainage depths of 150 cm and 50 cm respectively, it is found to be lower in a peat soil arable layer by a factor of 1.4, in subsoil by 1.2, and in sand by 3.3

Precipitation water entering the unsaturated zone in excess of the equilibrium moisture content is not retained by the soil, but percolates down and reaches the aquifer with a resulting rise in the groundwater level.

Generally, the moisture content (accumulated moisture) on drained peatland supporting agriculture is below field capacity due to evapotranspiration. As a result of this saturation deficiency with respect to the equilibrium state, a free storage volume is available in the soil which is capable of absorbing a certain quantity of water in the form of suspended moisture without any resulting rise in the water table.

In undrained peatland, the pores are almost completely filled with water and the free storage volume is negligible. The occurrence of available storage at periodic intervals corresponding to falls in the groundwater level is uncertain and rather small (10 to 20 mm) and may be neglected. Precipitation minus evaporation accounts for the surface runoff. The storage capacity of a drained but uncultivated peatland is not very large and for drainage depths of 100 cm amounts, on average, to 150 mm, of which 100 mm is concentrated in the upper 0.5 m layer (Lundin, 1970).

Agricultural use of a peatland which results in a cloddy soil structure combined with moisture consumption by plants contributes to the increase of available free storage. During periods of intense plant metabolism the soil is able to hold excess moisture which can later be utilized by plants. The soil acts as an underground reservoir for seasonal regulation of groundwater runoff. This reservoir contributes to regulation of the soil moisture regime and to stabilization of the groundwater level. Therefore, in well structured peat soils, the unsaturated zone, with its notable soil moisture capacity, is responsible for appreciable levelling of seasonal peaks.

In general, it can be said that agricultural use increases the soil moisture capacity in comparison with drained but non-cultivated peatland by 1.35 to 1.50 times for wet years and by 1.5 to 2.0 times for dry years.

Available soil moisture storage capacity is affected by the thickness of the peat formation. When the peat layer is 300 cm thick (Kossovskaya Experimental Station) the available soil moisture storage is less by a factor of 1.5 to 2.5 than that of a peat layer of 50 to 70 cm thickness (Polesskaya Experimental Station). At the latter station plots consisting of a peat layer 100 to 130 cm thick offer an available soil moisture storage capacity 1.5 times smaller than plots of 50 cm depth (Figure 22), other conditions being equal.

Available capacity is subject to seasonal variations (due to decrease of moisture stored in the soil) and increases from a value of 50 to 150 mm at the end of April to as much as 200 to 250 mm at the end of August depending on the level of the groundwater table and the soil moisture content.

Table 12a gives values of the available capacity in the upper 0.5 m layer or deficiency with respect to the equilibrium moisture storage for the normal drainage depth of 100 cm.

On the shallow peatlands of the Polesskaya lowland, which have a spring water storage of about 320-400 mm in the 0 to 50 cm layer, the minimum summer water storage varies from 120 to 175 mm for peat deposits 50 cm thick and from 200 to 250 mm for peat 100 cm thick, ie to temporary wilting point in the first case, and the point of interruption of capillarity in the second case.

In Byelorussia there are periods of heavy rainfall at the end of June and the beginning of July during which the precipitation of

Table 12a. Available moisture storage capacity of peat soil

Moisture condition	Water storage (mm) in the 0 to 50 cm layer	Available capacity mm
Total saturation	430	0
Field moisture capacity (the upper limit of the optimal range)	355	0
Interception of capillary movement (the lower limit of the optimal range)	250	105
Temporary wilting point	150	205

Table 11a. Equilibrium water storage in the upper 0.5 m layer

Item	Soil type	Total water holding capacity	from equation (2)	from equation (3)	Groundwater level, cm	Water Storage in 0 to 50 cm layer, mm	Water yield in 0 to 50 cm layer, mm
1.	Hypnum-sedge peat layer, Polesskaya Experimental Station R=30-25%	0.895	6.2	7.86	50	400	0.10
					100	287	0.32
					150	248	0.40
2.	The same station. Subsoil, forest-phragmites peat, R=40-45%	0.900	9.0	11.00	50	450	0.00
					100	372	0.16
					150	345	0.20
3.	The same. Transition layer between peat and underlaying sand. Ash content 90%	0.585	19.0	13.00	50	265	0.05
					100	230	0.12
					150	215	0.15
4.	Kossovskaya Experimental Station, arable layer, phragmites-sedge peat, R=45%	0.880	11.5	10.60	50	335	0.21
					100	275	0.33
					150	252	0.38
5.	The same, Subsurface horizons	0.880	38.0	34.20	50	430	0.02
					100	390	0.10
					150	383	0.12
6.	Cherebasovskaya area, ploughing horizon, phragmites-sedge peat, R=40-45%	0.830	8.5	9.6	50	365	0.10
					100	313	0.20
					150	286	0.26
7.	The same. Subsurface horizons.	0.910	9.4	12.00	50	457	0.00
					100	408	0.09
					150	383	0.14

Table 11a. (continued)

Item	Soil type	Total water holding capacity	from equation (2)	from equation (3)	Groundwater level, cm	Water storage in 0 to 50 cm layer, mm	Water yield in 0 to 50 cm layer, mm
8.	The same. Peat mineral soil.	0.40	-	-	50	170	0.06
					100	168	0.07
					150	130	0.14
9.	The same. Transition layer, ash content 86.7%	0.630	27.4	20.0	50	304	0.03
					100	286	0.06
					150	275	0.08
10.	Polesskays Experimental station						
	a) Fine sand	0.400	4.7	4.10	50	175	0.05
					80	140	0.12
					100	115	0.17
					150	77	0.25
	b) Cherabasovsky region	0.347	5.0	4.90	50	135	0.08
					80	90	0.17
					100	60	0.23
					150	30	0.29

a few days can amount to 80 or 100 mm. These occur at a time when the moisture content of peat soil approaches the point of interruption of capillary motion and in the case of grass cover in dry periods soil moisture content may well decrease to the point of initial wilting. Under such conditions the soil easily absorbs precipitation and stores it in the form of soil moisture, which will be gradually used by plants and will not contribute to runoff.

In wet years runoff generated by infiltration may occur if the available soil moisture storage is not sufficient to absorb snowmelt (80 to 100 mm) or in summer when field capacity is reached at the time of heavy rainfalls.

Such is the hydrological effect of bogs.

References

Korchunov, S. S., Mogilevsky, I. I., Abakumov, O. N., Dulkina, S. N. 1960. Izuchenie vodnogo rezhima osushenykh torfyanykh zelezhei. (Study of water régime on drained peat soils). Moscow Leningrad, Gosenergoizdat. Proc. (Trudy VNIITP, issue 17).

Korchunov, S. S., Mogilevsky, I. I., Abakumov, O. N. 1961. Opredelenie vlagokoeffitscientov metoim postoyannogo raskhoda na khnosti obraztsa. (Determination of moisture coefficients by the constant discharge method on a sample surface). Moscow. Gosenergoizdat (Trudy VNIITP, issue 18).

Lundin, K. P. 1970. Akkumuliruyushchaya emkost torfyanoi pochvy (storage capacity of peat soil). In "Melioratsiya i ispolzovanie osushennykh zemel. Minsk, Urozhai.

Lundin, K. P. 1964. Vodnye svoistva torfov. (Water properties of peats). Minsk, Urozhai.

DISCUSSION

W. H. Van der Molen (The Netherlands):- Peat soils are known to have a good property, namely, they can retain moisture for a long time. I think this is a reason for high cereal crop yields on peat soils. It is not quite clear from the paper whether the natural moisture in soil completely supplies plants with water.

What do you think about the drainage effect on moisture accumulation in soil and crop yields? How do you control water regime on peatland?

K. P. Lundin (Byelorussian SSR): - Compared to mineral soils peat soils contain a considerably larger quantity of moisture. But if we consider available productive water storage (except stagnant water) rather than total storage, this quantity is quite comparable in such different soils as peat, loam and clay.

As it was stated in the paper, moisture storage in a peat soil depends on the groundwater level after drainage. The experience gained in Byelorussia shows that deep peat soils (more than 1 m) do not require additional irrigation for agricultural use. Moisture deficit may only be found in occasional periods without rain. As to shallow peat soils, of a depth of 0.5 to 0.6 m, the total available soil moisture is practically consumed by the end of July. If no rain falls for a long time, then additional irrigation is necessary. The question of the best way of supplying water to the soil has not been answered as yet. Sprinkler irrigation is certainly the most productive method but also requires high investments. At present it may only be recommended for vegetables and pastures. As to subsurface irrigation, intensive investigations are being carried out at present at the Byelorussian Research Institute of Reclamation and Water Economy.

S. E. Vompersky (USSR): - I am interested by some details of your work on accumulated moisture dynamics using moisture potential. We obtained moisture potentials outside the range 0.7 to 0.8 atm. Have you used any other methods? Do the results agree?

K. P. Lundin (Byelorussian SSR): - Moisture potentials may be measured with ceramic gauges (moisture tensiometers) within a pressure range up to negative pressures of 0.5 atm. To measure high tensions a reference-standard filter paper method may be used. I should like to state that a capillary potential of up to 0.5 atm is quite acceptable in practice and that in most cases tensiometers are quite suitable.

S. E. Vompersky (USSR): - Can you say something about the effect of drainage on soil moisture condensation?

K. P. Lundin (Byelorussian SSR): - Prof. V. F. Shebeko has reported condensed moisture layers of about 25 to 50 mm in summer; this is probably a negligible contribution to the soil moisture balance of peat lands.

G. V. Bogomolov (Byelorussian SSR): - What is the role of condensation in moisture accumulation in the upper layers? Does drainage operate as a regulator? What is the proportion of condensation in soil moisture?

K. P. Lundin (Byelorussian SSR): - Prof. V. F. Shebeko is working on condensation of soil moisture. I have mentioned that he reported a condensed moisture layer of 25 to 50 mm. Laboratory observations involve heat transfer. As to drainage its effect has been described in some text books.

P. M. Zakrzhevsky (Byelorussian SSR): - What are the pore dimensions which cause the increase of storage capacity in shallow peat deposits?

K. P. Lundin (Byelorussian SSR): - In any type of soils, peat included, three categories of pores may be distinguished, namely:- macropores, interpores and micropores. In peat soils macropores dominate in moisture conduction, in drained peat soil micropores prevail. They affect moisture conductivity in a narrow range. Storage capacity in shallow peat deposits is mainly influenced by macropores.

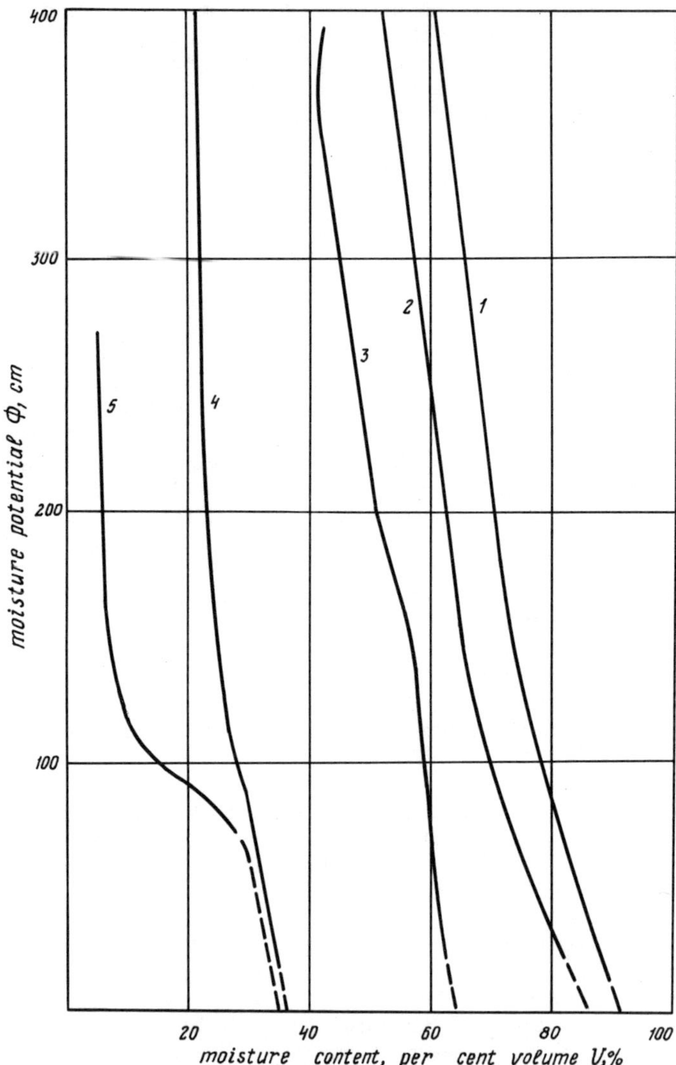

Figure 20. "Potential-moisture content" curves
1.2 peat soil, subsoil and arable layers,
3 transition layer; 4, peat sand;
5 fine sand

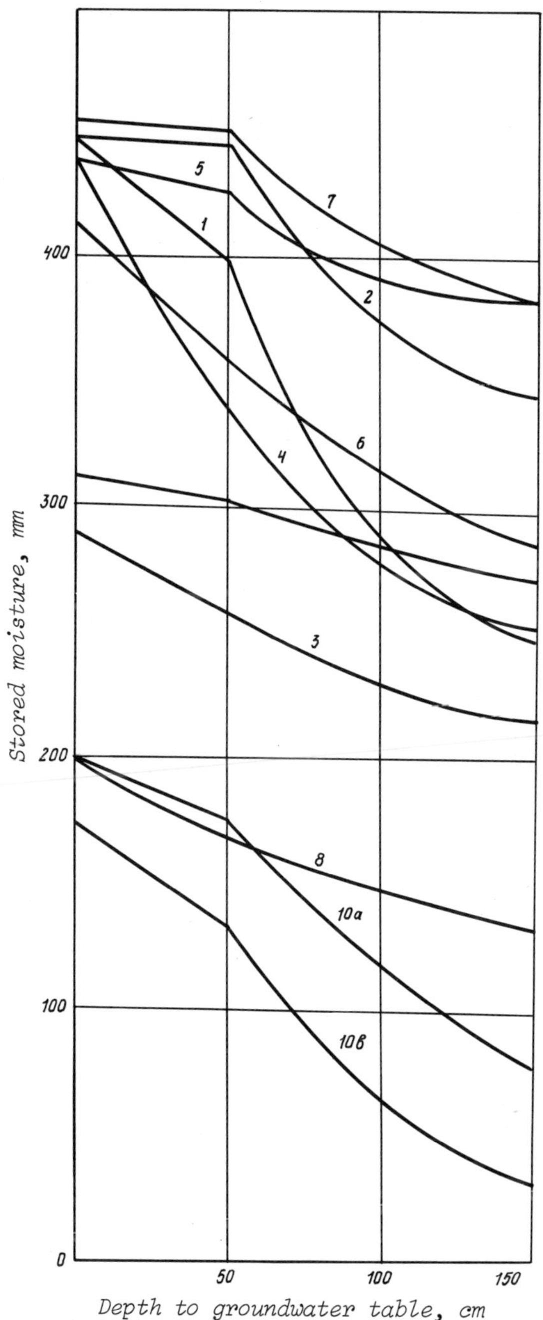

Figure 21 Equilibrium water storage in the upper 50 cm layer versus normal drainage rate
Curve Nos are in accordance with Table 1

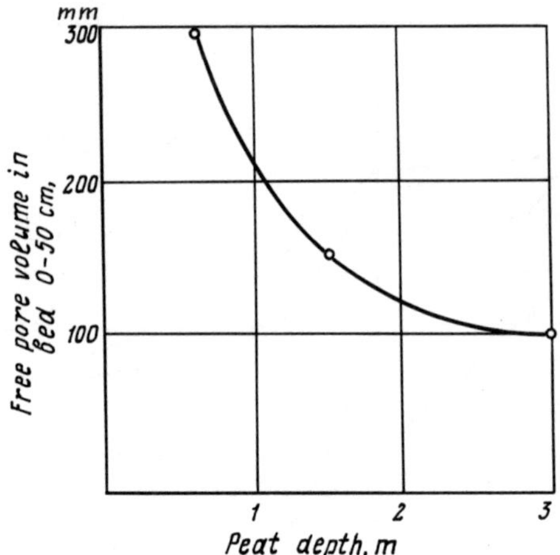

Figure 22 Available soil moisture storage capacity versus peat depth

HYDROLOGICAL FUNDAMENTALS OF BOG DRAINAGE

A. I. Ivitsky

Byelorussian Research Institute of Reclamation and Water Economy, Minsk, BSSR

SUMMARY: This paper considers the following problems:

The principles of reclamation by drainage;
The drainage rates (optimal groundwater regime) and new approaches to their determination;
The theory of drain spacing calculations for drainage and irrigation;
It allows groundwater dynamics and the required water regime of drained bogs to be estimated;
The method of calculation of the water yield of peat soil;
The method of calculation of drainage runoff from fen bogs;
The dependence of evaporation and transpiration coefficient on climate. groundwater table and crop yield.

FONDEMENTS HYDROLOGIQUES DE L'ASSECHEMENT DES MARAIS

RESUME : En partant des données fournies par des observations pluriannuelles et en s'appuyant sur des recherches théoriques l'auteur étudie les problèmes suivants :

Principes actuels qui président à l'assèchement par drainage ;
Degrés d'assèchement des marais (régime optimal des eaux souterraines) et nouvelles méthodes pour les déterminer en fonction de l'influence de la nappe souterraine et des facteurs climatiques sur la teneur en humidité et en air, en fonction aussi de l'influence de la nappe souterraine sur la présence d'oxygène et de gaz carbonique ;
Théorie du calcul de l'espacement entre les drains (canaux d'irrigation et d'assèchement), ce qui permet de connaître la dynamique des eaux souterraines et le régime hydraulique des marais drainés ;
Méthode de calcul de l'apport en eau des tourbières ;
Méthode de calcul de l'écoulement de drainage dans les marais d'assèchement ;
Influence du climat, de la nappe souterraine et de la culture sur l'évaporation et sur le coefficient de transpiration.

Certains des résultats obtenus par l'auteur sont publiés dans des manuels qui traitent d'hydrologie, d'hydrogéologie et de mise en valeur des terres et qu'on utilise dans la RSS de Biélorussie et dans les autres républiques soviétiques pour élaborer des projets de drainage de régions tourbeuses.

Reclamation by drainage should be designed and constructed scientifically according to the following principles (Ivitsky, 1966).

1. Reasonable and adequate utilization of water resources of the entire catchment area, taking into account the water requirements of the near future and those of a later date.
2. Maintenance of desired water levels in regulated rivers and the water regime in the entire catchment area.
3. The possibility of controlling the water regime and relevant air, heat and nutrition regimes in soil.
4. The extension of peat soil life.
5. The construction of durable and effective systems which combined both drainage and irrigation.
6. Comprehensive or multi-purpose solutions of the problems of water resources such as water transport, water supply to agriculture and industry, fishing, hydro-electric engineering, construction of recreation facilities.

To satisfy the above principles of modern drainage reclamation, it is necessary:

a. to control runoff by the construction of reservoirs. In this way it will be possible to retain the greater portion of spring runoff within a reservoir, and then release it for industry, irrigation, etc in the dry summer periods. This principle is being adopted in the drainage scheme for the Polesskaya low-land in the Byrelorussian SSR where it is planned to build 34 reservoirs.
b. reclamation projects should embrace not only the drained parts of the catchment, but its entire area. The combination of reclamation, hydro-technical, forestry and agricultural activities and techniques should be directed at the improved use of the water resources of the entire catchment area. These activities include:-

   -wide-spread construction of ponds on small rivers, streams, gorges and man-made water courses for the purpose of accumulating the greater portion of spring runoff;
   -measures for control of water erosion as evidenced by ravines and gullies in the catchment should be anticipated;
   -reafforestation throughout the catchment area that improves the conservation of water in the entire catchment area, rather than its individual parts;
   -snow detention, contour ploughing on light soils and deep ploughing and deep ripping (1 m depth) on heavy soils, suitable crop rotations, etc;

c. to build control gates to maintain a desired water level in small regulated rivers and ponds near settlements for everyday needs and recreation.
d. to construct drainage-irrigation systems for two-way operation creating an optimal moisture-aeration regime for crops,

both in excess moisture and drought periods;

e. to maintain the level of the groundwater table low enough to permit proper drainage;
f. to plan crop rotations on peat soils which include perennial grasses;
g. to drain shallow-deposit low-level bogs for use as meadows;
h. to apply pump drainage where hydrogeological conditions permit, including the use of groundwater for irrigation of drained lands and better control of soil moisture;
i. to use subdrainage, to stabilize the banks of open ditches in unstable soils; to use adequate methods of drainage.

In order to be able to design a combined drainage-irrigation system properly, it is necessary to determine the optimum soil moisture regime required for the successful growth of crops on the drained area.

Moisture and air content are basic soil characteristics. But in the case of a high groundwater table, which is a typical condition when bogs are drained for agricultural uses, there is a relationship between soil moisture, the water table level and climatic factors. Since the groundwater level can be defined more easily than the soil moisture content it is reasonable to express the optimal water regime for bog drainage conditions in terms of the optimal level of the groundwater table.

The drainage rate of peat soil (ie the optimal groundwater regime) can be determined in two ways: from the relation between the soil moisture content of peat and air capacity and the main factors affecting these, and from the aeration regime of the soil (Ivitsky 1962, 1966, 1971).

The following equations relating peat moisture content and air capacity with the level of the groundwater table and climatic factors, have been developed from long-term investigations (Ivitsky 1971).

$$W = 100\left[1 - A\,\frac{(H-h)^2\sqrt{H-h}}{e^{nH}}\,\sqrt[4]{\frac{\Sigma D + 1}{\Sigma N + 1}}\right] \qquad (1)$$

$$V = \frac{Ap}{100}\cdot\frac{(H-h)^2\sqrt{H-h}}{e^{nH}}\cdot\sqrt[4]{\frac{\Sigma D + 1}{\Sigma N + 1}}, \qquad (2)$$

where

W is the peat soil moisture content, % of total moisture capacity;
V is the air volume in the soil, % of porosity P;
H is the level of the groundwater table, m;
h is the distance from the soil surface to the point for which W and V values are required, m;
ΣD, ΣN is the air humidity deficit and total rainfall during the time from the beginning of the waiting period, as determined by the author's method (accumulation of spring positive mean daily air temperature, to a total value of 130°C) to the moment when W and V values are to be found;
e is the base of the natural logarithms;

n is 1.4, but according to experimental data may range from 1.3 to 1.5;

A an empirical factor determined experimentally. For low-level bogs on a permeable layer, A has an average value of 1.2 and may range from 1.1 to 1.4 depending on the nature of the evaporating surface. It is larger for crops with a high transpiration rate (grass) and smaller for crops with a low transpiration rate.

Expressions for the optimal level of groundwater table were derived after transformation of these equations. Groundwater levels of this type result in optimal moisture content and aeration of peat soils:

$$Hw = \frac{Bh + Cm + \sqrt{(Bh + Cm)^2 + (2C - Bh^2)(B - Cm^2)}}{B - Cm^2}, \qquad (3)$$

$$Hy = \frac{Bh + Sm + \sqrt{(Bh + Sm)^2 + (2S - Bh^2)(B - Sm^2)}}{B - Sm^2} \qquad (4)$$

where

H_w and H_y are the optimal groundwater levels for optimal moisture content and aeration in peat soils.

$$B = 2A_1 \sqrt[4]{\frac{\Sigma D + 1}{\Sigma N + 1}}, \qquad (5)$$

$$C = 1 - \frac{W}{100}, \qquad (6)$$

$$S = \frac{200V}{P}, \qquad (7)$$

A_1 = 1.1 to 1.2 for peat soils of low-level bogs on a permeable formation;

$$\begin{aligned} m &= 1.6 \text{ for } 0 < H \leqslant 0.5 \text{ m} \\ m &= 1.7 \text{ for } 0.5 < H \leqslant 1.0 \text{ m} \\ m &= 1.8 \text{ for } 1.0 < H \leqslant 1.5 \text{ m} \end{aligned} \qquad (8)$$

The rest of the notation is the same as in (1) and (2).

If optimal moisture content and aeration of a peat soil are known for each stage of growth of a given crop, the corresponding optimal groundwater level for the year under varying climatic conditions can be easily estimated and the optimal groundwater regime or drainage rates can readily be plotted by using equations (3) and (4).

Table 12 shows optimal groundwater levels (drainage levels) for dry and wet years estimated by equations (3) and (4). For optimal moisture content and aeration of a peat soil Kostyakov's data (1960) are taken.

The studies of aeration in soil show that there is a relationship between oxygen and carbon dioxide in a peat soil which depends the groundwater levels. The maximum oxygen and minimum carbon dioxide content are found in arable and subsurface layers of peat soils when the groundwater table is lowered to 120 or 130 cm (Ivitsky 1962).

Based on the two new methods described, and on the analysis of crop yields according to soil moisture content, nitrification and germination processes, the following levels for groundwater table for low-level bogs (Ivitsky 1966) are proposed (Table 13).

In the formula:-

$$H_b = \frac{W_s}{P_o} + H_n$$

H_b is the maximum groundwater level at the beginning of the maximum spring rise, cm;

W_s is the snow water to the moment of snow melt (excluding surface runoff);

P_o is the free porosity;

H_n is the permissible groundwater level for the maximum spring rise, cm.

The levels of the groundwater table in Table 12 show that the deepest values apply in wet years, and the lower ones, for the years when the soil moisture content is average. In dry years the lower values should be decreased by 10 to 15%. In order that the level of the groundwater table can be controlled so that it can be raised or lowered as required it is essential that the drainage network be adequate for this purpose. Thus it is necessary to determine the optimal distance between the drains for drainage and channels for irrigation. For this purpose the following theoretical formulae, have been developed which cover reclamation requirements for all natural conditions (Ivitsky 1971).

I to estimate the distance between drains for drainage:

$$E = \sqrt{\frac{kT\alpha R}{\delta Hp + N - V}} \qquad (9)$$

where

E is the drain spacing, m;

k is the permeability, m/day;

T is the time during which the groundwater level should be lowered from U to H along a line equidistant from two drains, days. Depending on the required drainage intensity, values of 5 to 15 days may be assumed;

α is a coefficient which accounts for the portion of the drain cross-section that is occupied by the base flow.

Table 12

Crops	Climatic conditions		Soil moisture content % of total moisture capacity	Drainage depth by equation (3), m	Optimal air capacity, % of porosity	Drainage depth by equation (4), m
	$\frac{D}{N}$	vegetation period conditions				
Grasses	5.0	dry	85	0.75	15	0.75
	1.3	wet	85	0.85	15	0.92
Cereals	5.0	dry	80	0.83	20	0.88
	1.3	wet	80	0.97	20	1.05
Row crops and	5.0	dry	70	1.01	30	1.09
Industrial crops	1.3	wet	70	1.22	30	1.35

Table 13

Crops	Level of groundwater table (cm)			
	Waiting period (mean for April)	Highest optimal value during vegetation period		Maximum value at the beginning of spring rise
		in bogs with shallow peat deposits	in cultivated bogs with deep peat deposits	
<u>Meadow grasses</u>				
Clover and grass	40	60 - 70	80 - 85	
mixture for hay; man-made pastures	50-60	70 - 80	85 - 95	
<u>Cereals</u>				
Oats;	50	70 - 80	90 - 100	$H_b = \frac{W_s}{P_o} + H_n$
winter cereals spring wheat and barley	60	80 - 90	100 - 110	or approximately 120 - 140
<u>Row crops and vegetables</u>				
Potato, sugar beet, maize, sunflower, cucumbers;	60-70	90 - 110	110 - 130	-
food roots, cabbage carrot, tobacco	60	85 - 100	110 - 120	-
<u>Industrial Crops</u>				
Hemp	70	100 - 110	120 - 130	-

At $\frac{F}{\alpha} \geqslant 3$ α is calculated from

$$\alpha = \frac{E}{E - \frac{8a}{\Pi} \ln \text{tn} \frac{\Pi d}{4a}} \qquad (10)$$

where

- α is the depth of the confining layer below the drain bottom, m;
- Π = 3.14; d is the external diameter of a drain;
- ln is the natural logarithm;
- tn is hyperbolic tangent symbol.

At $\frac{F}{\alpha} < 3$, α is defined by the more cumbersome expression after Averianov or by a specially prepared graph

$$R = \left[2t - H - u - 2h_o - \psi(H - u)\right]\left[2t - H - u + 2h_o + \psi(H - u) + 4\alpha\right]. \quad \ldots \quad (11)$$

where

- t is the drain depth, m;
- U,H is the groundwater level mid-way between drains at the beginning and at the end of the time T, m;
- h_o is the difference of elevation between the water level near a drain and the drain floor, m;
- ψ is the coefficient accounting for the nature of the fall of the groundwater level. When the depression curve drops such that it remains parallel to its original situation, as usually occurs in spring time, then cp = 1. If the depression curve drops in such a manner that its position near a drain remains unchanged, as occurs in the summer-autumn period when there is no back-water in the regulating network, then Q = O;
- N is the rainfall which, within period T, percolates through to the groundwater table, m;
- V is the evaporation which has its source in groundwater, during period T, m;
- δ is the water yield of peat soil which may be calculated with the help of the following formula*) (Ivitsky, 1966; 1971),

$$\delta = 0.116 \sqrt[8]{K^3} \left(\sqrt[4]{H_p^3} - \frac{U_p}{H_p} \sqrt[4]{U_p^3} \right). \qquad (12)$$

H_p, U_p are the estimated groundwater levels at the end and at the beginning of the time T, including their prescribed values (H,U) and water yield changes at different distances from the drain, m.

---

*) for mineral soils

The following analytical expressions are obtained:

$$H_p = H - \frac{t - H - h_o}{\ln\frac{E}{d}}, \qquad (13)$$

$$U_p = U + \frac{(t - U)(1 - \psi) + \psi(t - H) - h_o}{\ln\frac{E}{d}}. \qquad (14)$$

When drains are installed in a confining bed the ideal condition,
$\alpha = 0$,
$\alpha = 1$. Thus, formula (9) is valid for any depth of a confining bed.

2. If the drainage network is to be used for subsurface irrigation, where water from reservoirs or from rivers percolates through the drains then with a constant head H_o the distance between drains is given by:

$$E = 2\sqrt{\frac{kT\alpha(2H_o - h_2 - h_1)(2H_o + h_2 + h_1 + 4\alpha)}{\gamma\Pi(h \;\; - h\;) - 4(N - V)}} \qquad (15)$$

where

E is the drain spacing in metres for irrigation, the level of groundwater half-way between drains is raised from h_1 to h_2 by the head H_o measured from the drain floor in m during time T;

γ is the free porosity, it is approximately equal to the water yield δ.

The remaining notation is the same as for (9) above. For various hydrological and infiltration calculations in connection with drainage systems it is important to know the drainage runoff value.

Based upon 12 years of observations of the drainage runoff on the test plot "Sloust" and on 9 years of observation of the two test plots "Volma" the following formulae have been developed for the estimation of the maximum rate of drainage runoff from low-level bogs having a base flow*).

$$\bar{q} = \frac{\bar{S}\bar{R}}{t\sqrt{E + 1}} \qquad (16)$$

$\bar{S} = 2.7\sqrt{K}$ for bogs with phreatic groundwater feed,

---

*) L. B. Nizovtseva assisted in this work.

$$\bar{S} = 3.5\sqrt{K}$$ for bogs with phreatic and artesian feed, the units of K being m/day. (17)

$$\bar{R} = 0.043\sqrt{\bar{N}}\ , \qquad (18)$$

$$C_v = \frac{0,5}{\sqrt[6]{\omega}\ .\ \sqrt[3]{t^2}}\ , \quad C_s = 2C_v, \qquad (19)$$

where

$\bar{q}$ is the annual maximum rate of drainage runoff, l/sec. ha;

$\bar{S}$ is a factor characterizing the peat soil permeability;

$\bar{R}$ is a factor which accounts for the annual rainfall;

$\bar{N}$ is the annual rainfall, mm;

ω is the catchment area, ha;

C_v is a coefficient of variation;

C_s is a coefficient of asymmetry;

For the same purpose the following formula is also proposed:

$$q = \frac{2C_k\alpha(h - h_o)(h + h_o + 2\alpha)}{E^2}\ , \qquad (20)$$

where

q is the rate of drainage runoff, l/sec ha;

C is a scale factor; $C = 10^7$, if the permeability k is in m/sec;

h is the excess height of the groundwater level measured in metres above the bottom of the drain midway between the drains, m;

h_o is the excess height of groundwater level near the drain, measured in metres above the bottom of the drain;

The estimation of evaporation is very important for calculations of drainage systems. Based on observations and theoretical assumptions, the following relationships between evaporation and other important parameters (climate, groundwater level, yield) have been found (Ivitsky 1958).

a. for evaporation from peat soil without vegetation

$$V = \frac{\alpha_v D\ (1 + 0,1W)}{e^{\bar{n}H}}\ ; \qquad (21)$$

b. for evaporation from peat soil under timothy grass

$$V_T = V_n + 1,5\sqrt{P_y \Sigma D} \qquad (22)$$

c. for the transpiration coefficient of timothy grass

$$K_T = \frac{1,5\sqrt{\Sigma D}}{\sqrt{P_y}}, \qquad (23)$$

where

- V is the mean daily evaporation from peat soil without vegetation or total during the time T, mm;
- D is air saturation deficit, either mean daily, or, total during the time T, mm;
- $\bar{W}$ is the wind speed, m/sec;
- H is the distance of groundwater table from the surface, m;
- α_v is an empirical factor varying from 0.93 to 1.0;
- $\bar{n}$ varies between 1, 3 and 1.5
- V_T is the total evaporation for a soil carrying timothy grass, during the vegetation period, mm;
- V_n is the total evaporation from the same soil and at the same groundwater level, but with no vegetation, and during the same vegetation period, mm;
- P_y is the timothy grass yield (dry mass), in 100 kg/ha;
- D is the total saturation deficit of the air during the above vegetation period, mm;

According to A. A. Cherkasov the above expression (21) for the transpiration coefficient is valid for wheat in the Zavolzhic region, for cotton in Central Asia and for sugar beet in the Voronezh region (Chevkasov 1958).

References

Ivitsky, A. I. 1958. Osnovnye dostizheniya meliorativnoi nauki v oblasti proektirovaniya i raschetov osushitelnykh sistem v BSSR (Important advances in reclamation science in the field of designing and calculating drainage systems in the BSSR). In: Dostizheniya meliorativnoi nauki v BSSR, Minsk, Izd. Akad. Nauk BSSR.

Ivitsky, A. I. 1962. Novyi podkhod k opredeleniyu norm osusheniya melioriruemykh pochv (A new approach to the determination of drainage rates of reclaimed soils). Dokl. Akad. Nauk BSSR, 6(12).

Ivitsky, A. I. 1966. O printsipakh i sposobakh osushitelnoi melioratsii (On the principles and methods of drainage for land reclamation. Izv. Akad. Nauk BSSR. Ser. Selskokhoz. Nauk, No. 1.

Ivitsky, A. I. 1966. Obshchee uravnenie vodootdachi gruntov (General equation of groundwater yield). Dokl. Akad. Nauk BSSR, 10(11).

Ivitsky, A. I. 1971. Vlashnost i vozdushnyi rezhim torfyanoi pochvy v zavisimosti ot klimaticheskikh faktorov i urovynya gruntovykh vod (Moisture content and aerations of peat soils under different climatic conditions and groundwater levels) In: Nauchnye issledovaniya po gidrotekhnike v 1969 g., v.2, Leningrad, Energiya.

Ivitsky, A. I. 1971. K teorii rascheta osushitelnykh sistem (A contribution to the theory of drainage system calculation) Minsk, Urozhai.

Ivitsky, A. I. 1971. Uchet dinamiki vodootdachi gruntov pri raschetakh ponizheniya gruntovykh vod (Accounting for groundwater yield dyamics in the forecasting of lower groundwater levels). In: Nauchnye issledovaniya po gidrotekhnike v 1969 g., vol. 2, Leningrad, Energiya.

Kostyakov, A. N. 1960. Osnovy melioratsyi (Fundamentals of reclamation work). Moscow, Selkhozgiz.

Cherkasov, A. A. 1958. Melioratsiya i selskokhozyaistvennoe vodosnabzhenie (Land reclamation and agricultural water supply). Moscow, Selkhozgiz.

DISCUSSION

Prof. L. Wartena (The Netherlands): In your paper you mentioned some additional measures, particularly, you pointed to the necessity of including perennial grasses in crop rotations. I did not quite understand why this is necessary.

A. I. Ivitsky (Byelorussian SSR): Perennial grasses in a crop rotation are very important because they decrease the rate of decay of organic matter and require a smaller quantity of groundwater thus reducing washing out of decay products. In Gorki, Mogilev region, tile drainage was installed more than a century ago and organic matter still continues to decay.

SOURCES OF INCOMING WATER IN BOGS OF THE UKRANIAN PRIPYAT POLESSIE AS A RESULT OF RECLAMATION

V. E. Alekseevsky and K. P. Tereshchenko

Hydrology and Reclamation Expedition, Lvov, USSR.

SUMMARY: In the basin of the Pripyat in the Ukranian Polessie, hydrogeological factors, in addition to climatic and hydrological ones, appear to have a vital influence on the water supply of boggy regions. Hydrological factors must be considered since the artesian waters of the territory in question in certain regions contribute to the water supply of bogs. Naturally, the normal reclamation procedures in these regions do not yield the results that might be expected.

The overwhelming majority of boggy plains of the Pripyat basins in the Ukranian Polessie fall in the category of lowland bogs, mainly supplied with water from precipitation and surface runoff. However, groundwater may contribute to their water balance at the expense of both groundwater flow and groundwater outcropping where the aquifer is interrupted by river valleys. The same conditions hold for some of the upland bogs although precipitation is their principal source of water.

This paper presents a scheme of the Pripyat Polessie regions of Ukraine according to types of bog water supply and a diagram showing these different types of supply.

L'ALIMENTATION DES MARAIS EN EAU, DANS LA REGION DU PRIPET EN POLESIE UKRAINIENNE APRES LES TRAVAUX DE MISE EN VALEUR

RESUME : Sur le territoire du Pripet dans la Polésie ukrainienne ce sont essentiellement les facteurs hydrogéologiques, à côté des facteurs climatiques et hydrologiques, qui influencent l'alimentation en eau des marais. L'importance accordée aux facteurs hydrologiques tient à la présence de nappes artésiennes qui jouent un rôle direct dans l'alimentation en eau des marécages. Il est évident que les méthodes ordinaires de mise en valeur ne peuvent être efficaces dans cette région.

La plupart des marécages du Pripet en Polésie sont des marais de région basse alimentés par les précipitations atmosphériques et les eaux de surface. Cependant les eaux souterraines peuvent contribuer à leur alimentation au détriment à la fois de l'écoulement des eaux souterraines et des émergences lorsque l'aquifère est interrompue par la vallée d'une rivière. Il en va de même pour une partie des marais de région haute, encore que les précipitations atmosphériques demeurent la source principale de leur alimentation en eau.

Les auteurs fournissent un schéma de division de cette région selon les types d'alimentation en eau et un diagramme de l'alimentation en eau.

The Ukranian Pripyat Polessie is a vast boggy plain in the Pripyat basin which lies in the north-west of the Ukraine and includes the northern and central territories of the Volyn, Rovno, Zhitomir and Kiev regions. The northern boundary of the Pripyat Polessie conventionally follows the state frontier between the Ukranian SSR and the Byelorussian SSR. The western and eastern valley limits are the Zapadnyi Bug and Dnieper, respectively. In the south the boundary follows the northern slope of the Volyn Hills. The relief consists of a flat plain sloping north and north-east towards the valleys of the Pripyat and Dnieper, aqueogenetic types of relief being of primary importance in the modern surface structure. In the river valleys, besides wide flood plains, two subfloodplain terraces are generally present. Bogs of the area investigated are closely connected with new and fossil river valleys.

The highest points of the Pripyat Polessie are found in the Slovechansko-Ovruch Hills, which attain 320 m above sea level, whereas the lowest area in the Dnieper valley below the confluence with the Pripyat is 100-110 m above sea level. Thus, the total range of altitude is 200 m, which is relatively small. Generally, the relief of the northern part of the area does not exceed 25 m and has an average value of 25-50 m; only within the Zhitomir region does it exceed 50 m.

The Pripyat Polessie has a temperate continental climate with a warm moist summer and mild cloudy winter. Average monthly temperatures from December to March vary from -5° to -7°C except in the south-west where average temperatures for March are higher than 6°C, (Lutsk, Vladimir-Volynsky). Spring is a long period of changeable weather; the average air temperature in April is 10° to 11°C, in May it is 18°C and sometimes higher. The average temperature of the summer months (June-August) varies from 16°C to 18°C and upon intrusion of tropical air the maximum may reach 39°C. Autumn is lingering and dull, with drizzle. At the end of October the average daily temperature falls below 5°C and at the end of November it falls below 0°C. A minimum temperature of -39°C has been observed in this region (Vladimir-Volynsky) and a maximum of 39°C (Kiev). In general the lowest temperatures are observed in February and the highest in July or August.

The prevalence of Atlantic air is the controlling factor for precipitation which has an average total of 550-600 mm and is fairly uniformly distributed over the area. The considerable increase in rainfall towards the south may be attributed to the presence of the Volyno-Podolsk Hills. The maximum monthly precipitation usually occurs in July (61-106 mm). During the warm period of the year (April-October) the total precipitation is about 400 m whereas during the cold period (November-March) it varies from 140-200 mm. The relative humidity attains its maximum in winter and autumn and its minimum in spring: the values in November-December range from 84-89% (Korysten, Novograd-Volynsky) and 48-54% in May (Novograd-Volynsky).

The region is rich in rivers and lakes, with an average density of the river network of 0.22-0.27 km/km^2. The main waterways are the Pripyat and its rightbank tributaries: Vyzhevka, Turiya, Stokhod, Styr, Goryn, Sluch, Ubort, Slovechno, Uzh as well as the Teterev and Irpen wich flow directly into the Dnieper. All these rivers except the Pripyat run from the south-west towards the north-east. The valleys of most of the rivers, particularly within the Volynskoye Polessie, have wide, flat beds and very gentle slopes. Conversely, rivers flowing within the crystalline massif have deep valleys which are not very wide, and have steep slopes.

The majority of the lakes are concentrated in the remote north-western part of the Volynskoye Polessie (the Shatskie group of lakes). Here the lakes are mainly of karstic origin (Svityaz, Lyitsimer, Pulemetskoye, etc.); they are very deep and have well defined banks. Many lakes in the Volynskoye Polessie are located in river valleys, particularly in the Pripyat valley (Tur, Nobel, Tissabol, Lyubyazh). These are ordinary flood plain lakes with low sand banks, many being bog lakes. There are also lakes formed as a result of glacial erosion. The Zhitomirskoye Polessie has few lakes (the largest is lake Korsha) but the Kievskoye Polessie has even fewer.

The Pripyat Polessie has a complex geological structure because it is composed of three different geostructural regions of the Russian platform. These are: the northern part of the Galitsko-Volynskaya Depression (Volynskoye Polessie); the north-western part of the Ukrainian crystalline massif (Zhitomirskoye Polessie); and the north-western part of the Dneprovo-Donetskaya Depression (Kievskoye Polessie).

Each of these regions has a quite different geological history. The Ukrainian crystalline massif has been above sea level almost the whole of geological history and has had a continental climate, whereas the Calitsko-Volynskaya and Dneprovo-Donetskaya Depressions were often submerged. Consequently, crystalline formations play the main role in the geological structure of the massif, whereas sedimentary rocks are dominant in the depressions.

In the area under study deposits are evident which illustrate all ages of geological history of the Earth, from Pre-Cambrian to Quaternary. Among the main formations, the Pre-Cambrian, Cretaceous and Tertiary periods are particularly important as they are bedded above the erosion base and are responsible for the present day features of the relief. Quaternary deposits, which are widely distributed, are of almost equal importance.

Groundwater occurs in almost all the sedimentary formations, but of particular interest from the reclamation point of view are the water-bearing strata which directly contribute to the water supply of bogs. In the Ukrainian Pripyat Polessie such aquifers (in addition to a water-bearing stratum in Quaternary deposits whose participation in the water supply of bogs is obvious) are the following:

1. confined aquifer in upper Cretaceous deposits of the Volynskoye Polessie;
2. confined waters in fractured crystalline rocks of the Pre-Cambrian period and products of weathering in the Zhitomirskoye Polessie;
3. aquifers in the upper Cretaceous and Paleogenic deposits (Buchaksko-Kanevsky, Kharkovsky and Poltavsky horizons) in the Kievskoye Polessie.

The varied natural phenomena which lead to the formation of bogs (geomorphological, hydrological and climatic conditions, geological and hydrogeological characteristics) have resulted in an extremely non-uniform distribution of bogs in the Ukrainian Pripyat Polessie. In parts of the Volynskoye Polessie bogs occupy as much as 40 to 60% of the arable land; in the Kievskoye and Zhitomirskoye Polessie boggy areas are less extensive. Bog area,relief, depth of peat-bed and type are characteristics which vary with location.

According to the agricultural and silvicultureal classification of bogs and their association with morphological elements of the relief, the following types can be identified within the region studied:

a) Lowland bogs - occupying vast areas and generally associated with terraces of the Pripyat and its tributaries, only a small part being associated with lakes;
b) High bogs or Sphagnum bogs - in the form of isolated massifs. They are not numerous and are associated with depressions either in watersheds or between small sandy hills on ground-morainal plateaux;
c) Transient bogs - associated with flat depressions located either in ground-morainal landscape or between outwash plains.

About 80% of all bogs in the region are of the lowland type.

In the Ukrainian Pripyat Polessie there is intense reclamation of boggy and saturated land. About 700 000 ha have been drained and a further 300 000 ha approximately are to be drained in accordance with the current Five-Year Plan. Construction of drainage systems is accompanied by a dual control of soil moisture and aeration in reclaimed areas using the best methods and materials.

Selection of the most effective methods of reclamation of bogs in many respects depends on correct knowledge of the main sources of the water supply. Throughout the area, in addition to the usual climatic, hydrological and geomorphological factors, hydrogeological factors are also in operation. Their role is of particular importance in connection with artesian water-bearing horizons which in certain regions and under certain conditions may directly contribute to the water supply of bogs.

Thus, the problem is to define the groundwater component in the total water balance of bogs, for which it appears necessary to clarify the interrelation between free groundwater and the underlying confined aquifer.

The Ukrainian Scientific Research Institute of Water Engineering and Land Improvement has for many years conducted research into groundwater aspects of land reclamation. This has made it possible not only to classify all regions of the Pripyat Polessie according to the nature of the supply of water to the bogs (Fig 23) but also to understand the detailed pattern of flow of water to bogs (Fig 24). The presented pattern of supply (Fig 24) shows that within region I bogs are fed by precipitation and flood waters (Fig 24a) and by leakage from the confined aquifers (Fig 24b) where this water bearing stratum has been intercepted by river valleys.

Within region II highbogs are fed by precipitation (Fig 24a) and artesian water (Fig 24b) wherever there is no aquiclude to separate the water-bearing materials of the free groundwater body from the underlying confined aquifer. In certain areas of region II composed of loess deposits the water supply of the flood-plain (lowland) bogs is provided by rainfall and flood waters (Fig 24a).

In region III the water supply pattern depends on the fact that the bogs are located around the periphery of karstic lakes (Fig 24f), and so they receive a continuous flow of artesian water.

It can be seen that groundwater is a component in the water balance of all these bogs. Estimation of the magnitude of this component is of importance in all the cases considered and particularly when groundwater is the major source of water supply.

According to the data available on particular regions of the USSR (lowland bogs in the flood plain of the Yakhroma in the Moscow region), during the first years after land reclamation the total volume of groundwater flow to bogs attained 100 to 130% of the annual precipitation. This illustrates the necessity of a thorough investigation of the nature of groundwater resources. Disregard of this factor may result in undesirable consequences and sometimes may nullify the reclamation effort. Thus, in region III (Fig 23), instead of the expected result of a decrease of the groundwater level to the optimum value, reclamation measures caused further bogs.

Exploration and survey of areas presenting bogs mainly supplied by groundwater or boggy areas with different types of supply is essential for the correct design of reclamation systems and maintenance of the lands after reclamation. Detailed hydrogeological classification of the land to be reclaimed follows the initial survey. Planning of reclamation systems for the Pripyat Polessie, including hydrogeological investigation of the areas to be drained and study of the water balance of these areas is carried out by the Lvov Water Engineering Expedition, the Water Management Institute of the Ukrainian SSR as well as some other scientific research and design institutes. In this way the boggy and saturated regions will be transformed into high-yielding arable lands.

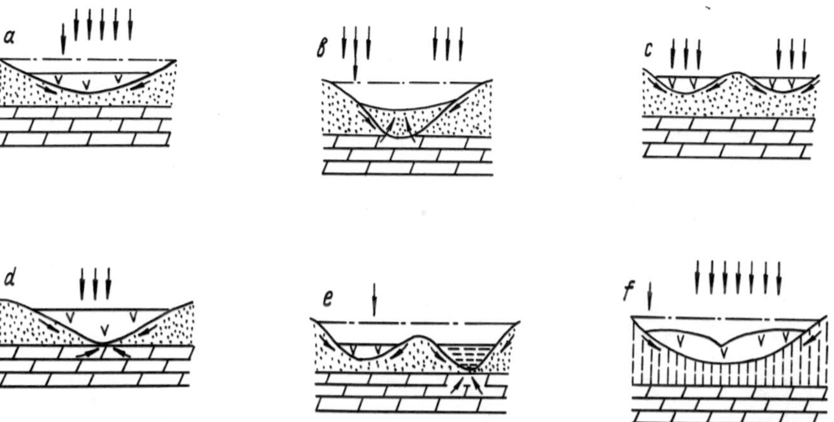

Fig 24. Different types of water supply patterns found in the bogs of the Ukrainian Pripyat Polessie.

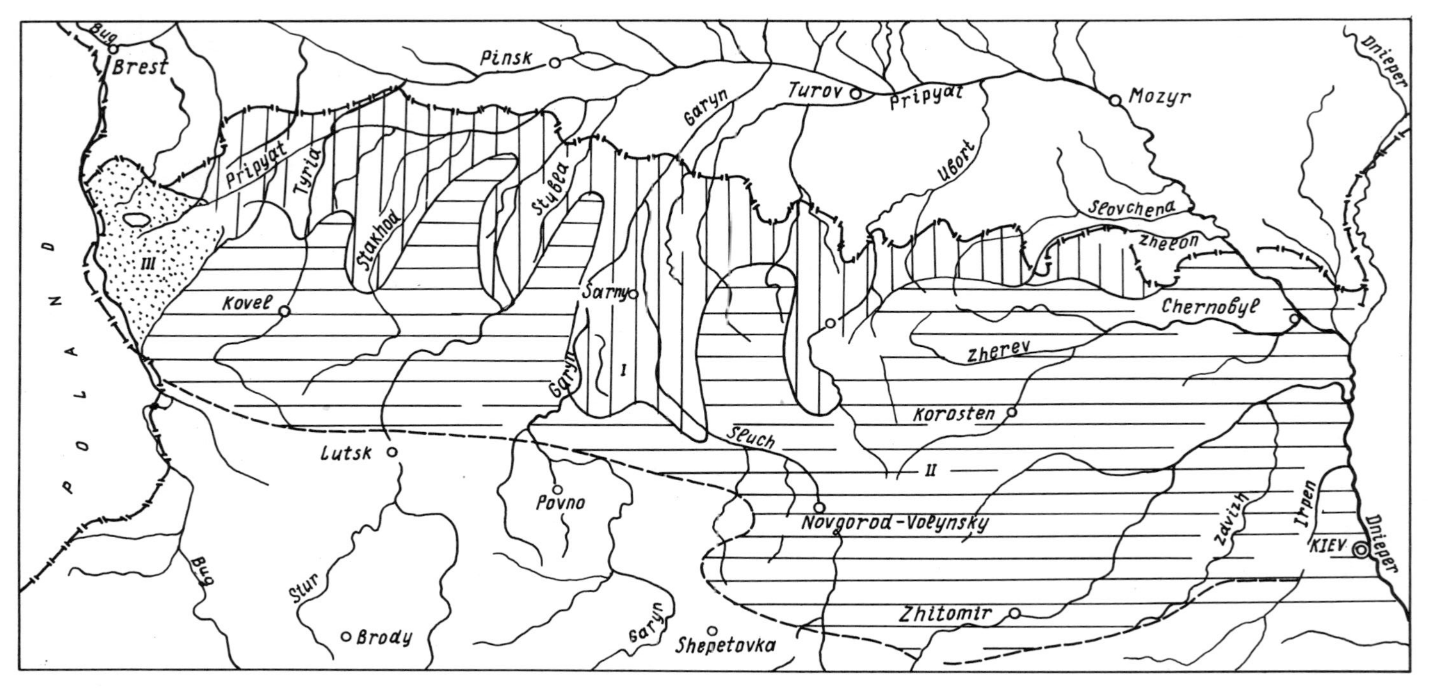

Fig 23. Bogs of the Pripyat Polessie in Ukraine

I. Bogs mainly fed by flood water
II. Small boggy areas receiving precipitation and groundwater
III. Bogs where artesian water predominates

DISCUSSION

P. F. Vishnevsky (Ukrainian SSR) - What can you say about types of water recharge in high bogs? What measures can you recommend for reclamation?

K. P. Tereshchenko (Ukrainian SSR) - The high bogs in Pripyat Polessie are closed basins with obstructed outlets and a slightly elevated relief. The situation is the second flood plain terrace of rivers and their tributaries, and these areas are very similar to other regions of Polessie. For example, in the Karsk region the bogs are located around the periphery of the lake. I think the most effective measure here would be vertical drainage which would lower the groundwater level. In the north-west part of the area reclamation works were begun, but they did not produce the desired result. We therefore conclude that available subsurface methods were inadequate for these particular conditions.

A. B. Diyachkov (USSR) - You said that artesian water recharge occurs in the region of the Karsk Lake but not in other regions. In these other regions is vertical drainage effective?

K. P. Tereshchenko (Ukrainian SSR) - Your question deserves serious attention. It is possible that pumping of groundwater from wells may be effective for drainage of bogs in other regions also.

V. M. Zubets (USSR) - You stated that reclamation works in the south-western part of the Karsk Lake did not give the desired effect. What are the soil types there?

K. P. Tereshchenko (Ukrainian SSR) - In the region studied there is a peat layer, thicker than 5 m, underlain by fine dust sands. Open drains made in this area appeared to be of low efficiency.

V. M Zubets (BSSR) - Can you specify the regions of Polessie to which you refer?

K. P. Tereshchenko (Ukrainian SSR) - This is the south-western part of the Pripyat Polessie.

A. A. Zhelobaev (USSR) - You spoke of the different types of water supply. Have you any quantitative values concerning the various sources of water supply to bogs?

K. P. Tereshchenko (Ukrainian SSR) - Unfortunately we have no values, as the studies of the types of water supply were started only recently. At present we have established a network of observation wells and a water balance station on the Pripyat. Here the object of the investigation is to estimate the contribution of artesian water. Groundwater supply is characteristic of all types of bogs in Polessie.

A. A. Zhelobaev (USSR) - You said that the greater portion of precipitation falls in summer. Does this imply that intensive drainage should be carried out in this period?

K. P. Tereshchenko (Ukrainian SSR) - In the period June-July precipitation totals about 150-200 mm, or half of the normal annual value. The period from April to May is a growing season, during which the groundwater level quickly falls to the optimal depth. In order to support normal plant life additional irrigation is required. Reclamation works enable agricultural work to commence earlier in the year.

G. H. Madissoon (USSR) - I did not quite understand your statement that studies of land reclamation are carried out.

K. P. Tereshchenko (Ukrainian SSR) - I spoke of drainage. Drainage implies the creation of optimal conditions or decreases of the groundwater level to the optimal depth. Drainage units of a single type cannot provide optimal conditions over the whole area. We do not undertake drainage in the north-western part of the region if the groundwater level is below the constant required depth; in our work we try to obtain the optimal groundwater level. In practice the systems are dual purpose; after reclamation irrigation is required.

INVESTIGATION OF RUNOFF IN DRAINAGE OUTLETS AND SUBSURFACE DRAINED SOILS IN THE LATVIAN SSR

A. A. Zivert, I. A. Rieksts and C. N. Skinkis

*Latvian State Design Institute of Water Economy,
Riga, SSR.*
and
*Latvian Research Institute of Water Enginerring,
Elgova, USSR*

SUMMARY: When designing drainage systems and drainage outlets, it is necessary to determine the expected hydrological regime during the operative life of the system. In the Latvian SSR research aimed at solving this complex problem follows two main directions: firstly, long-term observations are made of the hydrological regime of small and large drainage systems in the Republic under various soil and climatic conditions; secondly, methods of calculating runoff variations in large drainage outlets resulting from flow regulation are developed.

The procedure and technical requirements of the investigation of drainage system runoff are discussed in this paper and the main results of the observation of drainage, surface and total runoff from subsurface drained areas are presented. The basic principles of forecasting the expected runoff in drainage outlets are investigated and characteristics of the changes in the runoff from some of the largest land reclamation schemes in the Latvian SSR (the Lielupe river, the Aiviekste river and Lake Lubanas) are given.

ETUDE DU RUISSELLEMENT EN PROVENANCE DES RIVIERES RECEPTRICES ET DU SOUS-SOL ASSECHE PAR DRAINAGE EN RSS DE LETTONIE

RESUME : Avant d'élaborer des systèmes de drainage et de contrôle des rivières réceptrices, il faut déterminer quel sera le régime hydrologique pendant toute la période d'exploitation des installations projetées. A cette fin, en RSS de Lettonie, les recherches se font principalement dans deux directions : en premier lieu, on procède à des observations pluriannuelles du régime hydrologique des petits et des grands systèmes de drainage dans diverses conditions de terrain et de climat ; en second lieu, on élabore des méthodes pour calculer les variations que subit le régime d'écoulement des rivières réceptrices à la suite de leur régularisation.

Font suite à cette introduction : une analyse des méthodes et des moyens techniques mis en œuvre pour étudier le ruissellement des systèmes de drainage ; la présentation des principaux résultats obtenus en observant le ruissellement dû au drainage, le ruissellement dû à l'écoulement de surface et le ruissellement total dans les régions asséchées par drainage ; un exposé des principes essentiels sur lesquels sont fondées les prévisions en matière de ruissellement en provenance des rivières réceptrices ; une description des changements caractéristiques prévus pour le régime de ruissellement dans le cas de deux projets de mise en valeur parmi les plus importants de la RSS de Lettonie (rivières Lielupe et Aiviekste).

About 60% of the total area of the Latvian SSR suffers from temporary or permanent excess of water, about 5% of this area being occupied by bogs. Approximately 2 200 000 ha or 75% to 80% of the arable land is drained or needs to be drained. Over 900 000 ha of arable land and 300 00 ha of woodland have already been drained; from this total 700 000 ha has been reclaimed using tile drains.

When designing drainage or modifying drainage outlets, it is necessary to predict the future hydrological regime for their entire operating life. In order to solve this problem, two separate lines of hydrological investigation have been adopted in the Latvian SSR: methods are developed for predicting changes of conditions in the drainage outlet due to regulation and long-term hydrological observations are made of small and larger drainage systems with 300 mm diameter and larger offtake drains in different soil and climatic conditions.

Flood control of drainage outlets

Drainage and cultivation of marsh-ridden and periodically flooded areas are partly dependent upon flood control in the drainage outlets. All the principal measures that are taken in this direction, such as deepening and straightening of the channel, bank levees, and construction of flood-control basins, have a fundamental effect not only on the water levels in the area, but also on the discharge and particularly on the maximum discharge.

In order to predict the possible flood conditions of drainage outlets and taking into account changes which may result from hydrological measures, methods of calculation based on mathematical (in some cases hydraulic) simulation of flood propagation along the outlet channel have been derived. Simulation of observed floods under conditions holding before and after regulation, makes it possible to find the required adjustments.

For the simulation of an observed flood in a particular year, the following simplified calculation procedure for unsteady flow in open channels is suggested. Within the channel system considered, the nodal points of calculation (cross-sections) are selected such that they are only located at channel intersects or in places where a concentrated lateral inflow exists. Along channel reaches free of inputs the length of the reach for the calculation is determined with regard to the validity of linear interpolation of depth between adjacent cross-sections under the following basic assumptions.

1. The capacity of the channel system may be divided into individual parts, associated with the nodes, a single-valued relation existing for each of these

$$W_i = f(Z_i) \qquad (1)$$

where: Z_i is the level at the i-th node;
W_i is the volume of water associated with the i-th point.

2. For unsteady flow, the single-valued relation, valid for steady flow, between the hydraulic resistance modulus and the mean of the levels at the ends of the reach, may be used.

$$\phi_{i-1,i} = \rho\{\tfrac{1}{2}(Z_{i-1} + Z_i)\} \qquad (2)$$

where: $\phi_{i-1,i}$ is the hydraulic resistance modulus of the reach between the (i-1)th and i-th points;
Z_{i-1} and Z_i are the levels at the (i-1)th and i-th points.

Having made these assumptions, the differential equations for the i-th point may be written as follows:

The momentum equation (inertia forces being neglected)

$$Z_{i-1} = Z_i + \phi_{i-1,i} \cdot Q^2_{i-1,i} \cdot \text{sign}\, Q_{i-1,i} \qquad (3)$$

the continuity equation

$$Q_{i,i+1} = Q_{i-1,i} + Q_i - V_i - \frac{dW_i}{dt} \qquad (4)$$

where: $Q_{i,i+1}$ is the discharge between the i-th and (i+1)th points;
$Q_{i-1,i}$ is the discharge between the (i-1)th and i-th points;
Q_i is the lateral inflow at the i-th point;
V_i is the leakage to the groundwater reservoir (bank regulation) at the i-th point;
t is time.

For points of interflow or branching a further momentum equation is required.

By simultaneous solution of the differential equations for all the nodal points of a given channel system, with appropriate initial and boundary conditions, the unknown quantities Z and Q may be found. Relations (1) and (2) differ according to whether the channel system is regulated or not. The system of differential equations may be solved using a computer.

The above mathematical simulation technique is used to predict the possible flood flow of the two largest schemes of land improvement and water supply regulation in the Latvian SSR.

1. In the middle Lielupe large-scale levee works on flood lands and the construction of polders are planned; in the lower Lielupe possible ways of linking the river and Lake Babite are being considered. An example of the mathematical simulation of spring flood hydrographs is shown in Fig 25.

2. In the Lubanas lowland it is envisaged that the raising of levees around Lake Lunanas will create a large reservoir. In addition the Meiranu canal and other small-scale works are under construction. The general construction will be carried out in four stages of which three have already been completed. Land improvement and water supply regulation in the Lubanas lowland will considerably change the flood regime of the Aiviekste river. Fig 26 illustrates the maximum discharge of the Aiviekste after the

construction work has been completed. Analysis of the results of mathematical simulation of the flood conditions of the natural and possible future states shows that after completion of land improvement measures in the Lubanas lowland the flood water will be discharged at a higher rate and the maximum discharge of the Aiviekste river below the mouth of the Meiranu canal will increase by up to 40%.

Drainage and surface flows in areas drained by subsurface drainage

The hydrological regime of a soil is radically changed by drainage systems. A new component of runoff appears, namely a drainage flow that does not occur in bogs under natural conditions. Under the conditions of the Baltic region the contribution of this flow to the discharge is very large and is only somewhat less than evapotranspiration. On eight loamy soils and sandy loams in the Latvian SSR the average annual drainage flow ranges from 150 to 250 mm which is equivalent to about 25 to 35% of the annual precipitation. During wet years or in areas supplied by ground or ponded water this quantity may be considerably greater.

Consequently, hydrological properties of marsh-ridden areas calculated before draining cannot be used for drainage analysis purposes. The properties of drainage systems must be determined on the basis of the possible conditions after draining. In the Latvian SSR the calculation of drainage runoff is based on a comparison of hydrological systems. For this purpose, experimental drainage systems are constructed in areas having the most common types of soil offering various climatic conditions. The principal hydrological characteristics of the systems are then observed during a whole year. In many places observations have been made for the last 13 or 14 years and at Pëterlauki for more than 20 years.

In determining the design values of the drainage runoff, it should be noted that these vary considerably according to the drainage conditions, i.e. the depth and spacing of the drains, the drain design, especially tube diameter and filter types. For this reason each experimental plot includes several versions of drainage systems subject to different conditions of drainage. More than 200 such systems of various design have been installed in the Republic.

The optimal drainage rate for each experimental system has been established by analysis of groundwater level and soil moisture content data, whilst taking into account the economic characteristics of different drainage systems.

For new drainage systems the runoff is calculated by analogy with the experimental systems which have a long period of record and optimal values of the runoff parameters. Corrections are made for differences in climatic, soil, and other conditions between the area in question, and the experimental analogue.

The distribution of seasonal and long-term annual drainage flow is extremely nonuniform. Under natural rainfall the drainage flow distribution is of a clearly periodic type with sharp rises during the flood period, (Fig 27).

The greater part of the drainage flow occurs in spring (March to May) with an average 40-50% of the annual total. Occasionally the maximum drainage flow is observed in autumn or even in winter. In summer, with high evaporation and transpiration, the groundwater level usually falls considerably, 1.5 to 2.0 m and more, both on drained and undrained mineral soils subject to periodic saturation and there is no drainage flow except possibly after heavy rain. Consequently, in this season (June to August) the mean drainage flow only contributes from 3-8% of the annual total.

In peat soils of fen bogs the drainage flow is about 20 to 40% greater than that under the same climatic conditions in mineral soils. The hydrograph of flood drainage discharge is more uniform in peat than in mineral soils. On loam and sandy loam soils the maximum specific drainage discharge after rain sometimes reaches and even exceeds 4-5 l/s ha, whereas under similar conditions in peat it does not exceed 3 l/s ha.

The design value of discharge used for selecting the diameter of the drains is estimated as being less than the absolute maximum drainage discharge, thus allowing overload of the drains for a certain time.

It is found that such an overload of the system can be permitted for 2% of the year on mineral soils, and 3% on peat. From the long-term observations on experimental plots situated in different regions of the Republic, taking into account the main contributions to the maximum drainage discharge, a chart (Fig 28) has been produced for calculation of the design discharge under natural rainfall. If the water supply is of the mixed type, the specific discharge must be corrected accordingly.

Since 1963, because of the construction of large-scale subsurface drainage systems in boggy areas of 50-100 ha and more, new problems have arisen in the design of such systems. Thus, it became necessary to increase research work on drainage discharge. The following problems have to be solved:

Generalization of drainage flow in large subsurface systems, considering the differences of this process in subsystems.

Surface runoff generation on plots drained by subsurface drainage.

The latter is very important for the selection of rational means of removal of surface runoff from drained plots.

On five large drain systems continuous observations are being made of drainage flow and total (drainage + surface) runoff by means of weirs and automatic level recording facilities in wells and outlet ditches of the drain systems.

The hydrographs of the specific drainage discharge and total runoff of the 1966 spring flood at the Bërze state farm in the Dobeles region are shown in Fig 29. The drainage flow was measured in a large drainage system with a catchment area of 0.77 km^2 and the total runoff in the outlet ditch for an area of 3.68 km^2 with subsurface drainage. The relief of the experimental plot is flat; the soil is an alkaline, moderately heavy loam.

Fig 29 and Table 14 show that the surface runoff from areas with subsurface drainage is of very short duration; it occurs mainly in early spring when the soil is still frozen. In some years there was no surface runoff at all (1968, 1969).

Despite the short duration of surface runoff in these areas, the maximum specific discharge sometimes exceeds the maximum specific drainage discharge. The specific surface discharge used for calculating filter-absorbers and outlet drains for large drainage systems is 70 to 90% of the calculated drainage discharge. Since the surface discharge for certain years exceeds the design value it is necessary to provide special anti-erosion measures in large drainage systems.

Table 14. Characteristics of the drainage and surface runoff of areas subject to subsurface drainage, at the Bërze State Farm in the Dobele region.

Year of observation	Depth of runoff in spring, mm		Instantaneous peak specific discharge 1/s km^2	
	Drainage	Surface	Drainage	Surface
1966	75	31	60	361
1967	148	7	103	40
1968	140	0	74	0
1969	31	0	53	0
1970	67	12	87	99

In the Latvian SSR long-term observations of the hydrological regime of areas with subsurface drainage provide a scientific basis for the rational design of drainage systems to be used under similar conditions.

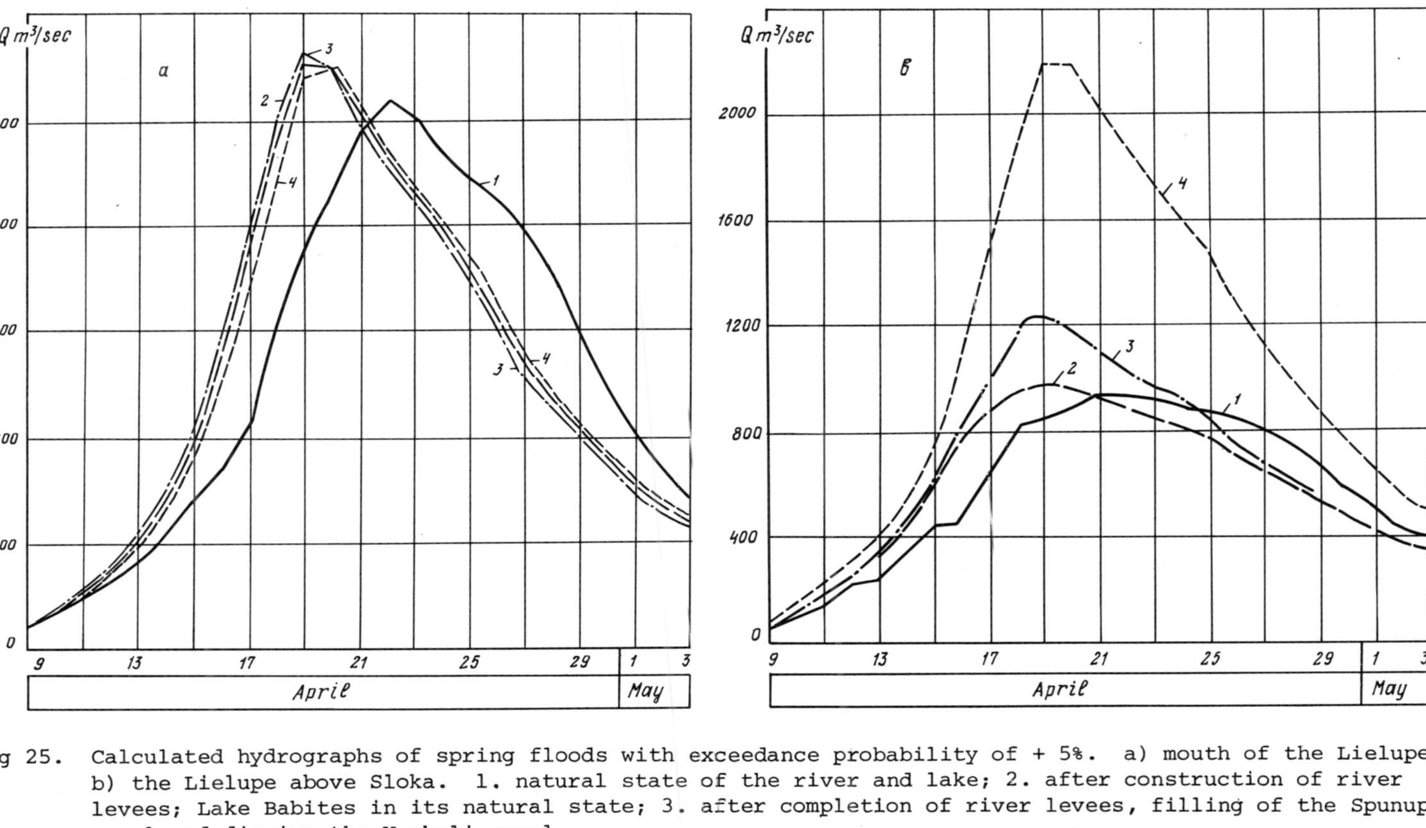

Fig 25. Calculated hydrographs of spring floods with exceedance probability of + 5%. a) mouth of the Lielupe, b) the Lielupe above Sloka. 1. natural state of the river and lake; 2. after construction of river levees; Lake Babites in its natural state; 3. after completion of river levees, filling of the Spunupe canal and digging the Varkali canal.

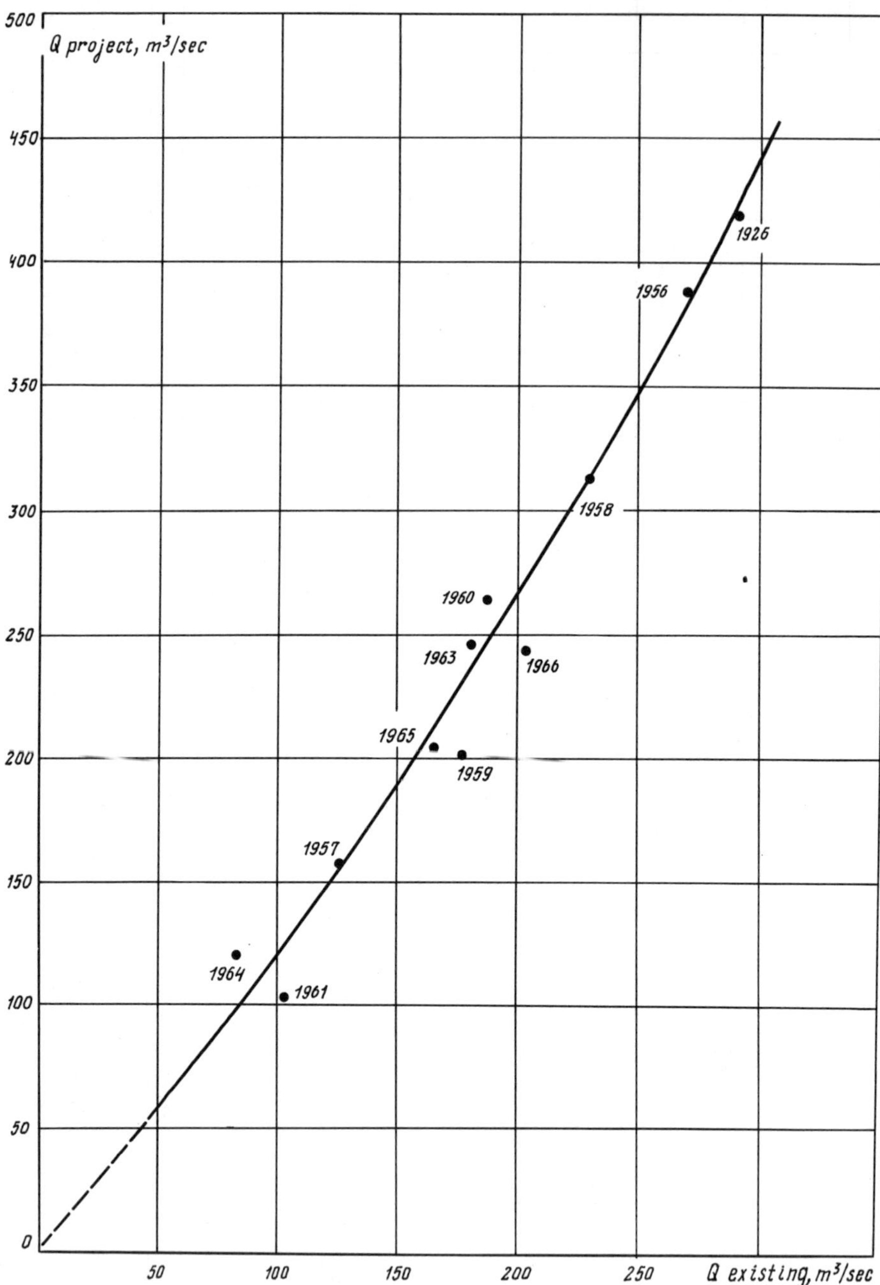

Fig 26. The relation between the calculated maximum discharge of the Aiviekste river below the mouth of the Meiranu canal in its present state (after regulation in the years 1927 to 1937) and its future state (after the second period of construction is completed).

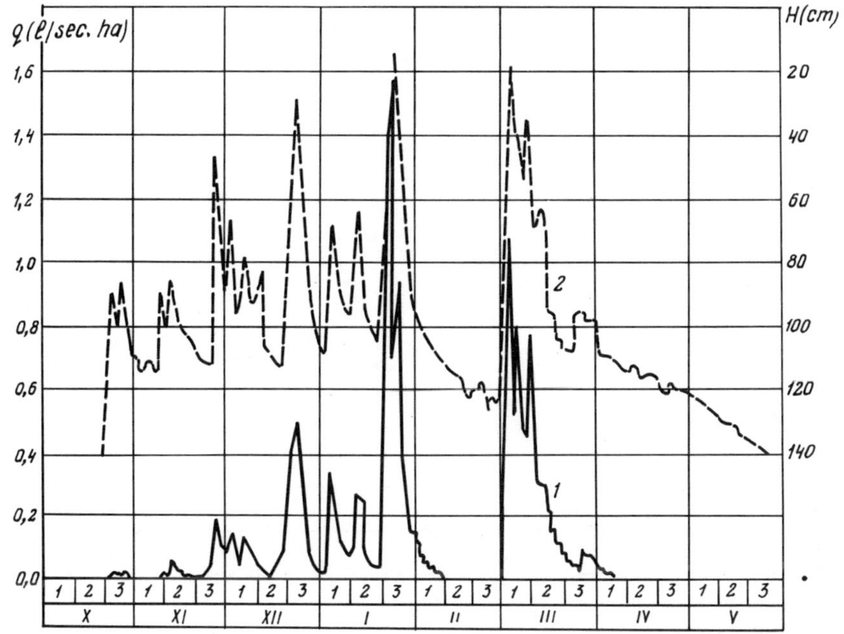

Fig 27. Hydrographs of the drainage runoff moduli (1) and ground-water levels (2) resulting from regular medium-depth drainage (1.2 m). The hydrological year 1958/59, Kolkhoz 'Kokness' in the Stuchka district.

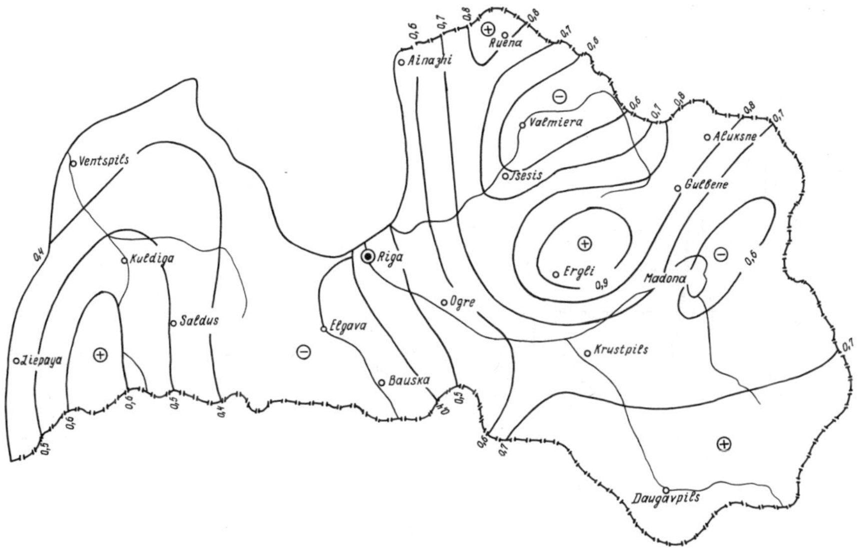

Fig 28.
The design specific drainage outflow (l/sec ha) for the Latvian SSR.

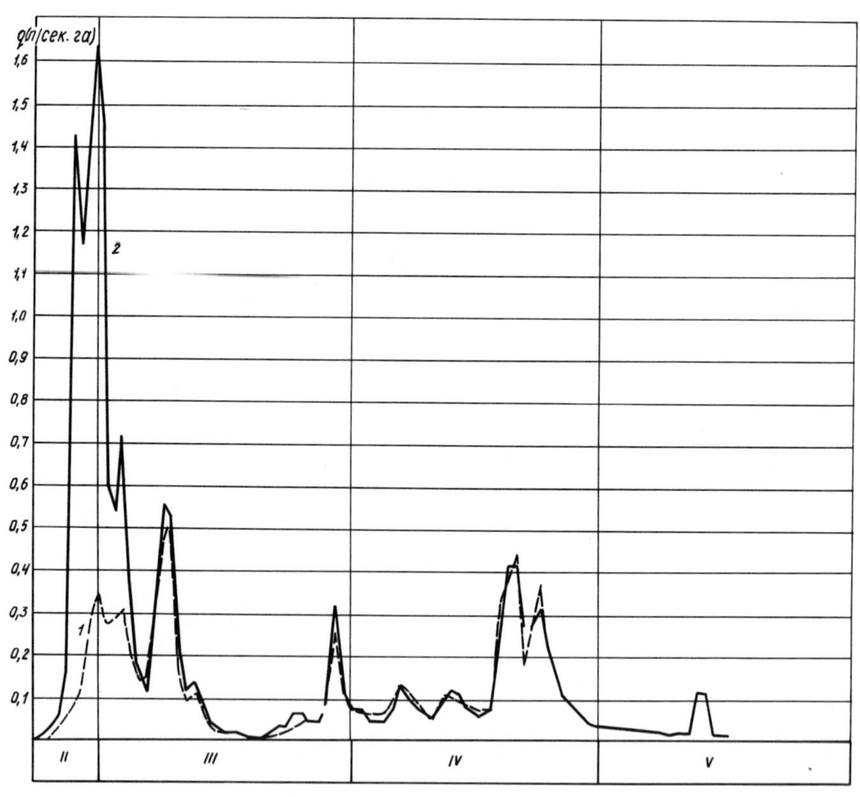

Fig 29. The mean diurnal moduli of drainage (1) and total (2) discharge in spring 1966 at the Bërze State Farm in the Dobeles district.

DISCUSSION

W. H. van der Molen (The Netherlands) - Do you account for streamflow variations and interaction between the streamflows from various drainage systems? If so, please give details.

A. A. Zivert (USSR) - In the particular case considered we do not account for the variations, as the main reclamation systems are located in regions with discharge ditches.

A. I. Zakrzhevsky (BSSR) - In Fig 29 one can see a solid line with a peak flow in spring. Is this peak a large value? Each year a large streamflow is observed during warm periods when water does not reach the drains, as described by Romanov. Is the soil in spring in the same frozen state as in winter, or does drainage occur ?

A. A. Zivert (USSR) - In the figure the streamflow data apply to 1966 and were characteristic for large surface watercourses. On the experimental plot this phenomenom was observed when snowmelt began very early, which explains why intensive surface flow was observed. In other years (the data are not given here) no surface flow was found.

WATER AND TEMPERATURE REGIMES OF LONG-CULTIVATED PEAT SOILS SUBJECT TO PROLONGED FREEZING IN THE RIVER VALLEYS OF THE MIDDLE TAIGA, THE KOMI ASSR

I. N. Skrynnikova

USSR

SUMMARY: This paper deals with the data obtained from investigations of the water and temperature regimes of long-cultivated peat soils subject to long periods of freezing in the river valleys of the Middle Taiga subzone of the Komi ASSR. It has been demonstrated that under the physiographical conditions of these regions, spring and autumn saturation of the drained peatlands is caused by surface water (autochthonous surface water above a frozen layer and allochthonous surface water due to subsurface or flood waters).

The paper also considers the relationship between the allochthonous and autochthonous sources of soil saturation for various combinations of weather conditions in the different regions of river valleys, their hydrothermal significance as well as the interrelation between the water and temperature regimes of soil in the continental regions of the Middle and Northern Taiga subzones of the European part of the USSR.

REGIME HYDRAULIQUE ET REGIME DES TEMPERATURES DES SOLS TOURBEUX CULTIVES DE LONGUE DATE ET SOUMIS A UN GEL SAISONNIER DANS DES VALLEES DE LA RSSA DES KOMIS (ZONE CENTRALE DE LA TAIGA)

RESUME : Cette étude présente les résultats obtenus sur des stations d'observation de régime hydraulique et de régime des températures. A partir de considérations géophysiques, l'auteur montre que, dans la région étudiée, la surhumidification de printemps et d'automne dans les tourbières drainées est due aux eaux de surface (les eaux de surface autochtones au-dessus de pergélisol et l'eau de surface allogène, résultant des eaux hypodermiques ou des crues).

L'auteur étudie la relation entre ces sources de saturation du sol autochtones et allogènes dans des conditions climatiques variées dans les diverses régions de vallées ; il expose la signification hydrothermique de ces phénomènes ; il montre combien il faut tenir compte des relations entre régime hydraulique et régime des températures du sol pour tous travaux de mise en valeur dans les ›ns continentales de la zone centrale et septentrionale des taïgas.

After draining, peat soils of the river valleys in the continental north-eastern regions of the European part of the USSR are most valuable for agricultural development.

In the continental regions of the Middle and, particularly, North Taiga with their short growing period and small sum of 'active' (> 10°C) temperatures, reclamation has a very great effect on both water and temperature conditions of the peat soil. This is very important for both soil formation and agricultural production. For instance, in the continental regions of the North Taiga drying of frozen peat soils may cause formation of soils with permafrost layers at a depth of 60-70 cm (Kochetkova, 1967).

The present paper contains the results of four years of continuous investigation of water-air and temperature regimes of long-cultivated peat soils in the Sysoly valley (Middle Taiga). This study is part of the complex research carried out from 1966 to 1969 by the Dokuchaev Soil Institute and Institute of Biology, the Komi Branch of the USSR Academy of Sciences.

Area studied

The investigations were carried out on a peatland situated in the Sysoly valley on the territory of the Vilgort Experimental Station. This peatland was drained in 1924-1927 by open ditches which were reconstructed after World War II. Surface and groundwater flow to Lake Elya-ty which in spring becomes connected to the Sysoly. The drained peat is sown with agricultural crops (perennial grasses, annual grass mixtures, root vegetables).

The experiments described were carried out on two arable plots with ditches 200 m long, 20 m and 50 m apart respectively, and on a third plot covered by natural vegetation.

The soil on two of the experimental plots has long been cultivated and on all three is subject to long frozen periods and is underlain by shallow peat. The arable layer (0 to 20 cm) is characterized by a high degree of peat decomposition (40%) and a well-pronounced cloddy or granular structure. This layer is followed (to a depth of 75 to 100 cm) by a sedge-bush-wood peat underlain by gleyed loam. At a depth of 40 to 60 cm there is an intercalated layer strongly enriched with wood residues.

Ash containing from 5.5 to 13 or 14% of mineral admixtures left by flood water composes 13-22% of the arable layer. The ash content of the subsoil is decreasing. The arable layer of the soil contains the following amounts of mineral elements which pass into 20%- HCl extract after ashing: CaO:- 3.5-4%; P_2O_5:- 0.34-0.36%; K_2O:- 0.07 - 0.23%; total nitrogen:- 2.20-2.52%. The soil contains small amounts of mobile nitrogen due to the low intensity of biochemical processes. The pH-value (1% KCl) varies from 4.8 to 5.4; exchange capacity equals 161-186 meg/100 g soil; calcium dominates among the absorbed cations.

Experimental methods

The groundwater levels were measured in a number of observation wells located both in the centres of the plots and near the ditches. Samples for determining the soil moisture content were augered once every 10 days (5-3 replicates) from June to October. The moisture content was determined thermally and soil temperature was measured by Savvinov-type angle thermometers and depth-thermometers on the 5th, 10th, 15th, 20th, 25th and 30th day of each month in the warm period of the year. During the autumn-winter and early spring periods of 1967-68 and 1968-69 depth-thermometers were used for measuring soil temperature. Groundwater levels and depths of soil freezing and thawing were also recorded. During the growing period the water and temperature conditions were investigated under two types of crops such as oat-pea grass mixture grown as green fodder and kuuzika (hybrid of swede and collard).

Climatic peculiarities and weather conditions during the investigation period

Table 15 contains some data characterizing the climate of the region under investigation. To distinguish more detailed climatic differences between the eastern regions of the Middle-Taiga and the weakly continental western regions, the Table includes comparative data obtained at the meteorological stations of Syktyvkar in the east and Petrozavodsk in the west but situated at the same latitude.

It can be seen from these figures that, according to climatic conditions, the growing periods in the west and east regions of the Middle Taiga are quite similar. These periods are short (100-104 days) with a low sum of 'active' (> + 10°C) temperatures (1360 for Syktyvkar and 1454 for Petrozavodsk). The fact that climate in the east of the subzone is more continental is demonstrated by a larger sum of negative temperatures (in Syktyvkar it is greater by 674°C. than in Petrozavodsk) and smaller total of solid precipitation. The severity of winter in the eastern regions manifests itself most clearly in the soil climate: the average and maximum freezing depths of loamy soils being 1.5 times greater than in the west. During mild winters with abundant snow, soils near Petrozavodsk do not freeze at all, whereas near Syktyvkar the minimum depth of freezing of loamy soil is 30 cm. There are no comparative data for peat soils but their soil climate is obviously governed by similar laws. The great 'reserve of cold' accumulated during winter is one of the main characteristics of the soil climate in the more continental eastern regions of the European part of the USSR.

The short growing period in the Middle-Taiga Subzone (especially in the eastern part) is to some extent compensated (for the development of farm crops) by high potential sunshine duration (19.5 hours per day in June and 18 to 19 hours per day in July) and low cloud cover. Thus, the daily increment of green matter of oats-pea mixture at the height of the growing period is 1 to 2 t/ha (data obtained at the Vilgort Experimental Station, 1967).

Table 16 contains some characteristics of the winter and summer weather of the region under investigation (Meteorological Station, Syktyvkar).

The table shows that the winter of 1965-66 was very cold and that of 1968-69 even colder. The winters of 1966-67 and 1968-69 were characterized by small amounts of snow, whereas the winter of 1967-68 was mild and had abundant snow; with total solid precipitation of almost 180% and sum of negative degree days 82% of their normal values. In spring the rivers attained high levels and flooded the drained peatland. The growing period was very warm and dry, having a mean temperature much higher than the normal value, whereas the growing periods of 1968 and 1969 were cold

Table 15. Some climatic data for western and eastern regions of the Middle Taiga

Geographical locality	Mean annual t °C	Annual precipitation, mm	Cumulated - ve degree days °C	Annual solid precipitation mm	Growing season, days	Sum of degree days for growing period °C	Depth of soil freezing, cm		
							Mean	Max	Min
Syktyvkar	0.32	500	1730	127	100	1360	83	149	30
Petrozavodsk	2.60	565	1056	176	104	1454	57	90	0

Table 16.

Hydrologic years	Cumulated - ve degree days °C	Annual solid precipitation, mm	Sum of degree days for growing period (June-August) °C	Total precipitation for period (June-Aug.), mm
1965-66	- 2065	195	1449	172
1966-67	- 1770	81	1490	101
1967-68	- 1425	228	1263	206
1968-69	- 2605	111	1138	206
Normal mean values	- 1718	127	1360	195

Table 16.

Hydrologic years	Cumulated -ve degree days °C	Annual solid precipitation mm	Sum of degree days for growing period (June-August) °C	Total precipitation for period (June-August) mm
1965-66	-2065	195	1449	172
1966-67	-1770	81	1490	101
1967-68	-1425	228	1263	206
1968-69	-2605	111	1138	206
Normal mean values	-1718	127	1360	195

and humid. It can be seen that during the years of investigation the weather conditions varied considerably.

In order to determine to which climatic period the years of investigation belong, we have calculated for various five-year periods the changes in the main climatic elements during the period 1898 to 1968 as observed at the Syktyvkar Meteorological station. We have also determined the total precipitation and the cumulated degree days not year by year but for periods of 'active' temperatures (> + 10°C) and plotted these values on a graph (Figure 30). It can be seen that climatic periods of 8 to 20 years duration can be clearly distinguished for both the cumulated degree days during the growing season and the summer precipitation. Attention is drawn to the opposition between the two parameters; periods of high rainfall coincide with low cumulative temperatures and vice versa. Similar comparisons made for some other continental regions have demonstrated the same relationship. The opposed character of these two parameters in the Komi ASSR reflects the prevalence of the southward circulation of air masses during cold summer periods when winds from the Arctic Ocean bring cold, moist air to the regions close to the Urals. The prevalence of latitudinal circulation of dry winds from the Mid-Russian Plain (in westerly and southwesterly directions) is responsible for warm, dry weather in summer. The period under discussion covers the end of a very warm, dry period (1966 and 1967) and the beginning of a period of cold summers (1968 and 1969) with high rainfall.

Soil moisture in cultivated humic peat

Like all peat soils those studied have a high moisture capacity; total moisture capacity (TMC) in the arable and subsoil layers being 85-86% and 89% by volume, respectively, whereas minimum moisture capacity (MMC) in these layers is 54 to 64% and 72%

respectively. Wilting point (WP) is somewhat higher than in the long-cultivated soils of the Southern-Taiga Subzone and attains 13 to 16% of the soil volume. Between MMC and WP we have calculated a further constant corresponding to the moisture content resulting in rupture of capillary bonds in mineral soils which we have called 'moisture content inside aggregate pores' (MCIAP), comprising 20 to 24% of the soil volume in the arable layer (Skrynnikova, 1961).

The large variation in the moisture content at a depth of 40-60 cm results from the presence of a large amount of wood residues in this layer. When strongly moistened or desiccated, the wood residues behave differently from the surrounding decomposed peat. If the groundwater table, having been at a high level for some considerable time is lowered, wood residues retain their moisture for a long time. After a long dry spell the pores of the wood residues as well as those of the surrounding peat are, on the contrary, filled with entrapped air (sometimes about 50% of the soil volume) which remains for a long time after a rapid rise of the groundwater table. Because of this we have defined three equal sub-intervals within the interval MMC to TMC to allow for the absorption of moisture by the wood residues (Figure 31).

Records of the groundwater level in observation wells show that during the period of investigation there was no permanent water table between the depths of 1 and 3 metres in the drained peat soils of the Sysoly valley. The saturation of soils was caused by surface water of two kinds: surface water above the frozen layer which appeared in the springs of 1967 and 1969 after severe winters with little snow, and surface water associated with the subsurface penetration and flooding of the peatland by the river flood waters after the snow-abundant, mild winter (spring of 1968). The diversity of surface water sources (precipitation, runoff, river floods,underflooding or flooding of the region) in addition to the above mentioned hydrophysical properties make it possible to give a quantitative description of the soil moisture content and water balance only for the upper 30-50 cm layer of soil. In the case of the lower horizons it is difficult to give more than a qualitative estimate.

From July to the beginning of September 1966 and from June to the 20th of October 1967 there was no groundwater in the soil above an aquiclude situated at a depth of 120 cm. Lenses of water lying above the frozen layer were detected at a depth of 35-55 cm only in the Spring of 1967 and these had disappeared by the end of July. The water which appeared in the lower soil horizons in the autumns of 1966 and 1967 was obviously allochthonous since it was known to originate in the rivers and lakes which submerged the reclamation network ditches after the autumn rain.

In 1966 evapotranspiration caused the soil in the root layer to dry to such an extent that by the end of July and August the minimum moisture capacity was attained and even the interval MCIAP to WP (in the 0 to 20 cm layer in August). Gravitational moisture in the lower soil horizons was retained only in wood residues. From the beginning of July in the drier summer of 1967, the entire soil profile dried to the minimum moisture capacity and gravitational moisture disappeared even from wood residues. The upper part of the arable layer was desiccated to a soil moisture level close to wilting point. Rainfall in September and October only replenished the soil moisture deficit in the upper horizons and failed to moisten the soil below a depth of 20 cm.

The allochthonous surface water which collected in October had disappeared by the beginning of March 1968. During the floods of 1968 the peatland was first underflooded and then flooded by the Sysoly river and lake Elya-ty. In May the entire soil profile was saturated but some of the wood residues located under the water contained a lot of entrapped air. From June to the middle of August the groundwater level fluctuated between 30 cm and 50 cm below the surface. By the end of August this water had disappeared but the entire soil profile still held moisture in excess of the minimum moisture capacity and rainfall in September rapidly raised the water table to within 30 to 50 cm of the surface again. The groundwater disappeared from the soil by the middle of February but the soil moisture content remained above minimum moisture capacity.

During the very cold winter of 1968-1969 with little snow, the soil froze down to a depth of 60 cm. Moisture from the lower unfrozen horizon migrated to the layer affected by freezing. When the soil was thawing, there appeared a lens of groundwater whose upper layer remained frozen but which partially percolated downwards into the frozen layer of soil until in July both groundwater horizons merged. During this cold and humid growing period the groundwater level fluctuated between 40 and 50 cm. The upper 50 cm of soil dried to the minimum moisture capacity only for a

Table 17.

Soil	Moisture storage (mm) corresponding to the following soil moisture levels defined in the text			
	TMC	MMC	WP	AMR*
Long cultivated, humic peat soil (experimental plots)	438	288	78	210
Strongly-podzolic loamy soil near Syktyvkar, Kononenko's data, experimental plot 1	216	162	39	123
The same plot, Kochetkova's data	265	153	46	105

* AMR - Active moisture range

few days in July. The arable layer was wetted to the total moisture capacity by rains which occurred at the end of September.

In Table 17 can be found the data on soil moisture in the 0 to 50 cm layer of long cultivated peat and strongly podzolic soils situated near Syktyvkar.

Comparison of these moisture storage values shows that the range of active moisture in peat soils is almost twice as great as that in podzolic soils, with the result that farm crops on peat soils do not suffer from moisture deficit even during the driest summer periods. To increase the moisture content in a 0.5 m-thick layer of peat soil from wilting point to minimum moisture capacity or to total moisture capacity, it is necessary to supply to the soil 210 mm or 360 mm of effective precipitation respectively. The corresponding figures for loamy, strongly-podzolic soils are 107 to 123 and 176 to 219 mm. Therefore, rainfall following dry spells first saturates the watershed soils and causes runoff, then percolates into reclaimed peat soils and continues to eliminate the moisture deficit. This explains the important role of allochthonous water in the autumn groundwater rise after dry summer periods on the cultivated peat soils of the middle and Southern Taiga in the European part of the USSR.

Temperature regime in cultivated peat soils subject to long frozen periods

The upper soil horizons remain frozen over a period of six to seven months, and accordingly are termed long-frozen soils. Complete thawing and warming of such soils to a temperature of + 5 to + 10°C at a depth of 10 cm occurs 22 to 27 days later than in the mineral soils in the vicinity of Syktyvkar. Floods play a very important role in warming these soils in spring time: in the spring of 1968 when the surface of the peatland massif was flooded by relatively warm flood water, the peat soil thawed by the 30th of April. According to long-term data, the earliest date for thawing of mineral soils is the 4th of May.

The isotherms in peat soils for the period of investigation have been plotted in Figure 32. These show that in 1967 and to a greater extent in 1969, following severe winters with little snow, frozen layers in the soil remained until the middle of June. Comparison of the data on precipitation and air and soil temperatures in the summers of 1967 and 1969 has shown that heating of the subsoil layers considerably lagged behind that of the air and was significantly influenced by percolation of concentrated rainfall in summer. The influence of water on the temperature regime of soils was quite evident throughout the years of study.

In Figure 33 the cumulative temperatures in different soil horizons in July and August are compared with the totals of solid precipitation and cumulative negative temperatures in the preceding winter periods. As can be seen from Figures 32 and 33, during the warmest but driest summer (1967), active temperatures did not go deeper than 60 cm and in July and August the subsoil layers were the coldest for the entire period of investigation. There is no correlation between the cumulative air and soil (below 60 cm) temperatures (Figure 33 a,b,c,d,) during the years of investigation, but the total solid precipitation in the preceding winter period is clearly correlated with the cumulative soil temperatures at depths of 80 and 160 cm in July and August. The highest temperature at those depths was observed in 1968 after a mild winter with abundant snow followed by spring flooding. In the autumn of 1969 the soil was saturated and the severe winter of 1969 accompanied by small amounts of snow resulted in deep freezing of soils, formation of and accumulation of a great 'cold reserve'. Hence, even the upper soil horizons were heated very slowly that year: the cumulative temperatures for July and August, show that at a depth of 20 cm the lowest temperature was observed in 1969.

Judging from the yield of pea-oat mixture, farm crops are very sensitive to the temperature of the air and arable soil layers. The greatest yields of green matter on the control plots (without application of fertilizers) were obtained in 1966 (22.96 t/ha) and in 1967 (20.87 t/ha), the lowest yield was in 1969 (7.78 t/ha).

Conclusions

1. The moisture and thermal conditions of drained and long cultivated peat soils are very closely related in the continental regions of middle and north Taiga Subzones.

2. Saturation of soils in spring and autumn is caused by superficial groundwater of two types. Autochthonous water which sometimes has a two-layer structure accumulated above a frozen layer plays a particularly important role in soil moistening after severe winters with little snow. The formation of allochthonous groundwater is mainly associated with subsurface percolation or flooding of drained soils in spring and less frequently in autumn by the overflow of drainage outlets. The role of the latter type of superficial groundwater is especially great in peat soils situated along river banks after mild winters with abundant snow, and insignificant when these soils are situated on the terrace. Deep winter freezing of soils accompanied by the accumulation of a large 'cold reserve' in the soil and formation of groundwater above a frozen layer, negatively affects peat soil heating especially in the lower horizons. Percolation of relatively warm river water into peat soil during floods has a great warming effect in spring. Such soils are also heated by percolation of warm summer precipitation to the deep soil horizons.

3. The close relation between the thermal and soil moisture conditions of peat soils in the northern continental regions should be taken into consideration when developing reclamation or agrotechnical methods. In order to improve these conditions, it is necessary to employ the following combination of methods:

(a) large spacings between drainage ditches (50-70 m) and regulation of drainage outlets (taking into account the warming effect of river water on soils);

(b) snow retention in early autumn;

(c) introduction of sand and clay into the arable layers;

(d) application of certain agrotechnical methods (sowing in furrows, levelling of reclaimed plots, etc.) to reduce the negative effect of superficial groundwater and improve soil heating.

References

Kochetkova, V.L. 1967. Osobennośti teplovogo rezhima torfyanykh pochv intinskogo promyshlennogo raiona. (Peculiarities of thermal regime in peat soils of the Intinskii industrial region). Trudy Komi filiula AN SSSR. No.11, Syktyvkar, Kimo Knizhnoe Izdatelstwo.

Skrynnikova, I.N. 1961. Pochvennye protsessy v okulturennykh torfyanykh pochvakh (Internal processes in cultivated peat soils) Moscow, Izd. Akad Nauk SSSR.

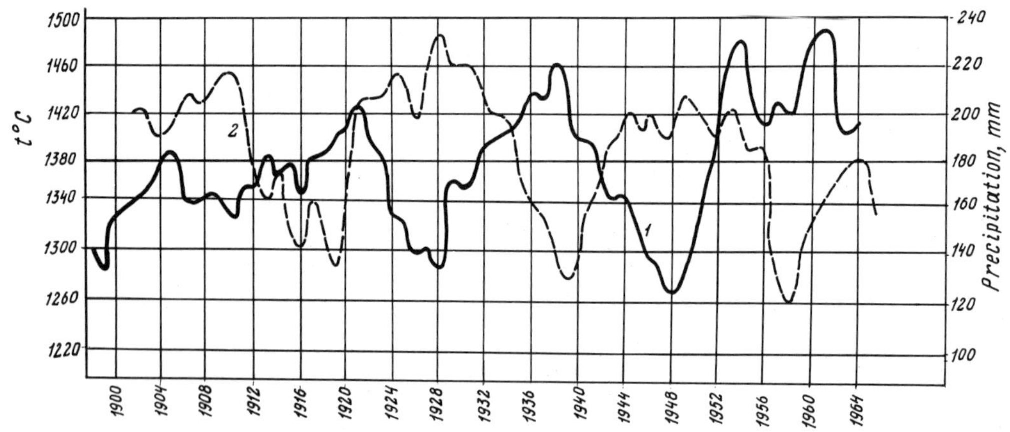

Figure 30. Cumulated temperatures and precipitation totals from June to August for arbitrary five-year periods (from 1897 to 1967)

——— cumulated temperature from June to August

- - - - - total precipitation from June to August

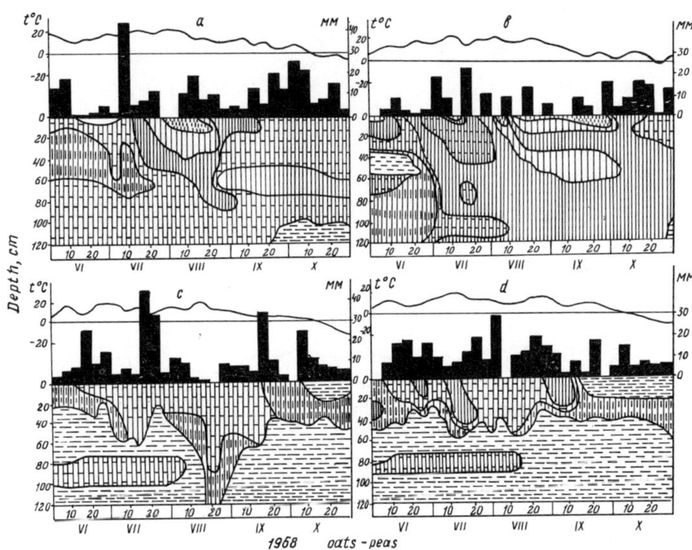

Figure 31. Chronoisolines of moisture content in long-cultivated peat soils subject to long frozen periods in the years 1966(a), 1967(b), 1968(c) and 1969(d).

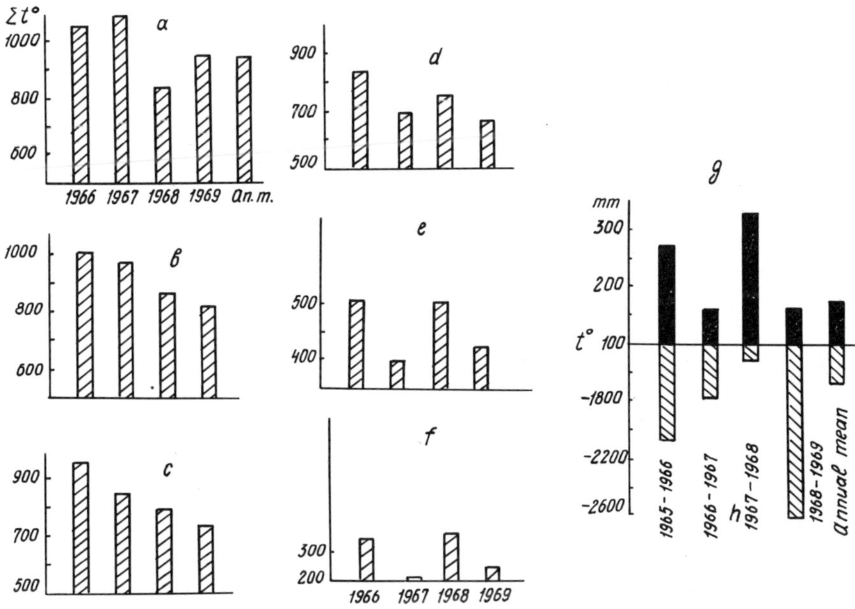

Figure 33. Correlation between cumulated degree days (from July to August) of the air (a), cumulated temperature of peat soil at a depth of 20 cm(b), 40 cm(c), 60 cm(d), 80 cm(e), 160 cm(f), cumulated negative temperatures total solid precipitation (g) during the years of investigation

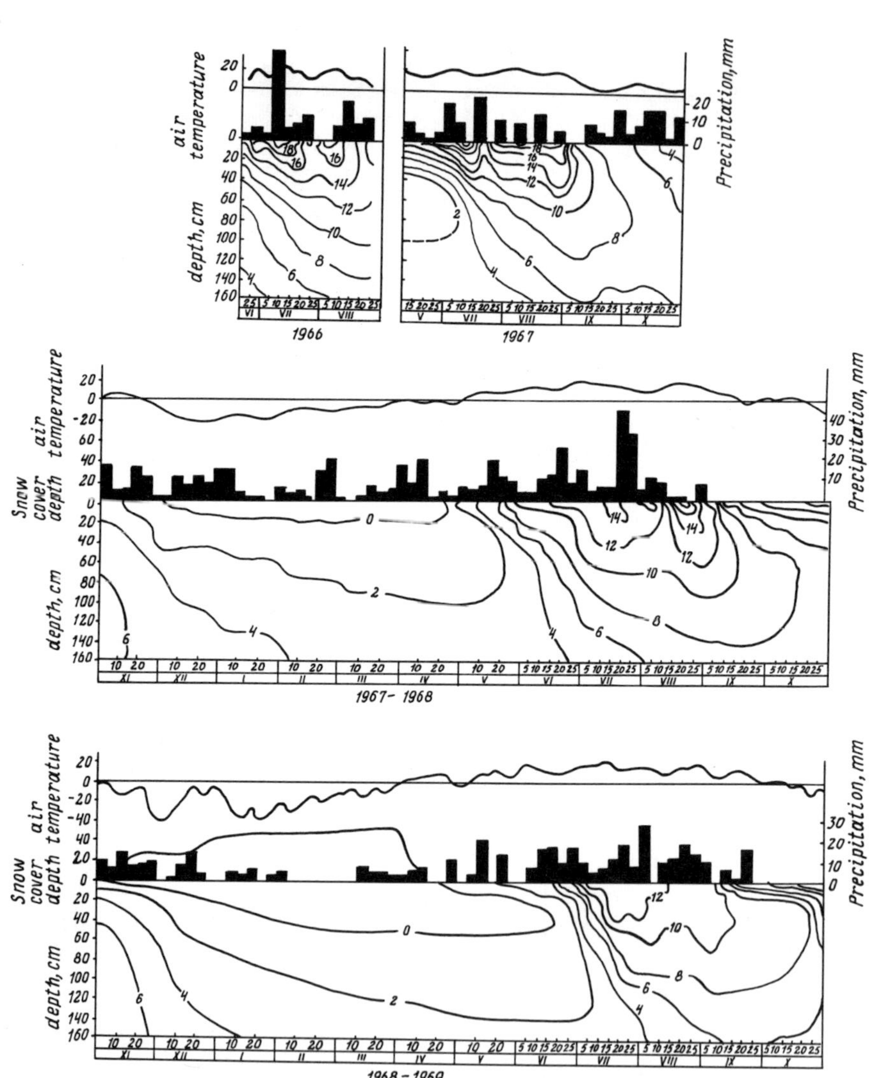

Figure 32. Isotherms in long-cultivated long-frozen peat soils

RESULTS OF COMBINED GEOLOGICAL AND HYDROLOGICAL INVESTIGATIONS IN THE BYELORUSSIAN POLESSIE AS A BASIS FOR ENGINEERING DESIGN OF MARSHLAND RECLAMATION SCHEMES

L. S. Panin, A. A. Zhelobaev and Yu. G. Bogomolov

State Design Institute for Water Economy,
Moscow, USSR.
and
Moscow State University,
Moscow, USSR.

SUMMARY: In this paper are examined the hydrochemical, geothermal, geophysical, paleohydrogeological and hydrodynamic methods of comprehensive investigation for land reclamation in the Byelorussian Polessie. The hydrodynamic method is found to be the most effective for studying the relation between aquifer water levels. The results of geohydrodynamic investigation and evaluation of the water balance components have made it possible to classify the region under consideration according to the nature of water supply to the bogs.

RECHERCHES GEOLOGIQUES ET HYDROLOGIQUES DANS LA POLESIE BIELORUSSE EN VUE DE L'ELABORATION DE PROJETS TECHNIQUES PORTANT SUR LA MISE EN VALEUR DES TERRAINS MARECAGEUX

RESUME : Cette étude passe en revue les méthodes hydrochimiques, géothermiques, géophysiques, paléohydrogéologiques et hydrodynamiques utilisées au cours de recherches complexes visant à la mise en valeur de la Polésie biélorusse. La méthode hydrodynamique fondamentale est l'étude du dynamisme des relations entre couches aquifères ; grâce aux résultats des recherches géologiques et hydrogéologiques et à partir de la mesure des divers facteurs qui entrent dans le bilan des eaux souterraines on a pu effectuer un classement des terres de cette région d'après les conditions d'alimentation en eau des marais.

The reclamation of water-logged regions requires general hydrogeological justification of improvement projects. Polessie is one of the largest reclamation areas in the USSR with very complex and variable natural conditions. Here, hydrogeological investigations have been considerably extended recently, in particular with respect to the behaviour of groundwater and its utilization for irrigation. At the same time hydrogeological classification of Polessie is proceeding for the purpose of differentiation of reclamation measures. Hydrogeological studies precede each of the three stages of the project: planning of reclamation schemes (long-term planning); engineering designs; and working drawings. At the engineering design stage a detailed study of the geological and lithological structure of the system is required. This involves the study of the hydrogeological conditions and a quantitative evaluation of the parameters involved in the design. A forecast is made of the future conditions keeping in mind the various uses to which the land may be put in order to satisfy the demands of the national economy. To solve these problems complex hydrogeological maps at scales of from 1:10 000 up to 1:50 000 have been prepared. Specific features of engineering works and their influence on natural conditions and peculiarities of hydrogeological conditions control the programmes and methods of investigation to be used as a scientific basis for reclamation work. Engineering measures for the reclamation of boggy regions in the Byelorussian Polessie are now reduced to the construction of horizontal drainage. This clearly necessitates a more detailed study of the upper layer of 20 to 30 m (or down to the first aquiclude), which is affected by such drainage. The upper 0.5 m of soil, in which the reclamation systems are constructed, and the effect of these measures is significant, is treated separately.

At the present time a number of engineering reclamation designs for the Byelorussian Polessie are under preparation. Studies in the western part of Polessie, within the South-east Brest depression are of particular interest. Here comprehensive geological, hydrogeological, geophysical, drilling and experimental programmes have been performed, to form a basis for engineering designs.

The boundaries of the regions are: in the north - the Mukhavets river and the Dnieper-Bug canal; in the south - the boundary with the Ukranian SSR; in the west - the USSR State Boundary; in the east - the Polessie saddle separating the Brest and Pripyat depressions. The territory represents a vast boggy plain, covered with forest and shrubs, dissected by a network of ill-defined rivers and drainage ditches. Three types of soil formation are found here; water accumulating, ashy and biogenic. The climate is moderately continental; rainfall varies from year to year (from 326.6 mm in 1963 to 728.4 mm in 1958), the average being 545.6 mm. The main geological process is swamping which currently affects about 40% of the region. In parallel with swamping, ashy processes occur in the area.

The crystalline basement, which is encountered at depth between 200 and 1000 m, presents a marked angular discordance with the overlying sedimentary cover which appears to be the upper structural horizon and is bedded by sedimentary rocks dating from the Protozoic to the Quaternary ages.

The rocks of Secondary and Tertiary ages are of particular interest in connection with the specific aims, such as choice and technical justification of drainage measures , of hydrogeological studies.

Jurassic deposits in the south-western and northern regions, which are marls and limestones with thickness ranging from 60 to 110 m, are pierced by holes. On Jurassic deposits offering stratigraphic discordance Upper Cretaceous rocks are bedded, which belong to Cenomanian, Turonian, Coniacian and Santonian stages. Deposits of the Cenomanian stage are represented by sands, sandstones and limestones, the thickness of which in northward and westward directions increases from 3 to 40 m. These deposits are overlain by a uniform marl-cretaceous mass attributed to the Turonian, Coniacian and Santonian stages. The lithology of these deposits reveals white chalk, sometimes intercalated with marl.

In the upper section (down to 10 to 15 m) the chalk is in a fluidized state. The thickness of the Cretaceous strata ranges between 66 and 124 m; deposits are highly karstic. On the cretaceous deposits are deposited rocks of the Middle and Upper Paleogene (Kiev and Kharkov Formation) composed of marine quartz-glauconite sand. These sands whose thickness varies from 10 to 15 m are extensive on the northern part of the area, but are found only in depressions of a supercretaceous relief in the south. At the end of the Paleogene era a new uplift occured causing a regression of the Paleogene Sea, washout of Paleogenic and Cretaceous deposits and formation of sandy-clay depostis of the Neogene and Anthropogen period. Neogene deposits (Poltava Formation) are found in the form of small patches consisting mainly of sands and clays, of a thickness not exceeding 14-20 m. Deposits of the Quaternary period represent a complicated parogenetic system of interrelated glacial, fluveo-glacial, alluvial, lacustrine and boggy formations. Three glacial complexes may be distinguished, namely, the Oka, Dnieper and Moscow formations with four corresponding horizons of interglacial deposits.

Depending on the geological structure and lithological composition, conditions of bedding and groundwater movement, several water-bearing systems may be distinguished, of which those of the Cretaceous and Quaternary contribute most to bog formation. To analyse the hydraulic interrelation between the above mentioned deposits, chemical, thermal, paleohydrogeological, geophysical and hydrodynamic methods have been used. This work has made it possible to classify the area according to conditions of water supply, and to evaluate the various elements of the water balance.

The chemical method is used to study the chemical composition of grounwater and its variation with distance from the supply region or with the depth of the aquifers. Salt content of groundwater on the Russian platform usually increases with the depth. A corresponding change in the chemical composition of groundwater takes place: hydrocarbonate waters are replaced by sulphate and then chloride waters. With such zoning, a water inflow of a different chemical composition results in the formation of zonal hydrochemical regions usually linked to hydrogeological inliers or tectonic distortions. Thus, for example, in the presence of hydrochemical zoning the appearance of a chloride water region within hydrocarbonate water is treated as evidence of deep water seepage. This chemical method which involves hundreds of chemical analyses of groundwater has verified the assumption that the total thickness of the sedimentary formation may be considered to be a zone of active groundwater flow. Three selective aquifers in the sedimentary formation were distinguished from chemical groundwater analyses, namely:

$$Q_{I-IV},\ Cr_2(t,Sn,Cm),\ I\text{-}S\text{-}CM\ Pt_3$$

An analysis of the distribution curve has shown that the distribution of ions in the upper Cretaceous aquifer is virtually insensitive to any influence and that water from all aquifers is mainly of calcium-hydrocarbonate composition. The salt content of groundwater generally does not exceed 1 g/l. Thus, a vertical chemical distribution showing the increase of salt content of groundwater with depth and the replacement of hydrocarbonates by sulphates and then by chlorides, was not generally observed for the Polessie region. Only the upper horizons of the Quaternary deposit are accumulations of chloride or sulphate waters found in certain areas, with the prevalence of certain cations due to domestic contamination, (Silin-Bekchurin, *et al*, 1971).

Since the chemical composition of groundwater and the content of microcomponents in the upper Cretaceous and deeper rocks are comparatively constant, the chemical method probably fails to interpret the groundwater movement.

A thermal method is used to reveal the general thermal background with the Brest depression, such as the thermophysical properties of the sedimentary cover deposits, geothermal gradients in boreholes and temperatures of rocks and water on the crystalline basement. When this method is employed, temperatures observed in boreholes are used as a measure of convective heat transfer by groundwater. During the research, (Bogomolov, 1970) 26 thermal properties of the sedimentary cover were determined (by applying an instantaneous heat source to samples with natural moisture content). Thermograms were obtained for ten boreholes and these were subsequently used for the calculation of a conductive heat flux density in four places (0.34 - 1.40 kcal/cm^2 s), thus verifying the ancient age of the depression base. The analysis of the results obtained shows that the area is located within the classic platform negative structure of the first order and that no additional heat sources should be expected here. Thus, the temperature drop between the base surface and neutral bed is very small (1 to 2°C) which makes the hydrothermal method inapplicable to the estimation of the interrelation of aquifers.

Geophysical vertical electric sounding (VES) has revealed a distinct spatial relation between washout zones in Cretaceous and Quaternary deposits, (Stikhov, *et al*, 1972). Investigations by the VES method support the suggestion that the lithologic composition of the Cretaceous deposits is non uniform and that this may result in considerable variations of permeability in the Cretaceous layer. Generally, the more sandy phases correspond to the crest of a structure edge. Therefore, the VES method allows one to deduce a qualitative description of the relationships between aquifers. In particular, the method suggests that this interrelation is most pronounced in the southern region of the Polessie. Concerning the paleohydrology of the area the principles of the proposed analysis will first be briefly discussed, (Bogmolov, 1970). The analysis is based on the isolation of temporal hydrogeological cycles and those structural material complexes which are the bodies where groundwater originates. Hydrogeological cycles are related to tectonic cycles (namely, the Caledonian, Herenynic and Alpine tectonic periods) with corresponding structural material complexes. Thus, temporal and spatial changes in the chemical composition of groundwater are considered. Absence of any strata in a strucural material complex of a given period does not distort the method: in this case the chemical composition of the water and rocks is considered in those deposits which are widespread at the present time. That part of the hydrogeological cycle which takes place before the sign of the oscillation changes from negative to positive is referred to as an eliasite state, whereas the state after the change of sign is said to be an infiltration state. Eliasite and infiltration hydrogeological stages are distinguished within tectonic periods or structures by completion of folding. Use of the paleohydrogeological method within the Brest depression (Bogomolov, 1970) has demonstrated that more than half of the geological history of the Phanerozoic period (52 to 55%) corresponds to infiltration stages. As to the area under study almost the whole Phanerozoic period was covered by infiltration which confirms a high degree of washing out of the sedimentary cover of the southern edge of the Depression. Comparison of the chemical composition of runoff and groundwater within the Brest Depression further strengthens this argument.

The main technique used for studying interrelations between aquifer water movements was the hydrodynamic method. After analysis of the hundreds of measured water head levels in the upper Cretaceous deposits and of the data on rock lithology and morphology of the region, maps of the water table contour were prepared showing the annual maxima and minima (from simultaneous measurements of groundwater levels in boreholes). Maps were also drawn up of the water table contour in the Quaternary mantle and of the piezometric heights in the upper Crétaceous deposits. In order to limit possible errors due to non-simultanity of measurements of artesian levels in the upper Cretaceous deposits, only the values measured in autumn were used, (Zhelobaev, 1971). Thus, zones have been defined and contoured where artesian levels in the Cretaceous deposits are either higher or equal to the free groundwater level. These zones were referred to as hydrogeological inliers or regions of direct hydrodynamic relation between ground and artesian waters, (Silin-Bekchurin, *et al*, 1971); here groundwater is most probably fed by artesian water. In accordance with the results of geological and hydrogeological research, the following equation was given by D. M. Kats and B. S. Maslov (1969):

$$q = \frac{K.\Delta H}{l}$$

where K is the permeability of the confining bed rock; ΔH is the difference between artesian and groundwater levels; l is the upper aquifer thickness with respect to a reference point at the ceiling of a pressure head horizon. Using the above equation contribution, (Zhelobaev, 1971), of the upper Cretaceous artesian water was calculated for lake Lukovo and found to be 0.0071 m per day (K=1.42 m/day, ΔH=2.0 m, l=400 m).

The knowledge gained by research and the evaluation of terms of the groundwater budget made it possible to classify the bogs of the territory according to their water supply conditions. Three types of bog were distinguished: those fed mainly by rainfall; those fed by both rainfall and groundwater; and bogs fed by artesian water. Here, ground or artesian waters were the main source of water supply to bogs in all cases when they provided more than 10% of the input term of the water balance. This classification of the regions seems to be the most sensible as it permits rational planning of reclamation of water-logged areas.

In the vicinity of bogs fed mainly by rainfall are found shallow layers (varying on average from 1.5 to 3.0 m) of lake loams Such zones covering 15% of the total area. Bogs of this type are associated with the deposits of residual lakes formed after lake Polesskoye had been emptied, and are connected with late Quaternary propyl rocks by Pra-Dnieper of the Ukranian crystalline massif (the region of the Dnieper gorges). The areas covered by individual bogs are not great, and range from a few hundred square metres to one or two square kilometres. On the input side of the water balance equation, accumulation of water due to groundwater discharge is less than 10% of the total input and is independent of rainfall. For example, in the case of the water balance for the 'Chernyany' region the total input due to groundwater flow for the 1967/68 hydrological year is 6% for a precipitation of 846.5 mm and 6.2% in the 1968/69 hydrological year for a precipitation of 461.3 mm.

The region of bogs fed by both rainfall and groundwater covers about 70% of the total area. Bogs of this type are associated with zones of waste and hollows of the late Quaternary period, early modern zones of waste and modern valley plains of rivers and lakes. Individual bog systems cover areas which vary from several tens of square metres to 20 or 25 km^2. The contribution of groundwater to the water balance is not constant. For example, in the case of the Barantsy region for the 1967/68 hydrological year, during which precipitation amounted to 465 mm, the input terms of the water balance were as follows: infiltration of rainfall, 93.2%; water originating in the aquifer (groundwater), 6.8%. In the 1968/69 hydrological year the total precipitation was 461.3 mm and the two input items amounted to 86.8% and 13.2%, respectively.

The third group of bogs, those fed by artesian water, lies in the southern part of the region where piezometric levels in the upper Cretaceous deposits exceed the level of the groundwater table. The Cretaceous aquifer is connected with the upper aquifer of the Quaternary deposits through the so-called 'hydrological inlier'. The presence of these inliers is predetermined by karst development in Cretaceous bed generally coinciding with the edge of the Brest Depression, (Makkaveev, 1969). This last circumstance explains why aquifers in Quaternary deposits are fed by waters from the Cretaceous layer.

According to the conditions of groundwater control and construction of vertical drainage wells, three regions may be distinguished within the entire area. The first is defined as the area where the upper aquiclude consists of loams of the middle-upper-Quaternary age and hydraulic conductivity does not exceed 10 m^2/day. This region appears to be unsuitable for supplying groundwater for irrigation and centralized urban water supply and for the construction of vertical drainage wells.

The second region was the largest area and covers the northern and central parts of the bog system, moraine clays and loams forming an aquiclude. The area can be subdivided into 12 zones with hydraulic conductivity ranging from 25 to 400 m^2/day. The zone with a hydraulic conductivity of 400 m^2/day is the most amenable to utilization of groundwater for irrigation, centralized water supply and the building of vertical drainage wells. The maximum predicted yield of wells in this zone is 50-70 l/s. In certain parts it is possible to build wells for supplying water to small farms having a water consumption of 20-30 l/s.

The third region which has no aquiclude in the upper Cretaceous deposits occupies the southern part of the territory. This may be divided into two zones, according to hydraulic conductivity, those with conductivities of 400 and 800 m^2/day, with rated maximum well yields of 60 and 120 l/s, respectively. Of the three regions this is the most promising for utilization of groundwater for irrigation, water supply and construction of vertical drainage wells. To evaluate such possibilities, pumping tests were carried out for a period of 60 days in the north-east near the village of Doroshevichi. After 30 days of pumping with a discharge of 110 l/s the level in the central well dropped by 14.78 m and in observation wells at distances of 100, 200, 500, and 750 m groundwater levels had decreased respectively by 3.05, 2.0, 0.68 and 0.26 m. This would therefore appear that the most rational drainage procedure for this region consists of decreasing the groundwater level in boggy areas and irrigation with fault waters of the adjacent high areas.

An analysis of the hydrogeological conditions shows that in the territory under consideration not only intense drainage of bogs is necessary but also irrigation of adjacent areas, particularly during dry periods in summer and autumn. In order to predict the effects of horizontal drainage and regulating networks, the territory is divided up into regions, subregions and zones. The regions are defined according to peculiarities of the geological and lithological structure to a depth of 5 m and permeability coefficients of the rocks in this layer. The depth to the uppermost aquiclude was chosen as the criterion for classification of subregions.

Maps have also been prepared showing areas where piezometric levels in the upper Cretaceous layer exceed those of the groundwater table. The construction of horizontal drainage networks, neglecting artesian supply, may be inadequate and in such cases the construction of vertical drainage wells is advocated.

REFERENCES

Bogomolov, Yu. G. 1970. Podzemnye vody Brestskoi vpadiny. Groundwater in the Brest Depression. Extended abstract of Candidate thesis, Moscow, Moscow State University.

Kats, D. M. and Maslov, B. S. 1969. Metodicheskie ukazaniya po gidrogeologicheskomy raionirovaniyu perenvlazhennykh zemel gumidnoi zony dlya tselei selskokhozyaistvennykh melioratsii. Methodological recommendations for hydrogeological zoning of saturated areas to be reclaimed in the humid zone. Moscow, Nedra.

Makkaveev, A. A. 1969. Raionirovanie Pripyatskogo Polessiya po usloviyam vodnogo pitaniya bolot. Subdivision of Pripyat Polessie according to bog water supply conditions. Moscow, Nedra (Voprosy regionalnoi gidrologii i metodiki gidroleologicheskogo kartirovaniya, vyp. 24).

Silin-Bekchurin, A. I. *et al.* 1971. Dinamika i gidrokhimiya podzemnykh vod Pripyatskogo Polessiya i ikh vzaimosvyaz s gruntovymi vodami v svyazi s zabolachivaniem (na primere Kazatskogo, Osipovskogo: Dneprovsko-Bugskogo massivov osusheniya). Dynamics and hydrodynamics of deep groundwater in Pripyat Polessie and the interrelation of these with groundwater in connection with swamping (examples of Kazatskii, Osipovskii and Dnieper-Bug drainage systems). Proc. of the 5th Sci. Conf., Faculty of Geology, Moscow State University, Moscow.

Stikhov, O. V. *et al.* 1972. Geologo-geofizicheskie aspecty vzaimosvyazi vodonosnykh gorizontov v predelakh yuzhnogo borta Brestskoi vpadiny (na pimere Osipovskogo, Kazatskogo i Dneprovsko-Bugskogo massivov osusheniya). Geological and geophysical aspect of aquifer interrelation at the southern edge of the Brest Depression (Osipovskii, Kazatskii and Dnieper-Bug drainage systems). Bull. Mosk. Obshch. Ispytat. Prirody, (6).

Zhelovaev, A. A. 1971. Gidrologicheskie usloviya yugo-vostochnoi chasti Brestskoi vpadiny i metadika gidrogeologicheskogo obosnovaniya tekhnicheskikh proektov melioratsii. Hydrogeological conditions of the south-eastern part of Brest Depression and methods of engineering justification of reclamation projects. Extended Abstract of Candidate Thesis, Minsk, Byelorussian Academy of Sciences.

DISCUSSION

F. R. Zaidelman (USSR) - In this paper you discussed only geological research whereas in practice the whole combination of geological, botanical, etc., research must be taken into account. Can your work be considered comprehensive in this respect?

A. A. Zhelobaev (USSR) - Before carrying out the design of reclamation projects we have to consider the aspects which have a bearing on the hydrogeological properties of the area. I would like to pursue this aspect as it is most complicated.

Design of reclamation systems is certainly a complex problem and not only soil science should be employed but also aerodynamics and other problems should be studied.

G. V. Bogomolob (BSSR) - In your paper you stated that for boggy areas the usefulness of vertical drainage for certain regions of Polessie has not been studied. Do you think this type of drainage may be used in the north zone where Cretaceous deposits occur near the surface?

A. A. Zhelobaev (USSR) - For the southern part of the area under consideration, vertical drainage systems should reach the lower layer of high conductivity. In this region vertical drainage may be successfully used. For the northern part no calculations have been made, but if vertical drainage wells are located in the active layer with coefficients of permeability of 10-15 m/day and thicknesses of 40 m and more, drainage will be effective. No values of the future decreases of the groundwater level in the northern zone can be given because of the lack of analysis.

THE CAUSES OF EXCESS SOIL MOISTURE AND ITS ROLE IN THE FORMATION OF BOGS IN ROMANIA

I. Enya and K. Sarta

The Institute of Meteorology and Hydrology, Bucharest, Romania

SUMMARY: The study of the causes of excess moisture is of great importance for planning and building hydraulic land reclamation structures and for developing land for agriculture. A network of 700 hydrogeological stations and 2500 observation wells has been set up to study the regime of deep groundwater under natural conditions and also during the building of structures. This network has been designed to study the behaviour of groundwater and to define suitable methods of preventing or reducing excess moisture with a view to eliminating water logging and salination of soils.

LES CAUSES DE L'EXCES D'HUMIDITE DU SOL ET SON ROLE DANS LA FORMATION DES TERRAINS MARECAGEUX EN ROUMANIE

RESUME : Il est très important de connaître l'origine des excédents d'humidité pour la construction des ouvrages destinés à l'assèchement des terres et pour la mise en valeur de celles-ci. Un réseau de 700 stations hydrogéologiques et de 2 500 puits d'observations a été établi pour étudier le régime des eaux souterraines profondes dans les conditions naturelles et aussi pendant la construction des ouvrages. L'étude du comportement des eaux souterraines permettra de mettre au point des méthodes appropriées pour prévenir ou réduire les excédents d'humidité et d'améliorer l'état des sols en éliminant leur saturation en eau et leur salinité.

The planning and construction of hydraulic and land reclamation structures for the development of agriculture require that the cause of rising groundwater levels be studied and that reclamation areas be classified to define measures for the prevention or reduction of excess soil moisture.

To study the regime of deep groundwater under natural conditions and during the process of building structures, a State hydrogeological network has been established comprising 700 first and second order stations and 2500 observation wells. The first order stations are located in valleys, 2 to 8 wells in a straight line at 10 km intervals; the second order stations are located on the catchments, one well per 16 km^2 area. Every third day the groundwater levels are measured in the wells to record both temporal and spatial variations.

Crop yields are known to fall not only where the water level reaches the soil surface but also where the level is less than 1 m below ground. The study of the causes of excess moisture is of great importance to reclamation schemes and the development of land for agriculture. Knowledge of these causes will help to solve the problems of boggy areas and soil salination.

Long term experiments and investigations carried out in different parts of Romania have enabled these causes to be divided into two classes; natural and artificial:

1. The natural causes apply to naturally waterlogged regions distinguished by various geomorphological features such as the following associations: a flood plain to terrace contact; a debris cone base; a foothills to lowland plain contact.

At the 'flood plain/terrace contact' the terrace aquifer forms springs which result in a boggy and saline flood plain where the grain size of the soil is fine and is deposited in gently sloping layers. In Romania such contacts are readily found in the lower reaches of the rivers Danube, Olta, Zhiu, Sereta and others.

The regions at the base of debris cones which form in foothill plains are also waterlogged areas. Such are the debris cones of Ardzhesha, Dymbovitsy, Prahove-Telyazhen, Buzeu, Muresha, Krishey, Somesha and others.

In these foothill regions of the Romanian and Western plains, waterlogging takes place particularly during the spring months, after snowmelt. In Buzeu, ground subsidence has resulted in the formation of layers of fine grained structures. This very gently sloping and dry zone is remarkable for its vast amount of groundwater storage, the level of which varies between 0 to 2 m below the surface and for its salt-marshes (Smeyen-Batogu).

Boggy regions are also found where loess deposits occur on clay layers. These are endoreic zones in which the groundwater level is near the surface and gives rise to swamps. Examples are found in the Leu-Rotunda plain and part of the Vlesiey plain (the zone of Proletaru-Shtefenesht-Afumats).

The presence below arable soil of loess deposits on deep and extensive clay layers has resulted in the formation of a ground stratum which assists the processes of waterlogging. Although loess is a consolidated deposit, when saturated with water, solution reduces its volume. This leads to the formation of undulations and depressions in which accumulate rain and snowmelt water bringing the groundwater level to the surface, with resulting boggy and saline conditions.

In wet years in these zones, the groundwater rises to inundate the land, which becomes useless for agriculture. This is illustrated by geological cross-sections of several zones in which excess moisture is due to the structure and stratigraphy, the rate of groundwater flow and the surface runoff (endoreic zones).

In the high Leu-Rotunda plain, there are clay formations of up to 30 m thickness under the arable soil which is, consequently, saturated with water. The zone lying to the west of Bucharest belongs to this category: under the arable layer there is a loess layer over clay. Areas where loess overlies fine sand may also be included in this class, as the groundwater discharge and influence of drainage are insignificant. Such regions are present on the terraces of Zhiuluy, Oltuluy, Kelerash, Breila and Visiru-Latinu.

The areas beyond the base of debris cones in the direction of groundwater flow are waterlogged because of the outflow of groundwater from the cones. These sand deposits are large groundwater reservoirs which raise the hydrostatic level (Krivina-Vynzhu Mare, the terrace of Sadove and also in the catchment of the Moshtishtya and Yalomitsy rivers).

Annual rainfall is another important factor which contributes to excess moisture, resulting in poor vegetation even when the groundwater level is less than 1 m deep. Measurement of precipitation, which varies temporally and spatially, and correlation with zone lithology, have made it possible to treat individually certain areas of the Romanian and Western plains.

2. The artificial causes of waterlogging are irrigation and the raising of embankments.

Our observations have shown that when the water table is more than 5 m below the surface, irrigation results in a reduction in the amplitude of seasonal variations in groundwater level. When the groundwater table is at lesser depths these variations increase as the level approaches the surface. A long-term study of groundwater conditions is being carried out in connection with the irrigation plan for the territory.

Several examples illustrate the constant rise of the groundwater level in irrigated territories.

Example I.

Intensive irrigation on the Kelerash terrace has modified the hydrological balance, worsened the agricultural conditions and created bogs and salination. Below the arable soil of the terrace there is loess, a formation into which water penetration is easier vertically than horizontally. Groundwater levels of 1965 and 1966 are higher in comparison with those of 1964 before rice irrigation had begun. Thus, rice cultivation on loess terraces underlain by fine-sand is inadvisable.

The flow of irrigation water from the Kelerash terrace towards the Trestik depression and the Geaku zone has resulted in creation of a bog. Before intensive irrigation, the groundwater level in this area was at a depth of 10 m, below the critical level, but the recorded rise since irrigation began amounts to several metres. If the resulting level is observed for a least 3 years and correlated with the volume of irrigation water and climatic factors, it may be possible to determine the causes and rate of the rise in level and to take appropriate action.

It has been established that the rise of the groundwater level is accompanied by an increase in the mineral content of the water, which after some time, may contain twice as much solid residues as initially.

Example II

Seepage losses from canals built in loess create a head which will lead to swamping as happened at the morphological contact between Karakal and Stoenesht terraces.

To ensure that the objectives are achieved, hydraulic land reclamation and water use plans should be implemented in accordance with the specific geomorphological conditions of each zone.

Irrigation rates must be determined with consideration for local conditions to avoid reduced crop yields because of excess moisture. Furthermore, to avoid bog formation and salination their causes must first be established.

Data received in the form of systematic measurements in the observation wells of the State hydrological network (which provides data for investigations of groundwater) were used to analyse the causes of waterlogging. Each area was classified according to hydrogeological conditions, and comparative maps were prepared as an aid to efficient land use.

Typical situations analysed were those of 1964 (a dry year after a long dry period); of 1968 (a dry year after several wet years); 1969 (a wet year after a dry one); and 1970 (a year of abundant precipitation).

The maps divide the territory into three groups: (a) areas with maximum groundwater level varying from 0 to 2 m; (b) those where the maximum level varies from 2 to 5 m; and (c) those areas where the groundwater level is below 5 m. These maps show that in the Rumanian plain the maximum groundwater levels of 0 to 2 m below the surface spread in 1964 over the morphological zones of 'flood-plain/terrace contact' along the Danube towards Benvase-Gostinu-Spantsov, to the south of the Buzeu river. The maps also show that the regions of dunes and debris cone discharge fell into groups (b) and (c) in succession.

The map of the maximum levels for 1968 shows the existence of zones in classes (a) and (b) in the territory to the south of Bucharest. For 1969, the base of debris cones and endoreic zones belong to class (a) if they are in areas with poor or no drainage; in comparison to 1968 category (b) is very extensive.

The map of maximum groundwater levels for 1970 shows that the number of zones in (a) has increased compared to those in (b). The Romanian plain has zones in class (a) even in dry periods; these are found either in areas where there is no surface runoff or at the base of debris cones which need drainage.

Rises in groundwater level can also be observed in regions where irrigation is out of control.

In the Western plain, the maximum levels in 1968 are characteristic of waterlogging at the base of debris cones (resulting from flow proceeding from the cone), in low-lying plains where-ever a layer of fine material is combined with a gentle slope, and in depressions. Areas of group (a) are particularly extensive in the depression of the plain and near the Muresh debris cone. This typical situation has been followed since 1968, the year when the measurements began in the network.

The distribution of maximum groundwater levels in 1969 shows a significant increase in category (a) compared to category (b).

In 1970, as a result of lower rainfall than in 1969, group (a) was contained within small areas (with the exception of the Arada and Arankey plains).

The first maps for the Romanian plain were plotted for the periods 1964 to 1968 and 1968 to 1970, and for the Western plain in 1968 and 1970. Their analysis shows that from 1964 to 1968, the rise in groundwater level in the flood plain area of the Rumanian plain varied from 15 cm to 1 m. In the plain to the east of Vynzhu Mare, falls in groundwater level of 21 cm to 74 cm were recorded in areas of steep slopes and coarse material. In areas subject to irrigation, the rise in groundwater levels varied from 50 cm to 1 m.

Comparison of the maps for 1968 and 1970 shows a rise in level in the irrigated regions of 0.5 to 1.5 m both in the flood plain and in the terrace land.

A comparison of the average annual levels for 1968 and 1970 in the Western plain (where the State hydrogeological network came into operation at a later date) shows rises of 0.5 to 1.1 m under irrigation, in the bogs at the foot of the debris cones and at the contact between the foothills and low-lying plains.

Such comparative maps are helpful in planning measures for the most efficient exploitation of the land and facilitate the timing of agricultural work.

GEOLOGICAL AND HYDROGEOLOGICAL CONDITIONS OF MARSHY AREAS AND THE FREEZING RATE OF SOILS

V. F. Kovalchuk, M. S. Karasev, V. S. Nabrodov, and Y. I. Litvinov

Far East Research Institute of Water Engineering and Reclamation, Vladivostok, USSR.

SUMMARY: In this paper the physiographical, geological and hydrogeological conditions which control the intensity and direction of the groundwater recharge of marshy areas in Primorie are described. These areas are classified according to the degree of natural drainage as intensely drained, drained, slightly drained and undrained. The individual features of the processes that result in water logging and formation of swamps are described for each of the regions. The relation of both ground and artesian waters with marsh water is examined in the case of the slightly drained and undrained regions. The regime of marsh water is defined. For the first time in Primorie considerable attention is given to the dependence of the depth of freezing of the ground on its moisture content.

CARACTERISTIQUES GEOLOGIQUES ET HYDROGEOLOGIQUES DE LA REGION MARECAGEUSE DE PRIMORYE ET RELATION AVEC L'INTENSITE DU GEL

RESUME : Cet exposé étudie les conditions géophysiques, géologiques et hydrogéologiques qui déterminent les caractéristiques de l'alimentation en eau du sol dans la région marécageuse de Primoryé. Les auteurs distinguent les zones intensivement drainées, peu drainées et non drainées. Pour chaque zone ils décrivent les particularités du processus qui mène à l'eau stagnante et à la formation de marais. Ils déterminent le type d'alimentation de ces marais et étudient les relations qui existent entre les eaux souterraines et artésiennes et l'eau des marais pour des zones peu drainées et pour les zones non drainées. Le régime hydraulique de ces marais est également examiné. Pour la première fois est étudiée dans cette région la relation entre le degré d'humidité des sols et la profondeur du gel dans ces sols.

Primorie is subjected to a monsoon climate and deep (over 2m) seasonal ground freezing caused by a shallow snow cover. The annual precipitation, observed over a period of many years, varies from 500 to 1000 mm, whilst the evaporation is of the order of 400 to 600 mm a year. About 80% of the normal annual precipitation falls between April and October, the maximum being observed in July-August.

Three large orographic regions may be distinguished on the territory in question: the West-Primorie plain, the mountainous regions of Sikhote-Alin and the piedmont of the West-Manchurian highland, (Anon., 1969).

The main features of the geological structure of Primorie are characterized by the development of ancient Paleozoic and Mesozoic structures forming a folded base and of the more recent Cenozoic deposits composing the mantle.

Paleogene and Neogene formations are known to be among the rocks of the mantle that form basins and volcano-tectonic structures: acid, middle and basic effusions, poorly cemented sandstone, aleurolites, argillites, brown coal beds, gavellites, having a total thickness over 1000 m. These deposits are overlapped by Quaternary sediments which can be divided into Early, Middle and Late Quaternary and modern sediments. According to origin there may be distinguished among these; alluvial, lacustrine-alluvial, lacustrine, marine sediments and formations of the slope series. Alluvial deposits are of a double-layer structure: in the top layer of the section are clays, loams, sandy loams 10-15 m thick; in the bottom layer, shingles, sands and gravels. Lacustrine and lacustrine-alluvial deposits are composed of clays and loams with thin interlayers of sandy loams and sands. The sediments of marine origin are represented by powdered sands, oozes and sandy loams. Formations of the slope series are composed of loams with rubble and gravel. The thickness of Quaternary deposits in the mountains does not exceed 3-4 m, while on the plains it attains 100 m.

Within the Primorie area, two hydrogeological structures of vast size may be distinguished: the Khanka artesian basin and the basin of fracture waters of the folded Sikhote-Alin system and the piedmont of the East-Manchurian highland, (Kladershcikov, 1968).

The Khanka artesian basin which covers the West-Promorie plain is a Cenozoic depression. Artesian interstratal waters are widespread in deposits of Paleogene and Neogene.

Both artesian and unconfined water is found in Quaternary deposits. Artesian water is confined to the Upper, Middle and Lower Quaternary deposits, the upper part of which is composed of clays and loams. Unconfined groundwater is present in the modern deposits of fluvial plains. It is also encountered on terraces above the flood plain where sands and sandy loams outcrop.

Because of the small thickness of the bed-rock and its discontinuity along the strike, the aquifers and complexes of pre-Quaternary rocks are hydraulically interconnected and connected to the aquifers in Quaternary deposits. Hence, the chemical composition of water in all aquifers is similar, being fresh water throughout with a mineral content of 0.1 to 0.5 g/l.

In the basin of fracture waters of the folded system, the groundwater is generally confined to the zone of jointing ledge rocks of various composition and age. Moreover, there is a number of small intermountain artesian basins with extensive development of interstratal artesian water.

According to the degree of natural drainage, the territory is divided into four zones; intensely drained, drained, slightly drained and undrained* . The intensely drained zone embraces mid-

* After D M Katz (1967), the criterion of natural drainage is the mean value of the base flow produced by the region in question, over a certain period.

mountain and low-mountain reliefs. Here groundwater recharge takes place at a depth of over 10 m. The land is not swamped.

The drained zone covers ridges separated by shallow river valleys and clays and loams to depths of 10 m and over are prevalent in the lithological composition of the deposits. The gradient of the ground surface is 0.02 - 0.04; the depth of the groundwater is over 3 m. The outflow exceeds the inflow. The recharge is attributable to precipitation and gradient. The land is subject to temporary excessive soil moisture because of intense summer rainfall and the low permeability of the argillaceous ground. This moisture is stored in the root layer during the second half of the vegetation season (July to September), which leads to a decrease of yields and sometimes to the loss of crops.

The slighlty drained zone includes fluvial plains, lacustrine alluvial, lacustrine and alluvial terraces within which movement and discharge of groundwater takes place. The depth of these varies between 0.5 m and 3 m, the gradient of the water table varies between 0.002 to 0.006. In this group the inflow exceeds the outflow. This zone also includes the areas in which water occurs at depths of over 3 m in a soil formation of heavy mechanical composition, 3-10 m thick.

The undrained zone is in the form of a narrow strip along the south-east shore of lake Khanka and also in the deltas of the rivers Lefu and Santakhesa. It is characterized by a groundwater depth of less than 1 m. The groundwater basin has a surface gradient less than 0.005 and the discharge is very small. The lacustrine-alluvial and lacustrine deposits consist of loams, sands and sandy loams.

Swamped lands and bogs are widespread in the slightly drained and undrained zones*. Swamped lands have the greatest extension on the West-Primorie plain. The main reasons for their development are insufficient permeability of the argillaceous soil and ground, abundance of rainfall, lack of slope which impedes runoff of rainfall. The mountainous relief of the surrounding area favours the runoff of water to the plain with the result that floods are frequent, which is one of the main reasons for the formation of bogs.

Swamps are mainly found on the West-Primorie plain. They are mostly low-land bogs extending along the south-east shore of lake Khanka, the right bank of the river Sungach and in the valley of the Lefu. The peat, composed of sedge-herbage and sedge-horestail species, has a thickness which varies between 0.8 and 3.0 m.

On the greater part of the plain, within the slightly drained zone, the peat bogs are underlain by impermeable clays and loams of a thickness varying from 1 m to 10 m and more. The swamps receive their water chiefly from atmospheric precipitation and runoff from the slopes. Watertight argillaceous deposits that form the inorganic bottom of swamps prevent any communication between the

---

* Swamps are the areas covered by a peat layer having a thickness when wet of over 30 cm; swamped lands are areas covered by a peat layer of a thickness less than 30 cm or wholly devoid of peat but permanently or durably saturated.

swampy water and the groundwater.

On the south-east shore of lake Khanka, within the undrained zone, peat occurs on fine-grained sands, sandy loams and light loams. Their permeability coefficients are 1; 0.5; 0.1 m/day respectively. Bogs are mainly supplied by groundwater which is close to the swamp surface.

Swamps are a groundwater-bearing horizon. In spring there is an increase in water supply on account of the thaw; in summer the swamp receives rainfall and runoff as well as water driven by strong winds from lake Khanka. In areas where the inorganic bottom consists of slightly permeable sandy loams and light loams, the swamps also receive artesian water. The artesian water supply mainly occurs in winter and equals 100 m^3/day per 1 km^2 of the area.

The amplitude of the annual variation of swamp water levels reaches 1.5 to 2.0 m. A slight rise in the level is observed during snowmelt in spring, then in May-June a steady fall takes place. An abrupt rise of the swamp level is observed in July - September, a time of heavy rains and flood, and swamps are covered by a layer of water 20 to 30 cm deep. Similar flooding of the swamp can occur during a period of strong winds which drive water from lake Khanka. In dry years the swamp water level falls and where the thickness of peat is small, the whole profile may dry out.

Swamps known by the name of 'mari' are widespread in the slightly drained zone in the valleys of the mountain rivers Beitsukhe, Alchan and Bikin in the north of Primorie. They are confined to the alluvial terrace of Upper-Quaternary age and slightly inclined towards the river bed. They are sphagnum bogs with small areas of sedge-hummocky bogs and 'erniks' (dry bogs, hummocky, covered with stunted arboreous vegetation). Here mostly larch grows and peat 0.3 to 0.5 m thick, but in some places up to 1.5 m, overlays impermeable heavy loams.

Besides 'maris' in the valleys of mountain rivers one may encounter small-sized sedge-blue joint, sedge-herbage near-terrace swamps and oxbow swamps occurring on sands, sometimes on pebble-beds, sandy loams and loams. They receive groundwater, surface runoff and a mixture of groundwater, rainfall and runoff water supply.

Nowhere in the world does the seasonal freezing reach such depths nor extend so far to the south, except in mountainous regions, as in the Far East. Soils which freeze seasonally are of great significance for agriculture. In the south of the Far East alone the areas with deep seasonal freezing occupy over 8 000 000 ha. Apart from deep seasonal freezing a peculiar feature of these soils is their late thawing which influences the vegation period.

A considerable part of the swamped land requires reclamation, but reclamation changes the soil moisture regime, which is one of the most important factors determining the dynamics and depth of freezing.

The hydrological conditions of Primorie control the groundwater recharge of the bog areas. Snow sublimates due to high solar radiation and negative air temperatures in winter, while in spring when air temperatures are still negative there is snowmelt above the seasonal deep frozen lower layers. Snowmelt water is not absorbed by the soil and runs off its surface.

The temperature of the soil continues to fall even when that of the atmosphere rises. Seasonally frozen soils in Primorie may be found at depths of 1.5 to 2.0 m even in August. This favours soil moisture storage and is of great importance for agriculture. Drainage of swamp and waterlogged areas would result in earlier ripening of crops, etc.

The dependence of the dynamics of frozen soils in reclamation areas on the moisture content of the soil was investigated at the experimental plots of the Khanka plain. Besides natural observations, investigations were carried out of the time changes in the temperature-moisture regime for similar conditions on the hydraulic intergrator of Prof. V. S. Lukyanov's system.

As an example, results are presented of estimates concerning inogranic soils of the Santakhes rice scheme. Since the most unfavourable conditions are of interest to practical workers, heat-transfer between the soil and the air for mean monthly temperatures of the coldest year registered for 40 years (Table 18) has been taken as the upper limit.

The lower limit is invariable temperature at a depth of 15 m equal to the mean annual temperature 0.8°C + 2°C, i.e. 2.8°C (after A. V. Stotsenko, 1952).

Table 18. Mean monthly temperature of the coldest year observed in 40 years.

Months	I	II	III	IV	V	VI	VII	VIII	IX	X	XI	XII
Temperature °C	-21.7	-17.5	-10.5	2.0	12.4	16.4	21.0	19.9	14.5	6.97	7.5	-26.9

To find the relation between the depth of the frozen ground and the soil moisture content, three types of argillaceous soils were considered. For each the frozen soil depth after reclamation was plotted against the moisture content and appropriate thermophysical properties were chosen, (Ushkalov, 1962). The results of calculation for one of the types of soil are given in Table 19. Similar results have also been obtained for other reclamations systems (Kaul, Melgunorka, Daubikha, Guberovo, Primorie, etc.).

The table shows that under the conditions of Primorie a decrease in the soil moisture content causes a considerable increase of the depth of soil subject to freezing.

Variations of moisture content change considerably the thermal conductivity and heat capacity of soils which in turn affects the growing period of plants.

Table 19. Estimate of freezing and thawing depth and thickness of frozen layer in clay soils of Primorie.

	Months								
	XI	XII	I	II	III	IV	V	VI	Moisture content %
1	0.08	0.28	0.52	0.80	1.03	1.11	1.07	0.80	
2							0.40	0.52	52
3	0.08	0.28	0.52	0.80	1.03	1.11	0.67	0.28	
1	0.11	0.45	0.85	1.12	1.46	1.64	1.52	1.20	
2							0.67	0.80	23
3	0.11	0.45	0.85	1.12	1.46	1.64	0.85	0.40	
1	0.26	0.8	1.55	1.81	2.35	2.40	2.38	2.25	
2							0.40	0.80	10
3	0.26	0.8	1.55	1.81	2.35	2.40	1.98	1.45	

1. freezing depth, m;
2. thawing depth, m;
3. thickness of the frozen layer, m.

REFERENCES

Anon. 1969. Geologiya SSSR (Geology of the USSR), vol. 32, Primorskii Krai. Moscow, Nedra.

Kats, D. M. 1967. Kontrol rezhima gruntovykh vod na oroshaemykh zemlyakh. Control for groundwater regime on irrigated lands. Moscow, Kolos.

Kladovshcikov, V. N. 1968. Usloviya formirovaniya i otsenka resursov podzemnykh vod Primorskogo kraya. Conditions for formation and estimation of underground water resources in the Primorie region. Extended abstract of the Candidate of Science Thesis, Moscow.

Stotsenko, A. B. 1952. Sezonnoe promerzanie gruntov Dalnego Vostoka vne oblasti vechnoi merzloty. Seasonal freezing of soils outside the permafrost zone. Moscow, Izd. Akad. Nauk, SSSR.

Ushkalov, V. P. 1962. Glubina i skorost ottaivaniya merzlogo osnovaniya, ikh predelnye velichiny i raschet. Depth and rate of melting of the frozen base, their limit values and calculation. Moscow. Gosstroiizdat.

METHODS FOR ANALYSING THE HYDROLOGICAL CHARACTERISTICS OF ORGANIC SOILS IN MARSH-RIDDEN AREAS

Don H. Boelter

Forest Service, US Department of Agriculture, Grand Rapids, Minnesota, USA.

SUMMARY: Organic soils, prevalent to marsh-ridden areas, have unique properties requiring special precautions when hydrological characteristics are measured. The high water content and low suctions make field sampling difficult. Laboratory analysis of water storage characteristics must be carried out with undisturbed and undried samples because changes in structure or bulk density and irriversible drying will alter these properties. The hydrological characteristics of peat materials in northern USA were found to be closely related to the degree of decomposition of the plant residues as measured by either bulk density or fibre content. Thus, the identification or classification of organic soils on the basis of the degree of decomposition of the peat material provide significant information about their hydrological characteristics and their role in the hydrology of marsh-ridden areas.

METHODES D'ANALYSE DES CARACTERISTIQUES HYDROLOGIQUES DES SOLS ORGANIQUES DANS DES TERRAINS MARECAGEUX

RESUME : Les sols organiques, qui prédominent dans les terrains marécageux, ont des propriétés spécifiques qui obligent à beaucoup de prudence lorsqu'on détermine leurs caractéristiques hydrologiques. La haute teneur en eau et la basse perméabilité rendent difficile l'expérimentation sur place. Pour procéder en laboratoire à l'analyse des caractéristiques du stockage en eau, il faut travailler avec des échantillons naturels dont les propriétés n'ont pas été altérées par des modifications de la structure, de la densité ou de l'humidité. Les études effectuées montrent que les propriétés hydrologiques des terrains tourbeux dans le nord des Etats-Unis sont liées au degré de décomposition des résidus de végétaux mesuré soit par la méthode des densités, soit par la méthode de la composition fibreuse. Il s'ensuit que l'identification ou la classification des sols organiques d'après le degré de décomposition de la matière tourbeuse revêt une extrême importance pour la connaissance de leurs caractéristiques hydrologiques et de leur rôle dans l'hydrologie des terrains marécageux.

Organic soils are common to marsh-ridden areas. To understand the hydrology of marshy areas, we must know the hydrological characteristics of the peat materials making up these soils. Because of their high organic component, they have unique properties that require special precautions to ensure that the measured hydrological characteristics represent the actual properties of the materials as they exist in the field.

There are an estimated 10 to 12 million hectares of organic soil in conterminous USA and more than 160 million hectares on the North American continent, (Davis and Lucas, 1959). The North Central Forest Experiment Station is studying the hydrology of forested peatlands in northern Minnesota, USA. Sampling and measuring techniques reported here were designed to evaluate the hydrological characteristics of peat materials and the role of organic soils in the hydrological regime.

Water storage

Organic soils develop under excessively wet conditions unfavourable for the decompostion of plant remains. Unless artificialy drained, they are saturated or nearly saturated at all times. The amount of water these soils contain and the quantity involved in water table changes are values needed by the hydrologist to evaluate the water resources of these marshy catchments.

To be useful to the hydrologist, all soil water contents should be expressed on a volumetric basis. This is especially true for organic soils becaue they have a wide range of bulk densities and there is no basis for comparing water contents on a per unit weight basis.

Field techniques

In addition to problems normally associated with the field measurement of soil water, some special problems are involved with organic soils because of the high water content and low water suctions. Under these conditions it is difficult to collect gravimetric samples which retain all their water as the sample is collected. Accurate gravimetric sampling of dense peat at lower depths is possible using a caisson (large cylinder) to prevent water from entering the excavation from porous surface horizons (Fig 34).

The best known indirect methods of measuring soil water content also present problems when used in organic soils. Resistance blocks are of little value at low suctions and the changes in suction are too small to be measured precisely with the normal tensiometer. The neutron scattering method has limited usefulness because of the hydrogen contained in the organic matter itself.

Laboratory techniques

Water storage characteristics can be measured in the laboratory using soil water extractors to reproduce natural conditions using samples collected in the field. Peat materials are difficult to rewet once dried, (Hooghoudt, 1950) and artificially drying or disturbing samples significantly changes the water storage characteristics (Boelter, 1964). Therefore, samples to be used in laboratory studies must be collected so that their natural structure is maintained and drying does not occur.

Undisturbed cores placed in single sample pressure cells, (Reginato and Bavel, 1962) are used to measure water retention at suctions ranging from saturation (0.0 bar) to 0.1 bar. Suction is applied by means of water columns. When equilibrium with the first suction has been reached, the complete apparatus and peat core are weighed and set up for extraction at the next higher suction. The oven-dry weight is not determined until after the final suction has been applied, making it possible to determine the water retention at a number of suctions with same core.

Undisturbed bulk samples are used to measure water retention at suctions of greater than 0.1 bar. Pieces approximately 1 cm thick are put on cheesecloth, saturated and placed in the pressure plate or pressure membrane extractor (Righards, 1949) using a slurry of peat or asbestos to ensure good contact with the porous plate or membrane. A capillary tube or outflow burette is used to indicate when equilibrium is reached, usually taking a week or 10 days.

All samples are dried at 105°C and the water content calculated from the loss in weight. Although there appears to be some charring, this also occurs at lower drying temperatures. It has been concluded that the dark material is a residue of organic substances which are brought to the surface of the sample by water and left as a residue when the water evaporates.

Some water retention values for several typical materials are presented in Table 20. Saturated water contents, given as total porosity, are large for all peat materials. However, at conditions less than saturation, striking differences in water retention are found.

Water yield

Hydrologists need to know the yield or quantity of water removed from the soil by gravity as the water level is lowered. Both field and laboratory methods have been used to calculate water yield. Heikurainen (1963) and Bay (1968) have determined water yield coefficients for bogs by relating the change in water table to the amount of precipitation. The quantity can also be calculated from laboratory-measured water storage characteristics. Vorob'ev (1963) tabulated a coefficient of drainage using the calculated water retention of various pore sizes and integrating water content changes for all horizons with given increments of water table change.

The water yield coefficients presented in Table 20 were calculated from laboratory-measured retention values and are the difference between the water contents at saturation and 0.1 bar suction. One-tenth bar (100 cm water) suction was chosen as the lower limit because studies indicate that all the gravitational water is removed at this suction. Therefore, these coefficients would represent the maximum amount of water that can be removed from specific peat material by simply lowering the water table. The values range from 0.08 to 0.86, much the same as coefficients of 0.10 to 0.67 reported for Finland, (Heikurainen, 1963). Any change in water table level in horizons of less decomposed peat (usually found in surface horizons) represents much more water than a corresponding change in the deeper, more dense peats.

Hydraulic conductivity

Hydraulic conductivity, a measure of the rate of water flow through saturated soil, also has important hydrological implications in peat catchments. The saturated hydraulic conductivity of northern Minnesota peats is measured using the piezometer method, (Boersma, 1965). A narrow tube is pushed to the desired soil depth and cleaned. Hydraulic conductivity is calculated from the rate of change in water level in the tube after water has been added or removed. This technique measures the horizontal hydraulic conductivity below the water table. Although investigators have reported that horizontal hydraulic conductivity is greater than vertical, (Colley, 1950) no difference has been found for northern Minnesota peats.

The hydraulic conductivity values have a wide range for peat materials (Table 20). Water movement is too rapid to measure using any available technique in the undecomposed surface or near surface horizons of sphagnum moss peat. By contrast, the hydraulic conductivity of more decomposed peat materials is lower than for clays and glacial tills and 15-30 minutes are required to get a measureable water level change.

Degree of decomposition

Differences in organic soil properties are due primarily to the degree of decomposition. As decomposition proceeds, large particles gradually break down into smaller particles, resulting in smaller pores and larger mass per unit volume. Although the degree of decomposition appears to be a key property of organic soils, it is not clearly defined and is therefore difficult to quantify. It is a relative quantity usually approximated by measuring one of the chemical or physical properties that change with advanced decomposition. Kaila (1956) studied several characteristics and concluded that volume weight (bulk density) is a basis for estimating the degree of decomposition in peats. More recently, Farnham and Finney (1965) used fibre content.

Bulk density

Bulk density (mass of oven-dry material per unit bulk volume) can be used not only as a measure of decomposition, but is also needed to calculate the water content on a volumetric basis. Because peat materials shrink when dried, their bulk densities must be calculated using the wet bulk volume. The values reported in Table 20 were measured using the oven-dry weight of a core of peat of known volume collected in the field at field water content.

Fibre content

The amount of fibres greater than a designated size will decrease with increasing decomposition. Fibre contents are measured by soaking a weighed sample (100-200 g) for 15 to 20 hours in a 5% solution of dispersing agent and then washing through sieves to collect particles larger than the designated minimum size. A gentle stream of water is used and care is taken to prevent rubbing or abrasion that could reduce the fibre size. The fibre collected from the sieves is oven-dried, weighed and compared with the

Table 20. Hydrological characteristics of several peat materials from organic soils in Northern Minnesota, USA.

Peat type	Sampling depth	Total porosity	0.1 bar H_2O content	Water yield coefficient	Hydraulic conductivity	Bulk density	Fibres 0.1 mm
	cm	%	%	cc/cc	10^{-5}cm/sec	g/cc	%
Sphagnum moss peat							
Live, undecomposed mosses	0-10	99.6	14.0	0.86	*	0.010	94
Undecomposed mosses	15-25	89.4	29.8	0.60	3810.00	0.040	87
" "	45-55	91.6	43.9	0.48	104.00	0.052	76
Moderately decomposed with woody inclusions	35-45	86.5	63.6	0.23	13.90	0.153	52
Woody peat							
Moderately decomposed	35-45	87.2	60.3	0.27	496.00	0.137	59
Moderately well decomposed	60-70	89.4	70.2	0.19	55.80	0.172	32
Herbaceous peat							
Slightly decomposed	25-35	86.8	30.2	0.57	1280.00	0.069	75
Moderately decomposed	70-80	87.0	73.9	0.13	0.70	0.156	40
Decomposed peat							
Well decomposed	50-60	82.4	74.8	0.08	0.45	0.261	25

* Too rapid to measure with techniques used.

oven-dry weight of the original sample which was calculated by using the measured water content of a duplicate sample.

Farnham and Finney (1965) arbitrarily selected 0.1 mm as the minimum fibre size in their analysis of the fibre content. Although Boelter (1969) reported a close relationship between fibres larger than 0.1 mm and bulk density, size fractions with minimum sizes of up to 2.00 mm were equally well related to bulk density and decomposition. Thus, the minimum size selected may not be critical.

Relationship of properties

The significance of degree of decomposition to the hydrological characteristics of peat materials is evident. Undecomposed peats contain many large pores that permit rapid water movement and are easily drained at low suctions yielding a large portion of their saturated water to drainage. The finer pores of the more decomposed peat permit only limited saturated flow, are not easily drained, and retain much more water at high suctions than do the undecomposed peats.

Regression analysis has been used to related the various hydrological characteristics to the degree of decomposition as measured by bulk density and fibre content, (Boelter, 1969). Curvilinear relationships were found for the regression of water retention values to both bulk density and fibre content (Fig 35). A significant linear relationship exists between the logarithm of hydraulic conductivity and both fibre content and bulk density. Using these relationships, one can estimate the hydrological characteristics of peat materials. Thus, the identification or classification of organic soils based on degree of decomposition provides significant information on the hydrology of marsh-ridden areas.

REFERENCES

Bay, R. R. 1968. Evapotranspiration from two peatland watersheds. 14th Gen. Assembly Int. Union of Geodesy and Geophysics, Switzerland: 300-307.

Boelter, D. H. 1964. Laboratory techniques for measuring water storage properties of organic soils. Soil Sci. Soc. Amer. Proc. 28. 823-824.

Boelter, D. H. 1969. Physical properties of peats as related to degree of decomposition. Soil Sci. Soc. Amer. Proc. 33. 606-609.

Boersma, L. 1965. Field measurements of hydraulic conductivity below a water table. In methods of soil analysis. Agron. Monogr. 9 (Part I). 222-223.

Colley, B. E. 1950. Construction of highways over peat and muck areas. Amer. Highways 29 (1). 3-6.

Davis, J. R. and Lucas, R. E. 1959. Organic soils, their formation, distribution, utilization and management. Spec. Bul. 425. Mich. State Univ. Agr. Exp. Sta., East Lansing, p 156.

Farnham, R. S. and Finney, H. R. 1965. Classification and properties of organic soils. Advances Agron. 7. 115-162.

Heikurainen, L. 1963. On using groundwater table fluctuations for measuring evapotranspiration. Acta Forestalia Fennica 76 (5). 5-16.

Hooghoudt, S. B. 1950. Irreversibly dessicated peat, clayey peat, peaty clay soils. Determination of degree of reversibility. Trans. 4th Int. Congr. Soil. Sci., Amsterdam 2. 3i-34.

Kaila, A. 1956. Determination of the degree of humification of peat samples. J. Sci. Agr. Soc., Finland 28. 18-35.

Reginato, J. and Van Bavel, G. H. M. 1962. Pressure cells for soil cores. Soil Sci. Amer. Proc. 29. 1-3.

Righards, L. A. 1949. Methods of measuring soil moisture tension. Soil Sci. 68. 95-112.

Vorob'ev, P. K. 1963. Investigations of water yield of low lying swamps of western Siberia. Soviet Hydrology. Selected Papers 1963 (3). 226-252.

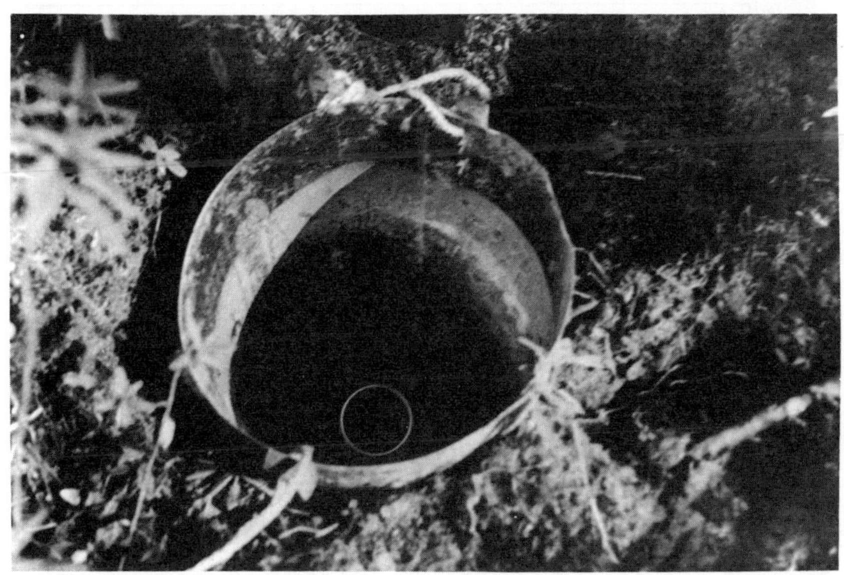

Fig 34. A large cylinder (caisson) was pushed into the organic soil to prevent free water from entering the excavation and to enable the collection of samples from below the water table. Core samples were obtained with the small cylinder inside the caisson.

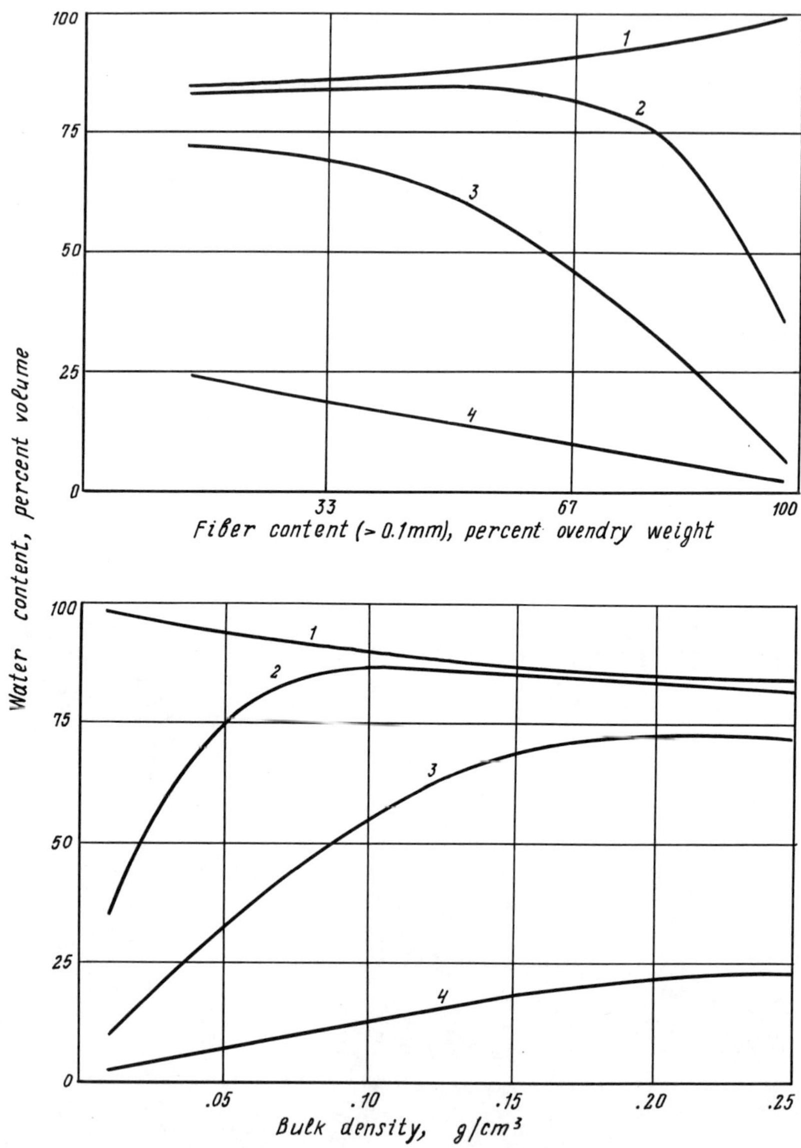

Fig 35. The relation of water content at saturation, 5000 mb, 100 mb and 1500 mb suction of fibre content (> 0.1 mm) and bulk density (Boelter, 1969).

1 - Water content, percent volume

2 - Fibre content (> 0.1 mm) percent oven-dry weight

3 - Bulk density, g/cm^3

DISCUSSION

Yu. N. Nikolsky (USSR) - What is the difference between the moisture content of individual peat samples after the suction process ceased at the pressure of 0.1 bar?

Don H. Boelter (USA) - The difference in the moisture content of individual peat samples is not large and increases with the pressure.

W. H. van der Molen (The Netherlands)- A volume weight may be a criterion of decay. On drained peat lands compression may increase the volume weight. Did you use the von Post scale? This field method is used in the Netherlands for the degree of peat decay which is closely connected with the hydraulic conductivity found by a similar procedure as described in your paper.

Don H. Boelter (USA) - I agree that mineral materials mixed with peat may considerably increase the volume of weight. I got samples with a low ash and mineral content.

I also agree that drainage favours an increase in the volume weight. I obtained the data for peat samples from undrained organic soils. I have not used the von Post scale.

ASPECTS OF THE HYDROLOGY OF BLANKET PEAT IN IRELAND

W. Burke

*Agricultural Institute, Kinsley,
Malahide Road, Dublin,
Ireland.*

SUMMARY: A hydrological study of drained and undrained plots on peat in the West of Ireland is described. Data are presented on precipitation, drain outflow, evaporation and water table levels. Typical summer and winter hydrographs are shown and the water balance for the area is discussed.

ASPECTS HYDROLOGIQUES DE QUELQUES TOURBIERES EN IRLANDE

RESUME: L'auteur étudie quelques aspects de l'hydrologie des tourbières dans l'ouest de l'Irlande. Il présente des données sur les précipitations, sur l'écoulement par drainage, l'évaporation et les niveaux des eaux souterraines. Il donne quelques hydrogrammes types pour l'été et pour l'hiver et traite du problème du bilan hydrique pour le secteur choisi.

Introduction

Some hydrological aspects of blanket peat were studied as part of a research project on the reclamation of peat for agriculture in Ireland. The research station is located at Glenamoy near the west coast (Fig 36). The prevailing winds are from the south-west and when they reach the west coast of Ireland they have a very high relative humidity. This results in very frequent and prolonged rainfall. It is mainly because of this rainfall pattern that the blanket peat has developed.

Physical properties of peat

The blanket peat at Glenamoy has very poor drainage properties. These have previously been described, (Burke, 1963). The underlying layers are highly decomposed and colloidal. H values (von Post) vary from 5 at the surface to 10 at about 3 metres. Moisture content of the undrained peat is approximately 90% by volume.

In undrained areas the water table seldom falls below 250 mm even in summer and the entire peat mass is often saturated in wet weather, particularly in winter. Under these conditions the volume of drained pores at the surface is very low, seldom exceeding 5%. The hydraulic conductivity of the peat is also very low; laboratory measurements gave values ranging from 18 to 130 mm/day, (Burke, 1963). Measurements made by Galvin and Hanrahan, (1967) showed a range of 10 mm/day on samples measured at a tension of 300 mm of water to 3 mm/day at a tension of 2100 mm.

Climate

Glenamoy lies in the cool temperate zone and has a highly oceanic climate, (Landberg *et al*, 1963). Annual rainfall is in the region of 1250 to 1500 mm falling on 250 to 270 days, fairly evenly distributed throughout the year. There are no distinctly wet or dry periods, but more rain falls on average between September and February than over the remainder of the year; the ratio is in the region 60/40. Snow and frost are almost unknown.

Objectives

The hydrological study described here was part of a general project set up to determine the drainage requirements of the peat for agriculture. The effects of subsurface drainage were compared with no drainage, on quantity and pattern of drain flow, to measure water table fluctuations and to estimate the water balance for the area. Other parts of the general project had shown that for adequate water table control, drains should be spaced at not more than 3.5 m and should be about 0.75 to 1 m deep for maximum efficiency. If large areas of this peat were reclaimed for agriculture, the hydrology of local streams and rivers would change and it was hoped that this study would enable some predictions to be made of the expected changes.

Methods and measurements

The major factor under study was the comparison of outflow from a drained area with that from an undrained one. Two similar plots, each of approximately 0.35 ha were selected. The vegetation is the typical peatland vegetation of the area consisting mainly of a sparse covering of grass, sedge and moss-type plants. One plot has drains installed at 3.5 m spacing and 0.8 m deep. The other had no drainage other than a perimeter drain approximately 150 mm deep. Its function was to divert all surface water originating within the area through a recorder. Precautions were taken in both areas to isolate, as far as possible, the recorded area from outside influences.

Water outflow from each area was measured by a V-notch weir equipped with a chart recorder. The recorders were calibrated individually *in situ*. Stand pipes were installed in the plots and water levels in the pipes were measured three times per week. The level of water in the stand pipes was regarded as the water table level. Precipitation was recorded with a recording raingauge, and evaporation measurements were obtained from lysimeters in a site about 400 m from the plots.

The following procedure was used in evaluating outflows. Flow equations were constructed from the calibration figures for different ranges of heads (H) on the V-notch. These equations were of the form:

$$Q/t = aH^b$$

where t = time, Q = quantity of water, H = height of flow in the V-notch and a and b are constants. The equations best suited to particular ranges of H were then selected by the following procedure. Regression equations, relating head on the V-notch (H) to measured flow, were computed for several ranges of H, e.g. 0 to 25 mm, 0 to 150 mm, 25 to 50 mm, 50 to 150 mm, etc. By comparing values calculated by means of these equations with values measured at frequent intervals, the equation best suited to different ranges of H were selected. The data on the charts were abstracted manually and processed as a series of values of H related to intervals of time. A computer programme was used to calculate daily flow by means of the regression equations used in conjunction with the data abstracted from the charts.

Results

Some results are presented for the period February to December 1968 inclusive. The calculation of exact volumes of flow proved to be more difficult than expected. Small pieces of vegetation and peat tended to foul the V-notch and water-borne peat tended to clog the entrance to the float chamber. However, by making exact measurements of flow at frequent intervals using a stop watch and a collecting vessel of known volume and by using these measured values to monitor the recorded flow, it is felt that reasonably accurate results were obtained. The data presented cover a period from winter (wet conditions) through summer (dry conditions) and back again to wet conditions, and are presented in summary form in Fig 37 and Table 21.

When the total hydrological picture as shown in Fig 37 and Table 21 is examined it is obvious that on the undrained area an approximate balance is achieved between precipitation, evaporation and runoff. Early and late in the year when evaporation is low and the peat is wet most rainfalls produce some runoff - the quantity depending largely on the rainfall amount and pattern of the previous days. In summer-time when evaporation is relatively high, heavy rainfall is required to produce runoff, and in general the amount of runoff is relatively small. In the drained area some runoff occurred on all days and the measured runoff for the eleven months was 60% (or 351 mm) greater than that from the undrained area. Flow was also much more uniform from the drained area and runoff hydrographs did not show the sharp peaks evident in the undrained area (Figs 38, 39, 40 and 41). In the Figures typical hydrographs of both areas are shown for summer (Figs 38 and 39) and winter (Figs 40 and 41). The buffering effect against

Table 21. Monthly totals of precipitation, outflow and measured evapotranspiration (P.E.).

Month	Precipitation mm	P.E. mm	Outflow (mm)		Surplus (+) or deficit (-) mm (Precipitation-P.E. outflow)	
			Drained	Undrained	Drained	Undrained
Feb	44.6	8.5	89.8	37.0	-53.7	-0.9
March	127.0	31.8	84.8	66.9	10.4	28.3
April	71.8	41.4	66.8	27.1	-36.4	3.3
May	72.4	73.6	52.9	19.1	-54.1	-20.3
June	76.1	93.6	27.8	3.3	-45.3	-20.8
July	31.8	82.2	13.7	0.0	-69.1	-50.4
Aug	98.2	80.9	39.5	15.0	-22.2	2.3
Sept	142.1	36.0	66.5	43.9	39.6	62.2
Oct	180.0	26.0*	158.1	102.7	-4.1	51.3
Nov	118.1	11.0	126.0	82.4	-18.9	24.7
Dec	135.8	0.0	138.9	121.2	-3.1	14.6
Total	1097.9	485.0	869.8	518.6	-256.9	94.3

* Estimated value

rapid runoff for the drained area was provided by the lowered water table with consequent increased temporary storage.

The effect of drainage on the water table is shown in Fig 37. In general, the water table level in the undrained area in summer was approximately the same as that in the drained area in winter, Mean depth of water table in summer was about 0.45 m in the drained area and 0.25 m in the undrained area. In winter the corresponding depths were 0.25 m and 0.1 m respectively.

Water balance

The assessment of an exact water balance for the area is extremely difficult. Apart from inherent difficulties in obtaining exact measurements of outflow and evaporation, great difficulties are encountered in allowing for changes in the water content of the peat itself. As the water content of the peat changes, the peat shrinks or swells. Large variations may occur in the amount of water stored in the peat and these will be apparent, partly as a reduction or increase in moisture content but largely as a lowering or raising of the surface. These changes in surface elevation are not uniform and in the Glenamoy peat the surface is very uneven. No attempt was made to measure this factor.

The water balance may be represented by the following equation:-

change in stored water (S) = Precipication (P) - Evaporation (E) - Runoff (R.O) - Losses to deep seepage (D), i.e.

$$S = P - E - R.O - D$$

For evaporation (E) the measured lysimeter value of potential evaporation under grass has been used and is therefore probably an over-estimate of actual evaporation. No account is taken of D although some deep seepage must occur, albeit small, as both peat and the subsoil have extremely low permeability values. For present purposes, P. E and D are assumed to have the same values for both plots. We can consider the equation on a monthly basis and also for the eleven months.

It is seen from Table 21 that in all months except March and September there was a net loss of water from the drained area while the only months in which a net loss was registered for the undrained area were May, June and July - the figure of 0.9 mm for February may be regarded as zero. Thus, there is a progressive de-watering of the peat by drainage but with some re-wetting in winter. The pattern in the undrained area is that excess surface water runs off in wet periods, evaporation dries the peat to some extent in summer but the original moisture status is rapidly restored by heavy rainfall.

If D is omitted, the equation S = P - E - R.O for the 11 month period has the following values (Table 21):

Drained area $S = 1097.9 - 485.0 - 869.8 = -256.9$ (1)

Undrained area $S = 1097.9 - 485.0 - 256.7 = 94.3$ (2)

These equations indicate that the reduction in water content in the drained area is 256.9 mm or 351.1 mm greater than in the undrained area. While this figure may be excessive for reasons stated above, it shows clearly the profound effect of drains in reducing the water content of the peat mass, and that this drainage effect is cumulative. It is now evident that much subsidence has occurred in the drained area and this subsidence is attributed to shrinkage caused by a reduction in the volume of water stored in the peat mass. Furthermore, the general lowering of the water table within the drained area, relative to that in the surrounding peat mass probably causes a small net inward seepage of water to the drained area through the peat mass beneath the isolation drain. The investigation is still in progress and new areas have been included to measure the effect of forest trees on the hydrology of the area.

The results also indicate that if widespread drainage is undertaken in the area, beneficial effects on stream and river flow will follow. Floods will be reduced in frequency and amount and summer flow of streams will be increased in the short term. Continuing measurements should reveal long-term effects.

REFERENCES

Burke, W. 1963. Drainage of Blanket peat at Glenamoy. Int. Peat Congress, Leningrad, 2. 809-817.

Galvin, L. E. and Hanrahan, E. T. 1967. Steady state drainage flow in peat. Highway Research Record, No. 203, 77-90. Highway Research Board, National Academy of Sciences, Washington, D.C.

Landberg *et al.* 1963. World maps of climatology. Springer-Verlag.

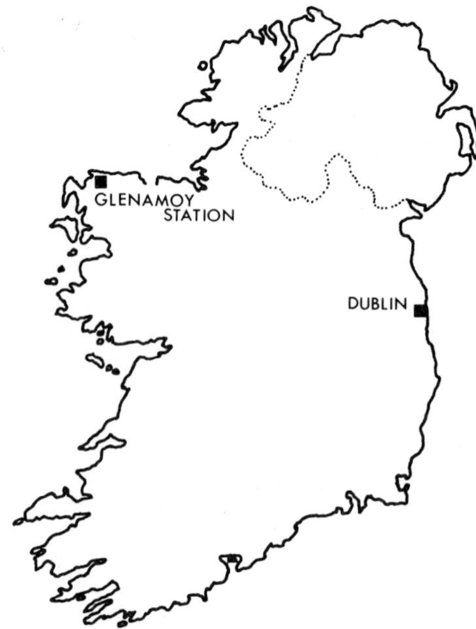

Fig 36. Ireland, showing Glenamoy Station.

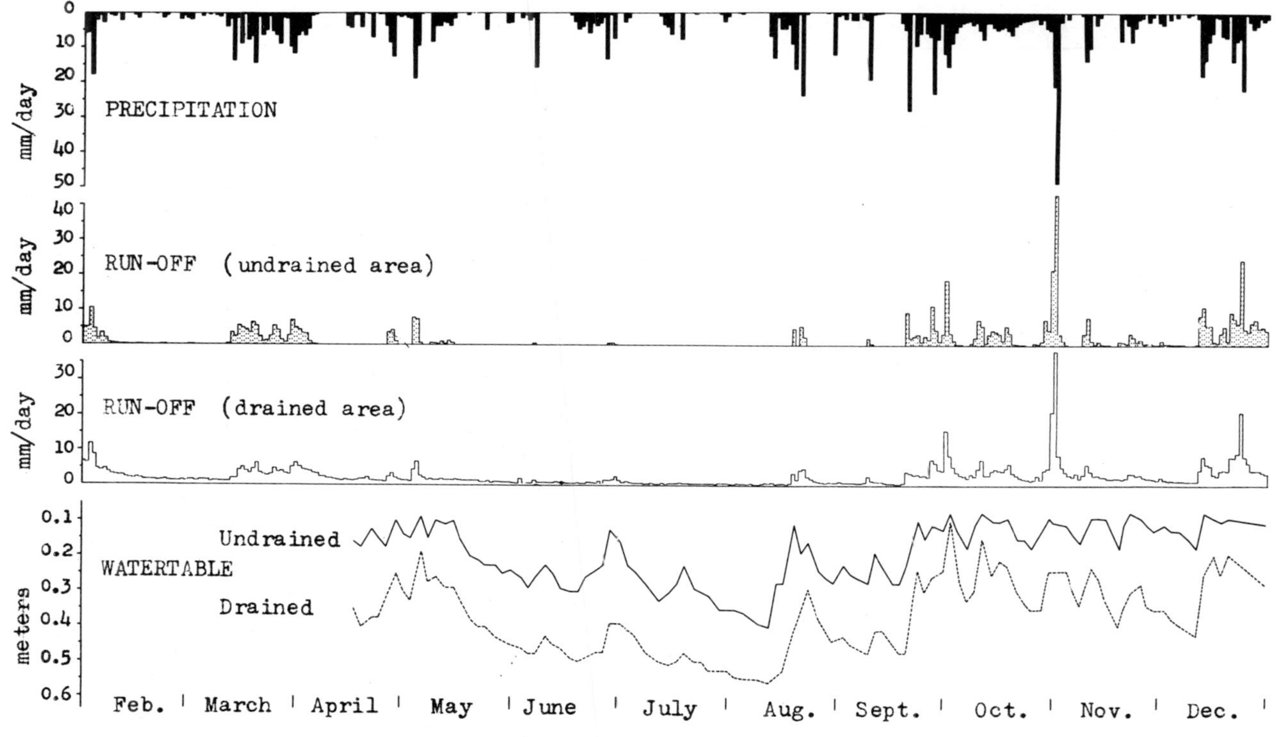

Fig 37. Hydrological pattern - Glenamoy 1968. 1 - mm/day; 2 - metres.

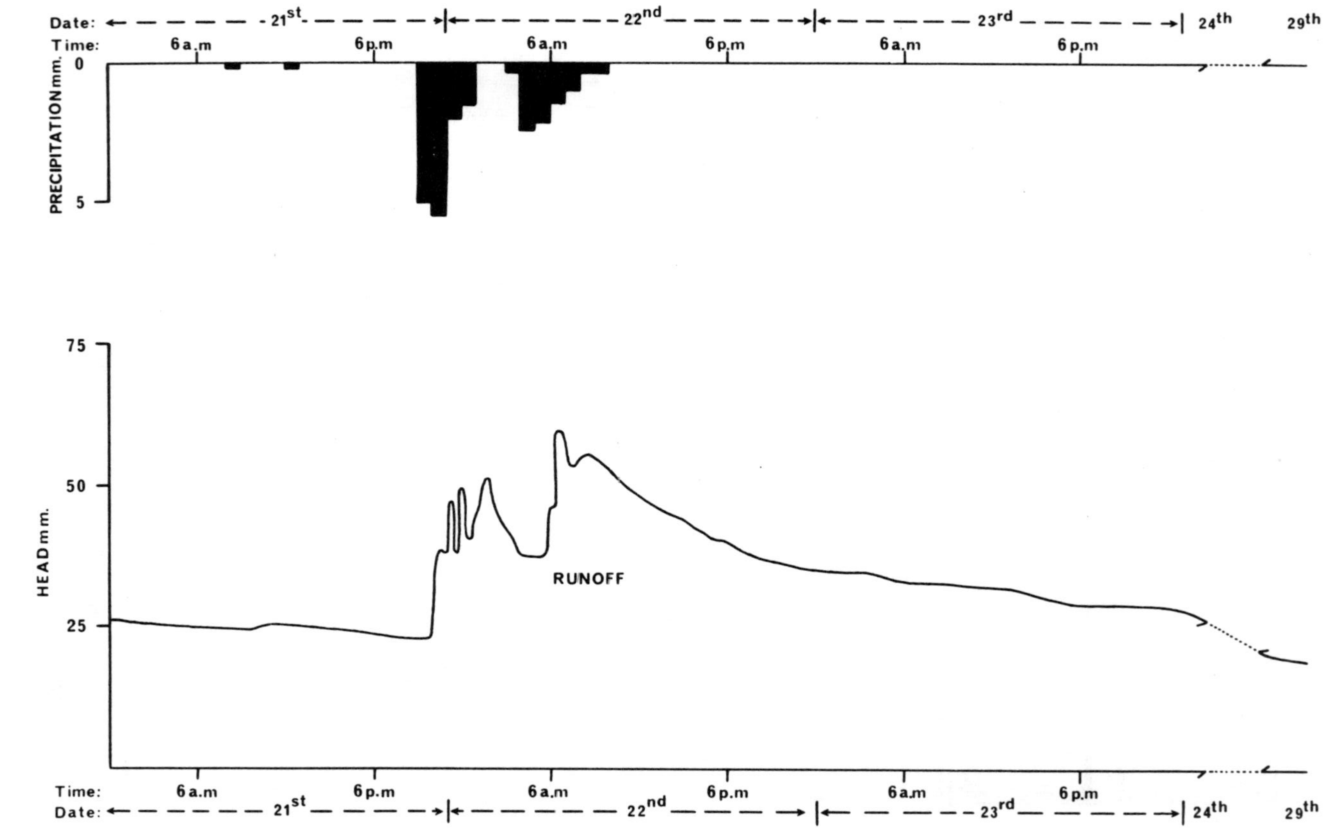

Fig 38. Hydrograph - Drained area - August.

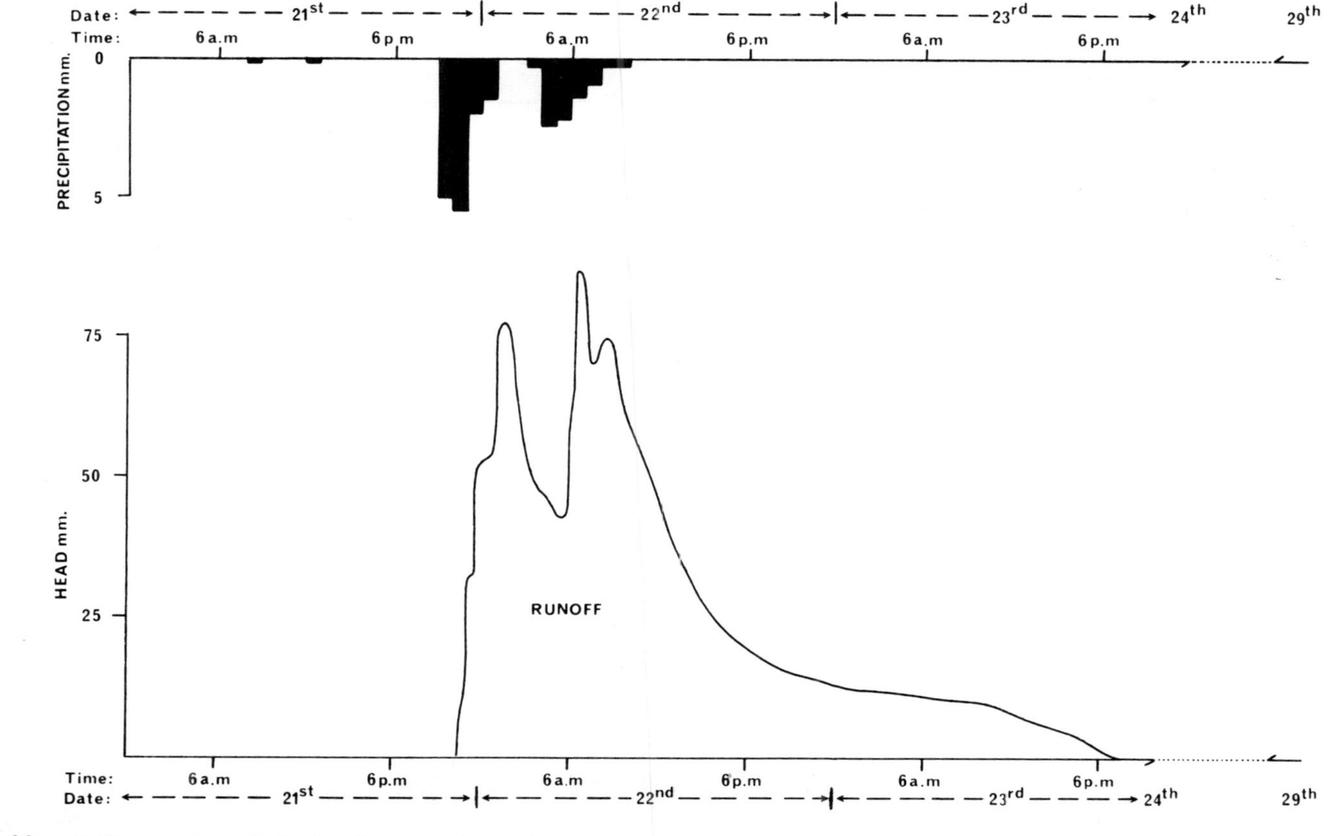

Fig 39. Hydrograph - Undrained area- August.

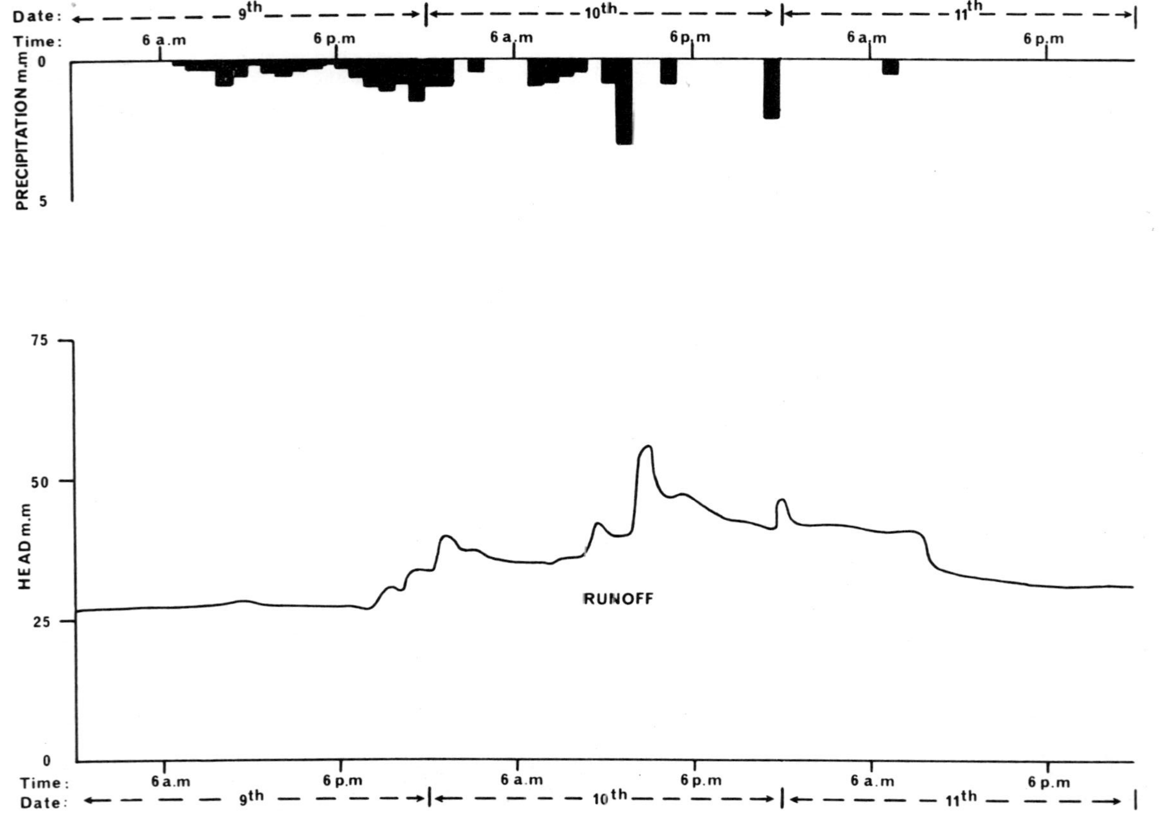

Fig 40. Hydrograph - Drained area - November.

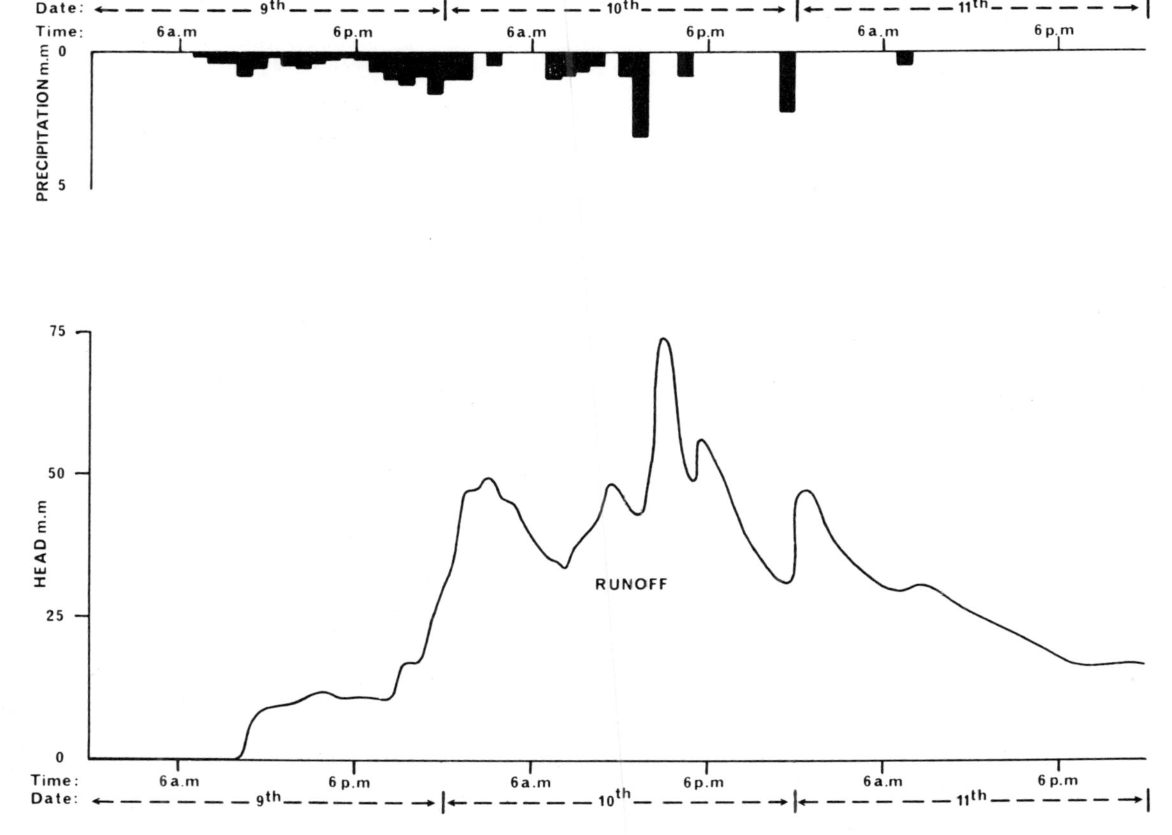

Fig 41. Hydrograph - Undrained area - November.

DISCUSSION

S. Mustonen (Finland) - In the paper it is suggested that peat soil is rather uniform. Is it really possible to assume this, as no calibration was made before the drainage arrangement?

W. Burke (Ireland) - Yes, this suggestion was made, and I believe it is correct.

SUBSIDENCE OF PEAT SOILS AFTER DRAINAGE

W. H. van der Molen

Agricultural University
Wagengingen,
The Netherlands.

SUMMARY: Considerable lowering of the soil surface occurs after drainage of peat marshes, due to subsidence.

Formulae derived by Fokkens are considered to be valid except in cases where oxidation of organic matter is an important cause of subsidence. A nomographical solution is given and the effects of compaction on the water balance are briefly discussed.

L'AFFAISSEMENT DES SOLS TOURBEUX APRES DRAINAGE

RESUME : Le drainage provoque un affaissement considérable des marécages tourbeux. Les formules de Fokkens sont généralement considérées comme valables sauf dans les cas où l'oxydation des matières organiques constitue un facteur important de cet affaissement. L'auteur cite la méthode nomographique et examine brièvement l'effet de la consolidation sur le bilan hydrique.

Considerable subsidence occurs after drainage of peat soils. Large areas of peat soils were drained and reclaimed in the western part of the Netherlands as early as the thirteenth century and have been used as grassland ever since. In the course of 700 years the soil surface has lowered several metres, thus impeding the original, though deficient drainage, towards the sea. To keep these lands in use, windmills and, since the beginning of the nineteenth century, pumping stations have been installed. Subsidence of peat soils is due to the following processes:

oxidation of organic matter;
shrinkage of the topsoil due to drying;
compaction of the subsoil due to the increased load.

In neutral or alkaline peats oxidation is the main process, but in acid peat soils shrinkage and compaction are the dominant processes. Many empirical and semi-empirical rules for predicting subsidence have been proposed. Recently Fokkens proposed a new approach based on a combination of methods used in soil mechanics with methods known in pedology.

Compaction of peat

The compaction of soft layers under the influence of an increased load is usually described by the Terzaghi equation:

$$\frac{\Delta Z}{Z} = \frac{1}{C} \ln(p_2/p_1)$$

where:
- ΔZ compaction (m);
- Z original thickness (m);
- C constant;
- p_1 stress due to initial load (gf/cm^2 or N/m^2);
- p_2 stress due to final load (gf/cm^2 or N/m^2).

Considering the properties of organic matter, mineral components and soil moisture, Fokkens derived expressions for 1/C in terms of soil properties for two situations:

peats which have never been subject to compressive forces;
peats which during their history of formation have been compressed (e.g. by a sand or clay cover sedimented on top of the peat, or by desiccation of the marsh during dry periods.

Uncompressed peats

For uncompressed peat, the final equation reads:

$$\frac{\Delta Z}{Z} = \frac{A_1}{A_1 + 0.62H + 38} \cdot \frac{\ln(p_2/p_1)}{25.3H/A_1 + \ln(p_2/p_1)}$$

where:
- A_1 original moisture content of soil in grams per 100 grams of dry matter;
- H organic content of soil in grams per 100 grams of dry matter.

Precompressed peats

If the peat has been subjected to a load during its formation, it may be assumed that the effect of such a load is reflected in its moisture content A_1 and that this moisture content is in equilibrium with the effective stress p_m caused by such a load. In that case only the final effective stress caused by the increased load enters the computation. In this situation, Fokkens proposed the following equation:

$$\frac{\Delta Z}{Z} = \frac{A_1}{A_1 + 0.62H + 38} \quad 1 - \frac{H/A_1}{0.0395 \ln p_2 - 0.066}$$

in which 0.0395 and 0.066 are two empirical constants. Even in natural peat-marshes a certain degree of precompression exists, mainly due to the occurrence of extremely dry periods. Therefore the equation is of general applicability and includes the equation for uncompressed peat as a special case. A nomograph for the compaction of precompressed peat is given in Fig 42 with the variables:

H humus content g/100 g dry matter;
A/H amount of moisture held per g organic matter;
p_2 final stress due to increased load (gf/cm^2).

Compaction due to increased load and desiccation

By lowering the water table the effective stresses acting on deeper layers are increased. A lowering of the water table of 1 m will increase the load by 100 gf/cm^2 or 9813 N/m^2. In deep peats, this increased load may represent an important contribution towards subsidence. In the upper layers, situated above the water table, desiccation causes even greater stresses. Though calculation of these stresses is not easy in granulated media, experience in the Netherlands leads to the conclusion that equating such mechanical stresses to soil moisture stresses gives good results. Such soil moisture stresses are often expressed on the pF scale (being the decimal logarithm of the soil moisture stress, expressed in cm of water). An average value for the root zone during dry spells is about pF 3.5, whereas in the upper 20 cm, values of pF 4.2 - 4.5 are reached. With such values for the moisture stress, the expected shrinkage may be calculated.

Effect on water balance

Peat subsidence will be accompanied by a release of water: in fact, lowering the soil surface by 10 cm liberates 100 mm of water, which is either evaporated or drained.

As subsidence of this order of magnitude is often encountered in newly drained peats, this irreversible change in moisture storage should be included in accurate determinations of the water balance of such areas.

REFERENCE

Fokkens, B. 1970. Ingenieur 83, B 23-28.

DISCUSSION

K. P. Lundin (BSSR) - How do you account for the subsidence time?

W. H. van der Molen (The Netherlands) - Our calculation gives the time of subsidence which is ordinarily short. In clay soils this time is longer than in peat soils.

V. M. Zubets (BSSR) - Can you give the subsidence value for each layer?

W. H. van der Molen (The Netherlands) - Subsidence may be calculated based on the peat thickness and we may introduce the existing load into the equation. Thus, near the surface subsidence will be larger, and in deeper layers the subsidence will be smaller. Similar examples may be found in Eggelsman's paper.

V. M. Zubets (BSSR) - Can you give any figures for subsidence during the 10 years after drainage?

W. H. van der Molen (The Netherlands) - The subsidence certainly depends on the peat bed thickness. The subsidence may reach 0.5 m and even 1 m for thick peat beds.

V. M. Zubets (BSSR) - During what time did such subsidence occur? What is the average subsidence for a year?

W. H. van der Molen (The Netherlands) - For about 50 years. The subsidence changes with drainage depth and thickness of a peat bed.

G. V Bogomolov (BSSR) - Does the type of underlying ground affect the subsidence value?

W. H. van der Molen (The Netherlands) - No difference in the subsidence is found if clay is not too soft. The type of underlying ground does not affect the subsidence. In my country subsoil is often very soft clay which decreases the subsidence rate.

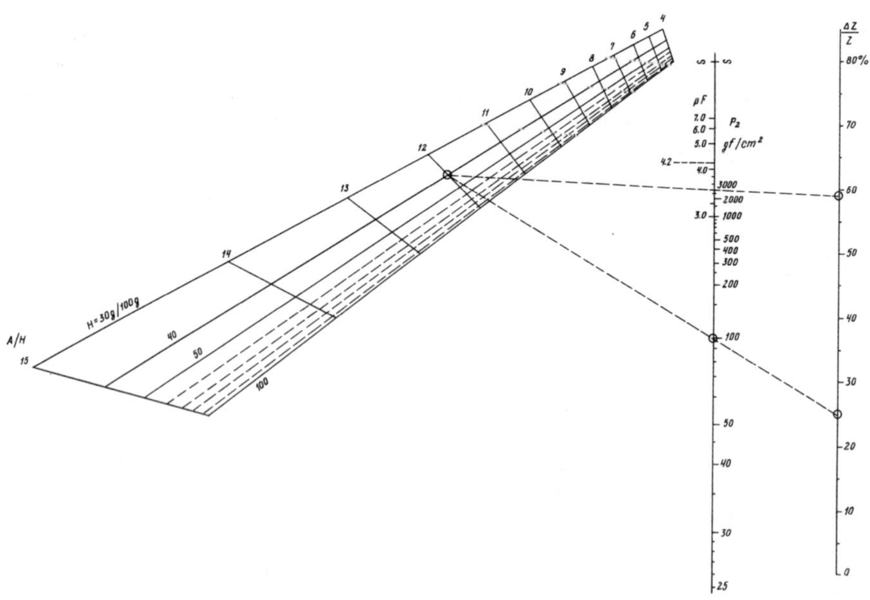

Fig 42. Subsidence of precompressed peat, (Fokkens, 1970).

THE HYDROLOGICAL REGIME OF WOODED MARSH-RIDDEN AND CUTOVER PEAT BOGS IN SOUTHERN BYELORUSSIA

V. K. Podzharov

Byelorussian Forestry Research Institute, Gomel, BSSR.

SUMMARY: The groundwater regime has a marked influence on the crop yield in marsh-ridden lands. The observed data on water table variations in various bogs over a 9 year period (1963-1971) are presented in this paper. They represent a variety of conditions that exist on natural marsh-ridden lands as well as in areas subject to reclamation measures. The most intense drainage of bogs never leads to overdrying of the soil in the root layer which would considerably decrease the growth of forest stands. Whereas summer submersion, although of very short duration, not only affects the growth rate of tress but also may result in their destruction.

The correlation data for groundwater levels in individual wells, under the various different conditions studied, are also given.

REGIME HYDROLOGIQUE DES TERRAINS MARECAGEUX ET FORESTIERS ET DES TOURBIERES A TOURBE TRAVAILLEE EN BIELORUSSIE MERIDIONALE

RESUME : Le régime des eaux souterraines est déterminant dans la croissance des plantes sur terrains marécageux. L'auteur fournit les données provenant d'observations effectuées pendant 9 années (1963-1971) sur les variations de la nappe souterraine dans diverses régions marécageuses du sud de la Biélorussie. Ces informations donnent un aperçu de la variété des régimes hydrologiques tant dans les terrains marécageux naturels que dans ceux qui ont subi des travaux d'aménagement. Cependant, le drainage le plus intense de ces marécages n'entraîne jamais le dessèchement de la zone radiculaire et une diminution considérable de la croissance des arbres alors que les crues d'été, même très brèves, non seulement affectent la croissance des arbres, mais aussi provoquent leur extinction.

Dans cet exposé, l'auteur analyse la dynamique des eaux souterraines dans quatre tourbières à tourbe travaillée. Il fournit aussi des informations sur les résultats obtenus par le creusement de puits séparés, en liaison avec la profondeur de la nappe souterraine et la distance des rigoles de drainage.

Land reclamation projects in the forest areas of south Byelorussia concern wooded bogs consisting of peat covered areas where the thickness of the peat is about 0.3 m (in accordance with the definition of International Land Cadastre). The main part of the area (about 600 000 ha) is covered by different stands and only 150 000 ha are free from forest. The mineral hydromorphic (boggy) soils are commonly covered by highly productive forest stands and no change of soil moisutre conditions is needed in these districts.

Groundwater conditions in natural bogs vary considerably, depending on the bog type, its category and thickness, on the soil formation and ground relief. Where these parameters are similar the groundwater regime depends on the extent of the bog as well as on the composition and evolution of the stand. It is affected directly by precipitation and summer climate. Data presented in Table 22 show a diversity of mean values of the groundwater level not only in various years but also in different bogs.

These data are least variable in high bog lands, where generally the water does not rise above the surface moss cover. The inflow of water to storage and surface flooding lasting until the middle of summer only occur along the low lying outer areas, which in rainy years may remain under water throughout the warm period. This, however, does not occur in high bogs but only in transitional or low-level bogs. The lower groundwater table and intensified dynamics are typical of areas of dense pine stands of bonitet V^b and of lower bonitets rather than of thin forests and forestless sites.

The variety of transitional bogs is much greater. In large areas, shallow saucer-shaped transitional formations may be encountered in lowland bogs where surface submersion occurs in spring as well as in wet years. This effect is most evident on the periphery of the bog where thalwegs and water courses appear; the central areas with a thick deposit of peat are more stable. Variations of groundwater level are always greater in areas with a shallow peat deposit and outer areas of the bog adjacent to dry land, generally covered by fairly productive pine and birch stands. Saucer-shaped bogs exhibit a particularly unstable groundwater regime. In wet years these store a large quantity of water which remains on the soil surface for a long time causing damage to growing stands.

Lowland bogs are characterized by wide variations of groundwater level. Even the mean length of growing period data for fluvial shallow-deposit bogs covered with black alder show highly significant fluctuations whilst somewhat smaller variations occur in large areas with a thick peat deposit.

Records on groundwater level for the Byelorussion Polessie show that the depth of subsurface water is subject to strong variations during the vegetal period and that it is slightly more stable in wet years. Under normal and especially under low precipitation the groundwater level falls appreciably causing overdrying of bogs at some tine in the summer. The hydrological conditions in large areas with a thick deposit are fairly stable, whilst the small, shallow saucer-shaped peat bogs are characterized by extreme variability of the groundwater level.

On this account shallow peat land requires special land reclamation treatment and care in the choice of methods which will alter the natural hydrological balance. The intensive land reclamation work in the Polessie region includes drainage of boggy forests, which is generally effected by open ditches with no flow regulation. This method of reclamation somewhat disturbs the natural balance. The data presented in Table 23 show that even good ditches cause no essential change in the water levels in thick deposit high bogs, but the forest growth is improved very

Table 22. Mean groundwater levels in various forest bogs of south Byelorussia.

Bog category	Plentitude of forest stands	Composition of forest stands	Bonitet	63	64	65	66	67	68	69	70	71
				Observation years (1963-1971)								
H i g h b o g s												
Raised	0.14	10C	V^b	22	25	10	23	21	20	22	13	27
"	0.22	10C	V^b	26	25	7	28	24	24	23	19	33
Slightly raised	0.54	10C	V^b	29	31	12	39	37	19	28	17	22
" "	0.70	10C	V^b	40	49	23	51	38	23	28	13	33
" "	0.80	10C	V^b	47	52	53	72	52	32	40	19	42
T r a n s i t i o n b o g s												
Bordered by dry lands	1.19	9CIB	IV	54	59	44	69	63	63	33	24	28
Shallow deposit	0.92	8C2B	IV	46	53	21	57	37	22	34	7	30
Medium deposit	0.84	9BIC	V^a	33	42	13	61	16	20	18	+2	9
Large, thick deposits	0.86	9CIB	V^a	23	33	8	8	13	17	15	+5	7
Saucer-shaped, lm deposit	0.23	8B2C	V^a	30	31	1	35	28	32	9	+12	+10
Saucer-shaped shallow deposit	-	-	-	31	33	10	37	29	31	16	+25	3
L o w l a n d b o g s												
Large, thick deposit	0.86	9CIB	V	34	41	12	15	15	24	17	+3	9
Shallow deposit	0.73	5B4O 10C	II	44	45	19	43	32	16	28	6	30
Fluvial bog	0.54	6O 4B	III	45	49	5	48	34	2	+13	+9	30

Note: Depth of peat on shallow deposit is less than 1 m, that of medium deposit varies from 1 to 2 m, and on thick deposits is more than 2 m.

slightly - from the V^b to V^c bonitet. Ditches that function well significantly lower groundwater levels on transition type bogs, and prevent submersion and excessive soil moisture in spring and wet years. On the contrary, under summer drought conditions, there is a fire hazard. It should be noted that the water table has never dropped below 1 m nor has the soil moisture content of the upper 15 cm layer fallen below wilting limits.

The groundwater level drops more in lowland bogs. The mean groundwater level during the vegetative period at a distance of 200 m from a ditch, regardless of the rainfall pattern in any season, is maintained at a depth of 68 to 75 cm. At a distance of 300 m from a ditch the water table is at a depth of 56 to 62 cm. The amplitude of the variation of level of the water table is 130 cm. Ofter the water table drops to a depth of 150 cm or more during summer droughts. This situation requires rather careful planning of drainage schemes and consideration of the trend towards increasing ditch spacing to 500 m. The depression curves are convex for all recently drained lands and only on lowland bogs in summer drought periods do they sometimes become concave as a result of increased transpiration of the forest stands. Silting results in a gradual decrease in the depth of drainage ditches which in turn leads to a progressive rise of the water table. In some areas the rate of silting may attain 2 cm per year, confirming the necessity of periodical cleaning of the ditches.

Cutover peat bogs represent a specific category of bog land, liable to significant modification by reclamation measures. Investigations have led to the classification of the whole range of cutover peat areas according to their hydrological conditions, feasibility and methods of their agricultural and forest development. Thus bogs may be arranged in the following four groups: submerged; low; middle and high fields.

These classifications are characterized by significantly different behaviour of the groundwater levels (see Fig 43). On submerged fields the water remains on the surface for 7 to 9 months every year, and the water table rises up to depths of 50 cm. Consequently, only bog vegetation can grow on such land.

The low fields are subject to short duration spring flooding and in exceptional cases may be submerged in summer. The groundwater level falls to 48-70 cm, following summer droughts, but at all times remains within the root layer. For practical purposes such fields are equivalent to undrained bogs.

The middle fields are characterized by a lower water table, whose depth varies between 60 and 145 cm during the growing period. At the time of flood peaks in spring the level may rise to 20 cm. The most favourable soil moisture and aeration conditions for vegetation are found in these areas.

Interuption of capillary movement towards the root layer is characteristic of the high fields during the summer months. The groundwater level drops to between 140 and 246 cm during the warm season and in spring it rises to 50 to 100 cm. The result is overdrying of the peat and consequently inner cracking and formation of cavities down to a depth of 1.5 m.

Annual variations of water table levels are greatest for high fields (150 to 170 cm); rather less for middle fields (95 to 105

Table 23. Mean depth (cm) of the water table during the growing period on forest bogs drained by single ditches.

Year	Ditch no.	Distance from drainage ditches, m								
		10	20	50	100	150	200	250	300	350
		High raised bog								
1966	63	44	42	40	24	23	24	22	21	19
1967	60	44	43	40	25	25	25	24	23	21
1968	60	32	31	29	20	19	19	19	18	17
1969	63	37	39	37	25	24	24	24	23	22
1970	62	33	33	31	18	16	16	15	14	13
1971	57	42	43	42	31	30	30	29	28	27
		Transition saucer-shaped bog								
1966	102	93	90	79	77	87	69			
1967	103	90	91	76	73	84	63			
1968	106	90	94	82	77	90	63			
1969	89	67	71	49	40	46	33			
1970	87	63	65	41	29	38	24			
1971	91	71	53	50	35	43	28			
		Typical transition bog								
1966	108	85	73	56	44	44	36	29		
1967	99	76	66	51	38	37	30	24		
1968	93	74	64	49	37	34	30	25		
1969	93	75	67	52	40	39	33	27		
1970	83	66	57	43	32	31	25	19		
1971	87	76	69	54	43	43	38	33		
		Lowland grass (black alder) bogs								
1966	90	86	80	74	72	73	75	71	62	57
1967	84	78	74	71	70	69	68	65	55	48
1968	87	79	76	71	68	68	70	68	58	47
1969	88	80	76	73	72	71	71	66	56	46
1970	91	83	79	75	71	71	70	65	56	60
1971	90	87	84	79	77	75	77	74	62	59

cm); and least for low fields (50 to 70 cm). Slightly larger amplitudes of water level fluctuation are observed on submerged fields (60 to 85 cm) and in ditches (59 to 72 cm). The variations in groundwater and ditch water levels are far from similtaneous. The correlation coefficient between the depth to the water surface in a ditch and the depth to the various bore holes does not exceed 0.60 ± 0.06.

The relation between groundwater levels in different fields

is characterized by the following correlation coefficients and regression equations

a) 3 years of observations

submerged low	r = 0.85 ± 0.03;	y = 25.98 + 0.89 x
submerged middle	r = 0.80 ± 0.03;	y = 91.44 + 1.23 x
submerged high	r = 0.76 ± 0.04;	y = 143.05 + 1.83 x
low middle	r = 0.76 ± 0.04;	y = 73.81 + 1.10 x
low high	r = 0.72 ± 0.04;	y = 115.62 + 1.65 x
middle high	r = 0.93 ± 0.01;	y = 10.45 + 1.46 x

b) during a warm period

submerged low	r = 0.87 ± 0.03;	y = 28.92 + 0.87 x
submerged middle	r = 0.84 ± 0.03;	y = 89.43 + 1.25 x
submerged high	r = 0.81 ± 0.04;	y = 137.21 + 1.88 x
low middle	r = 0.79 ± 0.04;	y = 65.97 + 1.17 x
low high	r = 0.78 ± 0.04;	y = 98.57 + 1.81 x
middle high	r = 0.93 ± 0.01;	y = 9.38 + 1.46 x

The highest correlation between groundwater levels is found for wells of similar depths within an homogeneous area. The relation between variations in level on middle fields at various distances from ditches (50, 100 and 200 m) is characterized by correlation coefficients of 0.96 to 0.97. The correlation coefficients between levels in submerged fields and any of the other three areas are considerably lower (0.81 to 0.87), and even less between low fields and either middle or high areas (0.78 to 0.79).

For all practical purposes it was found that groundwater levels and summer rainfall are not related. Extremely heavy and frequent rain may result in a rise of the water table, the rate of this rise decreasing as autumn approaches. In the presence of an impermeable layer as well as when two or multi-layer substrata are present the wells behave, generally, independently of each other. For this reason it is rather difficult to predict variations in the groundwater level even very near to a well.

The relation between water levels in rivers, channels and ditches and groundwater levels in various parts of the watershed is less well defined. Channel storage depends on catchment area and runoff conditions; the response to precipitation is always delayed.

The most important component of the hydrological balance of bogs, from the point of view of vegetation growth, is the soil moisture content in the root layer. Overmoistening of even the upper layers of the substratum is usually observed. Drainage measures, however, profoundly change the situation. When the water table falls below 100 cm, the capillary rise is interupted and the soil surface layer may suffer from excessive moisture extraction.

Observations of the soil moisture content in the root layer show that in dry periods, when the water table is rather deep (about 150 to 200 cm), as was observed on high fields, the top 10 cm layer dried out to below wilting point in sand, but also in peat. A similar situation was observed on middle fields when the water table was at a depth of 110 to 113 cm. In low fields the surface is not dried to such an extent even in periods of severe drought. However, the lower soil layers always contain sufficient stored moisture to enable most plants to survive during the drought periods. Bunchgrass species such as spear grass, meadow grass, fescue, etc. generally dry first; forest tree species lose only a part of their foliage.

Saturation of peat was observed after periods of prolonged rainfall, regardless of groundwater level. Such a condition may last for some time and is especially dangerous in late autumn prior to the fall of temperature in winter. These soils freeze through, and within the frozen layer a great number of macrocrystals of ice form; the ice crystals loosen the changeable peat substratum which results in a mass of segregated particles which break the roots of plants. Such peat assumes a cellular structure.

Ice melting and soil subsidence result in displacement of vegetation. Unfortunately, this was observed not only on poorly drained lands but also in quite well drained areas. Forests always grow better on mineral soils, which do not exhibit such mobility than on peat.

Analysis of the moisture regime of natural bogs, drained areas, and boggy regions affected by human activity suggest the following conclusions:

1) Saturation and especially submersion of soil are more harmful to forest growth than temporary lack of water;

2) Measures for forest drainage should be carried out selectively and in accordance with the peculiarities of the hydrology of bogs.

3) The effect of modern reclamation methods of boggy forests does not increase with time. On the contrary, with inadequate maintenance ditches rapidly become silted and overgrown with vegetation, with the result that in time the water table begins to rise once more.

DISCUSSION

V. M Zubets (BSSR) - What drainage norms to you recommend for forests?

V. K. Podzharov (BSSR) -For fen bogs, spring drainage norms of about 0.4 to 0.5 m may be recommended. We work on the assumption that if in spring the groundwater level falls to the required value, in summer it will be much lower.

V. M. Zubets (BSSR) - What tree varieties may be recommended for your drainage norms?

V. K. Podzharov (BSSR) - We do not classify forests in this respect. In theory, if this aspect is treated in every detail, some difference may certainly be found. For coniferous forests the

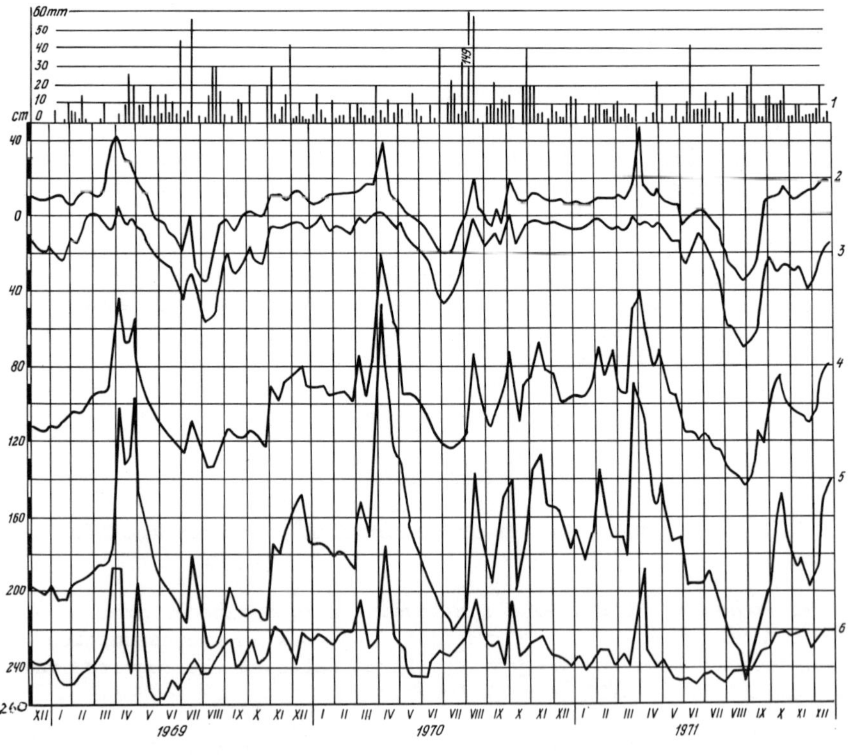

Fig 43. Groundwater level variations for different cutover peatlands. 1. Precipitation for five-day periods. 2. Flooded. 3. Low. 4. Medium. 5. High. 6. Main ditch.

drainage norms may be somewhat higher than for deciduous forests. But in practice it is impossible to account for this since the type of forest depends on the drainage network. Therefore definite conditions should be created first, and the tree varieties chosen accordingly.

P. I. Zakrzhevsky (BSSR) - The table gives the forest inventory but does not mention their replenishment.

V. K. Podzharov (BSSR) - The forest inventory is defined in the first year, then depending on the precipitation during the growing period it may change. In the present case, if we take the groundwater level in August or early September, the situation may be found to be different when the water level in ditches is higher than that between the ditches.

L. O. Heikurainen (Finland) - Your data seem to be very interesting. In Finland cases of over drainage are often encountered when the groundwater level is at a depth of more than 60 m. Have you met such cases in your work?

V. K. Podzharov (BSSR) - It is probably necessary in this case to take into account the types of bogs. Drainage may hardly be recommended for high bogs. They may be preserved in their original state as natural hydrological reserves. No over drainage is found on transient-type bogs and fens, although in particular growing periods, the levels fall to considerable depths. However, the water supply is quite adequate and crop yields do not decrease. In our conditions, saturation is more dangerous. A centimetre increase of the water level for the growing period results in a decreased crop yield. Therefore temporary over drainage is considered to be less dangerous than temporary saturation.

METHODS AND RESULTS OF INVESTIGATIONS OF THE HYDROLOGY, CLIMATE, HYDROGEOLOGY, SOILS AND VEGETATION IN MARSH-RIDDEN AREAS OF THE TEMPERATE ZONE

(Review Report)*

G. V. Bogomolov and D. M. Kats

Institute of Geochemistry and Geophysics,
Byelorussian Academy of Sciences,
Minsk, BSSR.
and
All-Union Research Institute for Water
Engineering and Reclamation,
Moscow, USSR.

Marsh areas are currently attracting world wide attention. In the temperate zone alone over 20% of the total area requires reclamation by drainage. The land has agricultural potential and as such is important.

In the temperate zone there are more than 350 000 000 ha of marshes. In the USSR marshes are encountered in 30 regions and territories of the Russian Federation, the Byelorussian SSR, and western regions of the Ukraine and Baltic Republics. The size of the marshes can be illustrated by the following figures: in the Byelorussian Polessie they occupy 6 000 000 ha, in the Ukrainian Polessie 4 000 000 ha, in West Siberia (Barabinsk Depression) over 11 000 000 ha and in the central regions of the RSFSR (Meshchersk Depression) more than 1 500 000 ha.

In the northern temperate zone over 30% of the marshes are fed by groundwater and more than 50% have a mixed recharge by ground and surface water.

In the Byelorussian and Ukrainian Polessie there are 11000 000 ha of marshes. More than 60% of this area is fed by groundwater from confined or unconfined aquifers.

In North America more than 100 000 000 ha of marshland are recharged by both ground and surface water. In Finland, France, the Netherlands, FRG and GDR groundwater in Quaternary deposits plays an important role in the formation of marshes.

Artesian water has an important function in the formation of bogs due to the hydraulic properties of peat. This provides an accumulation of chemical elements entrained by water at great depths in the form of organic soil materials. Such lands, after draining, contain in addition to organic materials and nitrogen, a large quantity of micro-elements which are essential to agricul-

---

* Some material used in this report have been provided by M. F. Kozlov (Byelorussian SSR), V. I. Moklyak (Ukranian SSR) and I. N. Skrynnikova (USSR).

ture. Recently the study of organic materials and the different components of soils, peat, sapropel, silt and water has occupied many scientists. Their work contains important data on the accumulation and leaching of iron, calcium, magnesium and other components within river basins in various regions of the northern hemisphere. The scientific investigations and development of marshlands have been considerably extended in the postwar period in many countries of the world. These investigations are carried out along three main lines:

a) Investigation of geological, hydrogeological, hydrological and climatic conditions and genetic peculiarities of soils and soil processes, particularly their water balance and thermal regime;

b) General inventory of soil and botanical resources;

c) Soil reclamation zoning and the preparation of projects for their development.

In the last two decades the modification and the effects of these changes on the physical properties and moisture balance of organic soils have been studied with the aim of identifying general principles.

In this paper a basis is given for chosing the type of drainage, taking into account the hydraulic and other physical properties of rocks and the hydraulic relationship between different water bearing horizons. Research has identified the evolution and changes in the hydrophysical properties of the moisture and temperature regimes in the non saturated zone of improved and developed soils in the different natural regions of the temperate zone. It is now possible to give a preliminary estimate of the effect of different reclamation measures and agricultural development on the moisture capacity of soils and runoff regime. The data obtained is of great importance for the purpose of forecasting the effect of reclamation and agricultural development on soils, vegetal cover and moisture regime under different natural conditions.

Current results suggest there is a need for a more thorough study not only of the physical properties of soils, but also of their moisture, aeration and thermal regimes. In virgin bogs and waterlogged soils the excess moisture reduces environmental, especially climatic, and local effects on soil processes. For example, the differences between lowland virgin peat soils of different natural regions are not very great. According to the data of I N Skrynnikova, drainage is followed by a sharp increase of environmental and local effects on soils which in turn effects their subsequent evolution.

The role of organic material in the fertility of reclaimed soil is extremely important. It is a source of plant nutrients, especially nitrogen, vitamins and growth stimulators, and it increases the moisture capacity of the soil.

Research has shown that the rate and nature of decomposition of organic material in reclaimed soils, as well as the fate of ash elements released during decomposition, depends primarily upon climatic conditions. The removal of the moisture reserve from marshy soils changes the nature and relationship between the sources of moisture and the moisture regime. Thus, after areas of the Northern and Middle Taiga subzones (which have thick poorly recharged layers of Quaternary deposits) were drained and reclaimed, the moisture for the reclaimed soils came from a variety of irregular sources.

The thermal regime is one of the limiting factors for the growth of crops and the development of biochemical processes in the north. The effect of reclamation on the thermal regimes of peaty and mineral soils is not the same. Whilst the removal of excess moisture from mineral soils gives rise to increased warming of such soils during the summer period and somewhat more intensive freezing in winter, the reclamation of peat soils adversely effects their thermal regime. In the continental regions of the north the less favourable thermal regime of peat soils after reclamation may even lead to the formation of soils with permanently frozen layers at a depth of 60 to 70 cm, or of a prolonged seasonal frost. A slow decomposition of peat in the Northern and Middle Taiga subzones results in a situation in which there is very little mobile nitrogen. Under these circumstance artificial fertilizing with NPK is necessary even though the long cultivated peats are naturally rich in nitrogen. In the Southern Taiga subzone the rate of biochemical processes is increased after the excess moisture is removed,and in parts of this zone where there are poor quartz sands, groundwater is the main source of nutrition for the vegetation.

For the above reasons it is acknowledged that the optimum type of moisture regime for the drained soils of the Southern Taiga and forest-steppes is a regime such that the capillary fringe reaches the upper layers of the tilled horizon. Such a moisture regime provides moisture for agricultural crops; it favours the hydrogenic accumulation of a number of elements, and soils do not become eluvial.

The methods of reclamation and the subsequent agricultural development affect the rate of biochemical processes. The deeper the groundwater table and the lower the capillary fringe in reclaimed soils, the thicker the upper horizon of soils involved in biochemical processes. This accelerates the decay of organic material.

The runoff hydrographs from drained soils can be modified in different ways. The canalizing of rivers in flood plains accelerates the surface runoff and consequently increases the peak discharges. On the other hand, a drainage network can retain a certain volume of spring runoff and consequently reduce the maximum discharge.

Generally, the features of any drainage system will produce individual effects in the runoff from that area. The simplest case is drainage by ditches with no sluices to control the discharge. In this situation, the runoff will reflect the efficiency of the network in removing excess water.

In the case of two-way regulation, moisture in the unsaturated zone is retained by storage reservoirs on tributaties, to be used later for irrigation. In striking a balance between the need to drain excess moisture and yet retain a surplus for use in dry periods, the runoff hydrographs will exhibit different features to thos of the simple drainage case.

Experience with drainage systems aimed at the two-way regulation of the moisture regime has shown that it is essential to have effective control of the level of groundwater in the areas to be drained.

Drained peat soils contain vast reserves of organic matter and nitrogen. Based on figures obtained from all over the world, crop yields from these peat soils are from 1.5 to 2 times higher than thos on mineral soils.

Reclamation has been practiced in many countires of the world for centuries. But the project for the development of the Polessie Depression, on the border between the Byelorussian and Ukranian SSR, is the greatest undertaking in hydraulic engineering in the non-chernozem zone of the Soviet Union. In this project control of the runoff will be achieved by construction of storage reservoirs and ponds, thus making it possible to store a considerable part of the runoff in reservoirs and to use it for irrigation during dry periods. It is planned to construct a two-way drainage-irrigating system ensuring the optimum water supply for agricultural crops. The papers presented earler in this volume by L. S. Panin, A. A. Zhelobaev and Y. G. Bogomolov and V. E. Alekseevsky and K. P. Tereshchenki were devoted to complex investigations which provide a scientific basis for development of marshlands.

Drainage projects of this magnitude require reliable and comprehensive field data on the environmental conditions of the area in question. These are often lacking in sufficient detail, and consequently, more geological, hydrogeological, soil, plant and engineering information is required for efficient planning. To facilitate the collection of the necessary data in the Soviet Union, manuals and specifications have been written defining planning requirements so that investigators can assemble the requisite facts.

In the USSR these requirements have been defined by the Ministry of Reclamation and Water Management which is co-ordinated with the Ministry of Geology. Their brief is to establish the engineering and economies of draining large areas and they consider that geological, hydrogeological and soil surveys at 1:200 000 are necessary. Given these surveys, potential reclamation areas can be selected, the basic engineering techniques chosen and preliminary estimates of cost can be made.

At this stage the geological, hydrogeological and soil surveys are repeated in the selected areas at the enlarged scales of 1:100 000 and 1:10 000 and investigations are made of the groundwater regime in areas adjacent to that selected, as well as the one chosen, to predict likely changes.

Gradually the planning enquiries focus in ever increasing detail on the selected area and on the installations and techniques to be used. These permit forecasts on water balance and groundwater regime of improved precision, and the likely costs of the project, leading finally to the preparation of contract specifications.

It cannot be overemphasised that the efficiency of planning depends entirely on the adequacy of the data available. In this connection detailed and accurate reporting are required on allied topics in both the area chosen for reclamation and its border zones.

The value of reports, diagrams, graphs and maps, as well as field notes, can often help to build up a comprehensive picture not only for a selected area but for interpretations of much larger zones.

Previously a wealth of data, such as described above, might have defied analysis, but the advent of computers, hydrological analogues and other simulation techniques enable it to be used to explore and test hypothetical situations, leading to accurate predictions on the possible changes.

TOPIC II

METHODS AND RESULTS OF WATER BUDGET CALCULATION (ANALYTICAL PROCEDURES, SIMULATION, ETC.): THEIR USE FOR DESIGN OF DRAINAGE AND IRRIGATION SYSTEMS IN THE TEMPERATE ZONE

METHODS AND RESULTS OF WATER BUDGET CALCULATIONS: THEIR USE FOR DESIGN OF DRAINAGE AND IRRIGATION SYSTEMS IN THE TEMPERATE ZONE

(Review Report)

V F Shebeko and K E Ivanov

Byelorussian Research Institute of Reclamation and Water Economy, Minsk, BSSR.
and
Leningrad State University, Leningrad, USSR.

Problems and significance of water balance calculations

Utilization of water resources in the general framework of the economy requires comprehensive analysis and the solution of many problems depends on the availability of data on surface and groundwater regimes. The water resources of an area are determined by complex physiographic processes, so quantitative estimation of these resources and forecasting of changes is inevitably associated with the establishment of patterns of behaviour of the individual components of the water balance. Amongst the numerous interrelated parameters affecting water balance we may distinguish the zonal features of a certain area, which are developed under the influence of global heat and moisture exchange, which determine the main natural patterns of water resources of a given area. Local factors (soils, geology, geomorphology, vegetation, etc.) introduce substantial variations into the general zonal process. They cause the natural differences of the hydrological regime within comparatively small regions. However, in spite of the high stochastic variability of water resources, and considerable differences in the moisture regime over the area, in the long-term, there can still be distinguished typical features and more stabilised characteristics of the moisture content for a certain natural zone. Examples might be: a definite relationship between the input and output components of the water budget; a characteristic combination in the intra-seasonal variation of individual components; patterns of formation and transformation of the water balance elements, etc. In this connection it is always possible to identify these patterns and to study the quantitative characteristics both of individual elements of the water balance and of water resources as a whole.

The methods used to investigate the water budget depend on the aim, but in all cases, practical problems of management of the hydrological regime are solved by taking into account the natural conditions of formation of input and output components of the water budget of the area. Economic activity mainly consists of distribution of individual components of the water budget in time and in space, in changing the ratios between individual elements over a certain time interval, in rational utilization of water resources and reasonable control of the water regime, etc. All these measures affecting the moisture regime should be appropriate to the natural formation of components of the water budget and to the interrelationship between intake and discharge components.

The possibilities of controlling the hydrological regime for more effective productivity are considerably limited. In the present state of knowledge, many input elements (precipitation, condensation, moisture exchange, etc.) cannot be controlled, and the extent to which the main output component of the water budget (evapotranspiration) can be influenced is determined by solar radiation and moisture regime. The study of the water balance of an area is thus an important hydrological and reclamation activity.

Water balance investigations are of great methodological significance, since in the simultaneous study of all input and output components, and of the laws of transfer and transformation of water over the area, whilst combining the components into a single budget, the accuracy and reliability of results is considerably improved. More accurate evaluation of water resources of an area and its hydrometeorological characteristics gives many scientific activities and practical measures a sounder basis. By studying components of the water balance, the effect of different aspects of economic activity (including water reclamation of waterlogged lands) on the moisture regime and water resources of an area can be more reliably determined and quantitatively estimated. The rational use of water resources and their development requires well-founded water economy budgets that determine available resources for use in a given region and evaluate the possibility of satisfying different national water consumption requirements. In many cases engineering problems are solved on the basis of water budget calculations; hydraulic engineering measures are designed for the regulation of runoff and of the mosture regime and the soil moisture regime of agricultural areas.

Significant results have been obtained in the USSR regarding the components of the water budget of a territory. Water balance methods are used to solve various problems, from small-scale investigations to establish patterns of moisture transfer in different layers of the unsaturated zone and rooting layer of the soil, etc., to water balances of catchments, large basins and regions. The methods used vary according to the aim of the investigation and the problems to be solved, but they are all based on the law of mass conservation.

Short description of papers

The following papers presented in Topic 2 throw light on various problems involved in the investigation, both of individual elements and of the whole water balance of marsh-ridden areas. Some of the papers discuss results of experimental investigations on the formation of elements of the water budget of bogs and marsh-ridden lands under different natural conditions, some papers give results of extensive investigations into the water and heat balance of swamped forests, while certain papers deal with hydrological investigations concerning elements of the water balance of marsh-ridden lands.

In the paper 'Water regime of the unsaturated zone in drained lands under irrigation' by S F Averiyanov, Yu N Nikolsky, N P Bunina, A I Uzunyan (USSR) an analysis is made of the water balance of the unsaturated zone of drained peat soils in a bog area in Byelorussia which are periodically moistened by sprinkling. The relationship between elements of the water balance in the unsaturated zone is established on the basis of the designed drainage rate, and the most useful drainage rate is determined.

Experimental data on the components of the water budget of undrained upland and lowland bogs in the forest zone of the European part of the USSR is discussed in the paper 'Water balance of swamps in the forest zone of the European part of the USSR' by L G Bavina (USSR). A comparison is made of the components of the water budget of bogs and adjacent river catchment areas, which has not revealed any significant differences between the long-term mean annual characteristics in water balances. Certain differences in the runoff are found for years with either heavy or light precipitation.

In the paper 'The water balance of lowland areas in north-west coastal regions of the FRG' by R Eggelsmann (FRG) results are given of investigations on the hydrological regime of peat beds in the north-west of the FRG. Characteristics of the water balance components are given and the effects of drainage and development on individual elements of the hydrological regime are determined.

The results of an experimental study of the components of the water balance of swamps in West Siberia drained by plastic drainage, are presented in the paper 'Hydrological regime and water balance of drained lands of the forest zone of West Siberia' by K D Vvedenskaya, I F Rusinov (USSR). The evaporation regime and drainage runoff are presented as tentative qualitative characteristics for the conditions of West Siberia.

The results of detailed investigations of elements of the water and heat balance in swamp forests of Byelorussian Polessie are given in the paper 'The water and heat balance of the forests of Byelorussian Polessie' by O A Belotserkovskaya (BSSR). These investigations are based on simultaneous observations in which the newest remote-control electrical measuring instruments were used. All the components of the radiation regime were studies, as well as those of temperature and moisture distribution along the vertical profile of the active forest layer to the upper limit of the displacement layer in the atmospheric strata nearest the earth and in the rooting zone of the soil. Observations were made of the runoff from afforested catchment areas and of groundwater regimes and precipitation. These observations have made it possible to characterize the heat and water balance of different types of forests and to draw conclusions regarding their hydrological role.

In the paper 'Mathematical simulation of runoff from small plots of undrained and drained peat at Glenamoy, Ireland' by J Dooge and R Keane (Ireland) it is suggested that use be made of mathematical models for investigation of the effect of bog drainage on runoff. Schematic models for simulating runoff from peat bogs in Glenamoy are considered, and the model parameters are determined on the basis that this method may be used for forecasting changes in the runoff after peaty areas have been drained.

In the second paper by the same authors entitled 'The effect of initial moisture content on infiltration into peat' the process of infiltration is shown to obey hydrogeological laws in general and conditions of the formation of the soil moisture regime in particular. Parameters entering into the Philip infiltration equation are determined in terms of initial moisture content and hydraulic conductivity of unsaturated peat soil. The moisture content distribution in the unsaturated zone and the process of infiltration in typical cases for different values of initial moisture content are predicted.

Some problems concerned with infiltration of precipitation water are treated in the paper 'A mathematical model of subsurface drainage in heavy soils' by F Feher (Hungary). A method is suggested for reclamation of heavy, periodically overmoistened soils using a system of compound drainage, and the hydraulic parameters of this system are presented. The system of drainage depends on the stage of percolation of precipitation water through the arable layer, on further seepage into mole drains located in the subsurface layer, then on the flow to the drainage ditch through filling filter to the drainage collectors.

In the paper 'Hydrological stability criteria and preservation of bogs and bog/lake systems' by K E Ivanov (USSR) the conditions, possibility of existence and stability of various forms of bog/lake systems are considered from the point of view of adaption of complex organogenous natural formations to changes in the environment, the main components of which are the intake of solar energy and the systems contributing to water recharge. An example is given of computation of the limiting values of morphological coefficients of external and internal drainage of systems depending on climatic factors, which for different latitudes make a limit for areas of steady-state systems when possible internal transformations take place in their structure. In the steady-state regions a rate of drainage may be found that does not cause degradation and decay of the system.

The methods used for investigating swamps and for organization of experimental work, hydrological and hydrophysical observations on swamps are considered in the paper 'Methods employed in the investigation of swamps with fixed stations and mobile surveys' by S M Novikov (USSR). Theoretical principles are discussed on which the methods of investigation are based.

Hydrological characteristics of a large marsh-ridden area are given in the paper entitled 'Distribution of annual natural stream flow in the Polessie lowland' by I M Livshits (BSSR), which presents results of investigations and estimation of natural uniformity of stream flow of the Polessie lowland, where catchments are generally waterlogged. Characteristics of the natural uniformity of the annual water regime of rivers of the investigated area are established and a particular method for determining parameters of frequency curves and uniformity values are given, both for investigated rivers and for those which are not investigated as yet.

The papers discussed at the Symposium cover only some of the water balance investigations of marsh-ridden areas in the USSR. This report also generalizes on other published methods and results of water balance investigations and their practical application to the reclamation of marsh-ridden lands.

Methods and results of water budget investigations

Calculations of the water balance of marsh-ridden areas may be made for various specific purposes, the data required and the method of calculation depending on the problem to be solved.

The main questions whose answers depend on water balance calculations are the following:

1. evaluation of total water resources and determination of possible rates of water consumption in complex use of water resources;
2. evaluation of the role of swamps and marsh-ridden lands as a natural factor in recharging of rivers, in the water balance of river catchment areas and stream flow;
3. evaluation of the effect of drainage and transformation of the water regime of marsh-ridden areas on runoff and moisture regimes, including their effect on the groundwater regime;
4. calculations and forecasting of the moisture regime of the area and moisture reserves in the rooting layer and in the plant nourishment layer on drained lands, to find an optimum drainage and irrigation rate for typical periods of an annual cycle and for years with average, light and heavy precipitation, also, to find sources of water supply and in the design of modern engineering methods of water reclamation;
5. control of the moisture content in drainage and irrigation systems and forecasts of the moisture content depending on expected meteorological conditions;
6. calculation of the moisture content in the surface layers and in the unsaturated zone of peat deposits and of marsh-ridden lands used for peat production and various types of construction, e.g. for roads, laying transport piping, erecting transmission lines, industrial and civil construction.

Calculations of the water balance have to be made for various types of bogs, and marsh-ridden lands with topography, vegetation, structure of peat deposits and, mainly with different rates of water flow and water exchange with the underlying ground and surrounding areas. Therefore, the structure of water balance equations and the values used should correspond to the features of the objects considered and to the solution of the problems stated for every particular case. In view of this it is more convenient to consider methods of calculation, and the data required, separately for each of the above problems.

Depending on the size of the swamps and marsh-ridden areas and the conditions of their formation in river catchment areas, three main types may be distinguished:

1. Units where swamps and marsh-ridden lands are located entirely within the catchment area, and have a concave or level surface sloping towards the beds of the hydrographic network. They are typical terrace or flood-plain type which in the zone of unstable moisture supply are, as a rule, represented by different types of bog facies.

2. Swamp systems that are located entirely within the river catchment area, but have a convex surface, so that the runoff from the swamp first reaches non-marshy parts of the catchment, before passing along the river network to the outlet section.

3. The system of swamps and marsh-ridden lands which are divides for adjacent river catchments.

In each of the above three cases we must distinguish catchments (with their marsh-ridden areas) which are completely drained by the river network from those which are incompletely drained. In the former case stream flow measured at the outlet of the catchment is the total runoff from marsh-ridden and non-marsh-ridden areas of the catchment. In the latter case, part of the runoff from marsh-ridden and non-marsh-ridden areas of the catchment areas is lost in infiltration to underlying deeper strata from which the groundwater flow reaches the river network downstream of the outlet from the catchment.

The latter case is most common and may be expressed as:

$$x = Ƶ + y_1 + y_2 + \Delta W \tag{1}$$

$$x = Ƶ_\beta + y_{1\beta} + y_{2\beta} + \Delta W_\beta \tag{2}$$

where:

- x is the precipitation;
- $Ƶ$ is the evaporation from the whole river catchment area;
- y_1 is the runoff from the catchment area measured at the outlet;
- y_2 is the loss to deep, underdrained layers of ground or groundwater recharge;
- ΔW is the change in moisture storage in the catchment area;
- $Ƶ_\beta$ is the evaporation;
- $y_{1\beta}$ is the runoff;
- $y_{2\beta}$ is loss to deep aquifers;
- ΔW_β is the change in moisture storage for the marsh-ridden part of the catchment area.

Equations in this form make it possible to solve the problems listed earlier in items 1, 2 and 3, with regard to calculation of the water budget of large regions: individual marshy river basins, territories embracing several river basins, marsh systems of the watershed type and bog systems lying within river basins. Calculation is then made on the basis of measurements and determination of individual components of the water budget.

Precipitation is determined on the basis of the observation data of the hydrometeorological network and their treatment in accordance with instructions of the Hydrometeorological Service; the runoff y_2 from the catchment area is the discharge measured at the outlet. Direct determination of component y_2 of the groundwater exchange for the whole catchment area presents great difficulties. This component must be calculated either as the residual term of the water balance equation, or it must be combined with the value ΔW; the sum $(y_2 + W)$ is also to be found as a residual term of the water balance equation (1). However, as groundwater exchange of marshy, unclosed river basins varies only slightly, it may be found from the balance equation composed for mean annual values, for which the change in moisture is $\Delta W \to 0$. Then

$$\bar{y}_2 = \bar{x} - \bar{y}_1 - z \qquad (3)$$

where bars over the terms denote a mean annual value.

With this approach the term ΔW may be determined separately as a residue of the equation:

$$\Delta W = x - y_1 - z - \bar{y}_2 \qquad (4)$$

In a number of cases, especially for small marshy catchments, the change in moisture storage may be determined directly from groundwater level measurements and the data on moisture reserves in the unsaturated zone for individual physiographic facies composing the catchment area in sections free of bogs and for bog facies (micro-landscapes) in the marshy section. The change in moisture storage in the catchment area in this case should include the change in groundwater volumes and also in volumes of moisture in the unsaturated zone on the non-marshy and marshy parts of the catchment.

Calculation of evapotranspiration from the catchment area presents considerable difficulties. However, a great deal of experimental and theoretical investigations of this component of the water balance have now been made, (Konstantinov, 1968; Romanov, 1962; Shebeko, 1965). It should be noted that little work has been done on investigation of moisture exchange in swamps and marsh-ridden lands with the adjacent areas. This problem should be investigated not only by hydrologists but also by hydrogeologists.

The independent solution of the water balance equation (2) for the marshy part of the catchment area, for which modern methods of calculating its components (worked out for undrained and drained swamps and marshy lands and determined by hydrophysical methods) may be used, facilitates considerably the calculation of the components of the water balance for the catchment area as a whole. In this case we may solve for any pair of values if we have a system of two equations with these components as unknowns.

In addition, when the marsh-ridden region for which the water balance is being calculated or investigated is a river catchment area (for example, in cases of watershed bog systems), the determination of the components of the water balance may be based on the methods in which determination of the stream flow at the outlet by hydrometric means is unnecessary. At the present time a number of hydrophysical methods of determining the stream flow and the water balance of natural (undrained) swamps have been developed. Runoff from bog systems $y_{1\beta}$ may then be calculated from filtration characteristics of the bog facies, from observation data on the groundwater levels on the bog and water yield coefficients, (Ivanov, 1957; Romanov, 1961), from hydrometeorological relationships from which the level of bog waters may be calculated on the basis of meteorological data, (Anon., 1971; Novikov, 1965), with subsequent use either of the filtration characteristics or of the water yield method.

Evaporation from swamps z_β is currently determined on the basis of a number of proposed methods and relations sufficiently reliable for calculations both of undrained and drained swamps. Calculation of the change in moisture storage ΔW_β is made either on the basis of direct changes in moisture content in the unsaturated zone or changes in the groundwater level, or on the basis of data on changes in the groundwater level and coefficients of water discharge for each layer (for undrained swamps).

The value most difficult to determine is water exchange of peat deposits in bog systems with underlying horizons $y_{2\beta}$. This term in the budget is determined either as the residue of the water balance equation or (for undrained swamps) from the relations used for groundwater exchange in accordance with the theory of hydromorphological relations. In this case, knowledge of types of bog facies and form of network of flow-lines obtained from aerial surveys is required to find the initial infiltration in undrained marsh-ridden lands, (Ivanov, 1965; 1967).

For estimates of the water balance a study is required of all the components. Any simplification of equations relating in-out and output moisture components, made for particular cases, should retain the physical meaning. In determining the unknown element as a residue of the water balance equation, an estimate of errors and reliability of results is essential. Only then can investigations on the water balance have the required scientific and practical value.

Water balance calculations acquire special practical significance in the development of marsh-ridden areas. They are necessary as a basis for reclamation methods, control of the water regime in soil horizons of the unsaturated zone, specification of required runoff in construction projects for drainage and irrigation systems, for forecasting moisture from expected meteorological conditions, and also for estimating changes in hydrological conditions on drained and adjacent areas in the transformation of the moisture regime as a result of water management works, (Anon., 1971; Globus, 1969). In this case wide use is made of the methods and results of calculations based on hydrophysical methods of determination of water balance components and the study of water exchange patterns in the unsaturated zone as a whole and on its boundaries (earth-soil surface, groundwater level).

Many calculations of moisture movement on drained areas are based on the potential theory of moisture transfer, using the measured potential and physical potential characteristics of the environment, (Globus, 1969; Luikov, 1954, 1969; Nerpin and Chudnovsky, 1967; Rode, 1965; Korchunov, 1960; Afanasik and Finskii, 1970; Afanaskii, 1971). A number of authors have proposed modifications of this method, (Romanov, 1961; 1962; Vorobiev, 1969), based on the determination of capillary potential, which finds application in the case of large-pore media in the zone of complete and partial capillary saturation. These methods of calculation may be used in narrow unsaturated zones when the distribution of capillary moisture is close to equilibrium.

In studying and calculating the soil moisture balance in given layers of soil, including the whole unsaturated zone, the moisture exchange procedure based on the potential theory is mainly used. A particular case of this is the potential movement of groundwater according to Darcy's law.

Investigations of the components of the water balance using evaporators, lysimeters, and other soil profile monoliths have become widespread in the USSR and elsewhere. Under such circumstances, water balance equations become considerably simplified, since the impermeable walls of the unit eliminate moisture exchange of the monolith with the surrounding layers of soil. In the general case

$$\zeta \pm \Delta W - E \pm d = 0 \qquad (5)$$

where:
- ζ is the precipitation on the surface of the monolith;
- E is the evaporation from the monolith;
- ΔW is the increase or decrease in moisture storage in the monolith;
- d is the discharge outside the monolith or inflow from the outside through special holes.

In solving practical problems of the water budget and water resources of water logged marsh-ridden areas, investigations on individual elements of the moisture regime and water balance of large river catchment areas in certain marshy regions is of great importance, since water management includes the comprehensive use of water resources. The results of important investigations along these lines have been published as follows: Voskresenskii, 1962; Bulavko, 1971; Ivanov, 1953; Kostyakov, 1951, 1961a, 1961b; Shebeko, 1965, 1970.

The use of water balance calculations in designing drainage irrigation systems

Waterlogged territories are little studied from a hydrological point of view. The effective use of such territories for agriculture requires a radical transformation of moisture conditions, thus engineering methods of controlling the water regime on such lands are vital.

The main task of reclamation is to ensure the dynamics of moisture reserves,water-air and heat regimes in the rooting layer of the soil within the optimum range and during the whole growing period of the crop instead of permanent superfluous moisture. For this purpose, drainage systems are constructed which remove the surplus water in periods of excess and make up the deficiency in soil moisture during dry periods. The direction and degree of influence on natural conditions of marsh-ridden areas depends on the hydrological regime, hydrological standards are thus the starting point in designing drainage-irrigation systems and estimating the water regime created under their influence. Among such standards may be chosen the conditions of formation and values of input and output elements of the water budget, maximum and minimum values and process parameters of particular categories of runoff, patterns of moisture exchange between the soil and the lower layers of the atmosphere and with the groundwater table. The laws of groundwater migration in the saturation zone have a dominating influence,(Averiyanov, 1956, 1957, 1960; Verigin, 1969; Aleinik and Nasikovskii, 1970; Polubarinova-Kochina, 1969;Vasiliev, 1970; Averiyanov, 1971)

Design of drainage-irrigation systems for marsh-ridden areas is constantly being improved as the science and practice of reclamation develop. In view of the high stochastic variability of factors controlling the moisture regime, the creation of optimum conditions is a complex problem, and dependence on the calculation methods in the planning of reclamation measures is increasing. At the present time the moisture regime is estimated and planned according to the hydrological situation for periods of maximum or fixed discharges on certain dates and for process characteristics of the water balance elements. Such planning makes it possible to forecast in detail the water level in canals and rivers as well as variations of the water table on a drained area and of the moisture storage in the rooting layer on waterlogged lands.

Often moisture conditions of an area are evaluated according to the position of the water table, about which there is ample information. This information is important; the water level is however a derivative of certain combined effects of the water balance elements and the groundwater dynamics, including the reclamation methods. Groundwater levels and levels of water in the drainage network are the outcome of the whole complex of elements of the water balance and the effect of engineering and agronomic measures. Therefore, the forecasting of the water balance requires detailed calculations. In addition, the moisture regime after reclamation measures have been carried out cannot be completely characterized by the water level alone. Reclamation provides for the optimal moisture content in the rooting layer of the soil. In this case detailed calculations are necessary of moisture exchange in the unsaturated zone and at the boundaries of the zone and of water exchange of swamps with the surrounding area, which have to be made on the basis of specific conditions of formation of input and output elements of the water budget. Water balance equations are also used to solve water reclamation problems. The design of a drainage-irrigation network and irrigation regime for agricultural areas should be based on relations and values of input and output elements in the water balance of the area considered, on moisture transfer in the unsaturated zone, on moisture exchange with groundwater tables and in the atmospheric layer just above the soil. Hydrogeological investigations are also essential for the underground recharge of swamps. When this data is incomplete problems of designing the moisture regime by reclamation measures may be solved only in particular cases or can be solved only partially. In many cases the suggested procedures of water balance calculations remain unfulfilled because initial hydrological, hydrophysical and agrophysical relations and standards have been incompletely studied.

The design of drainage-irrigation systems for swamps and marsh-ridden lands, is, in the USSR, based on data from water balance investigations. Design methods and water balance standards applicable to different regions have been worked out. Their application promotes improvement of water reclamation measures and has the following advantages.

If drainage systems are calculated only for particular extreme discharges in stream cross-sections, the water regime will be characterized for short time intervals only. Hydrological conditions in the network and moisture regime of the area for the main vegetative period remain unknown. The application of water balance calculations makes it possible to complete hydrological calculations in the network of canals and culverts by calculation of the water regime of fields under agricultural crops for the whole vegetative period.

Water balance calculations have made it possible to forecast the moisture regime and to design rationally all necessary reclamation layouts to ensure optimum conditions. In addition to extreme values of hydrological standards, the moisture regime on drained bogs and marsh-ridden lands should be based on process characteristics of input and output elements of the moisture balance, namely, intraseasonal runoff in years with typical moisture content or in calculated years; the precipitation regime, intraseasonal distribution of evapotranspiration by agricultural crops in years in which rainfall is not the same; standard values of moisture exchange in the unsaturated zone; infiltration and underground recharging of swamps and marsh-ridden areas.

As a result of designing drainage-irrigation systems and forecasting the moisture regime on the basis of water balance and hydraulic calculations, we establish with specified frequency the water reserves in the rooting layer of the soil for the years in question, the role of the water table regime in water flows and on fields under agricultural crops, deficiencies in soil moisture in the dry periods and necessary water discharges to provide the minimum allowable drainage rates during periods of excess moisture. These are subsequently used to find and design water sources for irrigation-drainage systems.

Specific methods of designing and estimating the irrigation regime may differ depending on the required accuracy, extent to which the initial data have been studied, features of natural conditions, methods and calculation procedures. Designs may be made for specific areas subject to typical natural conditions.

The general approach to designing irrigation regimes may be

demonstrated by using one of the accepted water balance methods for an evaluation of the level regime on fields under agricultural crops, internal specific discharges and moisture reserve regime in the rooting layer, (Shebeko, 1970).

In calculating the groundwater regime it should be remembered that in warm periods of the year this regime depends on moisture exchange at the soil surface and in the unsaturated zone, on precipitation seepage to the water table, flow of moisture from the water table to higher strata of the soil of the unsaturated zone, drainage by ditches and drains. Calculations are made for the whole unsaturated zone over successive intervals from any initial moment (from the end of the snowmelt, beginning of planting) with known initial conditions. The groundwater level H_k at the end of the interval will be found from:

$$H_k = H_{n-1} + \Delta H + \Delta H_2 \tag{6}$$

where: H_{n-1} is the groundwater level at the beginning of the interval considered, cm:

ΔH is the increase (+) or decrease (-) in the groundwater level under the influence of moisture exchange with the unsaturated zone;

ΔH_2 is the decrease in the groundwater level due to runoff and moisture exchange with adjoining catchments.

Moisture exchange with the unsaturated zone causes either an increase in the groundwater level ΔH during precipitation seepage (C, mm) or a decrease ($-\Delta H$) as a result of percolation of groundwater into the unsaturated zone (V, mm):

$$\Delta H = C \frac{K}{10} \tag{7}$$

$$\Delta H = -V \frac{K}{10} \tag{8}$$

where: K is the water rise coefficient.

Infiltration of precipitation to the water table will take place when, according to conditions of input and output, the accumulation of water approaches the maximum equilibrium moisture content (field capacity) W_v or exceeds it. In this case the difference between the maximum equilibrium moisture content in the unsaturated zone W_v and the actual moisture reserves W_a (or the accumulating capacity of this zone $W_{ak} = W_v - W_a$) is equal to zero, and as a consequence percolation of water from the groundwater table V = 0. Calculations are made according to the water balance equation:

$$C = W_{n-1} + \zeta_n - E + V - W_v \gtrsim 0 \tag{9}$$

If infiltration also takes place with a moisture content less than the maximum equilibrium moisture content W_V, when precipitation penetrates through larger pores and cracks but does not saturate the soil critically, then the term f_2W_V is introduced to the equation. This describes infiltration for a moisture content less than W_V. Then

$$C = W_{n-1} + \zeta_n - E + V - f_2W_v \geq 0 \qquad (10)$$

Here W_{n-1} is the water accumulation in the unsaturated zone at the beginning of the time interval considered, mm;

E, ζ_n is the evapotranspiration and absorption of precipitation on the soil surface, mm:

$$\zeta_n = \zeta - \zeta_{nc}$$

ζ is the rainfall, mm;

ζ_{nc} is the part of the precipitation which is discharged as surface runoff.

From equations (9) and (10) values of $C < 0$ and $C > 0$ may be obtained. The value $C \geq 0$ must be assumed. When $C < 0$, percolation V occurs. In the general case the value V is found from the following expressions:

$$D \leq W_v - W_{n-1} - \zeta_n + E \geq V \leq V_{max} \qquad (12)$$

where: V_{max} is the maximum possible percolation to the water table for a maximum accumulating capacity of the unsaturated zone and the given groundwater situation.

From such calculations the moisture regime in the unsaturated zone for an area with a given system of ditches and in years of normal precipitation can be characterized in detail. For individual particular cases calculations may be considerably simplified. For example, on developed bogs surface flow during the warm period of the year is zero, and $\zeta_n = \zeta$. Then $f_2 = 1$ may be assumed with a small error.

If, during seasonal variations in levels for certain periods, unfavourable conditions are created with higher water tables, it becomes necessary to discharge excess water to ensure the minimum drainage rate H_{min} for this period. The value of discharge h_c, mm for the considered time interval may be calculated from:

$$h_c = (H_{min} - H_k) \frac{10}{K} \qquad (13)$$

The value h_c determines the rational arrangement of drains and designed specific discharges which should be provided.

Water balance calculations may be made without calculating

drainage of groundwater by the designed network of ditches (without calculating ΔH_2). Then the required discharge h_c, which is somewhat overestimated, shows the specific discharge rates for the whole network of the drained area. The distribution of the total discharge of drainage systems for a particular season is obtained by adding groundwater flow components or the total lateral inflow.

Separate analysis of the components of the total discharge (internal and lateral inflows) is reasonable for the improvement of hydraulic engineering reclamation methods. The internal specific discharges mainly determine the necessary parameters of the regulating network (drainage), and the total runoff and intermediate inflow make up the design discharges for field canals and the peripheral network.

Water balance calculations for the whole unsaturated zone, as shown above, allow the water storage regime in this zone, the dynamics of groundwater and necessary internal specifice discharges from the drained area to be determined provided that the maximum groundwater level is not exceeded. Such characteristics permit the rational designing of drainage systems, a more detailed estimation of conditions for controlling the moisture regime to prevent overmoistening in wet years, or for controlling the groundwater levels in designing additional irrigation by sluices in dry years. But the water regime in the whole unsaturated zone alone does not completely describe the conditions for plants to be supplied with adequate moisture. For this purpose an analysis of moisture storage in the rooting layer is necessary. Moisture stored in the rooting layer on any date during the vegetative period can be determined from the water balance equation for this layer:

$$B = W_{spr} + \Sigma \Delta W + \Sigma(\zeta_p + \Pi_p) - \Sigma E \qquad (14)$$

where: W_{spr} is the design spring water stored in the rooting layer of initial thickness and at the beginning of the growing period;

ΔW is the change in water accumulation in the rooting layer with its depth from the previous to the subsequent time interval considered;

E is the evapotranspiration of crops;

Σ is the sum for the time from the beginning of the calculated period to the interval under consideration;

ζ_p, Π_p are design precipitation and groundwater flow from the lower layers.

The deficit (d) or excess (-d) of soil moisture in the rooting layer for a certain crop on any date of the warm period is found from the relation with the optimum moisture storage W_{op}:

$$d = W_{op} - B \qquad (15)$$

Thus, the water balance calculations for various combinations of input and output element of the water budget for a particular season characterize in detail the water regime in years with typical moisture conditions. This makes it possible to forecast the moisture regime when designing drainage-irrigation systems and to provide for rational reclamation measures, to use differential approaches to designing individual elements of the drainage network.

Reclamation methods used on marsh-ridden lands and engineering procedures depend on particular natural conditions of the moisture regime, formation of individual components of the water balance, seasonal and yearly distribution and combination of input and output elements and general characteristics of the water resources of the area. The division of the territory into reclamation regions is based on the water balance characteristics, geology of soils and climate. For these purposes results of investigations of the water balance of individual catchments of both small and large rivers, of bog and forest systems are used. Water balance investigations are also applicable for water management designs on redistribution and comprehensive us of water resources.

However, a number of problems have not as yet been sufficiently investigated for the great variety of natural conditions of marsh-ridden areas. This retards considerably the wide introduction of promising water balance methods. These problems are as follows:

1. Inadequate study of patterns of formation and magnitude of the individual water balance components, their interrelations and interdependence.

2. Conditions of water exchange between groundwater and the unsaturated zone, and also between different layers of the soil above the water table for various hydrogeological conditions.

3. Moisture extraction from the rooting layer plants under different soil, botanical and hydrophysical conditions, and for different soil moisture contents.

4. Water exchange of marsh-ridden areas with adjacent catchments, groundwater recharge of swamps and marsh-ridden lands.

5. Patterns of formation of individual components of the runoff from small marsh-ridden and reclaimed catchments, conditions of transformation of the runoff regime and its components, etc.

6. Improvement of investigation methods, development of physical and mathematical simulation procedures, etc.

In view of all these problems, more extensive and profound scientific investigations should be carried out for a better evaluation of water resources of marsh-ridden areas, the components of the water balance, and patterns of its formation for the purpose of adequate transformation of the water regime, finding the most effective engineering methods and construction and the rational use of water.

References

Afanasik, G. I. and Finskii, A. A. 1970 Metodika opredeleniya podpityvaniya korneobitaemogo sloya pochvy ot urovnya gruntovykh vod. (Methods for determination of inflow to a rooting layer versus groundwater level). Tezizy dokladov Nauchno-tekhnicheskoi konferentsii po melioratsii zemel Polesiya. Minsk, RVTsTsSU.

Afanskii, G. I. 1971 Model vlagoobmena v korneobitaemom sloe pochvy. (Moisture transfer model in a rooting layer). Sbornik trudov po agronomicheskoi fisike, issue 32 (AFI).

Aleinik, A. Ya., and Nasikovskii, V. P. 1970 Metody rescheta meliorativnogo dranazha v neodnorodno-sloistykh gruntakh. (Design methods for drainage in non-uniformly stratified ground). Kiev, Urozhai.

Anon., 1971 Ukazaniya po raschetam stoka s neosushennykh i osushennykh verkhovykh bolot. (Recommendation for runoff calculations from undrained and drained high bogs). Leningrad, Gidrometeoizdat.

Averiyanov, S. F. 1956 Filtratsiya iz kanalov i ee vliyanie na rezhim gruntovykh vod. (Filtration from canals and its effects on groundwater regime). Moscow, Izd., Akad. Nauk SSSR.

Averiyanov, S. F. 1957 Ob osushenii nizinnykh bolot. (On drainage of fen bogs). Nauchnye zapiski MIIUKH, vol. 19. Moscow, Selkhozgiz.

Averiyanov, S. F. 1960 O raschete osushitelnogo deistviya gorizontalnogo drenazha v usloviyakh napornogo pitaniya. (On calculation of draining effect of horizontal drains with pressure water recharge). Nauchnye sapiski MIIUKh, vol. 22. Moscow, Selkhozgiz.

Averiyanov, S. F. 1971 Rezhim osusheniya i metodika polevykh nauchnykh issledovanii. (Drainage rate and field investigation procedures). Otdelenie gidrotekhniki i melioratsii. Moscow, Kolos.

Bulavko, A. G. 1971 Vodnyi balans rechnykh vodosborov. (Water balance of river catchments). Leningrad, Gidrometeoizdat.

Globus, A. M. 1969 Eksperimentalnaya gidrofisika pochv. (Experimental hydrophysics of soils). Leningrad, Gidrometeoizdat.

Ivanov, K. E. 1953 Gidrologiya bolot. (Hydrology of swamps). Leningrad, Gidrometeoizdat.

Ivanov, K. E. 1957 Osnovy gidrologii bolot lesnoi zony i raschety podnogo rezhima bolotnykh massivov. (Elements of swamp hydrology of forest zone and water regime calculations of bog systems). Leningrad, Gidrometeoizdat.

Ivanov, K. E. 1965 Osnovy teorii mor fologii i gidromorfologicheskie zavisimosti. (Elimentary theory of swamp morphology and hydromorphological relations). Trudy GGI, issue 126. Leningrad, Gidrometeoizdat.

Ivanov, K. E. 1967 Teoriya gidromorfologicheskikh svyazei i ee primenenie. (Theory of hydromorphological relations and its application). In: Priroda bolot i metody ikh issledovanii. Leningrad, Nauka.

Konstantinov, A. P. 1968 Isparenie v prirode. (Evaporation in the nature). Leningrad, Gidrometeoizdat.

Korchunov, S. S., Mogilevskii, N. I., Abakumov, O. N. and Dulkina, S. M. 1960 Izuchenie vodnogo rezhima osushennykh torfyanykh zalezhei. (Study of water regime of drained peat deposits).

Moscow, Leningrad, Gosenergoizdat.

Kostyakov, A. N. 1951 Osnovy melioratsii. (Elements of land reclamation). Moscow, Selkhozgiz.

Kostyakov, A. N. 1961a Izbrannye trudy. (Selected works). vol. 1. Moscow, Selkhozgiz.

Kostyakov, A. N. 1961b Izbrannye trudy. (Selected works). vol. 2. Moscow, Selkhozgiz.

Luikov, A. V. 1954 Yavleniya perenosa v kapillyarno-poristykh telakh. (Transfer phenomena in capillary-porous bodies). Moscow, Gostekhizdat.

Luikov, A. V. 1969 Teoriya sushki kapillyarno-poristykh tel. (Drying theory of capillary-porous bodies). Moscow, Energiya.

Nerpin, S. V., Chudnovsky, A. F. 1967 Fisika pochv. (Soil physics). Moscow, Nauka.

Novikov, S. M. 1965 Raschet exhednevnykh urovnei gruntovykh vod na bolotakh po meteorologicheskim dannym. (Every-day calculations of groundwater levels on swamps using meteorological data). Trudy GGI, issue 126. Leningrad, Gidrometeoizdat.

Polubarinova-Kochina, P. Ya., Prazhinskaya, V. G. and Emikh, V. N. 1969 Matematicheskie metody v voprosakh orosheniya. (Mathematical methods in irrigation). Moscow, Nauka.

Rode, A. A. 1965 Osnovy ucheniya o pochvennoi vlage. (Elements of soil moisture science). Leningrad, Gidrometeoizdat.

Romanov, V. V. 1961 Gidrofisika bolot. (Hydrophysics of swamps). Leningrad, Gidrometeoizdat.

Romanov, V. V. 1962. Isparenie s bolot Evropeiskoi territorii SSSR. (Evaporation from swamps in the European part of the USSR). Leningrad, Gidrometeoizdat.

Shebeko, V. F. 1965 Isparenie s bolot i balans pochvennoi vlagi. (Evaporation from swamps and soil moisture balance). Minsk, Urozhai.

Shebeko, V. F. 1970 Gidrologicheskii rezhim osushaemykh territorii. (Hydrological regime of areas under drainage). Minsk, Urozhai.

Vasiliev, S. V., Verigin, N. N., Glizer, B. A., Razumov, G. A., Rudakov, V. K., Sarkisyan, V. S. and Sherkunov, B. S. 1970 Metody filtratsionnykh raschetov gidromeliorativnykh sistem. (Calculation methods for infiltration in water reclamation systems). Moscow, Kolos.

Verigin, N. N. 1969 Metod rascheta osusheniya stroitelnykh kotlovanov s pomoshchyu nesovershennykh skvazhin. (Design of constructional trenches with imperfect wells). Trudy soveshchaniya po voprosam vodoponizheniya v gidrotekhnicheskom stroitelstve. Moscow, Gosstroiizdat.

Vorobiev, P. K. 1969 Opredelenie vodootdachi iz torfyanoi zalezhi estestvennykh bolot. (Determination of water yield from peat deposits of natural swamps). In: Voprosy gidrologii bolot. Trudy GGI, issue 177. Leningrad, Gidrometeoizdat.

Voskresenskii, K. P. 1962 Norma i izmenchivost godovogo stoka rek Sovetskogo Soyuza. (Rates and variations of annual discharge of rivers in the Soviet Union). Leningrad, Gidrometeoizdat.

WATER REGIME OF THE UNSATURATED ZONE IN DRAINED LANDS UNDER IRRIGATION

S F Averiyanov, dec., Yu N Nikolsky
N P Bunina, A I Uzunyan.

Moscow Water Reclamation Institute, Moscow, USSR.

SUMMARY: This paper presents the results of the study of the soil moisture regime in drained lands under sprinkler irrigation.

The analysis of the experimental data obtained in situ has shown the change of the water balance of the unsaturated zone in low peat bogs when drained and maintained at the optimum level of soil moisture.

LE REGIME HYDRAULIQUE DE LA ZONE D'AERATION SOUMISE A L'ARROSAGE DES TERRES ASSECHEES

RESUME : Les auteurs présentent les résultats d'une étude sur les règles qui président à la formation du régime hydraulique de terres asséchées soumises à l'arrosage par aspersion.

L'analyse des données expérimentales obtenues *in situ* montre comment le bilan hydrique de la zone d'aération des tourbières de dépression varie avec la profondeur de l'assèchement et avec le degré d'humidité que requiert le sol.

Experimental evaluation of irrigation efficiency on drained lands and a study of the influence of the design moisture content and the water table depth on the moisture distribution in soil in the unsaturated zone has been made at the experimental plot 'Lesnoye' located on a lowland peat bog in the Minsk region. The peat depth was 1.5 to 3.0 m, its saturated hydraulic conductivity being 0.3 to 0.7 m per day. The porosity of the upper (0-50) cm layer varied over the range 0.82 to 0.90, the mean σ being 0.86. The moisture content of the soil at which hydraulic conductivity approximated zero was W_o = 0.45 (0.4-0.5) cm^3/cm^3. According to many authors (Averiyanov, 1949; Anon., 1969), the optimal range of moisture-content in lowland peat soils is (0.61 to 0.82) σ, which in the experimental plot corresponds to 0.50 to 0.70 cm^3/cm^3. The moisture mobility varies greatly within this range. With a moisture content of W = 0.7 the unsaturated hydraulic conductivity of the peatbog 'Lesnoye' was only 4 times less than the unsaturated hydraulic conductivity (Penman, 1968), whereas with W = 0.6 it decreased by a factor of 20. When analysing the experimental data this fact must be taken into consideration.

Soil moisture availability for plants and efficient use of irrigation water on drained lands depend on the groundwater level. The deeper the water table, the greater is the soil aeration and the more irrigation water is required to maintain the optimal soil moisture content.

Experiments and experimental procedure

Experiments with Lysimeters: Experiments on the water regime of drained peatbogs under irrigation were carried out using metal lysimeters filled with peat soil samples, 2.5 and 3.0 m deep and 1.5 m in diameter. On April 15, 1971 they were sown with the local spring wheat variety 'Minskaya'. The area around the lysimeters was also sown with wheat. The mean sowing density was 600 plants per sq m. During the growing period the level of the water table in the lysimeters was maintained at various depths (0.56, 1.0, 1.5, 2.0 m). During the growing period the lysimeters were watered by sprinklers so as to maintain the moisture content in the 0 to 50 cm soil layer within the optimal range of 0.55 to 0.70 cm^3/cm^3. The area surrounding the lysimeters was not irrigated. The soil moisture content of each 20 cm layer was measured by a neutron probe, with an accuracy of measurement of $\pm(0.01\text{-}0.02)$ cm^3/cm^3. The moisture balance equation for the unsaturated zone for the period under consideration is

$$\Delta\bar{W} = m_1 + m_2 - E_o \pm q \qquad (1)$$

where:

$\Delta\bar{W}$	change of soil moisture content in the unsaturated zone for the period under consideration;
m_1	precipitation;
m_2	amount of irrigation water;
E_o	evapotranspiration;
q	water exchange between the unsaturated zone and groundwater;
$+q$	flow from groundwater to soil;
$-q$	flow of soil moisture to groundwater.

The value q was found from the amount of water which had to be added to or removed from the lysimeter to maintain the level of the water table at the chosen depth. Precipitation was measured by standard devices such as the Pretiyakov precipitation meter and soil rain gauges GR-25. Evapotranspiration E_o from the lysimeter was determined from equation (1) with the measured values of $\Delta\bar{W}$, m_1, m_2 and q.

Experimental Plots: Experiments were carried out on four plots, 10 x 15 m, in addition to the lysimeter experiments. They were sown with a late variety of potato. The potato planted on June 4 and 5, sprouted between June 17 and 25. The moisture content in the 0 to 50 cm soil layer was measured by means of stationary gamma moisture meters installed at the plots ^{60}Co source with an activity of about 1.5 mg eqv. Ra was installed on a vertical bar at a depth of 50 cm. The sensing element was placed on the soil surface and moved periodically around the central bar. Moisture content of each soil layer was measured periodically by weighing. Watering was effected by a sprinkler nozzle CDA-2. Precipitation m_1 and irrigation rates m_2 were measured by soil rain gauges. The depth of the water table was systematically measured on the experimental plots, and evaporation was measured by a standard evaporimeter, GTI-500 at a depth of 50 to 100 cm. The moisture content of the 0 to 50 cm layer was maintained in the four plots within the following ranges: 0.50 to 0.55; 0.50 to 0.60; 0.55 to 0.65; 0.50 to 0.70 cm^3/cm^3. The moisture content of a fifth unirrigated reference plot varied from 0.40 to 0.60 cm^3/cm^3, whilst water table depth varied between 0.7 and 1.1 m.

Principal experimental results

Lysimeter experiments: The growing period of 1971 (May to September) was characterized by a severe drought. Precipitation during this period amounted to 217 mm only, whereas the normal precipitation for this time of the year is 340 mm. The probability of observing such low rainfall is 0.5%. Potential evaporation E (the greatest possible evaporation) according to a calculation based upon the temperature and humidity of the air (Shebeko, 1962) during the growing period of 1971 was equal to 465 mm. The probability level of this potential evaporation is 21% (Shebeko, 1965). The moisture deficit (m_1 - E) had a probability of 25%.

Table 24 shows details of the water table depth, soil moisture, irrigation volume, number of applications and crop yield for the lysimeters.

Each lysimeter received 5.8 mm of water on May 5, 1971. The following quantities of fertilizer were introduced at the same time: 1.5 t/ha of superphosphate; 0.25 t/ha of potassic salt; 0.08 t/ha of ammonia saltpetre. The highest crop yield capacity was obtained in lysimeters with a moisture content of about 0.55 to 0.65 cm^3/cm^3 in the 0 to 50 cm soil layer and the groundwater table at 1.0 to 1.5 m from the surface.

The lesser crop yield in lysimeters 3 and 4 drained to 20 m may be attributed to insufficiently accurate soil moisture control.

To analyse the variations of moisture distribution in drained peat soils under irrigation, the balance of soil moisture in the unsaturated zone is calculated for selected time intervals during the growing period. Table 25 shows the change of value of the water balance components for various drainage depths, (Anon., 1969). The data are given for the complete vegetative period (May 5 to August 18, 1971). In lysimeter No 4 where the grounwater table was only lowered to the assigned depth in June, the water balance of the unsaturated zone was calculated for the period between June 10 and August 18. In the table the water exchange q of equation (1) is the difference between the infiltration to the water table (q_2) and the flow from groundwater to the unsaturated zone (q_1) ($q = q_1 - q_2$).

Figure 44 illustrates the change in the water exchange between the water table and the unsaturated zone ($q/m_1 + m_2$), as the depth of drainage increases. When $\Delta \sim 1.2$ m, q = 0, so the water exchange between soil and groundwater is balanced; when $\Delta < 1.2$ m, $q > 0$ (flow of groundwater to soil); when $\Delta > 1.2$ m, $q < 0$ (flow from

Table 24. Summary of the principal data obtained with lysimeters containing peat soil and carrying wheat.

Lysimeter no.	Drainage depth Δ,m	Mean soil moisture content (0 to 50) cm $W_{mn}(W_1 + W_2)$		Total irrigation water applied mm	No. of applications	Crop yield t/ha
		May 5 to June 5	June 5 to August 18			
3	2.0	0.57 (0.50-0.67)	0.61 (0.53-0.74)	221.0	5	3.37
4	2.0	0.66 (0.53-0.75)	0.66 (0.53-0.75)	161.0	5	3.36
1	1.5	0.52 (0.44-0.60)	0.63 (0.56-0.69)	342.0	7	3.13
2	1.5	0.58 (0.52-0.69)	0.68 (0.62-0.75)	862.0	12	5.47
5	1.0	0.56 (0.54-0.59)	0.63 (0.55-0.74)	307.7	7	5.50
6	1.0	0.56 (0.51-0.62)	0.60 (0.55-0.63)	442.7	8	5.54
7	0.56	0.77 (0.74-0.79)	0.77 (0.75-0.78)	5.8	1	2.47

Table 25. Change in water budget components in unsaturated zone for various drainage depths. (same symbols as in equation (1))

Period	Lysimeter nos.	Δ m	$m_1 + m_2$ mm.	q_1 mm	q_2 mm	$q = q_1 - q_2$ mm.	$\Delta\bar{W}$ mm	E_o mm
May 5 to August 18, 1971	7	0.56	129	+447	0	+447	- 2	578
	5	1.0	431	+ 94	- 41	+ 53	+ 47	437
	6	1.0	566	+225	- 33	+192	+ 19	739
	1	1.5	466	+ 17	- 37	- 20	- 9	455
	2	1.5	986	+ 18	-221	-203	+ 45	738
	3	2.0	345	+ 18	-187	-169	-277	453
June 10 to August 18, 1971	4	2.0	385	-	-	-	-	-
	4	2.0	187	+ 5	-195	-190	-280	277

Table 26. Effect of drainage depth upon water budget components in the unsaturated zone of drained and irrigated peat carrying a potato crop.

Sections	Δ m	W_{mn}	$W_1 + W_2$	$m_1 + m_2$ mm	$\Delta\bar{W}$ mm	q mm	E_o mm	Crop yield t/ha
a	0.7 + 0.9	0.71	0.68 + 0.77	597	0	+450	147	35.0
b	0.7 + 1.0	0.62	0.50 + 0.64	313	+30	+132	151	31.0
c	0.8 + 1.0	0.65	0.55 + 0.70	457	+75	+236	146	10.0
d	1.2 + 1.4	0.57	0.50 + 0.62	146	-15	+ 22	139	28.0
e	0.8 + 1.1	0.50	0.40 + 0.55	43	-50	- 37	130	16.0

soil to groundwater).

Figure 44 also shows the relation between $E_o/m_1 + m_2$ and the depth of drainage Δ. In this case, too, when $\Delta \sim 1.2$, $E_o \sim m_1 + m_2$, which means that the sum of irrigation applications and rainfall is lost to evapotranspiration.

Experimental Plots: Table 26 shows the main results of the experimental evaluation of the influence of soil moisture content in the unsaturated zone of drained peat bogs under irrigation. Plots a, b, c, d, were watered in August and September of 1971. Plot e was not irrigated. The components of the water budget (Anon., 1969) were computed for August, the period of the most frequent watering.

W_{mn} is the mean soil moisture content in the 0 to 50 cm layer for the month: $W_1 - W_2$ is the soil moisture content within this layer. Precipitation in August amounted to 43 mm.

In Figure 45 the curves $m_2 = f(W_{mn})$ and $\frac{q}{m_1 + m_2} = f(W_{mn})$ have been plotted with the experimental values. The rise in the soil moisture content with increase of the irrigation rate is quite evident. Unsaturated hydraulic conductivity (Averiyanov, 1949), increases rapidly with the moisture content, therefore, moisture seepage to groundwater sharply grows, the water table being maintained at a constant level.

The curve $q/(m_1 + m_2) = f(W_{mn})$ is extrapolated to the value of the moisture-content which corresponds to complete saturation, assuming that $\frac{q}{m_1 + m_2} = -1$ when $W \sim \sigma$ ($\sigma = 0.86$).

Conclusions

When drained peat soils are irrigated with sprinklers, there exists a certain optimal depth of groundwater (for particular crops, soils, weather conditions) at which the total amount of precipitation and irrigation is spent on consumptive use and moisture exchange between the soil and groundwater is balanced. In the field experiments on a low bog in the Minsk region, this depth was equal to 1.2 m for a water consumption of about 400 mm (precipitation 100 mm, irrigation $\sim$ 300 mm); it is with this drainage depth that the greatest crop yield was obtained. A higher groundwater level results in increase of water consumption, but decreases the crop yield; a lower groundwater level increases irrigation rates (to maintain the optimum soil moisture content) and seepage to groundwater.

To increase irrigation efficiency, it is necessary to choose correctly the required range of soil moisture contents and the drainage regime. The choice may be made on the basis of crop requirements with respect to the water-air regime of soils, the calculation of the soil moisture regime in the rooting zone for typical conditions, as well as on the basis of economic feasibility.

References

Anon., 1969. Vodnyi rezhim selskokhozyaistvennykh rastennii. Water regime of agricultural plants. Moscow, Nauka.

Averiyanov, S. F. 1949. Zavisimost vodopronitsaemosti pochvogruntov ot soderzhaniya v nikh vozdukha. Effect of air content in soils on their moisture conductance. Dokl. Akad. Nauk SSSR, vol. 69, no. 2.

Penman, Kh. L. 1968. Rastenie i vlaga. Moisture flow. Leningrad, Gidrometeoizdat.

Shebeko, V. F. 1962. Vnutrigodovoe raspredelenie i obespechennost osadkov na territorii BSSR. Precipitation distribution and probability within a year in the BSSR. Minsk, Urozhai.

Shebeko, V. F. 1965. Isparenie s bolot i balans pochvennoi vlagi. Evaporation from swamps and soil moisture balance. Minsk, Urozhai.

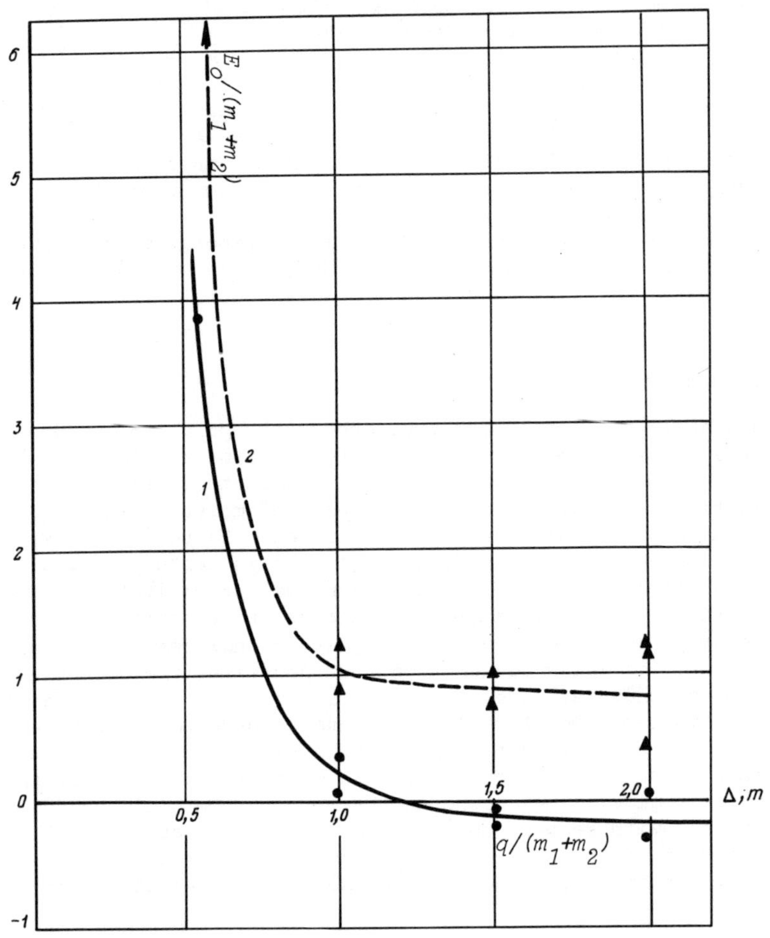

Fig 44. Effect of the drainage depth Δ on the value of the principal components of the water budget in the unsaturated zone of irrigated peatbogs.

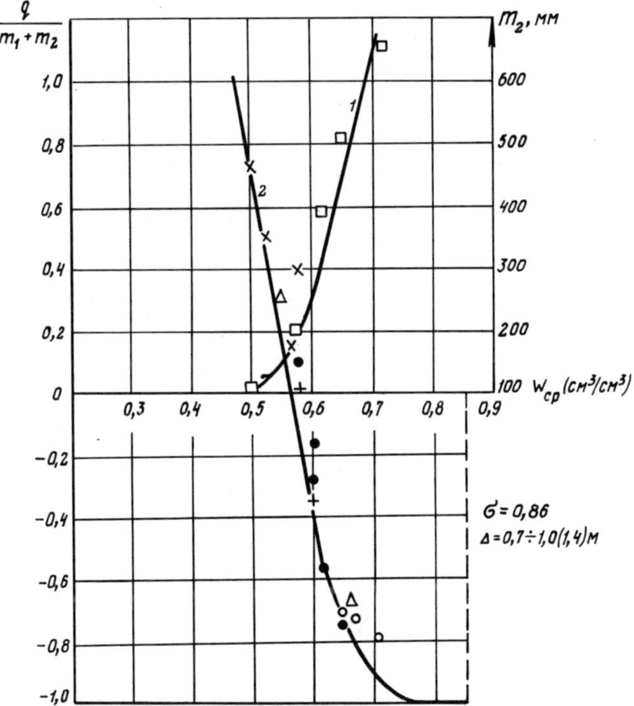

Fig 45. Effect of irrigation rate m_2 on moisture content W_{mn} of the 0 to 50 cm soil layer (curve 1) and a plot of W_{mn} versus the ratio $q/(m_1 + m_2)$ (curve 2).
o, ., Δ, +, x denotes the data for plots, a, b, c, d, e respectively;
□ denotes the data for $m_2(W_{mn})$.

DISCUSSION

L. Wartena (The Netherlands) - Did you find a high evaporation rate because the area surrounding the lysimeters was not irrigated?

Yu.N. Nikolsky (USSR) - The evaporation rate over the considered range of moisture content in a rooting layer is not high, therefore, it changes slightly with the groundwater level. The surrounding area was also irrigated. We tried to estimate the additional evapotranspiration but these values were found to be negligible (we applied the procedure conventionally used in the Soviet Union). I would like to make one more comment. For the analysis of the experimental data it is important to know the evaporation rate rather than compare these values with the evaporation rates which may be found on large areas.

THE WATER BALANCE OF BOGS AND FENS

(Review Report)

J. Dooge

University College Dublin,
Dublin, Ireland

Introduction

Objective and Scope Review: The objective of this paper is to sketch in the background for the more detailed papers presented under Topic 2 to this Symposium. Topic 2 is concerned with methods and results of water balance calculations and their use for projecting drainage or irrigation systems in the temperate zone. In order to give a balanced treatment of this theme, it is necessary to begin with some topics which might be thought more appropriate to Topic 1 of the Symposium. Similarly, to exemplify the use of water balances in the design of engineering schemes, it is necessary to anticipate to some extent certain subjects which fall under Topic 3.

A review of the literature on the water balance of marsh-ridden areas reveals work on the subject in a large number of countries. The bulk of the literature has appeared in Russian, English or German, and to a lesser extent in the Scandinavian languages, in Polish and French. As a theme paper on Topic 2 is being presented to the Symposium by Professor Shebeko and Professor Ivanov, I have not referred to the Russian literature in the detail that its importance deserves. Rather, I have confined myself to references to publications in the Russian language which are of special importance and to publications that have become available to a wide audience through translation. Accordingly, the material in this paper should be supplemented by the material dealt with by Professors Shebeko and Ivanov.

Following this introduction, this paper has been divided into four sections. Section 2 of the paper outlines the problem; it describes briefly the components of the hydrological cycle and the various forms of the water balance equation. It then deals with the classification of bogs and fens to which the water balance equation is to be applied. Section 2 ends with a discussion of the nature of peat which naturally influences strongly the hydrology of marsh-ridden areas such as bogs and fens.

Section 3 of the paper is concerned with the various moisture characteristics of peat which control the movement of water into, through and out of the peat deposits of bogs and fens. The most important of these are the waterholding capacity of the peat and its hydraulic conductivity. Other properties of hydrological importance are the shrinkage properties, the coefficient of storage which gives the amount of water lost during a fall in the water table, and its thermal properties which are of importance for the energy balance so fundamental in evaporation studies.

Section 4 deals with the individual components of the hydrological cycle which enter into water balances. In this section there are separate discussions about the evaporation from bogs and fens, the outflow from such areas and the water storage in both the saturated and unsaturated zones of bogs and fens.

Finally, Section 5 of the paper discusses the overall water balance and the relationship between the individual components. Reference is made to water balance studies in different countries. There is a brief discussion of the effect of human activity on changes of the water balance. The section ends with a brief note on the scope for future study and research.

Attention has been concentrated on basic principles and fundamental problems. Research results are only quoted as illustration of typical results and not as universal values. Further details of individual studies may be found in the references cited. It is hoped that the review may be useful in assisting scientific workers and engineers to compare their problems and results with those in other countries and to place their own work within a general framework.

Background to problem

Components of the Hydrological Cycle: The water in the atmosphere on the surface of the earth and within the earth's crust is known collectively as the hydrosphere. Though water occurs in many forms, the total amount of water in the hydrosphere remains constant. The never-ending circulation of water throughout the hydrosphere is known as the hydrological cycle.

The principal types of water storage and water transport with which we are concerned for individual catchment areas are shown schematically in Figure 46. Each of the boxes represents a type of water storage and the arrows between the boxes represent the transport of water from one form of storage to another. W_i represents the inflow of precipitable water into the atmosphere above the catchment area in question and W_o the outflow of precipitable water from the atmosphere above the area. P represents the precipitation of water from the atmosphere to the surface of the ground and E the evaporation of water into the atmosphere from surface storage on the ground. F represents the infiltration through the surface of the ground into the soil and T, transpiration of water from the soil through vegetation and its subsequent evaporation to the atmosphere. G represents the recharge of groundwater from soil moisture and C the capillary rise from groundwater to the soil. Q_o represents overland flow across the surface of the ground to the drainage network, Q_i the lateral flow of water through the unsaturated soil to the drainage network, and Q_g the groundwater flow (or base flow) which sustains the flow in the drainage network in the absence of precipitation. R represents the total runoff of the catchment which may be measured but which cannot, under ordinary circumstances, be divided into the various components of surface runoff, interflow and groundwater flow.

The UNESCO publication on 'Textbooks on Hydrology', (UNESCO, 1970) containes analyses and synoptic tables of the contents of selected textbooks which facilitate reference to various topics in hydrology and can be used to select a textbook which deals particularly with the problem of the hydrology of swamps.

A discussion of the methods of measurements of the components of the hydrological cycle is outside the scope of the present review. Internationally accepted methods of measurement are described in the WMO publication entitled 'Guide to Hydrometerological Practice', (WMO, 1965). The special problems involved in the measurement of these processes in swampy areas is dealt with in the 'Manual of Hydrometeorological Observations in Swamplands', (Anon. 1955), published by the Hydrometeorological Publishing House in Leningrad, and used as the basis of studies throughout the USSR. This publication became available to a still wider audience due to its translation by the Canadian Meteorological Service in 1961.

Water Balance Equations: A water balance may be calculated on various different scales, e.g. global, continental, regional, major or minor catchment, or on the scale of micro-relief of the landscape. Water balance studies were described by a number of authors in papers to the Symposium on World Water Balance held at Reading, England in 1970, (IASH/UNESCO, 1970).

In dealing with marsh-ridden areas, we are not concerned with water balance on so large a scale, but with such problems as the water balance of individual catchments, of marshy portions of catchment, or even the water balance of individual elements of micro-landscape.

In most water balance studies of catchments, the precipitation (P) and the runoff (R) are measured and the total evaporation (ET or E) is either measured or estimated. Accordingly a water balance equation for a stated period can be written for these elements as follows:

$$P = E + R + (S_2 - S_1) \tag{1}$$

where the figure for evaporation includes transpiration from the vegetation and where S_1 and S_2 represent the total storage in the catchment area down to the runoff point at the beginning and end of stated period. This total storage includes surface storage (SS), field moisture storage (FMS), groundwater storage (GWS) and channel storage (CS). For a more detailed water balance study, each of these elements of storage must be either measured or estimated or shown to be negligible.

Figure 46 and equation (1) refer to the case of a complete catchment for which the only inflow into the area is from precipitation. Where the water balance is required for only part of the catchment, it is necessary to modify both the figure and the equation. In Figure 47 which represents the modified diagram of the hydrological cycle, T_s represents the inflow of water to the area of study over the surface of the ground, T_i represents the inflow of the water into the area as interflow through the unsaturated soil, and T_g represents the inflow of water into the area as groundwater flow. As in Figure 46, Q_s represents outflow from the area as surface flow, Q_i outflow from the area as interflow and Q_g outflow from the area as groundwater flow. If i is used to denote the total flow into the area (i.e. $T_s + T_i + T_g$), then the water balance equation can be written as:

$$P + T = E + R + (S_2 - S_1) \qquad (2)$$

where the other terms have the same meaning as in equation (1).

A complete water balance study is strictly one in which every element of equation (2) is measured independently and substituted in the equation to see if the balance holds. Due to limitations in the data or to the expense of complete data acquisition, water balance studies reported in the literature are rarely complete in this sense. In many studies it is assumed that the storage term may be neglected. In long-term studies such an assumption is reasonable if one can be sure that the nature of the catchment area does not change during the period. In the cases of bogs and fens which have been drained, the assumption may give rise to error because the changes in the soil due to drainage do not occur all at once and may continue to affect the characteristics of the catchment for a large number of years. In many other studies, the water balance equation is assumed to hold and is used to determine the value of a component which has not been measured.

It is necessary to be clear in the case of all water balance studies as to exactly what the objective of the study is and what are the assumptions inherent in the methods adopted. It is helpful to write the water balance equation in the following form:

$$P + T - E - R - (S_2 - S_1) = b \qquad (3)$$

where the terms on the left hand side are the same as those in equation (2) and the term on the right hand side, b, represents the divergence of the water balance from zero due to the occurrence of errors in the measurement of one or more of the components; a small value of b in equation (3) is no guarantee of the absence of errors in the measurements, since some of the measurement errors may be compensating. If the water balance equation is written in the form of equation (3) then the various assumptions will become clear if in the study the equation is simplified by assuming b equal to zero, or S_1 equal to S_2, or I equal to zero.

In the above equations the balance has been taken for a limited catchment area in plan but for an unlimited vertical profile. If, for example, the groundwater storage is only measured down to a certain horizon, then a term would have to be included for deep percolation, which might be estimated or might be assumed equal to zero. For agricultural purposes, it might be desirable to construct a water balance of only the soil moisture in the unsaturated zone. In this case, it would be necessary to wirte the balance equation as follows:

$$F - E + T_i - Q_i - G - (FMS_2 - FMS_1) = b \qquad (4)$$

where F is the volume of infiltration, E the volume of evaporation and transpiration, G the amount of groundwater recharge, T_i and Q_i the inflow from and outflow to neighbouring soil layers in the unsaturated zone, FMS_1 and FMS_2 the field moisture storage at the beginning and end of the period, and b the error in the water balance. Still more detailed water balance equations can be written but in this paper we shall be largely concerned with the vertically integrated form of the balance equation given in equation (3) and with simplifications of it.

Mires, Bogs and Fens: One difficulty of discussing the hydrology of bogs and fens is the variety of nomenclature within individual languages and the difficulty of identifying the corresponding terms when translating from one language to another. The basic nomenclature adopted in this paper is that used and defined in the TELMA Questionnaire, (UNESCO, 1967) which corresponds to the terminology being increasingly adopted in English-speaking countries. For convenience of reference the appropriate part of the Questionnaire is reproduced as Appendix A.

The term mire is used to denote the category of peat formations in general. The main classification of mires is into the sub-categories of bogs and fens. The term 'bog' is confined to areas where the only inflow is direct rainfall and the term 'fen' used for all other mire complexes. Further subdivision is outlined in Appendix A and depends essentially on the morphology and hydrography of the mires. The effect of the latter factors on the ecology of the units is such that the different types of bogs and fens in fact support different types of vegetation.

An attempt is made in Table 27 to relate some of the more important of the English language terms in Appendix A to the corresponding terms in German, Russian and Swedish which represent three language groups which are well represented in the literature.

Table 27.

English	German	Russian	Swedish
mire	moor	болото	myr
fen	niedermoor	низинное болото	karr
bog	-	олиготворное болото	mosse
raised bog	hochmoor	верховое болото	-
blancket bog	terrain-bedeckendemoor	-	-

The terms of the above table do not correspond exactly but may be taken in each case as a good first approximation. Many other terms are of importance, e.g. a mire may be transitional (Ubergangsmoore) between fen and bog.

It is interesting to note that the basic classification of mires into bogs and fens may be related to equation (2) which is the equation of water balance of the mire unit. The mire is classified as a bog in cases where the inflow to the area from the surrounding area is negligible (i.e. I = 0 in the water balance equation). When the contribution of inflow is appreciable, so that both the terms P and T have to be included on the left hand side of equation (2), then the mire is classified as a fen. Bogs are said to be ombrotrophic because their vegetation thrives under heavy precipitation and are said to be oligotrophic because the nutrient supply is low. These conditions give rise to characteristic vegetation types such as Sphagnum which are more tolerant of conditions of acidity and scarcity of nutrients than competing types. The fens are said to be minerotrophic or eutrophic because of the supply of minerals by the inflowing water and are said to be rheophilous or sologenous because of the flow of water through the body of the fen. They are said to be topogenous where they occupy a depression and are fed by the surrounding groundwater table. The chemical quality of the water is also important in distinguishing between bogs in which the water is highly acid and fens in which the water is approximately neutral. Bellamy, 1968 has discussed the relevance of chemical quality to mire classification.

It is generally accepted that an area may be described as peatland (as distinguished from peaty soil) when the depth of the peat layer exceeds 300 mm when in an undrained condition. The corresponding depth in a drained condition is usually taken as 200 mm. The factors that result in the formation of bogs and fens or the conversion of a fen to a raised bog are outside the scope of the present paper, but useful general accounts are to be found in publications by Kulcznyski, 1949; Godwin, 1957; Gorham, 1957; and others.

There are, however, a number of features of peat formation which affect the water balance because of their influence on the movement of water either over or through the ground. An important concept is that of the active layer. This comprises the upper part of the mire surface formed by living plants together with that part of the plant residues which are still substantially undecomposed. There is a very marked decrease in permeability below the active layer and consequently a very marked decrease in lateral subsurface flow when the level of the water table falls below the active layer. The existence of an active layer in peat deposits and of this marked change in permeability results in a substantial amount of interflow when the level of the groundwater is between the bottom of the active layer and the surface. The water is unable to percolate downwards through the relatively impermeable peat layers but is free to move laterally through the permeable active layer.

Many bogs show a distinct demarcation line (known as the Grenzhorizont) separating the upper Sphagnum which is light in texture and colour from the lower Sphagnum which is darker in colour and more valuable as a fuel. In many cases the Sphagnum peats of both the upper and lower strataare built up in a lenticular fashion owing to the fact that a Sphagnum bog is built up as a mosaic of alternating hollows and hummocks in which the hollows are replaced by hummocks as time goes by and vice versa. This regeneration complex, as it is called, results in a deposit which is neither homogenous nor random and which therefore may exhibit marked changes in hydrological behaviour at different levels in the bog or in different parts of the bog. The classical profile of a raised bog shows a succesion from the bottom of layers of reedswamp peat, woody-fen (karr) peat, older Sphagnum and eriophorum peat (well humified) and finally a regeneration complex of younger Sphagnum (relatively humified).

The micro-relief of the surface of the bog is also important from a hydrological viewpoint. The existence of hummocks and of pools and of depression tracks on the surface of the mire complex, all affect the runoff process and the infiltration into the ground. Theoretical studies of the effect of the micro-relief on the hydrological regime have been reported by Ivanov in a number of books and papers, (Ivanov, 1957 and 1965).

The Nature of Peat: In a peat mire, the decomposition of plant remains by bacteria and fungi is slowed down by the acidity and the anaerobic conditions. The growth rate of the new plant material therefore exceeds its rate of breakdown and results in an accumulation of peat. The occurence, classification, properties and uses of peat have been studied from a number of viewpoints and have given rise to a vast literature. A good overview of these different approaches can be obtained by reference to the published Proceedings of the International Peat Congresses (1954, 1963 and 1968), and similar conferences.

Peat is a variable material and its properties change with the time. The drainage and cultivation of peat soil brings about irreversible changes in their physical and chemical characteristics. Discussion of these changes is outside the scope of the paper but reference may be made to specialist papers, (van Heuveln, 1960; Pons, 1960) or to books concerned with the reclamation of peat soil such as that by Skoropanov, (1961). A comprehensive review of the classification and properties of organic soil has been given by Farnham and Finney (1965).

The properties of peat of interest in hydrology vary greatly with the degree of decomposition. Accordingly comparisons can only be usefully made between the hydrological behaviour of individual mire units or complexes if the degree of decomposition of peat is recorded according to an objective and agreed method. One of the most widely used measures for the decomposition of peat is that due to van Post (1924). This method is based on a subjective evaluation of the material when squeezed in the hand. It takes account of the colour, amount of water, plasticity, etc., and is graded on a ten point scale. The first point of the scale H.1 corresponds to undecomposed peat and is applied when the water emerging from between the fingers is colourless and clear and no peat particles emerge through the fingers. For fairly decomposed peat, which is ranked as H.5, the water emerging between the fingers is very turbid and some peat mass emerges through the fingers. At this state the plant structure is still clearly recognisable. H.9 corresponds to almost fully decomposed peat in which almost all of the peaty mass emerges from between the fingers and the residue in the hand consists of resistant matter such as root fibres, wood, etc.

Kaila (1956) has proposed the designation of the degree of humification on the basis of the volume weight of air-dry ground peat expressed as a ratio of the volume weight for undecomposed surface vegetation. Kaila reviewed the methods previously proposed to determine the degree of humification and studied 220 samples from Finnish bogs by a number of methods. She found a coefficient of correlation of 0.8 between results obtained on a volume-weight basis and estimates based on the van Post scale. Kaila also discusses a number of methods used in Germany, including the Keppeler method, which is the basis of the standard used in the German Federal Republic, (Standards Committee, 1965). Other methods are based on the examination of such properties as the structure of the sample and the carbon-nitrogen ratio.

Moisture characteristics of peat

Water Holding Capacity: Peat is able to hold a relatively large quantity of water, particularly when it is in the original saturated condition and has not been subject to decomposition. These moisture capacities may be expressed in terms of moisture content on a dry weight basis or moisture content on a volume basis. When expressed in the former fashion, the figures reflect the high moisture capacity of saturated peat relative to saturated non-organic soil but these dry weight moisture contents are not so meaningful from the hydrological or agricultural viewpoint. In studying the water balance or studying the effect of drainage, water contents on a total volume basis are more easy to interpret.

The water holding capacity of peat depends greatly on its degree of decomposition. Grebenshchikova (1956) has given an interesting set of results for a number of different types of peat at different stages of decomposition. These show saturated moisture contents on a dry weight basis varying from a moisture content 4000% for a medium peat at 5% decomposition to a moisture content of 510% for an arboreal peat at 50% decomposition. Romanov (1961) who considers that the figures given by Grebenshchikova are somewhat high in absolute value but correct in relation to one another, gives a number of results from various Soviet authors on the maximum moisture capacity of Sphagnum of different types, of other plants and of peat. Results for peat from seven sites in various parts of the United States are contained in a paper by Feustel and Byers (1930). Their results show values of maximum moisture-holding capacities on a dry weight basis varying from over 3000% to under 300% and also indicate a decrease in moisutre capacity with increasing depth below the surface i.e. with increasing decomposition. Feustel and Byers also show the maximum moisture-holding capacity of their samples following air drying and re-wetting and report that in this case the maximum moisture capacity is between one-third and one-half of the moisture capacity of an original sample of the same type of peat which has not been air dried.

The lowering of the water table in the peat profile, either naturally or through drainage, results in moisture contents less than saturation in the upper layers of peat and a corresponding soil moisture tension. To describe the moisture-holding capacity of peat (or of any other soil) requires the determination of the relationship between moisture content and soil suction over the range of interest. The basic principles of the retention of moisture by unsaturated soil may be referred to in standard texts and publications on soil physics. Reference will only be made here to the paper by Peerklamp and Boekel (1960). These authors describe the physical principles involved and the main methods of measurement and discuss the effect of the organic element in the soil content on the moisture characteristic curves.

The determination of a soil moisture characteristic curve for a sample of peat gives rise to a number of experimental difficulties. If the results are to be applied to the in situ conditions, then great care must be taken to ensure that laboratory tests are carried out on an undisturbed sample. While the general shape of the soil moisture characteristic curve is very much the same for different samples of peat, the actual values obtained depend on the degree of decomposition of the peat. This may be illustrated by reference to the work of Boelter (1964, 1968). Table 28 shows some of his results for the moisture content of peat taken from different levels of a profile of a bog in Minnesota, and subjected to various suctions in the laboratory.

Table 28. Pertencage moisture content on a volume basis at different suctions.

Type of peat	Soil suction in bars					
	0.0[a]	0.005	0.035	0.1[b]	1.0	15[c]
Live Sphagnum	100	42	17	13	6	4
Undecomposed	97	64	40	31	16	9
Partially decomposed	87	84	75	64	35	21
Decomposed	83	82	78	72	44	22

a = saturation; b = field capacity; c = wilting point.

The results shown in Table 28 indicate that the moisture characteristic curves of undecomposed and decomposed Sphagnum peat are quite different. At saturation the moisture content decreases with the degree of decomposition but even for quite small suctions the position is reversed and the moisture content increases with the degree of decomposition. A fall in the water table of

50 mm would be sufficient to produce a suction of 0.005 bars and hence to reduce the moisture content of undecomposed peat appreciably i.e. from 97% to 64%, compared with the relatively small drop in the moisture content of decomposed peat from 83% to 82%.

Coefficient of Drainage: If there is no movement of water either upwards or downwards in an unsaturated soil profile, then the water at any level will be in equilibrium under the combined action of gravity and of the suction which exists in the soil water. The amount of the suction depends on the pressure drop across the curved interface which exists between the soil air which is at atmospheric pressure and the soil water in the unsaturated zone. This curvature is in turn relative to the structure of the soil, i.e. to the size and arrangement of the pores and cavities.

For the equilibrium condition, the soil suction at any point expressed as a negative head of water will be numerically equal to the height of the point above the groundwater table. The moisture content at this level will be the moisture content corresponding to this particular suction and may be obtained from the moisture characteristic curve for this particular soil. There is, therefore, a unique relationship for any given soil profile between the depth to groundwater level and the amount of water stored in the unsaturated zone above this water table under equilibrium conditions. If the water table is lowered, there will be a new equilibrium curve of moisture content versus depth below the surface and a new value of storage above the water table. The amount of water lost from a soil profile when the groundwater level falls by a determined amount is known as the water yield or coefficient of storage. It is normally expressed as a volume ratio.

The coefficient of storage is obviously of great importance in water balance studies where there are variations in the level of groundwater between the beginning and the end of the period under study. As might be expected, the coefficient of storage for peat varies greatly with the type of peat and the degree of decomposition. An examination of Table 28 above would lead us to expect that the coefficient of storage for live Sphagnum or undecomposed peat would be much greater than the coefficient of storage for decomposed or partially decomposed peat. This is borne out by such estimates as have been made of the coefficient of storage of peat.

Heikurainen and his colleagues, (Heikurainen, 1964), have reported field experiments and laboratory experiments in Finland on partially decomposed peat (H.3 to H.5). The differences in water content were determined by direct weighing and coefficients of correlation between water content and water level were almost always above 0.85 and coefficients exceeding 0.95 were quite common. This indicates that the concept of the coefficient of drainage is useful even when there is a departure from equilibrium conditions due to vertical movement of water in the soil profile. The field samples were peat cylinders 250 mm in diameter and consisted either of a single layer 200 mm high or a three layer sample 400 mm high. The variation of water level on the field experiments was from 80 mm below the durface to 470 mm below the surface and in the laboratory specimens from 200 mm from the sample surface to 275 mm from the surface.

The estimation of the storage coefficient from changes in water level and from recorded precipitation has been discussed by Ivanov (1957) and Vorobiev (1963). This in fact implies the use of a simplified form of the water balance budget in order to obtain the change in storage as a residual term. Boelter (1968) has suggested that a good approximation may be obtained for the storage coefficient by taking the difference between the moisture content at saturation and the moisture content at field capacity (taken arbitrarily a 0.1 bar). The range of values obtained for the coefficient of drainage in the different studies are much the same. They range from about 0.1 for decomposed peat, through 0.2 to 0.4 for partially decomposed peat, to 0.6 for decomposed peat and 0.8 for live Sphagnum moss.

Hydraulic Conductivity of Peat: The movement of water through porous media is normally assumed to take place in accordance with Darcy's Law:

$$Q = KAT \tag{5}$$

where Q is the volume rate of flow, K is the hydraulic conductivity, A is the cross sectional area of flow, and T is the slope of the hydraulic gradient.

It has not yet been confirmed beyond doubt that flow through peat does in fact take place in accordance with equation (5). Accordingly the hydraulic conductivity calculated on the basis of Darcy's Law may in fact be only an apparent hydraulic conductivity and may vary with the difference in head under which the flow takes place or may be a function of the pressure or the time elapsed since the beginning of flow or some other factor.

It is difficult to establish test conditions which will reproduce the circumstances of flow in the field in which we are inerested and all laboratory determinations of permeability suffer from the disadvantage that the peat is not in an undisturbed condition. Since structure is of great importance in the hydraulic properties of peat, the effect of disturbance is more marked for peat than for other soils. Boelter (1965) found that laboratory values of hydraulic conductivity were significantly higher than field values. Some of the results obtained by Boelter are shown in Table 29.

Galvin and Handrahan (1967) determined the permeability of blanket peat by pumping tests and found a value of 6.6 mm/day compared with an average laboratory result of 6.3 mm/day.

Even field measurements of hydraulic conductivity may not give reliable values because the methods used may not correspond to the conditions for which the conductivity value is to be applied. Tests may also be affected by the tendency, which appears in certain experiments, for the permeability to decrease with time after the commencement of the test before reaching a constant value.

Table 29. Hydraulic conductivity (m/day)

Peat	Horizontal		Vertical	
	field	lab	field	lab
Undecomposed	33	130	53	84
Partially decomposed	0.12	1.13	0.43	0.48
Decomposed	0.01	0.12	0.01	0.03

As shown in Table 29, the hydraulic conductivity of peat is very sensitive to the degree of decomposition of the peat and this effect is very marked in the case of the low degree of decomposition. Table 30 which is based on results by Sarasto (1961) shows the relative hydraulic conductivity for Sphagnum peats for various degrees of decomposition on the van Post scale.

Table 30. Variation of conductivity with decomposition (after Sarasto, 1961).

van Post no.	Hydraulic conductivity	
	Horizontal	Vertical
1	100	50
2	60	31
3	6	7
4	3	3
5	2	2
6	0.8	0.8
7	0.5	0.5
8	0.3	0.3

The results obtained by Sarasto for sedge peats are similar, though in that case the relative permeabilities are not so high for the undecomposed condition and somewhat higher for highly decomposed peat.

Dowling (1969) investigated the variation of hydraulic conductivity with types of peat and some average values based on his results are shown in Table 31.

In Table 31 the term 'partially drained peat' is used in reference to areas with open drains 1 m deep at 15 m spacing.

To speak of the hydraulic conductivity of peat without reference to the type of peat or the degree of decomposition is meaningless. Baden and Eggelsmann (1963) review the work done on the hydraulic conductivity of peat up to the year 1962 and report on over 2000 measurements of the hydraulic conductivity carried out by them on mires in north-west Germany. Their analysis also shows a decrease in hydraulic conductivity of one hundred fold for decomposed peat compared to undecomposed peat. Baden and Eggelsmann also plot the hydraulic conductivity against the percentage pore space to demonstrate that the wide variation in conductivity takes place when the peat is in a loose condition

Table 31. Variation of hydraulic conductivity with types of peat.

Peat Type	Degree of Drainage	Hydraulic Conductivity mm/day	
		Augerhole method	Piezometer method
Younger Sphagnum	undrained	303	149
Younger Sphagnum	partially drained	174	84
Older Sphagnum	partially drained	35	27
Reed	partially drained	107	166
Woody fen	partially drained	494	831
Woody fen	partially drained	752	644
Phragmites fen			

and that the permeability is relatively constant when it is in a denser condition.

Ivanov (1957) carried out a notable series of experiments on peat in a filtration flume and demonstrated differences in hydraulic conductivity of the order of a thousand fold at different depths in the peat layer. Ivanov suggested that the experimental data could be fitted by an equation of the type:

$$K(z) = \frac{A}{(z + 1)^m} \qquad (6)$$

where $K(z)$ is the hydraulic conductivity in cm/s and z is the depth from the surface in cm, while a and m are parameters for individual elements of the bog. If the variation of the permeability is expressed in the form of equation (6) or similar equation, it is a relatively easy matter to find by integration the mean hydraulic conductivity for a certain depth of flow. In this way the lateral flow in a peat profile can be estimated for different levels of groundwater above the bottom of the active layer. Romanov (1961) has described the application of this approach and also summarised the work of other Soviet writers on the subject.

Other Properties of Peat: Another important effect of lowering the groundwater level is the settlement of the peat surface due to shrinkage. If the new water table is relatively shallow, the shrinkage may be reversible to a large extent. However, if the lowering of the groundwater table is appreciable and prolonged, then irreversible shrinkage may take place. The changes in the soil which account for this irreversible shrinking have been discussed by Hooghoudt and others. Skoropanov (1961) gives some data in regard to Bielorussia, while Allison (1955) and Stephens (1956) have given the results of measurements on shrinkage of peat soils in the United States.

Prus-Chacinski (1962) has given a method for estimating the shrinkage of peat layer based on the original depth and the depth of the water table as follows:

$$h = aH^{1/3}t^{1/3} \qquad (7)$$

where h is the total shrinkage, H is the original depth of peat, t is the final depth to the water table after shrinkage, and a is a coefficient of shrinkage which depends on the moisture content. This is a modification of a method proposed by Ostromecki (1956) who adapted a formula by Kostiakov and checked it against measurements in Poland.

The thermal properties of peat are also of importance from the hydrological viewpoint since they influence both the heat exchange between the ground and the atmosphere (which may affect the extent of evaporation) and also the occurence of freezing (which may affect the amount of surface runoff). The most important properties with which we are concerned are the specific heat and the thermal conductivity. The coefficient of thermal diffusivity, which is the ratio of the coefficient of thermal conductivity and the specific heat per unit volume, is also worthy of attention not so much for its importance in hydrological processes as for the fact that it is sometimes measured and used to determine the thermal conductivity. These thermal properties depend on the moisture content of the peat and like most other properties are affected by its degree of decomposition. The specific heat per unit volume of peat in situ is largely made up of the specific heat of the moisture contained in it. A number of questions relating to these thermal properties are discussed in Chapter 3 of Romanov (1961). The temperature regime in bogs has been described by Chechkin (1965) and the effects of drainage on the temperature regime have been investigated in Finland by Pessi (1958). Schmeidl (1962) has reported on the temperature difference between the air and the ground surface of bogs in Southern Germany.

While the problem of the chemical quality of water in bogs and fens is not directly relevant to water balance studies, nevertheless it should be mentioned briefly here. The presence of ions of various types and the quantity in which they can occur is a useful indicator of the type of mire element with which we are concerned. The various methods of movement of water through a mire complex are of importance for an understanding of its complete hydrology. Apart from this, water quality is of indirect importance since the carriage of oxygen and of nutrients to various parts of the bog or fen have a considerable influence on the type of dominant vegetation in those parts of the complex. The type of vegetation may in turn affect the overall hydrological behaviour. The relation between water movement and vegetation has been discussed in a number of papers such as those on conditions in Sweden by Sjors (1948 and 1950), in Canada by Sparling (1966) and in Britain by Ingram (1967).

Components of the water balance

Evaporation from Mires: Having reviewed the nature and the basic properties of peat, we now turn to a more direct discussion of the water balance of bogs and fens. In the present section we will be concerned with the individual compenents of the water balance equation and with their relation to one another. Of all the factors in the water balance equation, precipitation is the one for which reliable measurements are most frequently available. In the following discussion it will be assumed that reliable measurements or estimates of precipitation are available. In the case of bogs, this leaves the terms for evaporation, runoff and changes in water storage to be determined. In the case of fens, the inflow term must also be measured or reliably estimated in some fashion.

As mentioned previously, a complete water balance would involve independent and reliable estimation of each of the factors involved which could then be inserted in the water balance equation and a residual obtained which would measure the overall accuracy of the balance. Relatively few such complete water balance investigations have been made and then only for the case of bogs since the inflowing water is very difficult to measure accurately in the case of fens. In most reported estimates of the water balance of bogs, one or more factors have been estimated or indirectly derived from other factors.

The evaporation term in the water balance equation includes vapour transfer to the atmosphere from free water surfaces, from bare soil surfaces and from transpiring vegetation. The measurement of this total evaporation presents a number of difficult problems. The general methods used for determining evaporation are given in standard reference works such as those published by WMO (1965 and 1966). The procedures approved of for swampland stations in the USSR are given in the relative sections of the Manual of Hydrometeorological Observations in Swamplands (1955). The various direct and indirect methods of measuring or estimating evaporation from bogs and the results obtained have been thoroughly discussed in books by Romanov (1962) and Sheneko (1965). A recent WMO report by Hounam (1971) on 'Problems of Evaporation Assessment in the Water Balance' gives an up to date review of the present position of the general problem. In the present paper, however, we will only be concerned with the application of various techniques to the problems of bogs and fens and with the particular results obtained in these cases. The direct measurement of evaporation from a vegetated soil profile can only be made on a small scale in a lysimeter. A lysimeter is a tank set into the ground filled with a volume of soil and covered by vegetation in a condition as representative as possible of the natural state. The use of a lysimeter involves the application of the water balance equation to the element of soil and vegetation within the tank in order to determine the evaporation as a residual term.

Accordingly, if the amount of water contained within the lysimeter varies, the differences must be measured and accounted for in order to get a true estimate of evaporation. This can be done in the case of a weighing lysimeter but it is difficult to construct a balance for weighing a lysimeter of large size. In the case of a drainage lysimeter in which the outflow is measured, either the change in storage must be measured or else the water level in the container maintained at a constant level. The latter arrangement is the more suitable in bogs since evaporation from a bog may vary sharply with the depth of the water table below the bog surface. The construction of two typical lysimeters is illustrated and the procedures for their observation described in detail in the USSR Manual of Hydrometeorological Observations in Swamplands (1955). Bay (1966) took advantage of the very low permeability of the lower layers of peat to use a bottomless caisson to measure evapotranspiration. Lysimeters are difficult to operate and more than one tank should be used as a check on the results. If lysimeters are to be used as the basis of predicting evaporation from a mire complex, then it is desirable to increase the number of lysimeters and to control the water table at different levels below the surface in a number of them.

In several studies, the evaporation measured in lysimeters has been correlated with indirect estimates of evaporation. These indirect estimates are based on such factors as the potential evapotranspiration estimated from climatic data, measured radiation (either total or net), or pan evaporation or some other variable which can be readily measured. This enables us to obtain reliable estimates of the evaporation from a mire complex in areas where direct measurements of evaporation have not been made in the past or where suitable lysimeters would be inconvenient or too expensive to install.

Examples of such indirect estimation of evaporation from bogs are available from a number of countries. Romanov (1962) proposed a linear relationship between evaporation from a bog (E) and the radiation balance of the bog (H) as follows:

$$E = aH + c \qquad (8)$$

where a and c are empirical coefficients. Bavina (1967) applied the method to a number of long term records of bog evaporation, determined the values of the parameters a and c and discussed their variation and their physical significance. Eggelsmann (1963) in Germany and Bay (1966 and 1968) in the United States correlated evaporation from a bog lysimeter with potential evapotranspiration. Both found the result to be highly significant in a statistical sense. Virta (1966) in Finland correlated the bog evaporation from lysimeters with a number of measures of the available energy and obtained high correlations in all cases.

Typical values for evaporation from undrained bogs and undrained fens and from drained bogs for the climatic conditions of the European part of the USSR are given by Romanov (1962). Results of a number of measurements and estimates for evaporation from bogs in western Europe have been given by Eggelsman (1964).

However, evaporation is frequently determined by local conditions and the transfer of measured or estimated values from one area to another may lead to serious error.

The fact that evaporation from the surface of a mire complex depends on the meteorological factors makes it inadvisable to apply a value of evaporation measured at one complex to another in a different climatic region. The fact that the rate of evaporation decreases appreciably when the depth of the water level below the surface is increased may make it inadvisable to transpose a measured value of local evaporation from one part of a complex to another where there is a different depth of water table below the surface.

Outflow from Bogs and Fens: In the case of bogs, we need to know the outflow from the bog in order to complete the water balance equation. In the case of fens, we need to know both the outflow from the fen and the inflow of groundwater (and possibly surface water) to the fen from the surrounding area. Hardly any measurements are available with regard to the inflow to fens and most of the literature is concerned either with a bog (which by definition receives no lateral flow) or with the larger catchment area of which the fen is a central part.

It is not possible in a general discussion to do more than refer to the problem of measuring or predicting the outflow from a mire. To understand or predict the hydrology of a mire completely, it would be necessary to have a good knowledge not only of the average outflow but also exceptionally high flows and exceptionally low flows. The long and extensive series of records of runoff from bogs and fens in the Soviet Union has enabled hydrologists in that country to devise standard methods for the estimation of maximum, average and minimum flows from bogs and fens. These standard procedures have been published in monographs edited by Ivanov (1963 and 1971), and will not be discussed further in this present paper.

The main points of interest in the hydrology of runoff from bogs and fens are the variations in runoff for different types of mire, the effect of topography and micro-relief, differences between the runoff from mires and from neighbouring catchments, the variations in runoff from year to year, the effects of drainage on the runoff, and so on. Some measured data are available on a few of these points for particular catchment areas. Except in the case of the USSR, sufficient information is not yet available to attempt to establish general principles which would give the basis for reliable predictions of outflow behaviour. Such information that is available is scattered throughout the literature.

The following are some examples of the type of studies which have been made outside the USSR. Reference should be made to the actual papers for details of the measuring procedures, the results obtained and the conclusions drawn. Conway and Millar (1960) in Britain measured the runoff from two small catchments of blanket bog for five years and from two other similar catchments for two years. Burke (1968) in Ireland published preliminary results for drained and undrained plots of blanket peat and showed graphs of continuously recorded outflow to illustrate the effect of drainage in smoothing the outflow. In Germany, Eggelsmann (1960) used four different methods to estimate the groundwater outflow from a raised bog and found them to give comparable results. Vidal (1962) also working in Germany, gave comparative results over a number of years for undrained raised bog and reclaimed and cultivated bog. Mustonen (1963) gives results for the effect of drainage on experimental watersheds in Finland. Results of the flow from bogs and swamp areas in various parts of the United States have been reported on by Vecchioli and others (1962), van't Woudt and Nelson (1963), Dingman (1966) and Bay (1969).

Groundwater Levels in Bogs: The measurement of the groundwater level of the various parts of the bog or fen is of the utmost importance in any attempt to study the water balance. The variations of the groundwater table are naturally a measure of changes in groundwater storage but they can also be used as a measure of the soil moisture storage between the surface and the water table. Both the evaporation and the runoff are affected by and may be related to the depth from the surface of the mire to the groundwater table. Thus the level of the groundwater table and its variations play a key role in all of the elements of the water balance equation (except precipitation which is taken as given). Apart from its key role in the hydrology of mires, the water level is also of great importance in regard to the vegetation and general ecology of the mire complex.

A number of studies have been made both from the hydrological and ecological viewpoint of fluctions in groundwater level. Thus Godwin and Bharucha (1931 and 1932) studied the seasonal and diurnal fluctuations in water level in a fen near Cambridge, England. Later Nicholson and others (1951) experimented with the control of the groundwater level in a part of the same fen complex. Burke (1961-2) in Ireland investigated the effect on the groundwater level of open drains 1 m deep spaced at 7.5 m, 15 m and 30 m. Holstener-Jorgensen (1966) discusses the effect of drainage and forest management of groundwater fluctuations in Denmark. Typical of studies reported on in the USSR are those described by Maslow (1961 and 1963) and by Shiskov (1969). Manson and Miller (1955) presented 15 year records of groundwater tables for undrained and drained forested bogs in Minnesota in the USA.

Effect of Groundwater Level on Balance Components: The water table level is closely connected with a number of the elements of the water balance equation. In a number of studies advantage has been taken of these relationships in order to study one element in terms of the other. Heikurainen (1963) has suggested that fluctuations in the groundwater table together with values for the coefficient of storage (determined either in the field or under laboratory conditions) can be used to estimate evapotranspiration in forested peatlands. Novikov (1964), on the other hand, has suggested that the water level in swamps can be determined by using the measured precipitation and estimates of the evaporation from climatological data together with the coefficient of storage. Both methods naturally depend for their verification on pilot studies the results of which are then used for prediction purposes. Ivanov (1957) and other workers in the USSR have studied the dependence of evapotranspiration on the depth of the water table below the surface.

The flow in the active layer of a mire depends on the depth of water in the active layer, the hydraulic conductivity and the inclination of the groundwater table. It seems reasonable to assume that the hydraulic conductivity and the inclination of the water table can both be expressed as functions of the groundwater level and accordingly that the total flow can be expressed as a function of that level so that the flow from a mire element can be expressed as:

$$q(W) = \int_0^W K(z) . \frac{dh}{dx} . z . dz \qquad (9)$$

where W is the level representative of the local groundwater table. If conditions are relatively uniform the runoff from a larger area can be expressed in terms of a representative water table as:

$$R = \Sigma q(W) = R(W) \qquad (10)$$

where W is either a representative or a mean water table level which has been found to give a stable rating curve with the mire outflow. This approach has been used in water balance studies by Ivanov (1959) and Virta (1966).

Considerations of soil physics suggest that for a given water level in a peat profile the amount of moisture stored in the unsaturated zone would be a function of the moisture holding characteristics of the unsaturated layers and of the rate of downward (or upward) movement of water throughout the profile. We would thus expect a fairly close relationship between water table level and moisture storage in the surface layers in the peat. This has been verified by a number of studies in which the moisture content of the peat profile was determined by sampling of the peat and measurement of its moisture content in the laboratory and the statistical correlation of the resulting figures of soil moisture content with the level of the water table. From experiments in Germany, Eggelsmann (1957) obtained a correlation coefficient of 0.895 and Heikurainen and others (1964) obtained a correlation coefficient of 0.95 for experiments in Finland. It is clear, therefore, that the concept of the coefficient of storage can be used to cover the storage of water in both the unsaturated and the saturated zone. In applying this approach to a mire complex we can define the storage coefficient as:

$$s(W) = \frac{dS}{dW} \qquad (11)$$

where s(W) is the storage coefficient expressed as a function of a representative water level W and S is the total amount of water storage in both the unsaturated and saturated zone for the given water level.

The overall water balance

Examples of Water Balance Studies: Water balance studies reported in the literature have been of many different types. Some of them have measured a few elements of the water balance and used empirical relationships in order to estimate the remainder. Others have measured directly or indirectly all of the elements except one and used the equation with the assumption that the data is perfect to derive the unknown element. Only in a few cases has a full water balance study in the strict sense been carried out, i.e. all of the elements measured directly or indirectly and inserted into the equation and the deviation of the result from zero recorded.

There is little point recounting in detail the studies of water balance which have been made. It is anticipated that the papers presented to the present Symposium will represent the latest and most accurate and most complete studies of the water balance of bogs and fens. Accordingly, only a brief mention will be made of a few studies in different countries which represent early attempts at work of this type. The papers cited provide a useful background for the study of the papers written for the present Symposium and for planning of future studies.

Water balance studies of one type or another have been carried out in the USSR over the past twenty years or so. Early studies of the subject were made by Romanov and Ivanov (1956). A number of studies by various workers in different parts of the USSR are summarised in chapter 6 of the book on the Hydrophysics of Bogs by Romanov (1961), and in the two books by Shebeko (1965 and 1970). Figures 48 and 49 taken from Romanov (1961) compare the observed groundwater level for a dry year and a wet year respectively with the groundwater level predicted on the basis of the evaporation, in turn predicted on the basis of the radiation balance, and the use of empirical curves for the variation of runoff and of the storage coefficient with groundwater level. Bavina (1966) has discussed the important question of the water balance of fens which involves the interchange of water between the fen area and the neighbouring mineral soil.

Studies in the USSR lead to the conclusion that the evaporation from bogs is less than that from non-marshy areas in the same climatic region but that the evaporation from fens is greater than that from non-marshy areas. Both the reduction in the case of bogs and the increase in the case of fens seem to be in the range of 5% to 15%. There is a corresponding increase in the runoff from bogs and a corresponding decrease in the runoff from fens. In cases where the runoff is small in comparison with the other elements of the water balance (namely precipitation and evaporation), then the percentage difference in runoff can be quite appreciable. There is less information on fens than on bogs but studies indicate that in wet years the fen receives more water from surrounding area and the underlying soil in April and May than it loses by runoff, but that in July and August the fen loses more water by runoff than it receives by inflow.

A notable series of investigations was carried out on the comparative water balances of an undrained raised bog and a reclaimed and cultivated raised bog near Hamburg in Northern Germany. The most detailed report on these investigations to date was published by Baden and Eggelsmann in 1964. A summary of the material was presented to the 3rd International Peat Congress (Baden and Eggelsmann, 1968), and a further short account was published in 1971, (Eggelsmann, 1971). Baden and Eggelsmann also found that the raised bog investigated by them had a slightly higher annual evaporation than neighbouring catchments free of peat areas, the difference in this case being about 5%. Vidal and Schuch have published over a period of ten years results on the average and extreme runoff and on groundwater levels in undisturbed and cultivated raised bogs in the Chiemsee area near Munich in Southern Germany (Vidal, 1962; Schuch, 1967).

In Britain, Robertson and others (1963) studied the rainfall and runoff for three years on a raised bog near Edinburgh in Scotland, and Chapman (1965) measured the precipitation, runoff, evaporation and water table levels in a raised bog in north east England. In Ireland, Burke (1972) has studied the water balance of blanket peat and is reporting some of the preliminary results to this Symposium. A number of studies of water balance have been made in Scandinavian countries. The publication by Virta (1966) reviews the work prior to 1966 and gives the results of a detailed study of three mires in Finland - a fen in Lapland in Northern Finland, a fen in Bothnia in Central Finland and a raised bog in Southern Finland. From the United States we have the work of Bay (1968a, 1968b, 1969) which discussed the water balance of a number of small bogs in Minnesota.

Changes in Water Balance: Many of the studies of water balance mentioned in the last section were concerned with the effect of the drainage or of cultivation of mires. While the effects of such human activities on the water balance of bogs and fens belongs more properly to third topic of this Symposium, it is as well to refer to them briefly here for the sake of completeness. Because all of the elements of the water balance equation vary from season to season and from year to year, a long record is required to establish the main features of the long term water balance of a mire complex. It is difficult to obtain a record of the required length which will be completely free from the effects of human activity. It is even more difficult to obtain reliable information on the effect of a change in drainage or of ground cover on the water balance of the mire.

Drainage schemes are referred to in the literature in terms of depth of drain, spacing of drains, height of groundwater table above level of drain, the drainage norm (i.e. the lowering of the groundwater level due to drainage), the method of drainage and similar factors. These factors are discussed in books dealing with the theory and practice of draining agricultural land such as Luthin (1957) and in publications which set out drainage standards and the drainage practices in a given country. Prus-Chacinski and Harris (1963) in Britain and Burke in Ireland give in their papers a brief comparison of drainage standards for peat soils in different countries. Guidelines for drainage design for both bogs and fens in Bielorussia are fully described in the manual by Ivitskii (1954), while the effect of drainage and supplementary irrigation on peat soils is discussed by Skoropanov (1961) in the light of results from many parts of the USSR and other European countries. Heikurainen (1964) has reviewed the effect of drainage on forest growth and the subsequent effect of the forest growth itself on the groundwater level. In order to relate the change in groundwater level due to drainage or afforestation to the water balance, it is necessary to determine the connection between groundwater level and the storage in the soil. This may require special studies of the actual storage in drained peat soils similar to the study by Wesseling (1968) which is related to a light clay.

Except for the recent book by Shebeko (1970) there are in the literature few detailed discussions of the full water balance equation for reclaimed soils. Nevertheless some general tendencies may be deduced from the studies which have been made. Studies made in the USSR (Romanov, 1961) indicate that drainage and cultivation increases the annual evaporation from raised bogs but does not appear to affect the annual evaporation from fens. It does, however, tend in the case of fens to redistribute the evaporation over the year, with a decrease of evaporation in April and May and an increase in June, July and August. This results in an increase in the spring discharge from drained fens. The above remarks apply to conditions in an average year. It would appear that in a very wet year the totals in the annual water balance are not affected by drainage but that there is a redistribution of evaporation and consequently of runoff throughout the year, though not to such a marked degree as in an average year. Romanov (1961) suggested that in dry years the runoff from drained bogs is about 100 mm less than from undrained bogs.

In the two areas studied by Baden and Eggelsmann (1964), the effect of drainage was to increase the average depth to the water table from 130 mm to 700 mm and to increase the average summer depth from 290 mm to 920 mm. The authors found that the effect of draining the raised bog and cultivating it as grassland was to increase the evaporation in spring and summer and to decrease it in autumn and winter as indicated in Figure 50. They found that after heavy rainfall or snowmelt the runoff was less extreme from the cultivated bog than from the undrained bog. It is interesting to note that in their studies it was found that the average of the yearly amplitude of the oscillation of the surface of the bog was 37 mm in the cultivated bog compared with 11 mm in the undrained bog. A comparison of the elements of the water balance for the undrained and for the drained and cultivated cases as determined by Baden and Eggelsmann (1964) is shown in Figure 51.

Scope for Future Research: This review of what has been done over the past twenty years in studying the water balance of mires reveals a good deal of information but not enough to constitute a consistent and comprehensive body of knowledge on the subject. Much remains to be done and the execution of this work efficiently and economically will require a good deal of thoughtful planning.

Because of the difficulty of measuring, or otherwise accounting for, the water flowing into a fen from the surrounding area of mineral soil, the water balance of such rheophilous mires presents a particular difficulty. For this reason most of the work which has been carried out to date has been in respect of the ombrophilous mires, i.e. of bogs. This tendency is likely to continue in the future. It is likely that we shall gain a reasonably complete understanding of the water balance of bogs before we do the same for fens. Nevertheless, it would be a mistake to concentrate on the study of bogs to an extent that we neglect the building up of necessary information in regard to fens.

Much of the quantitative information available in published papers suffers from the disadvantage that it is applicable only to the area of study and cannot be interpreted in terms of general principles. On the other hand, the topography of mires and the varying properties of the peat profile are so complex that the application of the principles of mathematical physics to a mire complex is extremely difficult. It appears to the author that a middle road will have to be adopted if real progress is to be made. The complexity of the topography and of the physical properties involved means that empirically derived relationships cannot be avoided. However, to be really useful and to be of significance in comparing one area with another, such empirical relationships must have a sound physical basis. Accordingly, all studies on the water balance of bogs or fens should take into account in some way the physical principles derived in the laboratory and in the field but procedures proposed as the basis of predictions of field conditions should be simple enough for practical application.

For example, we must take note of the fact that properties such as the moisture-holding capacity and the hydraulic conductivity are highly dependant on the degree of decomposition of the peat. Accordingly, no results should be reported in the literature without specifying the degree of decomposition of the peat in all layers of the profile in terms of some widely acknowledged scale. This creates a difficulty because there is not yet complete agreement on the way in which the degree of decomposition of peat should be measured and reported. Should we maintain the van Post scale, or use some measure of specific gravity, or a count of the number of fibres or some other measure? For some time it may be necessary to use and report in terms of a number of measures until one has been validated and generally adopted. As far as the water balance and hydrology of mires is concerned, what we require is a test that is as objective as possible and offers as good a discrimination as possible between peats with different moisture-holding and moisture-transmitting properties but is still convenient for field studies. While such a measure might not be ideal from the point of view of the botanist or the ecologist, it would serve the purpose of hydrological research. Comparing the various measures which have been proposed for the decomposition of peat from this particular viewpoint would be a fruitful field for research.

Again, we have the question of the relationship between the water table and the amount of water stored in the unsaturated zone. Work has been done which indicates that except following heavy rain or prolonged drought, the relationship is very stable. Useful work could be done in studying this relationship and devising ways in which corrections might be made to the relationship when conditions depart from the equilibrium. There is also great scope for research which would link laboratory and field measurements in regard to the relationship between evaporation and the depth of the water table below the surface. Pioneer work along these lines was done by Ivanov over fifteen years ago. What is needed now is fundamental work on the nature of the functional variation under different conditions.

Similarly we need development of a consistent body of knowledge about the flow of water through peat. This would in fact be adaption of standard groundwater hydraulics to the conditions in a peat soil where the permeability is highly variable and the soil liable to shrinkage. The overall parameters which characterise such movement represent an areal value which represents an average in some sense of the local point values determined by the usual field and laboratory tests. The relationships between the two types of parameters is another area for research.

Above all, we need full studies of the water balance equation. Work on the subject will not be complete until we have in respect of all areas of interest water balances in which all terms are measured independently and in which the resulting overall error is clearly indicated. While it is often necessary, due to deficiencies of data, to assume a static water balance and to derive one of the elements of the balance on the assumption that perfect balance exists, such a procedure gives us no notion of the accuracy of our measurements.

The attempt to find a middle path between an approach based entirely on mathematical physics and an approach based entirely on ad-hoc empiricism is likely to lead to the use of conceptual mathematical models in the hydrology of mires. Such an approach has resulted in considerable progress in dealing with other hydrological problems. If such conceptual models are based on reasonably sound physical principles, then the values of the parameters which give the closest fit to the experimental data should themselves have some physical significance. If the latter is the case, then it should be possible to obtain significant statistical correlations between these optimal values of the parameters and the topographical characteristics of the mires and the hydraulic properties of the peat. Such statistical correlations enable us to transpose results from one area to another and from one type of mire to another and from one stage of decomposition to another, and to take account of the effects of human activity on conditions.

Increasing attention has been given in recent years to the effect of human activity on the hydrological cycle as well as on other parts of the human environment (Boughton, 1970; Pereira, 1972). While our main concern is with the physical aspects of the water balance of bogs and fens, we should not neglect the fact that human activity often has unexpected results. In our search for hydrological understanding of agricultural development or forestry improvement, we must remember that a mire represents a special condition of ecological equilibrium and that disturbance to this equilibrium is not merely a question of hydrology but may involve quite complex biological consequences.

In conclusion, it may be said that the past twenty years have represented a herioc age in the investigation of the hydrology of bogs and fens and also that we stand now on the threshold of what may be a golden age of achievement in this particular field. We were lucky in having this Symposium as an occasion on which to take stock of the position, looking backward in satisfaction at what has been achieved and looking forward with eagerness toward the work that has yet to be done.

APPENDIX A

STANDARD NOMENCLATURE FOR MIRES, BOGS AND FENS

(Extracted from TELMA Questionnaire, Unesco, 1967)

A mire complex was described by Cajander in 1913, as 'a series of adjacent mires or different mire parts that are not completely separate but together form a more complex unit'.

The following is a tentative classification based primarily on the hydrology and topography of the mire.

The term Bog is used only for areas which receive exclusively precipitation water (rain falling directly on them) and the term Fen for all other areas. The former are said to be ombrotrophic or ombrophilous: the latter minerotrophic or rheophilous.

In oceanic climates, and also in some areas with high rainfall, the difference may be slight. In some extremely large mires 'minerotrophy' may be acquired by means of contact with subsoil rather than by inflow from the mineral surroundings. 'Minerotrophy' may also be due to decomposition of wind-bourne material.

A. BOG MIRE COMPLEXES

The larger part of the mire or mire complex consisting of areas directly or nearly exclusively fed by precipitation falling on the same areas. Smaller parts may receive other water.

1. Blanket Bog Complexes: Peat covering plateaux, slopes, depressions and valleys alike, up to a considerable degree of slope.

2. Raised Bog Complexes: Surface of the peat largely higher than the margin.

    a. Domed - i.e. convex and usually highest near the centre.
    b. Crested - i.e. sloping sideways from a central crest.
    c. Ridge shaped - i.e. forming a long ridge or hog's back.
    d. Plateau-like.

3. Eccentric Bog Complexes: Containing one or more peat masses that have their highest point or points near the margin and usually slope unilaterally; sometimes fan or saddle-shaped.

    a. Lowland and upland types with moderate relief.
    b. Montaine and sub-alpine type, with stronger relief.

4. Flat Bog Complexes: Bogs with insignificant differences in the level of the peat surface.

    a. Developed on large, almost horizontal surface.
    b. In basins, kettle holes and similar depressions.

B. FEN MIRE COMPLEXES

The whole mire or the larger parts of it receiving some influx of water that has earlier been in contact with mineral soil. The 'minerotrophy' of 5 and 6 is open to question.

1. Horizontal or Nearly So:

    a. Flooded permanently.
    b. Flood plain or lake-side inundation fens, flooded for a long period by river or lake water.
    c. Flooded only for a short period, e.g. at snow thaw, but retaining water for a long period or over the whole year ('retention fen'). May cover flat surface, valley floors or basins.
       N.B. This type of fen is extremely variable, e.g. sedge, scrub, tamarack, alder vegetation, acid to alkaline, pure peat to nearly mineral soil, etc.
    d. Floating fen ('schwingmoor').

2. Aapa Fen: Typically only slightly sloping, often very complex mires with structures such as string-like elevated parts and wet terraces, ('Rimpis', Flarks') forming patterns that may be related to water flow and seasonal soil frost.
   'Mixed mires' mostly belong to this category.

3. Sloping Fens: More appreciably sloping, developed in less flat topography than 1 and 2 and often in areas with a higher rainfall. Fed by seepage rather than a distinctly localised outflow of spring water.

N.B. This type occurs as 'flushes' within blanket bog or together with eccentric raised bog or on slopes beneath spring fen, etc.

4. Spring Fens: May be convex or sloping; formed directly over springs or spring areas.

5. Palsa Mires: With high permanently frozen mounds, the areas between which are only seasonally frozen.

6. Polygon Mires: Permafrosted mire terraces, etc., which are frozen throughout the year.

Types 2, 3, 5 and 6 are regional but the others are widespread.

C. OTHERS

These may be intermediate or special types, or those not covered by the above system.

References

Allison, R. V. 1955. The influence of drainage and cultivation on subsidence of organic soil under conditions of Everglades reclamation. Proc. Soil and Crop Science Society of Florida, Vol. 16. pp. 21-31.

Anon. 1955. Nastavlenie gidrometeorologicheskim stantsiyam i postam, V yp. 8. Gidrometeorologicheskie nabludeniya na bolotakh. Instructions for meteorological stations and posts, No. 8. Manual of hydrometeorological observations in swamplands. Leningrad, Gidrometeoizdat.

Baden, W. and Egglesmann, R. 1963. Zur Durchlassigkeit der Moorboden. On the permeability of marshland. A. Kulturtech Flurberein, vol. 4, pp. 226-254.

Baden, W. and Egglesman, R. 1964. Der Wasserkreislauf eines Nordwestdeutschen Hochmoores. The hydrological cycle of a raised bog in Northwest Germany. Schriftenreihe de Kuratorium fur Kulturbauwesen. vol. 12, pp. 1-155.

Baden, W. and Egglesmann, R. 1968. The hydrological budget of raised bogs in the Atlantic region. 3rd International Peat Congress, Quebec. Proceedings, pp. 206-211.

Bavina, L. G. 1966. Vodnyi balans nizinnykh bolot Polesskoi nizmennosti. Water balance of lowland bogs in the Polesie lowlands. Trudy GGI, issue 135.

Bavina, L. G. 1967. Utochenie raschetnykh parametrov ispareniya s bolot po materialam nabludenii bolotnykh stantsii. Refinement of parameters for calculating evaporation from bogs on the basis of observations at bog stations. Trudy GGI, issue 145, pp. 69-96.

Bay, R. R. 1966. Evaluation of an evapotranspirometer for peat bogs. Water Resources Research, vol. 2, no. 3, pp. 437-442.

Bay, R. R. 1968a. Evapotranspiration from two peatland watersheds.

IASH General Assembly of Berne. IASH Publication No. 78, pp. 300-307.

Bay, R. R. 1968b. The hydrology of several peat deposits in Northern Minnesota, USA. 3rd Internation Peat Congress, Quebec. Proceedings, pp. 212-218.

Bay, R. R. 1969. Runoff from small peatland water sheds. Journal of Hydrology, vol. 9, pp. 90-102.

Bellamy, D. J. 1968. An ecological approach to the classification of European mires. 3rd International Peat Congress, Proceedings, pp. 74-79. Quebec.

Boelter, D. H. 1964. Water storage characteristics of several peats in situ. Soil Sci. Soc. Amer. Proc. vol. 28, pp. 433-435.

Boelter, D. H. 1965. Hydraulic conductivity of peat. Soil Sci. 100. pp. 227-231.

Boelter, D. H. 1968. Important physical properties of peat materials. 3rd International Peat Congress, Quebec. Proceeding. pp. 150-154.

Boughton, W. C. 1970. Effects of land management on quantity and quality of available water. Water Research Laboratory Report No. 120, University of New South Wales.

Burke, W. 1961-1962. Drainage investigation on bogland. The effect of drain spacing on groundwater levels. Irish Journal of Agric. Research 1, pp. 31-34.

Burke, W. 1968. Drainage of blanket peat at Clenamoy. 2nd Int. Peat Congress, Leningrad. Proceedings, vol. 2, pp. 809-817, Edinburgh.

Burke, W. 1972. Aspects of the hydrology of blanket peat in Ireland. Paper to the present Sumposium.

Chechkin, A. S. 1965. Temperaturnii rezhim bolot. The temperature regime of swamps. Trudy GGI issue 126.

Chapman, S. B. 1965. The ecology of Coom Rigg Moss, Northumberland. III. Some water relations of the bog system. J. Ecol., vol. 53, pp. 371-384.

Conway, V. M. and Millar, A. 1960. The hydrology of some small peat covered catchments in the Northern Pennines. J. Inst. Wat. Engrs., vol. 14, pp. 415-424.

Dingman, S. L. 1966. Characteristics of summer runoff from a small watershed in Central Alaska. Water Resources Research, vol. 2, No. 4, pp. 751-754.

Dowling, P. C. 1969. The hydraulic conductivity of peats. Thesis for degree of M. Eng. Sc., University College, Dublin.

Egglesmann, R. 1957. Zur Kenntnis der Zusammenhange zwischen Bodenfeuchte und oberflachennahem Grundwasser. Data on the connection between soil moisture and groundwater level. Die Wasserwirtschaft, 283-287.

Egglesmann, R. 1960. Uber den Unterirdischen Abfluss aus Mooren. On the underground outflow from bogs. Die Wasserwirtschaft, vol. 50, No. 6, pp. 149-154.

Egglesmann, R. 1963. Die potentielle und aktuelle evaporation eines seeklima-hochmoores. The potential and actual evaporation of an Atlanctic raised bog. IASH General Assembly of Berkeley. IASH Pub. No. 62, pp. 88-97.

Egglesmann, R. 1964. Die Verdunstung der Hochmoore und deren hydrographischer Einfluss. The evaporation from raised bogs and their hydrographic influence. Deutsche Gewasserkundliche Mitteilungen, vol. 8, No. 6, pp. 138-147.

Egglesmann, R. 1971. Uber den hydrologischen Einfluss der Moor. On the hydrological influence of a bog. Telma 1, 37-38. Hanover.

Farnham, R. S. and Finney, H. 1965. Classification and properties of organic soil. Adv. Agron., vol. 17, pp. 115-162.

Feustel, I. C. and Byers, H. G. 1930. The physical and chemical characteristics of certain American peat profiles. Tech. Bull. No. 214, US Dept. Agric.

Galvin, L. F. and Hanrahan, E. T. 1967. Steady state drainage in peat. Highway Research Record No. 203, pp. 77-90. National Research Council, Washington.

Godwin, H. 1931. Studies in the ecology of Wicken Fen. I: The groundwater level of the fen. J. Ecol., vol. 19, pp. 449-473.

Godwin, H. 1956. The history of the British flora. Cambridge University Press.

Godwin, H. and Bharucha, F. R. 1932. Studies in the ecology of Wicken Fen. II: The fen water table and its control of plant communities. J. Ecol., vol. 20, pp. 157-191.

Gorham, E. 1957. The development of peatlands. Quart. Rev. Biol., vol. 32, pp. 145-166.

Grebenschikova, A. A. 1956. O vlagoemkosti torfov. The water capacity of peats. Pochnedenie, No. 9, p. 102.

Heikurainen, L. 1963. On using groundwater table fluctuations for measuring evapotranspiration. Acta. Forestalia Fennica, vol. 76, No. 5, pp. 1-16.

Heikurainen, L. 1964. Improvement of forest growth on poorly drained soils. International Review of Forestry Research, vol. 1, pp. 39-78. Academic Press.

Heikurainen, L., Paivanen, J. and Sarasto, J. 1964. Groundwater table and water content in peat soil. Acta forestalia fennica, vol. 77, No. 1, pp. 1-18.

Holstener-Jorgensen, H. 1966. Influence of Forest Management and Drainage on Groundwater Fluctuations. Int. Symp. on Forest Hydrology. Proceedings, Pergamonn Press, pp. 325-333.

Hooghoudt, S. B. *et al.* 1960. Irreversibly drying peat soil in the west of the Netherlands. Versl. Landbook. Onderz, vol. 23, No. 66, 308.

Hounam, C. E. 1971. The problems of the evaporation assessment in the water balance. Reports on WMO/IHD Projects, No. 13, WMO No. 285, Geneva.

IASH/UNESCO. 1970. Reading Symposium on World Water Balance. Proceedings. Volume I and II. IASH Publications No 2, 92 and 93.

Ingram, H. A. P. 1967. Problems of hydrology and plant distribution in mires. J. Ecol., Vol. 55, pp. 711-724.

Ivanov, K. E. 1957. Osnovy gidrologii bolot lesnoi zony. Principles of bog hydrology for the forest zone. Leningrad, Gidrometeoizdat.

Ivanov, K. E. 1963. Gidrologicheskie raschety pri osushenii bolot i zabolochennykh zemel. Hydrological calculations in connection with the drainage of bogs and marsh-ridden lands. Leningrad, Gisrometeoizdat.

Ivanov, K. E. 1965. Teoriya gidromorfologicheskikh svyazei v bolotnykh massivakh. Fundamentals of the theory of swamp morphology and hydromorphological relationships. Trudy GGI, issue 126, pp. 5-47.

Ivanov, K. E. *et al.* 1971. Ukasaniya po raschetam stoka s neosushennykh i osushennykh verkhovykh bolot. Directions for the computation of runoff from drained and undrained highland bogs. Leningrad, Gidrometeoizdat.

Ivitskii, A. I. 1954. Printsipy proektirovaniya zakrytogo arenazha v Belorusskoi SSR. Principles of underdrainage design in the Belorussian SSSR. Minsk, Ozd. Akad. Nauk BSSR.

Kaila, A. 1956. Determination of the degree of humification of peat samples. J. Sci. Agr. Soc. Finland. Vol. 28, pp. 18-25.

Kulcznyski, S. 1949. Peat bogs of Polesie. Mem. Acad. Cracovie Classe Sci. math. et nat. Ser. B, no. 15, pp. 1-356.

Luthin, J. N. 1957. Drainage of agricultural lands. Amer. Soc. Agronomy, Monograph no. 7, Madison, Wisconsin.

Manson, P. W., Miller, D. G. 1955. Groundwater fluctuations in certain open and forested bogs in Northern Minnesota. Minn. Agric. Exp. Sta. Techn. Bull., vol. 217, pp. 3-29.

Maslov, B. S. 1961. Vodnyi rezhim torfyanykh pochv v letnii period v usloviyakh Meshcherskoi nizmennosti. Moisture regime of peat soils in summer in the Mescher lowland. Pochvovedenie, no. 3, pp. 48-59.

Maslov, B. S. 1963. Vodnyi rezhim torfyanoi pochvy v vesennii period. Water regime of a peat soil in Spring. Pochvovedenie, no. 10, pp. 73-82.

Mustonen, S. 1963. On effect of drainage on hydrology of peatland. Rakennustenkiikka, No. 3, pp. 204-209.

Nicholson, H. H., Al Derman, G. and Firth, D. H. 1951. An experiment in the control of the groundwater level in a fen peat soil. J. Agric. Sci., vol. 41, pp. 149-162.

Novikov, S. M. 1964. Raschet urovennykh rezhimov verkhovykh bolot po meteorologicheskim dannym. Computation of the water level regime of undrained upland swamps from meteorological data. Trudi GGI, issue 112, pp. 5-32.

Ostromecki, J. 1956. Projektowanie profilu podluznego rowow i drenow w torfowiskach z uwz glednieniem osidiana. Design of ditches and drains in peat, with reference to shrinkage. Roczniki nauk rolniczych, 71-F-3, pp. 627-671, Warsaw.

Peerklamp, P. K. and Boekel, P. 1960. Moisture retention by soils. Versl. Meded, Comm. Hydrol. Onderz. TNO. The Hague.

Pereira, H. C. 1972. Influence of man on the hydrological cycle: guidelines to policies for the safe development of land and water resources. Status and trends of research in hydrology 1965/74. IHD Studies and reports in hydrology, No. 10, UNESCO, Paris.

Pessi, Y. 1958. On the influence of bog draining upon the thermal conductions in the soil and in the air near the ground. Acta. Agric. Scand., vol. 8, pp. 359-374.

Pons, L. J. 1960. Soil genesis and classification of reclaimed peat soils in connection with initial soil formation. Trans. 7th International Congress, Soil Sci., vol. 4, pp. 205-211.

Proceedings of 1st International Peat Congress, Dublin, Ireland. July 1954. Published by Bord na Mona, Dublin.

Proceedings of 2nd International Peat Congress, Leningrad, USSR 1963. Published in 2 volumes by HMSO, Edinburgh.

Proceedings of 3rd International Peat Congress, Quebec, Canada, Ugust 1968. Published by National Research Council of Canada.

Prus-Chacinski, T. M. 1962. Shrinkage of peatlands due to drainage opererations. J. Inst. W. Engrs., vol. 16, pp. 436-448.

Prus-Chacinski, T. M. and Harris, W. 1963. Standards for lowland drainage and flood alleviation and drainage of peatlands with special reference to Crossesns scheme. Proc. I.C.E., vol. 24, pp. 177-206.

Robertson, R. A., Nicholson, I. A. and Hughes, R. 1963. Runoff studies on a peat catchment. 2nd International Peat Congress, Leningrad, 1963. Proc., vol. 1, pp. 161-166. Edinburgh, 1968.

Romanov, V. V. 1956. Gidrofizicheskie metody rascheta vodnogo balansa bolot. Hydrophysical methods of calculating the water balance of bogs. Pochvovedenie, vol. 8, pp. 49-56.

Romanov, V. V. 1961. Gidrofizika bolot. Hydrophysics of bogs. Leningrad, Gidrometeoizdat.

Romanov, V. V. 1962. Isparenie s bolot Evropeiskoi territorii SSSR. Evaporation from bogs in the European Territory of the USSR. Leningrad, Gidrometeoizdat.

Romanov, V. V. and Ivanov, K. E. 1956. Vodnyi Balans dolinnykh bolot Poles'ya i ego bozmozhnoe preobrazovanie. Water balance of valley bogs in Polesie and its possible transformation. Trudy

Konferentsii po melioratsii i osvoennyi bolotnykh i zabolotchennykh pochv. Minsk, Izd. Akad. nauk BSSR.

Sarasto, J. 1961. Tutkimusksia Rahka - Ja Saratuspeiden Vedenlapaisevyydesta. (A study of the permeability to water of different types of peat). Suo. vol. 14, pp. 32-35.

Schuch, 1967. Ergebnisse Abfluss- und Grundwasserbeobachtungen auf einer unberuhrten bzw, kultivierten Hochmoorflache in den sudlichen Chiemseemooren im Abflussjahr 1965. Comparative results of observations on water discharge and groundwater on virgin and cultivated bogs in the southern mires of the Chiemsee in 1965. Bayer Land, Jb. vol. 47, pp. 351-364.

Shebeko, V. F. 1965. Isparenie s bolot: balans pochvennoi vlagi. Evaporation from bogs and the balance of soil moisture. Minsk, Urozhai.

Shebeko, V. F. 1970. Gidrologicheskii rezhim osushaemykh territorii. Hydrological regime of reclaimed area. Minsk, Urozhai.

Schmeidi, H. 1962. Kleinklimatische Vergleiche in Moorgebieten. Wetter und Leben, No. 14, pp. 77-82.

Shishkov, K. 1969. Water movement and use of peat soils in lowland swamps of the USSR. Wageningen Symposium on Water in the Unsaturated Zone. Proc. vol. 2, pp. 682-693. UNESCO/IASH. Paris.

Sjors, H. 1948. Mire vegetation in Bergsladen. Acta phytogeogr, suec. 21, pp. 277-290.

Sjors, H. 1950. On the relation between vegetation and electrolytes in north Swedish mire waters. Oikos. 2, 241.

Skoropanov, S. G. 1961. Osvoenie i ispolzovanie torfyanobolotnykh pochv. Reclamation and cultivation of peat bog soils. Minsk.

Sparling, J. H. 1966. Studies on the relationship between water movement and water chemistry in mires. Canadian Journal of Botany, vol. 44, p. 747.

Standards Committee of German Federal Republic. Peat for Horticulture and Agriculture. DIN 11542.

Stephens, G. 1956. Can we save our organic soils? J. of Soil Conservation , vol. 22, No. 3.

UNESCO. 1970. Testbook on Hydrology. IHD Technical Paper in Hydrology No. 6, Paris.

van Heuveln, B., Jongerius, A. and Pons, L. J. 1960. Soil formation in organic soils. 7th Int. Congress of Soil Science. pp. 195-204. Madison, Wisc. USA.

van Post, L. 1924. Das genitische System der organoganen Buildungen Schwedens. The classification system of organogenic formations in Sweden. Inter. Soil Sci. Congress, Helsingfars. Proceedings. pp. 287-304.

van't Wouldt, B. D. and Nelson, R. E. 1963. Hydrology of the Alakai swamp. Kauai, Hawaii. Hawaii Agric. Exp. Sta. Bull. vol. 132.

Vecchioli, J., Gill, H. E. and Lang, S. M. 1962. Hydrological role of the great swamp and other marshlands in the Upper Passaic River Basin. AWWA Journal, vol. 54.

Vidal, H. 1962. Ergebnisse Vergleichender Abfluss- und Grundwasserbeobachtungen auf einer unberuhrten bzw. kultivierten Hochmoorflachen in den sudlichen Chiemseemooren im Abflussjahr, 1961. Comparative results of observations on water discharge and groundwater on virgin and cultivated raised bogs in the southern mires of the Chiemsee in 1961. Bayer, Landw. Jb, vol. 39, pp. 819-835.

Virta, J. 1966. Measurement of evapotranspiration and computation of water budget in treeless peatlands in the natural state. Commentationes Physico-mathematicae, vol. 32, No. 11, pp. 1-70.

Vorobiev, P. L. 1963. Issledovanie vodootdachi nizinnykh bolot Zapadnoi Sibiri. Investigations of water yield of low lying swamps of western Siberia. Trudy GGI, issue 105.

Wesseling, J. 1968. The relation between rainfall, drainage discharge and the depth of water table in tile-drained land. Netherland Journal of Agricultural Science, vol. 6, No. 1.

WMO. 1965. Guide to Hydrometeorological Practices. WMO Publication No. 168. T.P., 82, Geneva.

WMO. 1966. Measurement and Estimation of Evaporation and Evapotranspiration. Technical Note No. 83, Geneva.

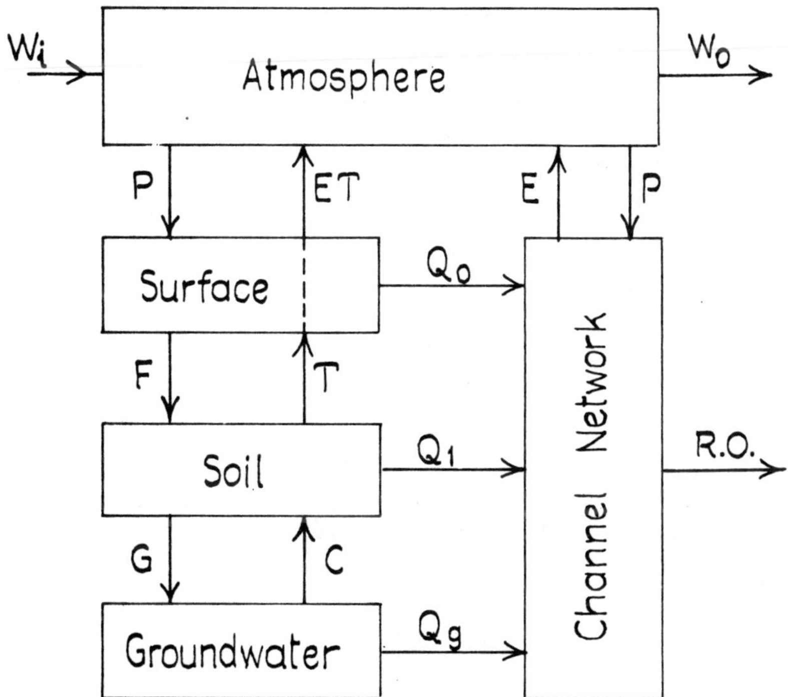

Fig. 46. Hydrological cycle for catchments.

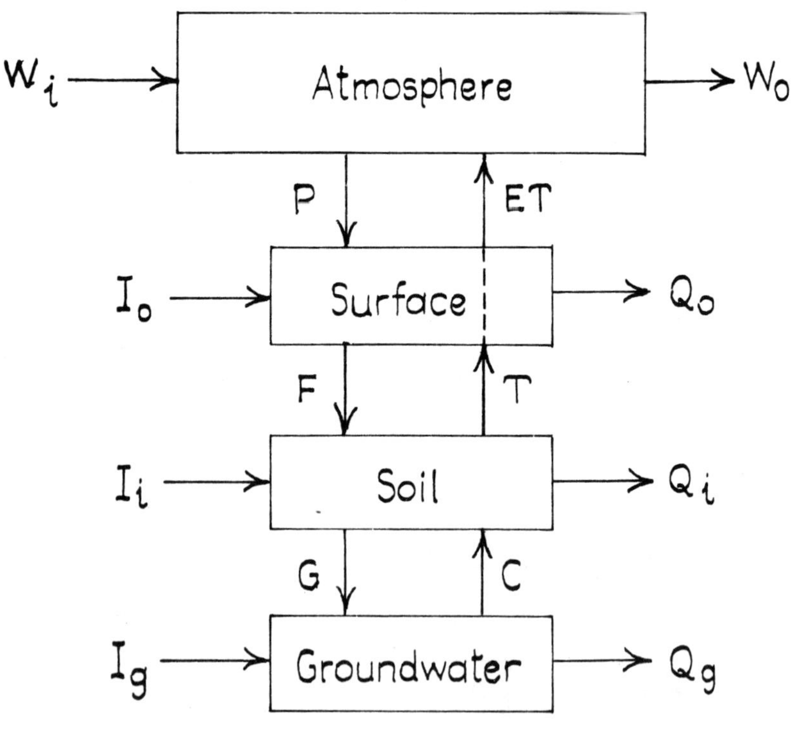

Fig 47. Hydrological cycle for element of catchment

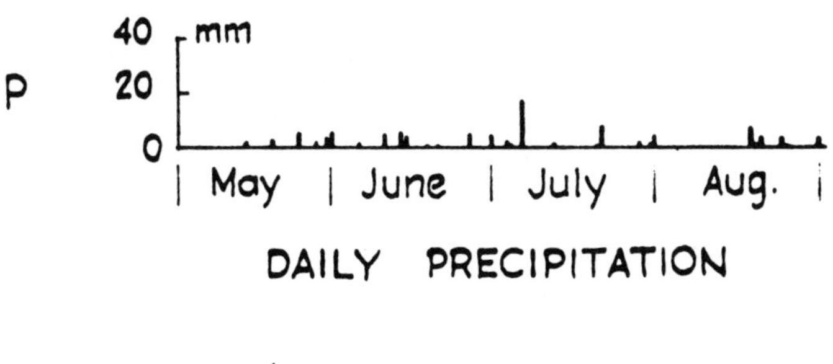

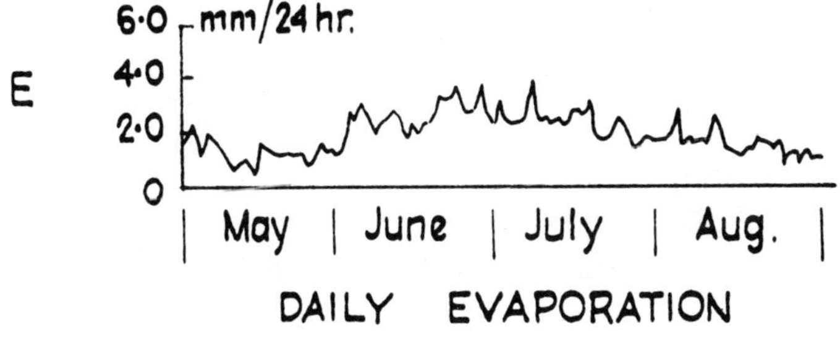

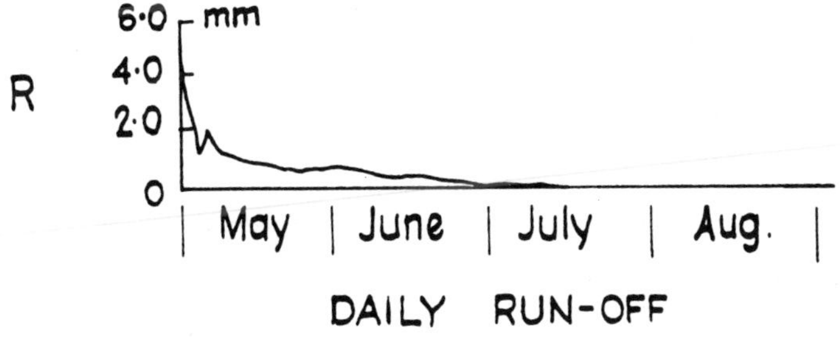

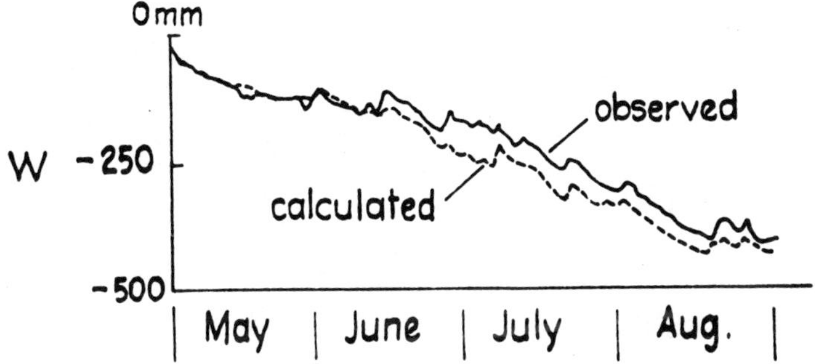

Fig 48. Observed and calculated water table levels for 1950 (dry year)

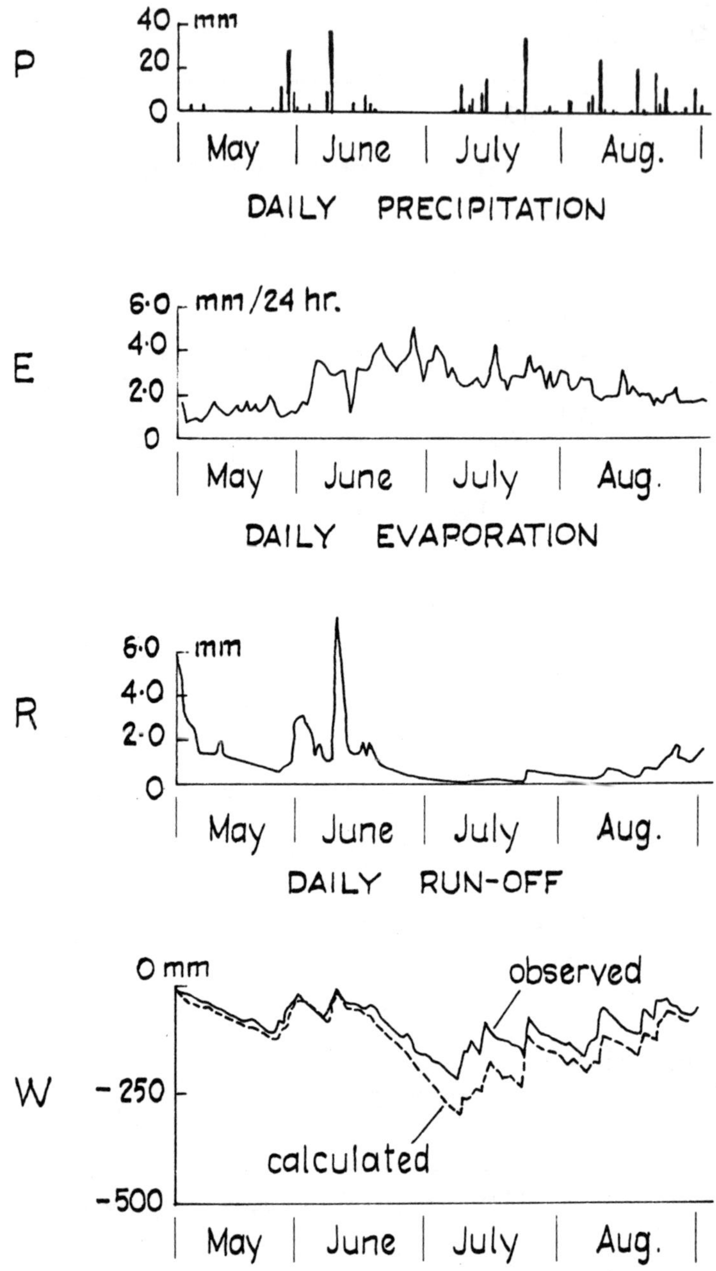

Fig 49. Observed and calculated water table levels for 1953 (wet year)

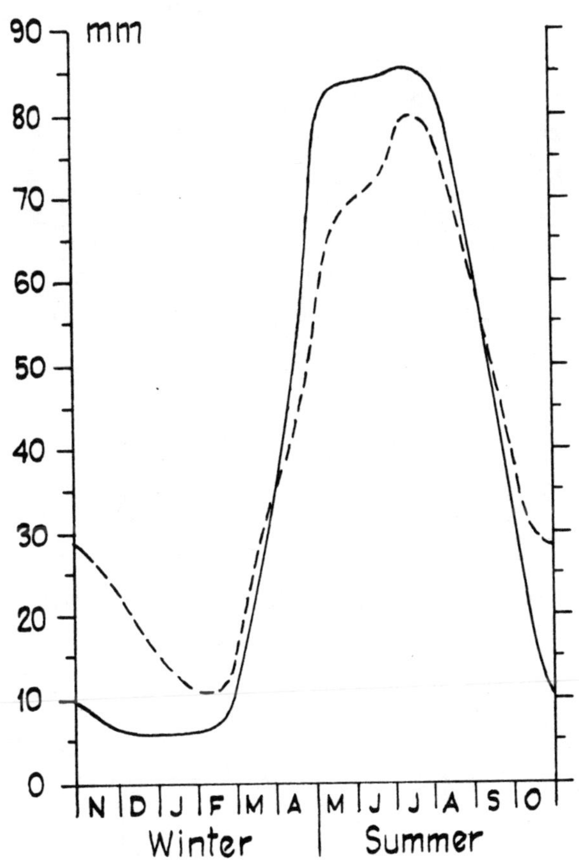

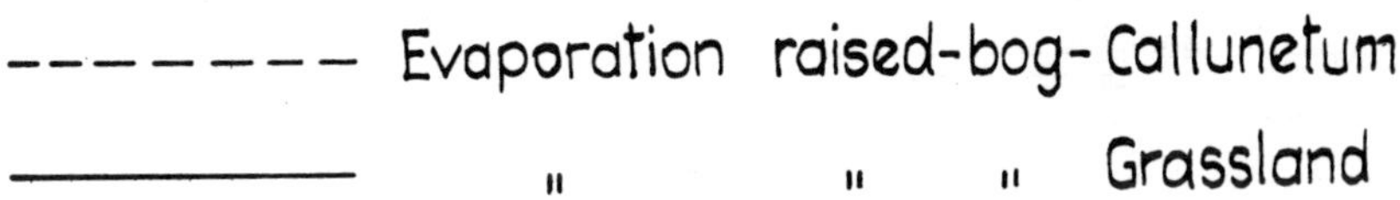

Fig 50. Yearly motion of the evaporation of the waste, heather-covered raised bog and the raised bog grassland.

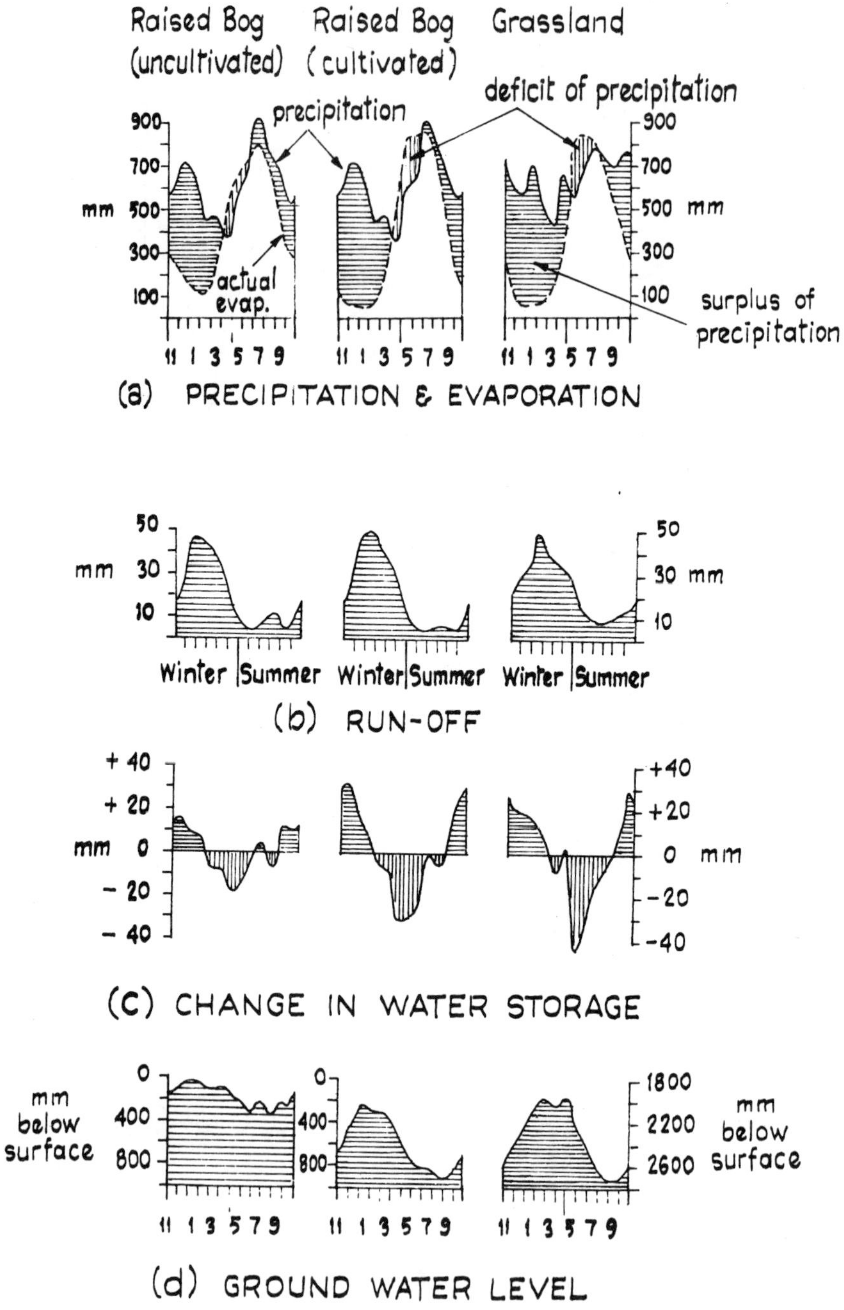

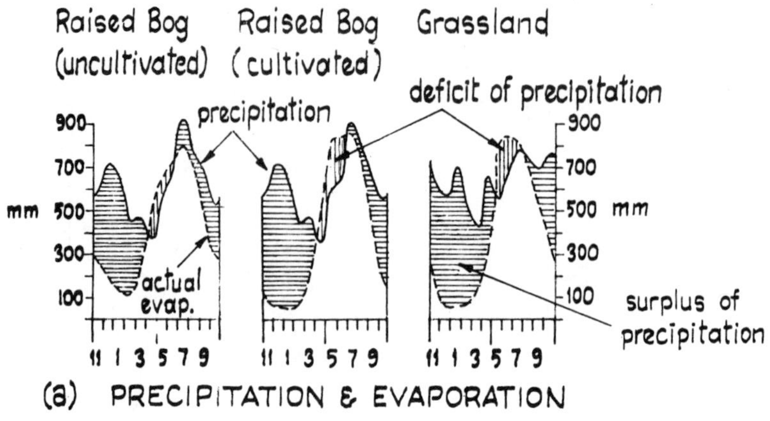

Fig 51. Example of water balance (Baden and Eggelsmann).

DISCUSSION

K. P. Lundin (BSSR) - Have you studied the capillary zone, particularly that of peat soil?

J. Dooge (Ireland) - I would like to say first that division into capillary and non-capillary layers is artificial; when studying suction processes we deal with a compound, three-layer material - solid, liquid and gaseous. Therefore, we cannot divide the layer or pores into capillary and non-capillary. The relation between the moisture content and hydraulic conductivity must be constant.

V.F. Shebeko (BSSR) - A physical approach to pore distribution in a peat soil allows us to distinguish a capillary zone. A reclamation approach to capillary flow in soils, however, reveals the existence of a zone in which capillary suction sharply decreases and the moisture conductivity coefficient sharply changes. Therefore, our point of view is different. I think that any soil may be divided into capillary and non-capillary layers. Do you agree with this view?

D. Dooge (Ireland) - I don't think this division is reasonable. I think this is a matter of convenience rather than of a physical approach.

THE EFFECT OF INITIAL MOISTURE CONTENT ON INFILTRATION INTO PEAT

R. Keane and J. Dooge

University College Dublin,
Dublin, Ireland.

SUMMARY: Laboratory data on the moisture characteristics and the unsaturated permeability of two types of peat are used to determine the appropriate values of the parameters in Philip's general equation for infiltration. A computer program which utilises the data to predict the moisture profile and the quantity of infiltration at specified times is described. A comparison is shown of the amount of infiltration during the first hour for a range of initial moisture contents. The predicted infiltration in the first hour is seen to be a maximum for an initial moisture content of about 85% on a volumetric basis.

L'EFFET DE L'HUMIDITE INITIALE SUR L'INFILTRATION DANS LA TOURBE

RESUME : Cette étude utilise les données obtenues en laboratoire sur les caractéristiques de l'humidité et sur la perméabilité de deux types de tourbe pour déterminer les valeurs des paramètres de l'équation générale de Philip pour l'infiltration. Le traitement par ordinateur des informations ainsi obtenues permet de calculer le profil d'humidité et le taux d'infiltration. L'auteur procède à une étude comparée du degré d'infiltration pendant la première heure pour diverses teneurs initiales en eau. Pendant la première heure l'infiltration calculée est maximale pour un volume d'eau initial de 85 %.

Introduction

In the course of a study on hydrology of peat carried out at the University College, Cork, Ireland (Keane, 1972), the results from some laboratory investigations were used as the basis of a theoretical study of infiltration into peat. Predictions were made of the shape of the moisture profiles and the amount of infiltration for various initial values of the moisture content.

PROPERTIES OF PEAT SAMPLES

The peat samples used in the study were taken from two areas: (A) from a raised bog at Lullymore in Co. Kildare in the midlands and (B) from a highland blanket bog at Nad in Co. Cork near the south coast. Because of limitations of space, only the results for the Lullymore peat are reported in detail in this paper.

The moisture characteristic curves (i.e. the relationship between soil suction and moisture content) were derived firstly on a dry weight basis and then transformed to a volumetric basis. In order to cover as wide a range of soil suction as possible, use was made of four types of apparatus. Suction plates were used for the pF range from 0 to 2.5, an oedometer was used over the pF range from 2 to 4, pressure membranes were used over the pF range from 3 to 4 and vacuum desiccators were used over a range from 4.5 to 7. Figure 52 shows the drying curve and four typical rewetting curves for Lullymore peat.

The measurement of the unsaturated permeability of the peat samples gave rise to difficulty. The values in the present study were obtained by passing water through samples which were being consolidated at an extremely slow speed in a triaxial testing apparatus. The main purpose of these tests was to investigate this new technique for the measurement of unsaturated permeability. The results have not been confirmed by an independent method but the values obtained seem reasonable. The results for a sample of Lullymore peat consisting largely of young Sphagnum is shown in Figure 53.

INFILTRATION EQUATION USED

Most numerical solutions of the infiltration problem use the diffusion formulation:

$$\frac{\partial m}{\partial t} = \frac{\partial}{\partial z}\left(D\frac{\partial m}{\partial z}\right) + \frac{\partial k}{\partial z} \quad (1)$$

where m is the moisture content, z is the elevation, D is the hydraulic diffusivity and k is the permeability, (Bear, *et al*, 1968). This equation can be reformulated in terms of the rate of penetration of the moisture profile as

$$-\frac{\partial z}{\partial t} = \frac{\partial}{\partial m} \cdot \left(D\frac{\partial m}{\partial z}\right) + \frac{\partial k}{\partial m} \quad (2)$$

Philip (1957) solved the problem of ponded infiltration into a semi-infinite column of soil at a given constant initial moisture content by expressing the solution in the form:

$$-z = at^{1/2} + bt + ct^{3/2} + dt^{2} + \ldots\ldots\ldots \quad (3)$$

which he showed to be a valid and convergent solution except at high values of t. The assumption of a constant initial moisture content throughout the profile implies an initial rate of infiltration equal to the unsaturated permeability (K_o) corresponding to the initial moisture content.

A series of algebraic manipulations can be used to get a series of equations which can be solved for the values of the parameter a, b, c, d, etc., in terms of the hydraulic diffusivity (D) and the permeability (K) and their variation with moisture content (m). For a given initial moisture content, the volume of infiltration is given by

$$F = At^{1/2} + (B + K_o)t + Ct^{3/2} + Dt^2 \; .... \qquad (4)$$

where K_o is the permeability corresponding to the initial moisture content and where the parameters (A, B, C, etc.) are obtained by integrating the values of the corresponding parameters in equation (3) with respect to moisture content over the range from the initial moisture content to saturation.

COMPUTER PROGRAM

A computer program was written in Fortran IV for the IBM 1130 computer to solve the above formulation of Philip for a given initial moisture content and for given variations of soil suction with moisture content and given variations of permeability with moisture content. For a given set of input data, the program computes and prints out the values of the parameters a, b, c and d in equation (3) at specified moisture contents, plots on a single graph the moisture profile at initially specified time values, plots the total quantity of water infiltrated at specified intervales during the first hour, prints out in tabular form the total quantity of water infiltrated and the rate of infiltration at each minute during the first hour of infiltration.

The computer program was run for some of the data obtained in the laboratory on the peat samples from Lullymore and Nad. The rewetting curves were run for both types of peat from initial pF values of 4, 3.5, 3.0, 2.5, 2.0 and 1.5 to saturation. The time chosen as the basis for the curves of a, b, c and d versus moisture content was one hour. The number of data points used to describe the moisture properties of the peat vary from 16 to 60 depending on the nature of the moisture characteristic and permeability curves. The volume of the results is too large to be presented in a paper of this type but a selection of typical results for the Lullymore data are described below.

PREDICTED INFILTRATION INTO DRAINED PEAT

Figure 54 shows the values of the parameters in equation (3) for ponded infiltration into Lullymore peat with an initial moisture content of 19% on a volumetric basis (i.e. a pF of 4.0). For this low moisture content, the values of a and b predominate, while the values of c and d are completely insignificant and cannot be plotted to the same scale. For a higher initial moisture content (i.e. a lower initial pF value) the c and d curves become significant. For some higher initial moisture contents the influence of the c and d terms render the moisture profile unstable at times greater than about 3 hours. This indicates that the Philip equation in its full form is no longer valid for infiltration at these times into Lullymore peat at these high initial moisture contents. The moisture profile following infiltration for three hours is shown in Figure 55 for four initial volumetric moisture contents, namely 19% (pF = 4.0), 53% (pF = 3.0), 71% (pF = 2.5) and 84% (pF = 2.0). These cases correspond to the four wetting curves shown on Figure 52. It can be seen that only in the case of the very low initial moisture content does the water profile resemble the type of profile which is usually accepted as typical for non-organic soils.

EFFECT OF INITIAL MOISTURE CONTENT

From a hydrological viewpoint, the main interest is the quantity of water which will be infiltrated. Figure 56 is based on the computer outputs and shows the predicted quantity of water infiltrated into completely drained Lullymore peat in the first hour as a function of the initial moisture content. The calculation takes account of the infiltration corresponding to the initial unsaturated permeability. The relationship is similar for other durations. For example, the relationship between the amount infiltrated during the first minute and the initial moisture content shows a curve of similar shape with the amount of infiltration being about 10% of that shown on Figure 56. The maximum amount of predicted infiltration is seen to correspond to an initial moisture content of about 85%. This corresponds to an initial pF value of 2.0. For higher initial moisture contents, the infiltration is less because of the lower value of soil suction. For lower moisture contents the infiltration is less because of the sharp falloff in permeability.

For the case of the peat samples from Nad in Co. Cork a similar result was obtained and this is shown in Figure 55. The optimum initial moisture content for maximum infiltration was again found to be about 85% (pF = 2.0) and the curve has the same general shape.

REFERENCES

Bear, J., Zaslavsky, D. and Irmay, S. 1968. Physical principles of water percolation and seepage. Arid Zone Research XXIX. UNESCO.

Keane, R. 1972. The hydrology of peat. Thesis for degree of Master of Engineering Science. University College, Cork, Ireland.

Philip, J. R. 1957. Numerical solution of equations of the diffusion type with diffusivity concentration dependent. Australian Journal of Physics, 10.

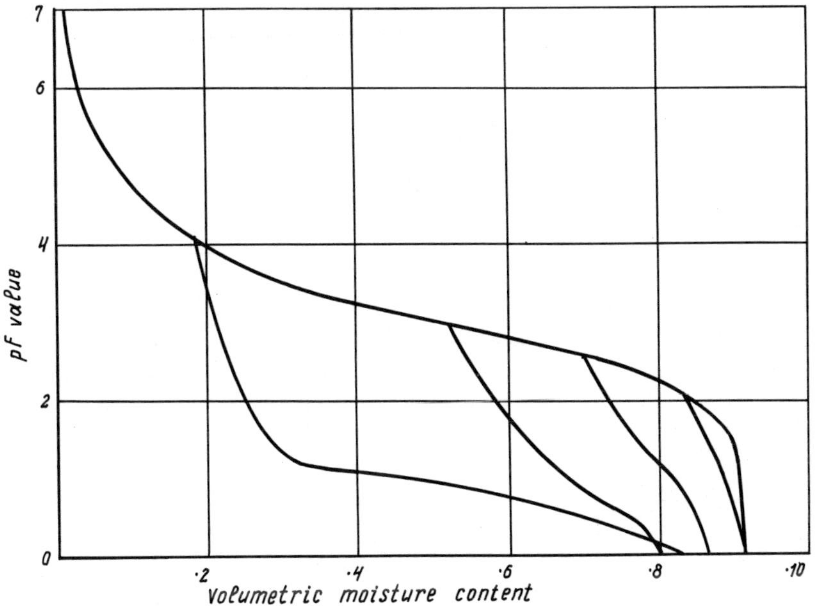

Fig 52. Moisture curves for Lullymore peat.

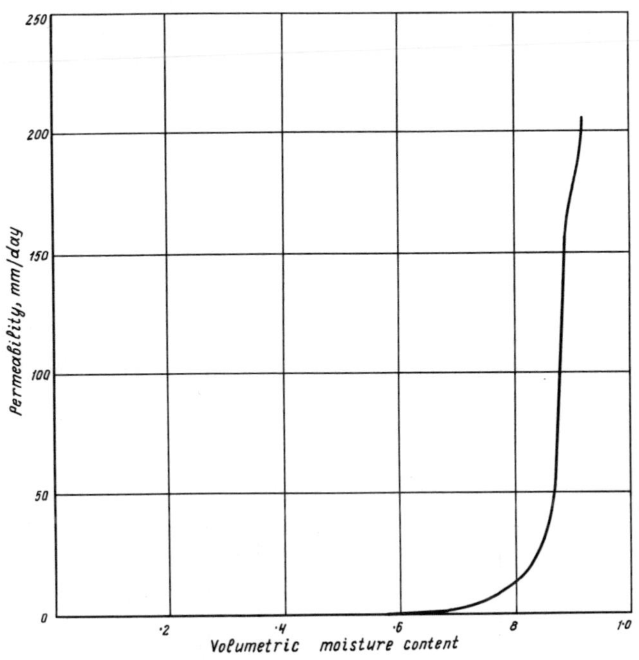

Fig 53. Permeability of Lullymore peat.

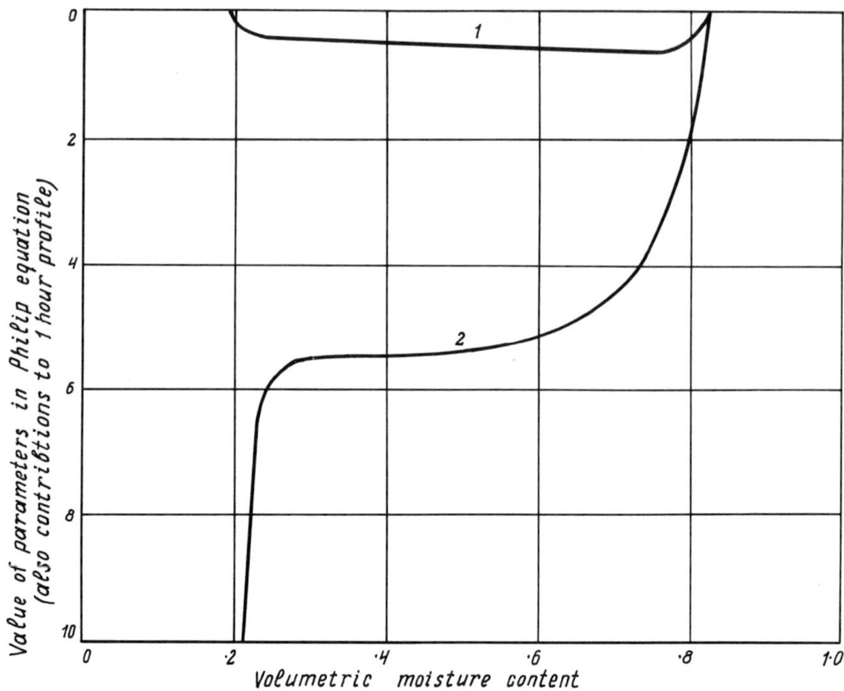

Fig 54. Parameters in Philip equation for Lullymore peat (initial pF = 4.0).

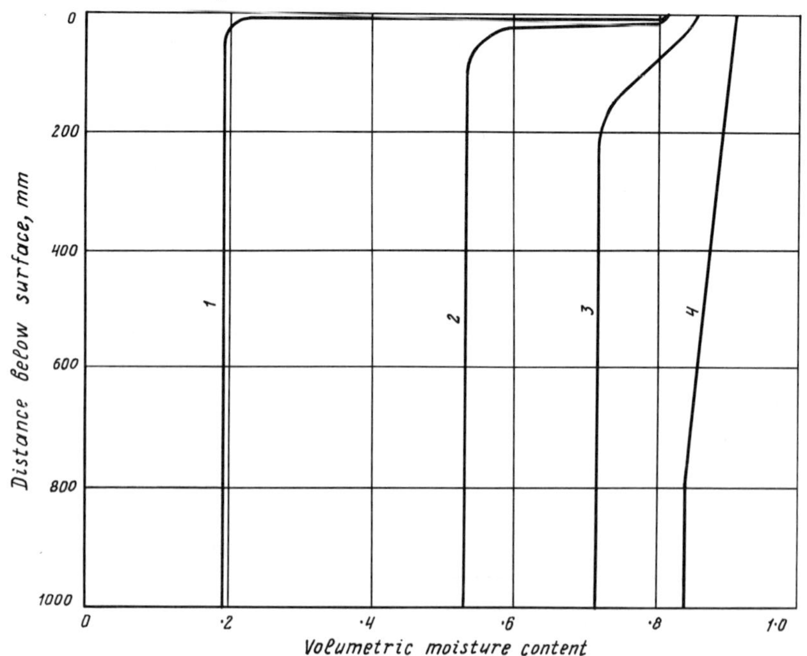

Fig 55. Predicted profiles for infiltration into Lullymore peat.

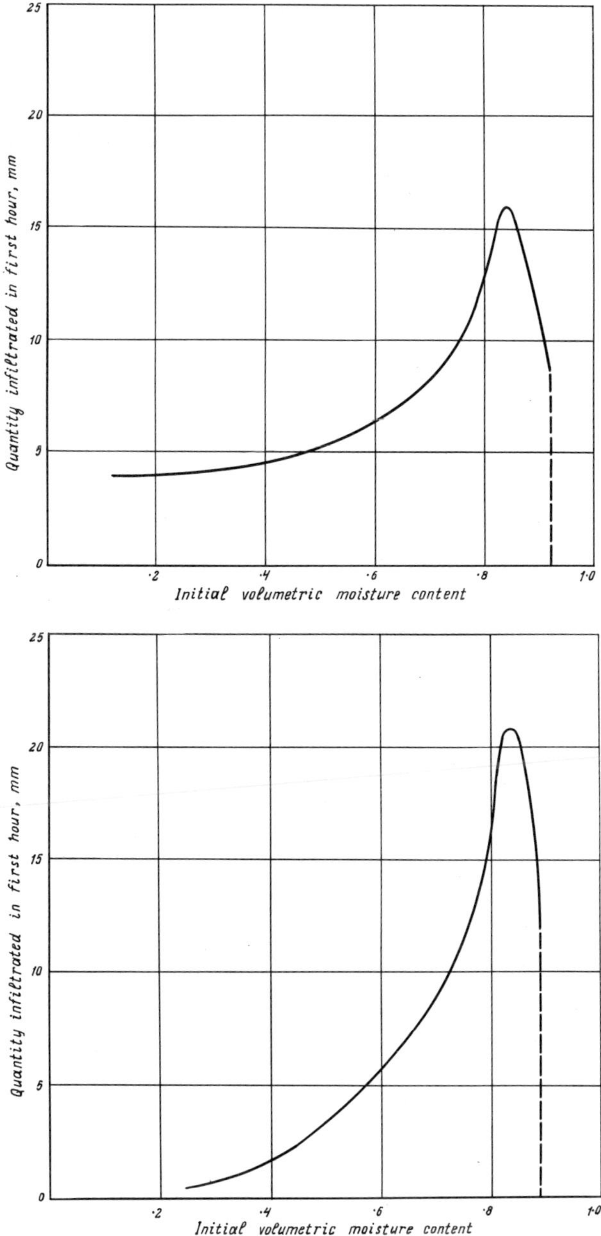

Fig. 56 Predicted infiltration for various initial moisture contents, into

(a) Lullymore peat (upper diagram)
(b) Nad peat (lower diagram)

DISTRIBUTION OF ANNAUL NATURAL STREAMFLOW IN THE POLESSIE LOWLAND

I. M. Livshits

Byelorussian Polytechnical Institute, Minsk, BSSR.

SUMMARY: An analysis is made of the existing and proposed characteristics of the annual streamflow distribution in the rivers of the Polessie Lowland. The correlations are examined and regression equations are obtained which relate to various characteristics of the streamflow distribution of Polessie and adjacent regions.

The characteristics suggested by the author are examined. Based on the analysis of the effect of the physiographic factors of Polessie on the normal annual values and variability of the distribution characteristics, formulae are presented for the indirect determination of the parameters of their finite distribution curves.

REGIME NATUREL ANNUEL DES RIVIERES DE LA PLAINE DE POLESIE

RESUME : L'auteur présente une analyse des caractéristiques connues et des caractéristiques supposées du débit naturel annuel des rivières de la basse plaine de Polésie. Il étudie les indices de corrélation et établit les équations de régression se rapportant aux diverses caractéristiques du débit des rivières de la Polésie et des régions limitrophes.

L'auteur montre que les valeurs caractéristiques du débit telles qu'il les propose sont utilisables et raisonnables. En partant de l'analyse de l'influence des facteurs physiques et géographiques de la Polésie sur les valeurs normales du débit sur les variations des caractéristiques, il établit des formules pour la détermination indirecte des paramètres des courbes de distribution limitées de ces caractéristiques.

The choice of principles and methods to change the hydrological regime of swamp lands depends essentially on the annual natural streamflow regulation of the individual units.

Annual streamflow variations which primarily depend on the processes due to the earth's revolution around the sun, are characterized in certain physiographic regions by appropriate regular phases have different patterns of internal streamflow distribution.

The streamflow distribution depends on the basin area, climatic conditions,number of lakes, swampiness, the river bed and flood land control, vegetation and other natural factors.

The effects of these factors on the streamflow distribution have been studied by D.L. Sokolovskii (1968), V.G. Andriyanov (1960), Yu.P. Burneikis and B.V. Gailyushis (1965), I.N. Livshits (1970) and some other workers.

In projects concerned with the simultaneous utilization of water resources of the Polessie region such as streamflow control by storage reservoirs, interbasin streamflow distribution, runoff utilization under natural conditions, etc., estimation of the natural streamflow distribution is of great importance.

The shape of the models of individual phases of the streamflow in Polessie is well defined but they are found to begin irregularily in different years, their duration differes in different years and constituents of individual phases are variable.

Annual natural flow distribution of the Pripyat, the main river in Polessie, is also considerably affected by its east-west orientation and asynchronous inflows of flood waves from the north and south effluents.

For every hydrological station, annual deviations from the 'standard model of the annual streamflow behaviour are random because of many interconnected factors which affect the process. Thus, variations of streamflow distribution characteristics over many years may be described as a stochastic process.

Most of the physiographic factors which affect the annual streamflow distribution have a similar effect on normal annual characteristics.

ANALYSIS OF THE EXISTING METHODS OF ESTIMATION OF ANNUAL STREAMFLOW DISTRIBUTION

The available methods for the estimation of annual streamflow distribution may be roughly divided into two groups. The first is based on the ratios of the discharge parameters, the second is based on the ratios of individual areas of the discharge hydrograph or the frequency curve.

The following characteristic ordinate ratios are suggested in the hydrological literature (the first group):

$$\frac{\bar{Q}a}{Qmin},\ \frac{Qmax}{Qmin},\ \frac{Q180}{Qmin},\ \frac{Q58}{Q308},\ \frac{Q90}{Q270},\ \frac{Qmax}{\bar{Q}a},\ \frac{\bar{Q}a-Qmin}{Qmax-Qmin},\ \frac{Qmax}{Q180},\ \frac{\bar{Q}a}{\bar{Q}},\ p(\bar{Q}a),$$

where: Q_{max}, Qmin, $\bar{Q}a$. Maximum, minimum and meann annual discharges, respectively;

Q58, Q180, Q307. Discharges of the corresponding diurnal frequency (0.16, 0.50, 0.84 fractions of a year);

$\bar{Q}$ The model discharge;

$p(\bar{Q}a)$ The mean annual discharge frequency.

The quantity ρ and Cr (Babkin, 1970) belong to the second group of distribution characteristics.

The quantity ρ (the natural uniformity coefficient after D.L. Sokolovskii) may be found from the discharge hydrograph or the frequency curve expressed in fractions of the mean annual discharge and annual period duration so that

$$\rho = \rho(1) = \int_0^1 p\,dk$$

For complete uniformity of distribution of the streamflow within a year, $\rho = 1$ and it decreases as the annual uniformity decreases.

In practice, for the rivers that are considerably regulated by lakes its value rises to 0.90, and for the semi-desert zone it falls to 0.15. For the rivers of Polessie $\bar{\rho}$ ranges between 0.40 and 0.75.

The quantity Cr which is the natural uniformity coefficient after P. M. Dmitrovsky expresses the volume of an imaginary storage reservoir necessary for total annual regulation (Babkin, 1970).

For a single-peak hydrograph $\rho = 1 - Cr$. With several peaks present $\rho < 1 - Cr$. For the rivers of Polessie $1 - Cr/\rho$ ranges between 1 and 1.10 and in the vast majority of cases it is below 1.05. Thus, both ρ and Cr are equivalent for estimating the quality of consumed and utilized water resources of Polessie.

Investigation of the relation between $\bar{\rho}$ and the characteristics of the first group is of practical interest. Table 32 furnishes the correlation coefficients between ρ and the uniformity characteristics of the first group for 126 hydrological stations of Polessie and the adjacent regions.

Note 1. In calculating the correlation coefficients, the ratios of the low streamflow rates to the high ones are adopted, which are illustrative for annual uniformity.

2. Characteristic ordinates are taken from the 'mean' frequency curves of the daily discharges.

3. $\bar{K}$ are obtained by graphical differentiation of the daily discharge frequency curve.

Correlation between the uniformity characteristics of the first group and $\bar{\rho}_3$ is essential throughout. However, for practical use the following regression equations may be recommended:

$$\bar{\rho}_3 = 0.88\,\bar{K}_1 + 0.41 \quad (1)$$

$$\bar{\rho}_3 = 4.26\,\frac{\bar{K}_1}{\bar{K}_0} + 0.49 \quad (2)$$

$$\bar{\rho}_3 = 0.55\,\frac{\bar{K}_{0.75}}{\bar{K}_{0.25}} + 0.35 \quad (3)$$

The parameters entering into equations (1) through (3) are published in Hydrological Annuals (Q_a, Qmin, Qmax, Q90, Q270) and the above equations allow easy and straightforward calculation of $\bar{\rho}$. Simplicity of calculations of streamflow uniformity characteristics is essential for the study of their variability over many years and for generalization of the probability distribution parameters over large areas. Attempts to find alternative estimation methods resulted in simpler characteristics having a clear physical meaning.

PROPOSED CHARACTERISTICS

The analysis of the annual streamflow distribution in Polessie has revealed that low discharges are dominant within a year (or according to D I Kocherin 'rivers with a low water regime are dominant within a year'). For this type of river statistical discharge distribution is such that the mode is less than the median which in turn is less than the arithmetic mean, and the low discharge frequency is considerably smaller than the high discharge frequency. The deviations of the high discharges from the mean value are essentially larger than the low discharge deviations from the mean discharge (positive asymmetry).

The studies have shown (Livshits, 1970), that the median to annual discharge ratio $\frac{Q_{0.5}}{Q_a} = K_{0.5}$ may serve as an annual uniformity characteristic. For rivers largely regulated by lakes $K_{0.5}$ may be as large as unity and for rivers in semideserts it falls to 0.005.

The characteristic $\bar{K}_{0.5}$ is closely related with $\bar{\rho}$. For 78 hydrological stations of Polessie and the adjacent areas the correlation coefficient between normal annual $\bar{K}_{0.5}$ and $\bar{\rho}$ is found to be 0.99, (Livshits, 1970). The regression equations are:

$$\bar{\rho} = 0.65\,\bar{K}_{0.5} + 0.24 \qquad (4)$$

$$\bar{K}_{0.5} = 1.54\bar{\rho}_3 + 0.37 \qquad (5)$$

Figure 57 illustrates this correlation for 126 hydrological stations of Byelorussia and the Upper Pridneprovie.

It should be noted that a very close relationship between $\bar{K}_{0.5}$ and $\bar{\rho}$ is revealed for various physiographical stations. So, for 41 hydrological stations in various regions of the USSR with different water content, swampiness, vegetation, geology and soil types, the correlation coefficient of 0.98 ($\bar{K}_{0.5}$ and $\bar{\rho}$ after B. G. Andriyanov) is obtained. The regression equation is

$$\bar{\rho} = 0.63\,\bar{K}_{0.5} + 0.25 \qquad (6)$$

For the diverse conditions of the Caucausus r = 0.90 is found from the data of 122 hydrological stations. For Lithuania r = 0.94 is derived from the data of 67 hydrological stations. As reported by O. Dub (1954) a correlation coefficient between $\bar{\rho}$ and $\bar{K}_{0.5}$ is found for the rivers of Slovakia. The data obtained at 67 hydrological stations enabled him to distinguish and correlate nine types of daily discharge frequency curves peculiar to different typical streamflow regimes. The regression equation is

$$\bar{\rho} = 0.63\,\bar{K}_{0.5} + 0.26 \qquad (7)$$

and r = 0.99

In table 32 correlation coefficients are given for the relations between the annual distribution characteristics of the first group and $\bar{K}_{0.5}$. They differ slightly from the correlation coefficient of the annual distribution characteristics of the first group $\bar{\rho}$.

The following regression equation may also be recommended for practical use:

$$\bar{K}_{0.5} = 1.20\ \bar{K}_1 + 0.28 \tag{8}$$

$$\bar{K}_{0.5} = 4.50\ \frac{\bar{K}_1}{\bar{K}_0} + 0.41 \tag{9}$$

$$\bar{K}_{0.5} = 0.87\ \frac{\bar{K}_{0.75}}{\bar{K}_{0.25}} + 0.19 \tag{10}$$

Table 32.

Specific coefficient	K_1	$\frac{K_1}{K_0}$	$\frac{K_1}{K_{0.5}}$	$\frac{K_{0.84}}{K_{0.16}}$	$\frac{K_{0.75}}{K_{0.25}}$	$\frac{1}{K_0}$	$\frac{1-K_1}{K_0-K_1}$	$\frac{K_{0.5}}{K_0}$	-K	p(1)
$\bar{\rho}$	0.90	0.79	0.71	0.72	0.72	0.56	0.46	0.72	0.51	0.93
$\bar{K}_{0.5}$	0.92	0.66	0.70	0.74	0.67	0.59	0.45	0.73	0.52	0.90

The correlation coefficients calculated for annaul $K_{0.5}$ and ρ from the data of 39 hydrological stations are found to vary between 0.79 and 0.99, the values for 25 stations being greater than 0.90. The parameters of the regression equations also appear to be fairly close to those of equations (4) and (5).

Thus ρ as a parameter of uniformity of the annual streamflow rate may be replaced by the ratio of the median discharge $Q_{0.5}$ to the mean annual $Q_a (Q_{0.5}/Q_a = K_{0.5})$. The value of $K_{0.5}$ may be obtained both for a particular year and for an 'averaged' year by a simple procedure as the initial data on Q180 and Q_a are reported in the publications of the Hydrometeorological Service.

The uniformity of the streamflow rate within a year may be characterized not only by the ratio of the median discharge to the arithmetic mean value, but also by the ratio of the harmonic mean ($\bar{K}_h = \bar{Q}_h/\bar{Q}_a$) and the geometric mean ($\bar{K}_g = \bar{Q}_g/\bar{Q}_a$) to the arithmetic mean. Both ratios possess the same properties as ρ and $K_{0.5}$. The upper is unity, which corresponds to a complete uniformity of discharges within a year. In this case the discharges may be used within their whole range for a year. However, these properties do not also include the calendar distribution of the streamflow and their calculation is more tedious.

For the rivers of Polessie the correlation coefficient between

$\bar{K}_g$, $\bar{K}_{0.5}$ and $\bar{K}_h$ appeared to be large ($r > 0.90$). This may be attributed to the stability of the statistical discharge distribution within a year. Thus, $K_{0.5}$ may be adopted as the annual streamflow uniformity parameter.

The above parameters of the natural streamflow uniformity characterize the averaged annual streamflow rate.

Variation of streamflow uniformity parameters is a stochastic process and may be treated using the mathematical statistical methods which are also applicable to the study of variations of other hydrological characteristics of homogeneous phases.

When studying probable oscillations of ρ and $K_{0.5}$ we have to deal with the problem of the generalization of the variation of the finite distribution curves. The upper limit of ρ, in the case of imaginary complete uniformity of the streamflow, is unity. The lower limit approaches zero. The upper limit of $K_{0.5}$ for diurnal discharge frequency curves with a positive asymmetry is unity. In extreme cases of rivers with a strongly regulated streamflow $K_{0.5}$ may be above unity, although only very slightly (for the Svir River near the settlement of Bogachevo $K_{0.5} = 1.03$). The lower limit of $K_{0.5}$ may approach zero.

Probable variations of ρ and $K_{0.5}$ may be calculated by the method developed by S N Kritsky and M F Menkel for plotting the finite frequency curves (1956). To do this, we shall pass from the paramaters σ and χ to the parameters of the auxiliary curve related with the variable χ by the equation $\chi = 10^{-Av}$ or $v = -\frac{\lg\chi}{A}$. As χ changes from 0 to 1, the variable v ranges from $+\infty$ to 0. This corresponds to the binomial distribution curve at $\bar{v} = 1$ and $Cs = 2Cv$. The transition from the ordinates of the frequency curve to those of χ may follow the equation $\lg\chi = Av$. In the transitions, probabilities of exceeding the initial curve (Pv) are replaced by an addition to 100% as $\chi = \infty$ corresponds to $p = 0$ and $\chi = 0$ to $p = 100$.

Thus, to plot the distribution curves $K_{0.5}$ and ρ calculation of the two parameters $\bar{K}_{0.5}$ (or $\bar{\rho}$) and $\sigma K_{0.5}$ ($\sigma\rho$) is quite sufficient. Plotting the exceedance probability curves for these quantities is considerably simplified as the distribution of $K_{0.5}$ (or ρ) does not require determination of the asymmetry factor.

The values of $\sigma K_{0.5}$ and C_{vk} calculated for 65 hydrological stations of Polessie and the adjacent regions range between 0.90 and 0.24, and 0.12 to 0.82 respectively.

INDIRECT CALCULATION METHODS

Rivers which have not been studied in detail require indirect methods for the determination of the annual natural streamflow uniformity.

As $K_{0.5}$ is a fairly rational uniformity parameter, regular behaviour of the mean and standard values is examined and generalized for $K_{0.5}$. The transition to the relationships for ρ is practically possible.

In 1949 from the hydrological data of the rivers of Polessie the present author established the formula which allowed determination of the normal annual $K_{0.5}$ with no direct observations (Dub, 1954). The formula is verified by the recent data, (Livshit, 1970).

The formula looks like this

$$\bar{K}_{0.5} = aF^{0.08} \tag{11}$$

where: F **is the drainage area, km^2;**
a is the parameter which describes the units of the physiogrpahic factors for the region under investigation, (Livshits, 1955).

Examinations of the data from various regions of the USSR have shown that parameter a increases with the water content of the area. At present, on the basis of the accumulated data of 151 hydrological stations of Polessie and adjacent regions the parameters entering into formula (11) have been recalculated. The exponent appeared to be stable and no essential changes in the shapes of the isolines a are found for Polessie. For the basin of the Pripyat the above has been confirmed by the results of Babkin (1970) by calculation of individual effects of each of the above factors on $\bar{\rho}$ and consequently, on $K_{0.5}$, which certainly does not imply the absence of these effects on the relation between the median and mean annual discharges. Their joint effect is accounted for by the physiographical parameter a.

Several versions of the study on the generalization of regularities of $K_{0.5}$ variations for long term periods have been made. Assuming that the annual streamflow uniformity is related to the water content, the correlation between variation factors of $K_{0.5}$ and q_a was examined. The linear correlation coefficient is found to be 0.80 for 39 stations; the regression equation obtained is

$$C_{vK_{0.5}} = 0.91\ C_{vq_a} - 0.04 \tag{12}$$

The relation between $C_{vK0.5}$ and the normal annual discharge $\bar{q}_a$ was found to be inverse, as was expected (r = - 0.55), in basins with a high water content the amplitude of the variations of the individual elements of hydrographs decreases. The correlation coefficient between $C_{vK0.5}$ and $\bar{q}_a$ is practically significant, but the regression equation for two variables does not prove the accuracy required for practical designs. The correlation between C_{vk} and $\bar{K}_{0.5}$ appears to be closer. For these series with the same number of terms, the correlation factor r = - 0.87 is found. The inverse relation may be attributed to the fact that $K_{0.5}$ increases with the basin and water content of the region considered. The same factors should probably operate to decrease variability of annual values of $K_{0.5}$.

The correlation between $\bar{K}_{0.5}$, $\bar{q}_a$, and $C_{vK_{0.5}}$ found by the multiple regression method is expressed by

$$C_{vk} = 0.81 - 0.72\ \bar{K}_{0.5} - 0.03\ \bar{q}_a \tag{13}$$

the correlation coefficient between $\bar{K}_{0.5}$ and $\bar{q}_a$ being 0.49 and multiple correlation factor R = 0.91. Referring to equation (11), we obtain

$$C_{vK_{0.5}} = 0.81 - 0.72F^{0.08} - 0.03\,\bar{q}_a \qquad (14)$$

As was demonstrated above, the relation between $\bar{\rho}$ and $\bar{K}_{0.5}$ expressed by formula (4) may be considered to be functional.

The author's investigations of the materials of the same hydrological stations in Polessie have also revealed a close relation between the variation factors of ρ and $K_{0.5}$. The value r = 0.86 is obtained and the linear regression equation is

$$C_{v\rho} = 0.29\,C_{vK_{0.5}} + 0.07 \qquad (15)$$

Thus, there exists a possibility of passing from the finite distribution curve $K_{0.5}$ to the corresponding curve ρ.

REFERENCES

Andriyanov, V. G. 1960. Vnutrigodovoe raspredelenie rechnogo stoka. Streamflow distribution within a year. Leningrad, Gidrometeoizdat.

Babkin, V. I. 1970. Issledovanie vnutrigodovoi zaregulirovannosti stoka rek Belorusskoi SSR i Verkhnego Pridneproviya v zavisimosti ot fiziko-geograficheskikh faktorov metodom mnozhestvennoi korrelyatsii. Study of annual streamflow uniformity of rivers in the Byelorussian SSR and Upper Pridneprovie as a function of physiographic factors by multiple correlation method. Sbornik rabot po gidrologii, No. 9. Leningrad, Gidrometeoizdat.

Burneikis, Yu. P., Gailyushis, B. V. 1965. Pokazateli vnutrigodovoi neravnomernosti stoka rek Litovskoi SSR i ikh zavisimosti ot mestnykh fiziko-geograficheskikh factorov. Parameters of annual streamflow non-uniformity of rivers in the Lithuanian SSR, and their dependence on local physiographic factors. Trudy Academii Nauk Litovskoi SSR, ser. B, 2(41), Kaunas.

Dub, O. 1954. Vsenecha hydrologia Slovenska. Bratislava, SAV.

Kritskii, S. N. and Menkel, M. F. 1956. O dvoyakoogranischennoi krivoi raspredeleniya veroyatnostei i primenenii ego k gidrologicheskim raschetam. On finite probability distribution curve and its application to hydrological calculations. In: Problemy regulirovaniya rechnogo stoka, issue 6, Moscow.

Livshits, I. M. 1955. Obespechennost sutochnykh raskhodov rek Polesiya. Frequency of daily mean discharges of rivers in Polessie. Trudy Instituta melioratsii, vodnogo i bolotnogo khozyaistva BSSR, vol. 6, Minsk, Izd. Akad. Nauk BSSR.

Livshits, I. M. 1970. Ob otsenke vnutrigodovoi estestvennoi zaregulirovannosti stoka v usloviyakh Belorussii. On estimation of annual natural streamflow uniformity in the conditions of Byelorussia. 4th Lithuanian Conference on Meteorology, Hydrology, Agricultural Microclimatology, Vilnius.

Sokolvskii, D. L. 1968. Rechnoi stok. River streamflow. Leningrad, Gidrometeoizdat.

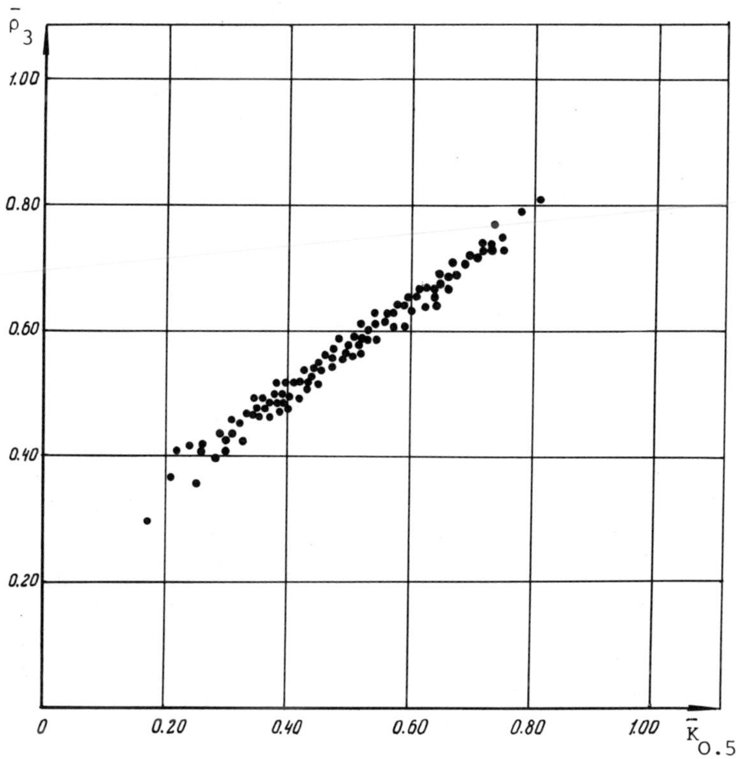

Fig 57. Correlation between $\bar{\rho}_3$ and $\bar{K}_{0.5}$.

DISCUSSION

B. V. Fashchevsky (USSR) - Does the annual streamflow change in a region under drainage? If so, what is the direction of the change?

You have suggested a new criterion of annual natural streamflow uniformity. What is its advantage compared to the number ρ suggested by Sokolovsky?

I. M. Livshits (BSSR) - In our works we have not found a dependence on drainage. This may be explained by the fact that there are no areas which may cause radical changes which affect the whole river regime.

Plotting of distribution curves following Sokolovsky is very laborious work involving the plotting of hydrographs for each year. As to the median coefficient, its values are published and using this coefficient does not require tedious treatment of data. Not only the median coefficient may be used but also a geometric mean with respect to an arithmetic mean.

METHODS EMPLOYED IN THE INVESTIGATION OF SWAMPS WITH FIXED STATIONS AND MOBILE SURVEYS

S.M. Novikov

State Hydrological Institute, Leningrad, USSR.

SUMMARY: The paper considers the methodological principles of hydrological and hydrophysical observations and investigations of undrained swamps of the USSR, as well as theoretical fundamentals of the hydrology of swamps.

Guidelines for fixed and mobile research work in swamps are presented. The different aspects of the two approaches are discussed and shown to be complementary. The combination of fixed stations and field surveys for the study of the hydrological and thermal regimes of waste areas of swamp is considered to be essential.

Emphasis is given to the value of aerial photography for the hydrological survey and landscape study of swamps.

METHODOLOGIE DE L'ETUDE HYDROLOGIQUE DES MARAIS A L'AIDE DE STATIONS ET DE MISSIONS SUR LE TERRAIN

RESUME : L'auteur expose les principes méthodologiques sur lesquels sont fondées l'observation et l'étude hydrologiques et hydrophysiques des marais non asséchés ainsi que les bases théoriques de ces études.

L'auteur met l'accent sur l'utilisation de la photographie aérienne, il expose les principes d'organisation des postes d'observation et des stations expérimentales et les règles qui régissent le choix des secteurs à étudier. Il décrit les programmes d'études et d'observations de ces postes et stations. Il procède de même pour les recherches qui sont effectuées par des spécialistes en mission sur le terrain.

Il expose les différences principales entre ces deux méthodes de recherches, il montre ce qu'elles ont de commun et en quoi elles permettent d'étudier le régime hydrique et le régime thermique des marais sur de vastes territoires.

The total area of swamp and marshland in the USSR exceeds 200 000 000 ha. Although swamps occur in all vegetation zones of the country with the exception of semi-deserts and deserts, they are more prevalent in the forest zone. In some regions (Bylorussian Polessie, Meshcherskaya Lowland, The Ob basin, etc.) 30-50% of the area is swamp, and in some catchments swamps constitute 70% of the area. The wide range of climatic, hydromorphological and hydrogeological conditions explains the great variety in the structure and hydrological regime of swamps.

The main hydrological features that distinguish swamps from dry land and determine to some extent the methods of research, are the following: the presence of a single aquifer, extending from the top to the bottom of the swamp; division of the peat deposit over the depth into the active and the inert layers; runoff from the swamp in the form of subsurface flow uniformly distributed over the swamp surface; similarity of groundwater slope and swamp surface slope; simultaneous groundwater level fluctuations at any point of the swamp area.

The classification of swamp landscapes with respect to hydrology has been developed by K. E. Ivanov (1957) and E. A. Romanova (1961) on the basis of classification of swamp phytocoenosis proposed by S. N. Tyuremnov (1949). In this classification , swamp micro-landscape is taken to mean a portion of swamp area which is homogeneous with respect to the vegetation, the surface micro-relief and the hydrophysical properties of upper layers of peat deposits.

Since the type of vegetation on the micro-landscape corresponds to the environment, the main factors of which are water and mineral supply and water exchange rate, every type of micro-landscape has its characteristic hydrological regime.

Investigations of the peat deposit corresponding to various swamp micro-landscape (Ivanov, 1957; Romanova, 1964; Lopatin, 1949) have revealed great differences in the physical properties of the active and inert layers of peat. In particular, it was fould that the coefficient of permeability can vary widely. The high hydraulic conductivity of the active layer of the peat deposit (the layer in which the water level fluctuates), formed of moss and little decomposed peat, compared to the low hydraulic conductivity of the inert layer, formed of well decomposed peat, gives rise to many hydrological processess in natural swamps. The active peat layer, despite its insignificant depth, is responsible for all the main peculiarities of hydrological regimes of swamps.

Every swamp mirco-landscape has definite hydrological characteristics. These include mean long term water level (Z); active layer depth (Z_o); the relation between permeability coefficient (K_z) and swamp water level $K_z = f(Z)$; relation between water yield coefficient (ξ_z) and swamp water level $\xi_z = f(Z)$; the slope of the swamp surface within the micro-landscape; mean long term discharge; relation between the coefficient α and swamp water level $E = \alpha_{R\sigma} = f(Z)$, (where E is the evaporation and Rσ is the net radiation. Every type of swamp micro-landscape has quite definite values and forms of the above parameters and relations.

Swamp micro-landscapes of the same type wherever they are situated, have the same hydrological characteristics. This may be attributed to the fact that the peat vegetation which formed the peat of the active layer, (or the majority of it) is the same as the vegetation which exists today. The upper peat layers, in similar micro-landscapes, generally have the same composition (Abramova, 1954; Boch, 1958; Romanov, 1961; Romanova, 1960). Thus it follows that the vegetation of swamps serves as a rather good indicator of their hydrological regimes. This property makes it possible to apply aerial survey methods to the study of the hydrological regime of swamps.

Aerial photographs have become a useful tool for the hydrological and hydrophysical investigation of swamps. Methods of typological interpretation of aerial photographs of swamps (Galkina, 1953; Galkina *et al*, 1949) have made it practical to study the landscape structure of swamps and their hydrological regime on a large scale. The introduction of such methods to practical hydrological research has opened up a new phase in the study of swamps and marshlands. The application of aerial survey to research has radically changed the methods of studying swamps and favoured a more rapid investigation of inaccessible areas.

Interpretation of aerial photographs of swamps is based on the relation between vegetation of swamps, their geomorphological position and their hydrological regime. Photo interpretation of swamp micro-landscape types and boundaries is a rapid and simple matter. The photographs also indicate the lasw of distribution of micro-landscapes over the swamp area and make it possible to establish the limits of swamp areas, the character of the hydrographical network and the water content of the various parts of swamps. Interpretation is followed by the preparation of a detailed typological map of the swamp on which are distinguished the types of micro-landscapes, their prevalence and relation to different parts of the swamp area. In conjunction with the typological map, the photographs are used to plot lines of flow of both surface and subsurface water. This map gives a clear picture of the flow pattern of the swamp, and is used when computing the runoff from swamps.

A typological map, prepared with the assistance of aerial photographs, is a basic tool for organizing stationary and mobile investigations of swamps. The map is used to ensure the most rational and scientific layout of fixed posts for the observation of the hydrological and thermal regime of the swamp area as well as to select field survey courses correctly, taking into account the peculiarities of the typological structure of the swamp area under study.

A typological map is also essential for the study of the hydrological regime of ungauged swamps. This is of the utmost importance in the case of extra large swamp areas when in is practically impossible to apply the usual surface methods of investigation. In the case of the Soviet Union with its vast swamp areas, a method has been developed for hydrological and hydrophysical studies of natural swamps, which combines both surface investigations and aerial surveys. The essence of this method consists in the examination and detailed study of the hydrological regime of selected swamp micro-landscapes. The results, concerning the characteristics of the hydrological and thermal regimes and the hydrophysical properties of the peat deposit in the active layer or the relation between these characteristics and factors which control them, are extended, with the help of aerial surveys, to anologous micro-landscapes in uninvestigated swamps of the zone under study.

When the location of observation posts in swamps in planned, the factors considered are: its landscape structure; the conditions of geomorphological position and water supply to the swamp; distribution of micro-landscapes within the swamp area; the direction of swamp water flow; the boundaries and relief of the swamp area; and the distribution of peat deposits depths. Hydrological and meteorological variables that change with the type of swamp micro-landscape must be measured in every type of micro-landscape. The variables that do not change or vary only within an acceptable limit, are measured at one or several points only, selected according to the peculiarities of the swamp. The places selected for studying the hydrophysical properties of the active layer of a peat deposit and the chemical composition of swamp waters must be co-ordinated (with respect to their elevation) with the points selected for the observation of water levels.

The work programme of special swamp hydrological stations, organized by the Hydrometeorological Service of the USSR, includes observations of groundwater levels of swamps and adjoining dry lands, the channel inflow and outflow of swamps, evaporation, freezing, thermal distribution in the peat deposit and climatology. The programme also includes experimental investigations of the hydrological properties of peat, such as its permeability and capillary properties, water yield and moisture content. In the programme of investigation a special place is reserved for the quality of water in the hydrographic network connected to the swamp.

Mobile investigations of swamps are designed to obtain initial data on the swamp structure (swamp relief, surface microrelief in different micro-landscapes, stratigraphy and depth of peat deposit) and the internal hydrographic network (depth of lakes and streams, flow, velocity, water level variations, etc.). Preparation for mobile surveys begins with the collection and study of all available information concerning the swamp area in question, interpretation of aerial photographs and plotting of a typological map of the swamp. This map is the basic document used in the preparation of the final programme of field work such as the selection of surface survey routes and where hydrological observations are to be made of the swamp, its lakes and water courses. The results of surface observations are used to prepare a more detailed and accurate typological map.

When preparing the programme of observations for the swamp area, one should adhere to the following principles: the survey routes must pass through all main micro-landscapes and at least one of the routes should pass through the genetic centre of the swamp area or follow the line of the greatest slope of swamp surface. When examining swamps located in river catchments, the survey routes should be planned to start near one of the bordering streams, cross the watershed divide and terminate at the stream which defines the opposite border of the swamp.

Geobotanical surveys of the swamps are carried out in summer when the water level is low. Topographical surveys of the swamp are made to determine the slopes of the micro-landscape and the height of the convex parts in relation to its borders; the depth of peat deposits is measured; samples of peat for laboratory determination of botanic composition and the degree of decomposition are taken by auger.Other undertakings are: detailed geobotanic description of the micro-landscapes; calculation of the number of stumps per unit area of micro-landscapes; micro-topography of the bog surface in different micro-landscapes; description of the hydrographic network within the swamp; periodic measurement of the water level; ground control of the typological map of the swamp area.

The planning of obserservations of hydrological thermal regimes of swamps and the experimental study of hydrophysical properties of peat deposits during mobile investigations is based upon the same principles as those which govern fixed investigations. Both the programme and the method of observations and experiments are similar to those employed at specialized swamp stations.

The main endeavour at fixed stations in swamps is the study of the hydrological and thermal regimes of the processes of water exchange with the environment.

Long- term fixed observations of hydrological and thermal behaviour lead to an understanding of the principal laws which govern the various hydrological and hydrophysical processes that occur in swamps, and to the development of methods for calculating the hydrological variables with the help of meteorological data, obtained by the network of meteorological stations, (Novikov, 1964).

Mobile surveys are mainly aimed at studying the landscape structure of swamp areas, the stratigraphy of peat deposits and peculiarities of the internal hydrographical network layout. Short hydrological and hydrophysical investigations are mainly carried out in this case to study peculiarities of the hydrological behaviour of particular swamp areas, as well as to check design specifications prepared on the basis of observations at fixed stations.

Thus, the fixed stations are used to study changes with time of the components of the hydrological and thermal regimes, whilst mobile investigations reveal changes in space of these components. Only the application of both these methods of investigation in parallel can produce satisfactory results in vast swamp areas.

We have seen that fixed stations and mobile surveys are complementary. Data obtained at fixed stations is extrapolated to wide areas of swamp with the help of observations obtained during the course of field surveys. Aerial photographs, after adequate ground control, furnish information on the more inaccessible areas of the swamp.

REFERENCES

Abramova, T. G. 1954. O svyazi mezhdu rastitelnym polrovom bolot i stroeniem verkhnikh sloev torfyanoi zalezhi. On the relation between vegetal cover of swamps and structure of the upper peat bed. Uchenye Zapiski LGU, No. 167, Ser. Biol. Nauk, issue 34.

Boch, M. S. 1958. K voprosu ob ispolzovanii rastitelnogo pokrova kak indikatora stroeniya torfyanoi zalezhi. Vegetal cover as evidence of peat bed structure. Vestnik LGU, No. 3, Ser. Biol, Nauk, issue 1.

Galkina, E. A., Gilev, S. G., Ivanov, K. E. and Romanova, E. A. 1949. Primenenie materialov aerofotosyomki dlya gidrograficheskogo izucheniya bolot. Application of aerial surveys to the hydrological investigation of swamps. Trudy GGI, issue 13.

Galkina, E. A. 1953. Ispolzovanie aerofotosyomok v bolotovedenii. Air survey in swamp investigations. Botanicheskiy Zhurnal, vol. 38, No. 6.

Ivanov, K. E. 1957. Osnovy gidrologii bolot lesnoi zony. Fundamentals of swamp hydrology in forest zones. Leningrad, Gidrometeoizdat.

Lopatin, V. D. 1949. O gidrologicheskom znachenii verkhovykh bolot. On the hydrological effect of raised bogs. Vestnik LGU, No. 2.

Novikov, S. M. 1964. Raschety vodnogo rezhima i vodnogo balansa nizinnykh bolot i ryamov yuzhnoi chasti Zapadno-Sibirskoi nizmennosti. Calculations of hydrological regime and water balance of lowland bogs and mires in the south part of the West-Siberia Lowland. Trudy GGI, issue 105.

Romanova, E. A. 1960. O svyazi mezhdu rastitelnostiyu, verkhnimi sloyami torfyanoi zalezhi i vodnym rezhimom verkhovysk bolot Sverozapada. The relation between flora, upper peat bed and water regime of raised bogs in the North-west. Trudy GGI, issue 89.

Romanova, E. A. 1961. Geobotanicheskie osnovy gidrologicheskogo izucheniya verkhovykh bolot. Geobotanical fundamentals for hydrlogical investigations of raised bogs. Leningrad, Gidrometeoizdat.

Romanova, E. A. 1964. Tipy bolotnykh mikrolandshaftov kak pokazateli vidov torfa v verkhikh sloyakh nizinnykh bolot. Bog microlandscape types as indicators of peat types in upper peat beds of lowland bogs. Trudy GGI, issue 112.

Romanov, V. V. 1961. Gidrofizika bolot. Hydrophysics of swamps. Leningrad, Gidrometeoizdat.

Tyuremnov, S. N. 1949. Torfyanye mestorozhdeniya i ikh razvedka. Peat deposits and their prospects. Moscow, Leningrad, Gosenergoizdat.

DISCUSSION

Don H. Boelter (USA) - Have you found a large difference between the active layer thickness and that of vegetation?

S. M. Novikov (USSR) - I mentioned that in the present work a relation between water engineering characteristics, water yield, evaporation and vegetation is found. That means that the vegetation is an indicator of bog occurrence. The known hydrophysical properties of a micro-landscape may probably be applied to the study of any bog system.

WATER BALANCE OF SWAMPS IN THE FOREST ZONE OF THE EUROPEAN PART OF THE USSR

L. G. Bavina

State Hydrological Institute, Leningrad, USSR.

SUMMARY: This paper considers the main problems involved in the investigation and calculation of the water balance components of swamps. The investigations have been carried out in the forest zone of the European Territory of the USSR where swamps are widespread on account of specific physiographical conditions.

The swamps of the forest zone may be divided into two groups according to their hydrological regime: upland swamps (raised bogs) and lowland swamps (flat or valley bogs).

The paper presents results of calculation of water balance components in typical upland and lowland swamps. It then examines the relationship between the hydrological regime of swamps and that of rivers by comparison of the run-off from swamps as calculated by the water balance equation and the runoff from river basins.

LE BILAN HYDRIQUE DES MARAIS DANS LA ZONE FORESTIERE EUROPEENNE DE L'URSS

RESUME : L'auteur traite de l'étude et du calcul des éléments du bilan hydrique des marais. Les recherches ont porté sur la zone forestière du territoire européen de l'URSS où les marais sont très répandus en raison de conditions physico-géographiques spécifiques.

D'après les caractères du régime hydrologique, les marais de la zone forestière sont divisés en deux groupes : marais hauts et marais bas.

L'auteur donne les résultats du calcul des éléments du bilan hydrique d'après les données des observations des marais hauts et des marais bas. La relation entre régime des marais et régime des rivières est faite par comparaison entre l'écoulement provenant des marais (calculé par l'équation du bilan hydrique) et l'écoulement provenant des bassins versants des rivières.

Temperate climate, excessive moisture combined with flat relief and abundant groundwater near the soil surface create favourable conditions for the formation of swamps in the forest zone of the European part of the USSR.

The study of the water balance of swamps is of great importance for the establishment of the relationship between the hydrological regimes of swamps, rivers, lakes and groundwater. Such knowledge is necessary for the selection of correct methods for drainage and reclamation of swamps and for the evaluation of the effect of complex reclamation undertakings on the hydrological regime of the whole area.

Swamps differ in their vegetation, depth and properties of the peat deposit, conditions of water supply, etc. There are a great number of rather strict classifications of swamps which depend on various characteristic features. The type of hydrological regime of a swamp is determined by three principal features, (Ivanov, 1957): 1) geomorphological conditions of location, which determine the source of water supply; 2) surface relief, which determines the conditions of runoff; 3) vegetation cover and its distribution over the swamp.

The great variety of swamps found in the forest zone exhibit sharp differences in the above features but may be divided into two main groups, i.e. upland swamps and lowland swamps. The most characteristic features of upland swamps are their location in river catchments, the convex shape of their surface, their dependence on precipitation, and the oligotrophic vegetation. In the case of lowland swamps the following features are most characteristic: their location in a depression, river valley or flood plain, a concave or flat surface, mixed water supply (precipitation, surface and underground inflow from surrounding dry land), eutrophic vegetation.

The diverse properties which differentiate upland and lowland swamps explains largely the different roles which they play in hydrological processes. Upland swamps are primary regulators of atmospheric moisture and an independent source of water supply to streams, while lowland swamps mainly transform groundwater flow and surface runoff, and may therefore be called secondary regulators of water flowing to rivers and lakes.

For a long time, the views on the role of swamps in streamflow patterns were contradictory. This was because of a lack of direct and systematic observations of the components of the water balance of swamps. Now long-term series of measurements of water balance components of swamps are available and the analysis of such measurements for two typical (upland and lowland) swamps in the forest zone of the European part of the USSR is presented here.

The water balance equation for swamps may be written as follows:

$$P = E \pm \Delta W + Q_2 - Q_1 + G_2 - G_1 \qquad (1)$$

where:

- P precipitation;
- E evapotranspiration;
- ΔW change of water storage in the swamp;
- Q_1 surface inflow to the swamp;
- Q_2 surface outflow from the swamp;
- G_1 groundwater inflow to the swamp, through the mineral banks or through the bottom of the swamp;
- G_2 groundwater outflow from the swamp, through the mineral banks or through the bottom of the swamp.

Depending on the type of swamp, the scope of the investigation and its practical aims, equation (1) may be refined with respect to its components. Some water balance components may be either differentiated or combined. Thus the last four components are often combined and the result is called the total water exchange or actual runoff from the swamp (S).

$$S = Q_2 - Q_1 + G_2 - G_1 \qquad (2)$$

For upland swamps with no surface or groundwater inflow, the value $S = Q_2 + G_2$ is called the total runoff from the swamp.

Equations (1) and (2) yield

$$S = P - E \pm \Delta W \qquad (3)$$

With equation (3) it is possible to evaluate the total water exchange (S) when the other components are known. Such computations were made for a long period for both upland and lowland swamps using observations of precipitation, evaporation and water levels.

Precipitation P was measured by raingauges. For the winter period a 10-15% correction for wind effect was introduced into the records of the gauges.

Evaporation E was measured by evaporimeters GGI-B-1000 and by the heat balance method, (Romanov, 1961). Winter evaporation from the snow cover was not measured; it was assumed to be approximately equal to 25 mm taking into account the duration of snow cover, (Bavina, 1967).

Due to the high porosity of the upper layers of peat and the presence of the water table close to the surface, the water content in the unsaturated zone of swamps is as a rule close to field capacity. Hence the variation of water storage in swamps (ΔW) was computed from the data on water level fluctuations (ΔH) and water yield (μ):

$$\Delta W = \Delta H \mu$$

The water yield was determined under field conditions, (Vorobiev, 1963). Observations were carried out in all the principal swamp micro-landscapes*; the results were then averaged for the whole swamp area taking into account the area of each micro-landscape.

The upland swamp, where the observations were carried out, is situated in the northern part of the forest zone of the European part of the USSR. It is a typical swamp, which has reached the stage of development of a convex moss bog. The central part of the swamp is covered by Sphagnum and bushes with scattered pine trees and the slopes are occupied by ridges and hollows. At the periphery of the swamp, Sphagnum-grass and Sphagnum-sedge micro-landscapes prevail.

---

* A swamp micro-landscape is a part of the swamp area which is occupied by similar plant associations and has homogeneous surface microrelief and hydrophysical properties of the active layer, (Ivanov, 1957).

The total runoff (S) was computed by equation (3) using monthly values of the water balance components for the warm season (May to October) and total values for the cold season (September to April) which include snow accumulation in winter and snowmelt and melt water runoff from the swamp in spring.

Mean results for the period 1955 to 1966 as well as results for typical dry (1960) and wet (1962) years are presented in Table 33. As can be seen in this table, the total runoff from the swamp both in the wet year and during the 12 year period of observation, represents about one half of the discharge component of the water balance, whereas in the dry year it is only one third. The greater portion of the runoff occurs during the spring period in the form of snowmelt (57% in the wet year, 76% in an average year and 85% in the dry year). During the warm season rainfall gave rise to runoff in May, August, September and October. In June and July of the wet year the runoff was 20 mm or 9% of the total for the warm season.

Table 33. Water balance components for an upland swamp (mm).

Balance component		Periods								
		Sept to April	May	June	July	Aug	Sept	Oct	May to Oct	Annual
					Mean annual values					
Precipitation	(P)	378	49	64	80	89	80	88	450	828
Evaporation	(E)	49	89	109	96	67	38	8	407	456
Change of storage	(ΔW)	45	-70	-26	-5	12	17	23	-49	-4
Total runoff	(S)	284	30	-19	-11	10	25	57	92	376
					1960 (dry year)					
Precipitation	(P)	289	26	69	73	82	32	44	326	615
Evaporation	(E)	50	103	105	103	67	34	7	412	469
Change of storage	(ΔW)	42	-67	-5	-12	16	-4	4	-68	-26
Total runoff	(S)	197	-10	-31	-18	-1	2	33	-25	172
					1962 (wet year)					
Precipitation	(P)	447	60	119	103	148	132	56	618	1065
Evaporation	(E)	58	76	101	92	65	39	10	383	441
Change of storage	(ΔW)	54	-47	16	-7	37	-21	7	-15	39
Total runoff	(S)	335	31	2	18	46	114	39	250	585

Negative values of S in summer months result from errors of measurement of components of the water balance. The mean error is 10%, the most significant errors occurring in dry months, when the water level of the swamp is low. In this case the errors are caused by neglecting the drying of the upper layers of peat below field capacity.

Evaporation from upland swamps varies very little from year to year and primarily depends on the value of solar radiation at the swamp surface. Table 34 presents variation coefficients C_V of monthly evaporation as measured by evaporimeters, and of monthly sums of radiation balance.

The calculated total runoff value from the swamp were compared with the runoff from three river basins. The swamp system under study was located on their divide(all basins are afforested (about 70%) and slightly marsh-ridden (6%)). The area of basin No. 1 is

Table 34. Coefficients of variation of monthly evaporation and radiation balance for an upland swamp.

	May	June	July	August	Sept
Evaporation	0.12	0.10	0.09	0.16	0.11
Radiation balance	0.12	0.10	0.10	0.16	0.11

89 km^2, that of basin No. 2 is 293 km^2 and the area of basin No. 3 is 390 km^2. The result of the comparison presented in Table 35, show that for a long-term period and within the limits of accuracy of calculation (±10%), runoff from the swamp and from the river basins was the same under similar physiographic conditions. The correlation coefficients vary between 0.85 and 0.90.

In dry years (1960 and 1964) runoff from the swamp was less than that from the river basins, and in wet years (1962 and 1966) it was much greater.

Let us now examine the total water exchange calculated for a lowland swamp. The swamp is situated in the southern part of the forest zone of the European part of the USSR. It is a typical flood plain swamp: the area near the river terrace is mainly occupied by forest; the central area is covered by sedge-hypnum and reedgrass-birch associations and the area adjacent to the river is occupied by reed and willow.

Computation of total water exchange was made for the central part of the swamp on the basis of monthly values of the water balance components. Positive values of total water exchange indicate that the outflow (total outflow) from the swamp, including runoff and infiltration into the mineral soil, are greater than the inflow; negative values imply that the inflow from the dry land or underlying mineral soil is greater than the outflow.

Within a year, three periods may be distinguished according to the variation of the water exchange (Table 36).

In the autumn-winter period (November to February) the inflow is generally greater than the outflow. During this period the increase of storage is mainly attributable to groundwater inflow. The excess of outflow over inflow in 1957 can be explained by early snowmelt (in February). Spring(March to April) is characterized by the predominance of outflow from the swamp; although when the snow cover is small and winter water levels are rather low, as was the case in 1954 and 1960, actual runoff from the swamp may not occur. During the summer (May to October) water exchange in some months is insignificant; in all years the outflow from the swamp into the underlying sand layer exceeds the inflow.

The total annual water exchange is positive, the mean value for nine years being 119 mm.

The values of the total water exchange (actual runoff from the swamp) were compared with runoff values from a river basin having an area of 1450 km^2, 50% of the area being covered with swamps and 37% with forest (Table 37).

Runoff from the swamp is in general agreement with runoff from the swampy catchment even on a yearly basis (within the limits

Table 35. Swamp runoff compared with runoff from three watersheds(mm)

	1955	1956	1957	1958	1959	1960	1961	1962	1963	1964	1965	1966	Mean
Swamp	422	317	484	424	300	172	363	585	360	216	315	558	376
Basin (1)		366	374	367	358	195	305	429					342
Basin (2)	500	338	428	391	302	215	344	472	335	305	331	426	366
Basin (3)	480	321	407	380	296	200	298	449	409	258	316	441	354

of computation error). The difference between 1960 and 1961 can be explained by the fact that precipitation late in 1960 caused runoff from the swamp in December 1960 and from the river basin in January 1961. In dry years runoff from the swamp is less than that from the river basin, and in extremely dry years (1954) the water exchange of the swamp may even be negative.

During normal or moderately wet years the values of runoff from swamps do not significantly differ from the values of runoff from river basins in the forest zone of the European part of the USSR. In dry years and dry seasons the runoff from swamps is less than that from river basins, and lowland swamps may even delay a part of the transit runoff from the surrounding dry lands.

REFERENCES

Bavina, L. G. 1967. Utochnenie raschetnykh parametrov ispareniya s bolot po materialam nablyudenii bolotnykh stantsii. Refinement of design parameters for a bog evaporation on the basis of bog station observations. Trudy GGI, issue 145.

Ivanov, K. E. 1957. Osnovy gidrologii bolot lesnoi zony. Fundamental hydrology of bogs in the forest zone. Leningrad, Gidrometeoizdat.

Romanov, V. V. 1961. Gidrofizika bolot. Hydrophysics of swamps. Leningrad, Gidrometeoizdat.

Vorobiev, P. K. 1963. Issledovanie vodootdachi nizinnykh bolot Zapadnoi Sibiri. Water yield from fen bogs of West Siberia. Trudy GGI, issue 105.

Table 36. Water exchange between lowland swamp and its boundaries (mm)

Year	Periods		
	Autumn-winter (Nov-Feb)	Spring (March-April)	Summer (May-Oct)
1953-1954	7	-53	48
1954-1955	-2	61	70
1955-1956	-95	137	77
1956-1957	166	42	51
1957-1958	-185	207	157
1958-1959	-28	56	8
1959-1960	23	5	28
1960-1961	69	99	18
1961-1962	-4	61	53
Mean	-5	68	56

Table 37. Swamp and watershed runoff compared (mm).

	Year									Mean
	1954	1955	1956	1957	1958	1959	1960	1961	1962	
Swamp	-33	122	167	177	233	111	134	57	104	119
River basin	20	108	146	134	245	110	100	94	102	117

HYDROLOGY OF MARSHY PONDS ON THE COTEAU DU MISSOURI

Wm. S. Eisenlohr, Jr

SUMMARY: The Coteau du Missouri in central North Dakota, USA, is covered with glacial drift that contains hundreds of thousands of marshes. These marshes have an area less than 0.6 km^2. Their water surfaces are continuous with the groundwater table, which follows the land surface at a shallow depth. Some marshes contain large quantities of emergent aquatic vegetation (hydrophytes). Evapotranspiration computed by the mass-transfer method showed that seasonal evaporation from marshes clear of hydrophytes was close to published regional values. Evapotranspiration from ponds with hydrophytes was about 15% less, owing to the reduction in evaporation caused by the hydrophytes. All the ponds are in till; groundwater moves vertically through this till with relative ease near the ground surface, but with difficulty at depth; it can move laterally only with difficulty at all depths. As a result, most water from precipitation is dissipated locally.

HYDROLOGIE DES BASSINS MARECAGEUX DU COTEAU DU MISSOURI

RESUME : Le Coteau du Missouri, dans le centre du Dakota-Nord (Etats-Unis d'Amérique), est recouvert d'une moraine qui comporte plusieurs centaines de milliers de petits marécages. La superficie moyenne de ces marécages est inférieure à 0,6 km^2. Leurs eaux sont le prolongement de la nappe souterraine. Certains de ces marécages contiennent un grand nombre de plantes aquatiques émergées (hydrophytes). L'évaporation saisonnière des marécages sans hydrophytes est à peu près la même que la valeur régionale publiée ; l'évapotranspiration des marécages avec hydrophytes est inférieure à 15 % environ, cette moindre évaporation étant due à la présence des hydrophytes. Tous les marécages étudiés sont situés dans un sol erratique. L'eau souterraine, près de la surface du sol, se déplace verticalement avec une relative facilité, mais avec difficulté en profondeur. Latéralement son déplacement est difficile à toute profondeur. Il en résulte qu'une partie considérable des précipitations se dissipe localement.

INTRODUCTION

The Coteau du Missouri is a rolling upland area (covered with glacial drift) in the prairies of central North Dakota. The area has little or no integrated drainage systems, and there is an almost complete absence of permanent streams. The area is covered with hundreds of thousands of depressions, formed as a result of glacial processess, known as prairie potholes. A prairie pothole is merely a depression in the prairie, capable of holding water, and with an area generally less than 0.6 km^2. Although they are known locally as prairie potholes, they will be referred to in this paper as marshy ponds or marshes in keeping with the Symposium. The water table is at shallow depth and is continuous with the water surface of those marshes that contain water for several months. Many marshes contain large quantities of emergent aquatic vegetation. The area in which this study was made is only a small part of the prairie pothole region that covers about 800 000 km^2 in north-central United States and south-central Canada. It is the most important breeding area for migratory water fowl in North America

The glacial drift in the study area is mostly unsorted ice-laid rock debris known as till. This till was left by Late Pleistocene glaciers, and consists of about equal parts of sand, silt, and clay, plus a small percentage of gravel. The clay size fraction is dominantly montmorillonite. The till has a very low permeability for water moving laterally, but for vertical flow the permeability can be high near the land surface owing to vertical joints and cracks that are common there. The soils overlying the till are very permeable, probably as a result of prairie grass that develop root systems more that a meter deep.

Average annual precipitation ranges from about 500 mm in the southeast to about 400 mm in the northwest; average annual lake evaporation is about 800 mm. Thus average annual precipitation is less than 60% of the annual lake evaporation. There is, therefore, a strong tendency for marshy ponds to disappear during periods when precipitation falls only in the form of scattered light showers. However, the difference between precipitation and evaporation is such that a wet year with precipitation concentrated in a few intense storms, or heavy snowmelt, will produce enough basin inflow to assure the persistence of a pond for several years.

BASIN INFLOW

The water supply of a marsh comes from precipitation directly on the marsh and basin inflow. Basin inflow as used here includes runoff (channel flow), overland flow, and seepage inflow. As seepage inflow is believed to be small, and drainage channels are uncommon, overland flow was considered as the only significant source (with a few recognized exceptions) of inflow and thus it is the key factor in maintaining a marshy pond. Basin inflow was measured as the rise in the water level of a pond less the amount of precipitation on its surface. Even though marshy ponds could be called 'outcrops' of the water table, the fact that they are above the ground increases the opportunity for evapotranspiration. Also, the low permeability for lateral movement of groundwater precludes significant replenishment from that source.

Overland flow can occur only when the precipitation rate exceeds the infiltration capacity of the soil. It is convenient to describe the soil as saturated when this occurs. The infiltration capacity of the soils was high when the soil was dry, but if the soil had been wetted by a previous rain so that the montmorillonite clays became wet and expanded, or if the soil was frozen, the infiltration capacity became very low. Therefore, the amount of basin inflow provided by a given amount of precipitation is very unpredictable. For example, at a typical marsh, the highest May to October precipitation produced less basin inflow thant the lowest May to October precipitation. Also the highest November to April basin inflow occurred when the precipitation for that period was less than normal. At another marsh, the basin inflow in 1963 was 80 mm and in 1964 it was 600 mm.

Despite the unpredictable variability of the water supply, a large percentage of the marshes hold water throughout the breeding season (June to August) every year.

EVAPOTRANSPIRATION

In the study, evapotranspiration losses were measured for the May to October period from 1960 to 1964, using the mass-transfer method developed for the project, (Eisenlohr, 1966a). The results, for ponds clear of vegetation, were very close to published regional values, (Kohler, Nordenson and Baker, 1959).

An objective of the study was to determine the effect of emergent aquatic vegetation (hydrophytes) on evapotranspiration losses. It is generally accepted that at the height of the growing season, transpiration by hydrophytes plus evaporation from the water surface will exceed evaporation from a clear water surface. However, hydrophytes, by their presence, reduce evaporation by shielding and sheltering the water surface. Thus, in vegetation the wind speed at the water surface was generally less than a third of the speed over open water. They also shelter the water surface from incoming radiation that would otherwise be available for evaporation. The vegetation absorbs this energy for its own use, especially for transpiration.

The effect of vegetation in reducing evaporation is cyclic. As the plants grow above the water surface in the spring they progressively stop the wind and shelter the water, so that evaporation should be progressively reduced reaching a minimum at the time of maximum growth, which was assumed to coincide with maximum height. Evaporation should then continue at the minimum rate until the vegetation has died and starts to break down as a result of weathering by wind and snow. As the dead stalks are broken down, evaporation can be expected to increase, reaching a maximum just before new growth starts the following spring.

The net effect of these processes made the total seasonal evapotranspiration losses about 15% less than the evaporation lossess from a nearby similarly situated pond clear of vegetation. The growing season in North Dakota is short, however; only about 120 days. It could be expected that in other regions with longer growing seasons, an opposite relationship would exist.

The transpiration rate of hydrophytes, for the period during which plants were fully active, was found to have a highly significant correlation with the emerged height of stem. By means of this relationship, transpiration was computed separately from evaporation using the mass-transfer method, (Eisenlohr, 1966b). During the decline of activity in the autumn, a significant correlation was found between transpiration and moisture content of the hydrophytes, (Eisenlohr, 1969).

A seasonal (May to October) water budget for a typical marshy pond containing hydrophytes is:

Evaporation	330 mm
Transpiration	200 mm
Net seepage outflow	150 mm
Total decrease	680 mm
Precipitation	440 mm
Basin inflow	80 mm
Total increase	520 mm
Computed decrease in storage	160 mm
Measured decrease in storage	160 mm
Residual error	0

As no term in this budget is a residual, the budget could not be expected to balance. The fact that it does is purely fortuitous.

SEEPAGE OUTFLOW

In making the mass-transfer computations of evapotranspiration, seepage outflow from a pond has to be considered as constant. The rate seldom exceeded 3 mm per day per m^2 of water surface. Variations in seepage outflow rates were noted, but the variations were significantly small in relation to evapotranspiration rates for little error to be introduced by assuming a constant rate of seepage outflow.

No satisfactory explanation of the variation in seepage outflow rates was found until after evapotranspiration calculations had been made. At that time a computer program became available from which it was determined that there was a significant correlation between seepage outflow rate and the rate of decline in water level in a nearby observation well.

It has been shown that outflow seepage from a marshy pond filled with hydrophytes is very much greater than a pond clear of vegetation. It is known that some species of hydrophytes require bare soil for germination. Therefore, it could be expected that marshes with higher seepage outflow rates would become dry more often, and thus allow more frequent opportunities for the germination of hydrophytes, than a marsh with lower seepage outflow rates that remains clear.

GROUNDWATER

The water table was at shallow depth in the study area. A water table is that surface within the zone of saturation at which the pressure is atmospheric. It follows the contours of the land surface rather closely and in places it is very steep. This is because there is little lateral movement of groundwater through the till. There were large fluctuations in vertical movement of the water table caused by vertical flow of water near the land surface. Water from precipitation reached the water table rapidly through the permeable soils and then returned to the atmosphere through transpiration by the vegetation, evaporation from the soil surface, and possibly as water vapour moving upward through the soil from the water table. Diurnal fluctuations in the water table were noted in relation to the demands of evapotranspiration, (Sloan, 1972).

A special situation existed near the shores of the marshy ponds. There is evidence that groundwater can move laterally without difficulty for several metres from the pond. This can be explained, at least in part, by layers of drift from which much of the fine material had been removed, probably by ancient wave action. Also, a line of observation wells extending away from a marsh shows that the water table slopes away from the pond. This results from evapotranspiration losses from the ground surface being replaced by water from the marsh. Although there is evidence that a very small portion of seepage outflow moves vertically downward from the marsh, it is now believed that most of the seepage outflow moves laterally to be dissipated by evapotranspiration on the shore. As this is the same process that lowers the water table generally it explains the correlation between the seepage outflow rate and the rate of decline of the water table. It also explains why Millar (1971) found that the length of shoreline per unit area of water surface is a meaningful factor for comparing seepage outflow rates. In view of these factors it is concluded that almost all the water received from precipitation is dissipated locally.

The most important effect of groundwater movement is the way the direction of flow controls the quality of water in the marshes. It has been emphasized that the lateral movement of groundwater is slow, especially in relation to vertical movement. This does not mean, however, that there is no lateral movement and is is this slow lateral movement that controls the quality.

WATER QUALITY

All water entering a marsh contains dissolved solids. The only effective ways for these solids to be removed are by overflow and seepage outflow because evaporation and transpiration remove only pure water. Most marshy ponds overflow rarely if at all. If there is also no outflow seepage, the solids are concentrated in the pond and the water becomes brackish to briny. If there is good seepage outflow, the waters may remain rather fresh, depending on the amount of dissolved solids in the inflowing waters. An intermediate quality of water indicates an intermediate relationship between inflow and outflow. To a limited extent, the concentration of dissolved solids is dependent on the volume of water in a pond, which varies with season. Annual variations in any one pond were not great, and there was no systematic variation in salinity with geographical location. In ponds that had large concentrations of dissolved solids, the dominant salt was almost always sodium sulphate.

VEGETATION

Many species of hydrophytes grow in the marshes. Some genera found frequently were: *Alisma; Carex; Scholochloa; Scirpus* and *Typha*. Most hydrophytes have a limited tolerance to water of a quality much different from the optimum value for their growth. This intolerance is so marked that the species of hydrophytes growing at any one time in a marsh are excellent indication of the quality of the water in which they are growing; Stewart and Kantrud (1972) discuss this in detail.

Vegetation in any marsh was usually composed of several species, often grouped into distinct plant associations related to a range of salinity to such an extent that it was possible to infer the salinity present. Where species related to different ranges of salinity were growing together the ranges usually overlap and the salinity indicated is that of the overlap. Vegetation is such a good indicator that in several places it was used to identify an area where much fresher water, from a spring or seepage inflow, was entering a more saline marsh.

ACKNOWLEDGMENTS

The work described in this paper was a project of the US Geological Survey. It was first reported in a contribution to the International Hydrological Decade, (Eisenlohr, 1965). The present paper is based on the author's final report, (Eisenlohr *et al*, 1972), which contains complete descriptions of the work and more than 40 bibliographic citations. A general report (Eisenlohr and Sloan, 1968) has also been published.

REFERENCES

Eisenlohr, W. S. 1965. Hydrology of prairie potholes in north-central United States. Int. Assoc. Sci. Hydro. Bull. 10(3), pp. 49-50.

Eisenlohr, W. S. 1966a. Determining the water balance of a lake containing vegetation. Int. Assoc. Sci. Hydro. Publ. 70, Symp., Garda 1966, vol. 1, pp. 91-99.

Eisenlohr, W. S. 1966b. Water loss from a natural pond through transpiration by hydrophytes. Am. Geophys. Union Water Resources Res. vol. 2, no. 3, pp. 443-453.

Eisenlohr, W. S. and Sloan, C. E. 1968. Generalized hydrology of prairie potholes on the Coteau du Missouri, North Dakota. US Geol. Survey Circular 558.

Eisenlohr, W. S. 1969. Relation of water loss to moisture content of hydrophytes in a natural pond. Am. Geophys. Union Water Resources Res. vol. 5, no. 2, pp. 527-530.

Eisenlohr, W. S. *et al.* 1972. Hydrological investigations of prairie potholes in North Dakota, 1959-68 US Geol. Survey Prof. Paper 585A.

Kohler, M. A., Nordenson, T. J. and Baker, D. R. 1959. Evaporation maps for the United States. US Weather Bureau Tech. Paper 37.

Millar, J. B. 1971. Shoreline-area ratio as a factor in rate of water loss from small sloughs. J. Hydrol., Amsterdam. vol. 14, no. 3/4, pp. 259-284.

Sloan, C. E. 1972. Groundwater hydrology of prairie potholes in North Dakota. US Geol. Survey Prof. Paper 585c.

Stewart, R. E. and Kantrud, H. A. 1972. Vegetation of prairie potholes, North Dakota, in relation to quality of water and other environmental factors. US Geol. Survey Prof. Paper 585D.

THE HYDROLOGICAL REGIME AND WATER BALANCE OF DRAINED LANDS OF THE FOREST ZONE OF WEST SIBERIA

E. D. Vvedenskaya and I. F. Rusinov

West Siberian Branch of All-Union Research Institute of Water Engineering and Reclamation, Tyumen.

SUMMARY: The paper gives the results of three years of study of drainage flow on the first experimental site in West Siberia where plastic drains were used in thick peatlands.

Observations to date of runoff, groundwater, evaporation and soil moisture enable both quantitative and qualitative statements to be made about the drainage flow. Detailed studies of the dynamics of spring drainage have shown that it is similar to the surface flow from experimental plots. The data obtained have confirmed the feasibility of drainage under the severe conditions found in Siberia and may serve as preliminary recommendations for the design of plastic drains in the forest zone of West Siberia.

REGIME HYDRAULIQUE ET BILAN HYDRIQUE DES TERRAINS DRAINES DANS LA ZONE FORESTIERE DE LA SIBERIE OCCIDENTALE

RESUME : Cette étude donne les résultats de trois années de recherches sur l'écoulement de drainage de deux sites expérimentaux de tourbières puissantes en Sibérie occidentale où, pour la première fois, les drains utilisés étaient faits de matière plastique.

En observant le ruissellement, les eaux souterraines, l'évaporation et le dregré d'humidité du sol, on a pu déterminer qualitativement et quantitativement les caractéristiques de cet écoulement. L'examen détaillé de la dynamique de l'écoulement de drainage au printemps a montré son analogie avec l'écoulement superficiel des petites surfaces (parcelles expérimentales). Les données obtenues ont confirmé la possibilité de procéder à des drainages dans les conditions climatiques rigoureuses de la Sibérie. Elles peuvent servir à formuler des recommandations préliminaires pour le projet de drainage avec des canalisations en matière plastique dans la zone forestière de la Sibérie occidentale.

To study the feasibility of using closed plastic drains and to test their operation under the conditions of the souther forest zone of West Siberia, two experimental sites on peatlands (sites I and II) were selected in autumn 1968 and 1969, respectively, by the West Siberian Branch of All-Union Research Institute of Water Engineering and Reclamation. Site I is a 3 ha plot on the Salairsky State Farm, 35 km north-west of Tyumen. Site II is a 14 ha plot in the Svobodny Trud Collective Farm, 100 km north-west of Tyumen. Drains (50 mm diameter) were installed with a DPBN-1,8 mole drainage machine designed at the Institute.

Nine drains were installed on site I, at a spacing of 20 m, depth 0.9 m and slope 0.003. The drains discharged to the main canal.

Drains on site II were spaced at 20, 30 and 40 m, with 6 drains for each spacing. The slope was 0.002, the depth 0.9 to 1.1 m and the drain length 150 m. The drains were brought to an open collector which discharged to the main canal.

Both experimental sites were located in the Tarmany peat area, where soil was low moor, poorly decomposed and 1 to 2 m thick. The medium and youger peats were underlaid by loam with a sandy loam layer. Inflow to the bogs was from groundwater and by direct precipitation.

Investigations were conducted on both sites throughout the year. The quantity of drainage water, precipitation, air temperature and humidity, the accumulation and melting of the snow cover, soil moisture content, soil temperature, groundwater levels and evaporation were measured.

Three years of study has yielded preliminary information on the distribution, quantity and drainage volumes and on other components of the water balance of the drained area.

The climate varied throughout the experimental period. During the 1968-1969 winter, air temperature fell to -52°C, when the frozen layer was 900 mm thick. During the two subsequent years, minimum temperatures did not fall below -45°C and before snowmelt occurred the depth of frozen soil had reached 700 mm. Soil water melted both from below and from above, at an average rate of 1 mm per day and 8 mm per day, respectively. Complete melting had taken place between the end of July and the beginning of August, but some drainage occurred in the frozen layer.

Throughout the 3 years, the drainage system operated well during spring, summer and autumn. In winter when the water table was low, there was no flow in the drains. No deformation of the drainage system was observed.

In the spring, flow in the drains begins as soon as the air temperature exceeds 0°C. In 1969, 1970 and 1971 this occured on April 7, 1 and 12, respectively (the average date is April 10). Because of backwater submersion of drain outlets, the flow from the drains was measured on only 5 to 13 days. In summer and autumn, when either a high water table prevailed at the sites or during high intensity rainfall, the drainage system again did not operate.

The patterns of spring and summer drainage differ depending on their origin. The spring drainage regime is similar to that of surface runoff from experimental plots, whereas the summer-autumn drainage regime ressembles that of the total flow in the drainage ditches as it is controlled mainly by the groundwater level. Because of these differences of behaviour the drainage flow was analysed separately during both the spring and summer-autumn periods.

In 1971 a second spring flood occurred which affected the specific discharge. In addition the value of spring flow was considerably underestimated on account of submission of drain outlets. The duration of spring submission in the period from 1969 to 1971 was 6, 46 and 23 days, respectively.

Table 38. Drainage flow data at the Salairsky State Farm, Tyumen Region.

year	duration			specific discharge l/s ha			runoff volume m^3/ha
	start	end	No. of days	mean	maximum daily mean	maximum	
			Spring-summer				
1969	April 7	April 21	13	0.05	0.22	2.22	56.2
1970	April 1	April 5	5	0.07	0.14	1.25	30.2
1971	April 12	April 13	5	0.03	0.05	0.16	13.0
	April 19	April 22					
			Summer-autumn				
1969	June 25	Aug 31	68	0.0128	0.0332	0.0437	75.3
1970	May 21	Sept 29	131	0.0234	0.0773	0.2620	264.0
1971	May 15	Dec 10	178	0.0023	0.0290	0.0400	35.3

In 1971 the drainage flow was measured in November and part of December. Because the origin of this flow is similar to that which occurrs in summer-autumn, it was included with the latter. Moreover, during the summer season, the drainage system operated for over a month under backwater conditions on account of heavy and prolonged flood levels in the drain outlet structure. For this reason it was not possible to measure the runoff.

The present analysis of the annual distribution of the drainage flow does not reflect the true relationship because the flow during periods of backwater was not considered. It is, however, possible to suggest that the highest summer-autumn discharge occurs in July, and that the runoff coefficient for the summer of 1970 was 0.47.

The three year maximum specific spring discharge had a mean value of 1.22 l/s ha, while that of summer-autumn had a value of 0.12 l/s ha. The average value of two years of summer-autumn discharge on site II (where no runoff was observed in spring as the outfalls were submerged) was 0.3 l/s ha. Under the conditions found in southern forest zones of West Siberia, the drainage system must, therefore, be desi ed for spring flows.

During, thereore, the summer-autumn period, the drainage discharge decreases with time when the water table is high and increases after intensive rainfall. No discharge occurs with a low groundwater table.

The spring hydrograph shows a distinct diurnal cycle. A rise in discharge accompanied the onset of snowmelt to give a peak flow between 3 and 5 p.m. As the snow melts, the maximum discharge increases, the time of the peak gradually approaching noon which is the time when the maximum temperatures and solar radiation occur. At the end of the runoff period the peak generally occurs between 11 a.m. and 1 p.m. The hydrograph of spring drainage flow is similar to that of spring surface flow from small watersheds and has a triangular or trapezoidal shape (Figure 58). The latter shape is a function of the rate of snowmelt and is not due to surcharging of the drainage pipes.

Similarity between spring drainage and surface flows may be explained by the fact that both originate from snowmelt. Surface melt water enters the drains through cracks and through the frozen soil. However, after backfilling and compaction the cracks in the top layer practically disappear. The work of Shalabanov (1902), Komarov (1957), Kuznik (1954), Pavlova (1970) and others show that frozen soils do absorb meltwater.

The results of a number of studies of the rate of evapotranspiration and its distribution over the forest zone of West Siberia have been published. The evapotranspiration from drained lands naturally differs from the general regional data, as has been shown by the author during the course of a comparative study of evaporation.

Evapotranspiration from drained lands sown with a vetch-oats mixture at the Salairsky State Farm was estimated by three independent methods. These were the water balance of the cropped lands, the water balance of isolated soil blocks (GGI - 500 - 50 soil evaporation pans), and a calculation based on climatic data, using the methods of Konstantinov (1963) and Mezentsev (1969).

Table 39. Evapotranspiration at drainage Site I (Salairsky State Farm).

Months	Evapotranspiration during the growing season, mm								
	Long-term data				Data for 1970				
	Budyko	Oldekon	Tyurk	Mezentsev (map)	Romanov	Mezentsev	Konstantinov	Water balance	GGI-550 evaporation pans
May	95.6	38.4	39.4	85	111.1	75.1	57.8	-	-
June	98.0	44.8	45.9	96	122.0	81.0	79.1	122.5	-
July	79.2	57.6	57.5	75	110.2	64.2	74.0	173.4	144.2
August	60.0	47.9	45.3	59	80.7	49.1	48.0	134.1	123.8
September	38.7	34.4	33.2	35	45.4	30.8	32.0	48.8	33.0*
Total	371.5	223.1	221.3	350	469.4	300.2	290.9	(478.8)	(301.0)

* September 1 to 17 only

Calculations of the evaporation from soil pans were made using the following equation:

$$E = x + 0.02\ (P_1 - P_2) - y \qquad (1)$$

where: E is the amount of water evaporated from the soil block, mm;
x is the precipitation, mm;
P_1 and P_2 are the antecendent and subsequent weights of the block in the evaporation pan, g;
y is the flow into the bottom of the pan, mm.

The water balance equation on cropped areas for the summer period takes the form:

$$x + \Pi = E + f + \Delta W + y_s \qquad (2)$$

where: x is precipitation, mm;
Π is the flow from lower layers, mm;
f is the infiltration to the layer below that considered, mm;
ΔW is the difference between initial and final moisture content in the soil layer considered, mm;
y_s is the surface runoff, mm;
E is evapotranspiration, mm.

During the summer there is no surface runoff from the cultivated bogs, and hence $y_s = 0$. Because the thickness of the soil layers under consideration is 1.5 m, it can be assumed (Shebeko, 1965) that the total summer rainfall will normally be held in this layer, hence f = 0.

The normal evapotranspiration for the experimental site was also determined by the formulae of Budyko (1948), Konstantinov (1963), Oldekop (1911), Tyurk, Romanov (1962) and from the map given by Mezentsev (1969). The results obtained are summarized in Table 39.

The results indicate that evaporation pans give an average 12% underestimate of evapotranspiration. This difference is considered to be sufficiently insignificant to confirm the suitability of pans for the calculation of evapotranspiration. The other methods are not applicable to the calculation of evapotranspiration from drained peat lands of the forest zone of West Siberia. Design formulae which take into account local features such as groundwater level, soil moisture capacity and mechanical composition of soil give better results. Thus, Romanov's method gives a difference for the months of June to September of only 25%. The distribution obtained by using the methods of Tyurk and Oldekop follow more closely the distribution given from the water balance data; the maximum evapotranspiration is found to occur in July.

The heat-balance method gives results within 1% of those given by the balance methods in drained peat lands.

The first data obtained on the principal components of the water-balance of drained lands in the Southern forest zone of West Siberia demonstrated that further investigation is needed. This work is now underway and will form the basis for the design of more advanced reclamation systems in West Siberia.

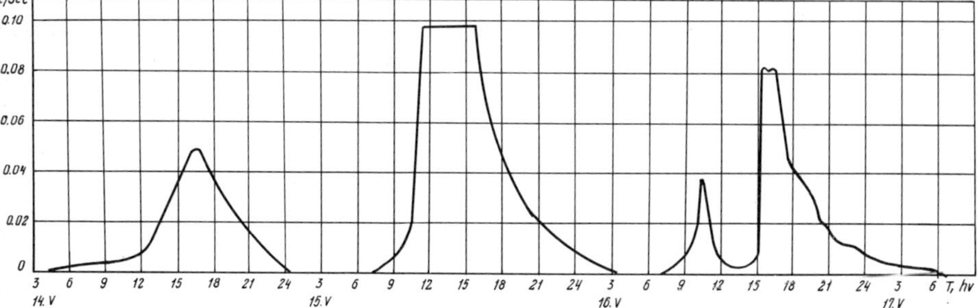

Fig 58. Hydrograph of spring drainage flow, the Salairsky State Farm.

REFERENCES

Budyko, M. I. 1948. Isparenie v estestvennykh usloviyakh. Evaporation in natural conditions. Leningrad, Gidrometeoizdat.

Iozopaitis, A. V. 1971. Issledovaniya gidrologicheskogo rezhima malykh rek Litovskoi SSR s basseinami osushennymy goncharnym drenozhom. Hydrological investigations of small rivers in the Lithuanian SSR with basins drained by Ceramic drainage. Vilanyoi.

Komarov, V. D. 1957. Laboratornye issledovaniya vodopronitsaemosti merzloi pochvy. Laboratory studies on the permeability of frozen soil. Trudy Tsentralnogo Institutal Prognozov, issue 54 Moscow.

Konstantinov, A. R. 1963. Isparenie v prirode. Evaporation in nature. Leningrad, Gidrometeoizdat.

Kuznik, I. A. 1954. Nekotorye fakticheskie dannye o vodnofizicheskick svoistvach merzlykh pochv i poverkhnostnom stoke talykh vod. Some data on the hydrophysical properties of frozen soils and surface runoff due to snowmelt. Trudy po agronomicheskoi fizike, issue 7.

Mezentsev, V. S., and Karnatsevich, I. V. 1969. Wlazhnennost Zapadno-Sibirskoi ravniny. Moisture content in the West Siberia Plain. Leningrad, Gidrometeoizdat.

Oldekop, E. M. 1911. Ob isparennii s poverkhnosti rechnyka basseinov. On surface evaporation from river basins.

Pavlova, K. K. 1970. K voprosu o fazovom sostave vody i teplophizicheskikh kharakteristik merzlogo torfa pri izuchenii infiltratsii. On phase composition of water and thermal properties of frozen peat in seepage studies. Trudy GGI, issue 182. Leningrad, Gidrometeoizdat.

Romanov, V. V. 1962. Isparenie s bolot evropeiskoi territorii SSSR. Evaporation from marshes in the European part of the USSR. Leningrad, Gifrometeoizdat.

Shalabanov, A. A. 1902. Propuskaet li vodu merzlaya pochva? Is frozen soil permeable to water? Pochvovedenie, No. 3.

Shebeko, V. F. 1965. Isparenie s bolot i balans pochvennoi vlagi. Evaporation from swamps and soil moisture balance. Minsk, Uroshai.

Tyurk, L. Balans pochvennoi vlagi. Soil moisture balance. Leningrad, Gidrometeoizdat.

DISCUSSION

N. M. Fedorova (USSR) - What is the thickness of the layer melted from below and what reason can you suggest for melting?

E. D. Vvedenskay (USSR) - The soil is melted from below and from above. The melting rate from below is smaller (1 mm per day) than the rate of melting from abouve (8 mm per day). The thickness of the layer melted from below is 100 mm. Melting from below is mainly caused by heat flux from groundwater.

I. P. Kalyzhnyi (USSR) - You have observed a drainage flow in spring. Have you investigated water absorption and infiltration in frozen peat soils?

E. D. Vvedenskay (USSR) - We have not made direct measurements of infiltration into frozen soils. Indirect conclusions concerning infiltration may be made from the good correlation obtained between the drainage flow and air temperature.

THE WATER AND HEAT BALANCE OF THE FORESTS OF BYELORUSSIAN POLESSIE

O. A. Belotserkovskaya

Byelorussian Research Forestry Institute, Gomel, BSSR.

SUMMARY: This paper gives the results of investigations of the components of the water and heat balances, and of the productivity of marsh-ridden and dry land forests of Polessie. The research programme was carried out at the Experimental Station of the Forest Research Institute, situated in the marsh-ridden forest near Tomichevo. The types of forest investigated are representative of all woodlands of the Polessie region.

The experimental procedure and results are described in detail. The relationship between productivity of the forest stand and radiation, the transfer of heat and moisture in the active forest layer, and transpiration of the forest stand are discussed. The input and output components of the water and heat balances are analysed for all types of forest investigated as well as for the entire watershed.

ETUDE DU BILAN HYDRIQUE ET DU BILAN THERMIQUE DANS LES FORETS DE LA POLESIE

RESUME : L'auteur traite de l'étude et du calcul des éléments du bilan hydrique et du bilan thermique, ainsi que du rendement de forêts situées dans les terrains marécageux et arides de la Polésie. Le programme de recherches a été réalisé à la station expérimentale établie en 1967 par le Centre de recherches forestières de Biélorussie dans le massif forestier et marécageux de la commune de Tomichevo. Le sol et les caractéristiques hydrologiques de cette région sont typiques des forêts de la Polésie.

L'auteur décrit en détail les méthodes et les données utilisées. Il montre les variations du rendement du peuplement forestier en fonction du rayonnement, de l'échange de chaleur et d'eau dans la strate active de la forêt et de la transpiration végétale. Les éléments du bilan thermique et du bilan hydrique sont analysés pour chaque type de forêt étudié ainsi que pour l'ensemble du bassin versant.

Intensive land reclamation is now underway in the woodland district called the Byelorussian Polessie. This requires comprehensive investigations of the natural conditions, including the hydrological characteristics. Because Polessie comprises the highest proportion of forest land (about 31%) in Byelorussia and that this area is nearly one half marsh-ridden, a detailed knowledge of both water and heat balances in wetland forests and in woods adjacent to drainage areas is of great scientific and practical interest.

Every year during the growing season a complex research programme involving measurements of the components of the water and heat balances and investigations of forest stand productivities is undertaken at the Hydrological Forest Experimental Station of the Byelorussian Forest Research Institute. This Institute was established in 1967 near Tomichevo in the catchment of the marsh-ridden Kaptsevichi-Petrikovo forest.

This marsh-ridden area is due to be reclaimed as woodland between 1972 and 1974. From the hydrological, geomorphological and pedological viewpoint this catchment is typical of all woodlands of the Polessie region.

The investigations were carried out both in marsh-ridden woods with a high water table, the level of which varies in spring from 200 to 40 mm (pine, sphagnum and birch-cowberry forests), and in dry woods with either a low water table varying from 1100 to 2000 mm below the surface during spring (pine, heather and pine-lichen) or a high water table between +200 and zero mm below ground, in the spring (middle-aged and young pine, bilberry). To obtain reliable data on the water and heat balances, special forest experiments are required in which the following extensive and complex measurements are made simultaneously:

1. Measurements of incoming and reflected radiation at two metre intervals from the soil surface to the tree tops and beyond to the upper limit of the turbulent layer. These measurements are made in all the above species on wooden masts, which are sited to avoid disturbance of the surrounding forest vegetation and of the understorey relief. Remote measurements of the radiation components are also useful in this respect.
2. Measurements of biomass temperature, air temperature and humidity in the active layer are made following the same procedure as above.
3. Measurements in the unsaturated part of the root layer, of the surface and bulk temperatures of the soil, its moisture content and the heat flux into the soil are made at depths of 20, 50, 100, 200, 300, 400, 500, 800 and 1000 mm.
4. Runoff measurements in the forest and measurements of streamflow at the exit from the river basin are under study.
5. Measurements of variations in groundwater level in different types of forest.
6. Measurement of precipitation above and below the canopy, wind speed along vertical profiles, cloudiness and general climate.

Observations are made at 07.00, 08.30, 10.00, 13.00, 16.00 and 19.00 hours and hourly throughout the day during periods of intensive observation.

The investigations made use of new electron remote measuring instruments to measure radiation, soil temperature, temperature, albedo and humidity; all instruments were manufactured especially for the Byelorussian Forestry Research Institute for investigations of the gradient of these variables in the forest environment. After quality control of the data obtained, the values of turbulent heat transfer, evapotranspiration, evaporation and transpiration are estimated. Turbulent heat transfer is calculated as the residual term in the thermal balance and by the eddy diffusion method. Evapotranspiration is estimated from the thermal balance formula given by Budyko and the heat balance equation:

$$R_b - Q_a - P - LE - LPh = 0 \qquad (1)$$

where: R_b is the surface net radiation;
Q_a is the heat flux into the active forest layer (biomass and soil) ($Q_b + Q_h$);
P is the turbulent heat flux to the atmosphere;
LE is heat lost to evaporation;
LPh is heat lost to forest vegetation photosynthesis.

Evaporation is also estimated by the heat balance method but with reference to the active soil layer (Belotserkovskaya, 1969).

$$\pm E_\phi = \frac{10(\Delta Q_h - c\rho . h_l . \Delta t)}{L} \quad \text{mm/hr} \qquad (2)$$

where: ΔQ_h is the difference between the values of heat flux to the upper and lower boundaries of the layer under consideration, $cal/cm^2.hr$;
$c\rho$ is the specific heat of the layer, $cal/cm^3.^oC$;
h_l is the layer thickness, cm;
Δt is the temperature rise of the given layer during a given time interval (min., hour, day, etc.), ^oC;
L is the latent heat of evaporation, cal/g.

All the terms in the above formula are obtained by direct measurements made in the forest.

Transpiration is estimated by the difference from equations (1) and (2) and is therefore obtained without disturbing the biological processes in the forest stand.

The agreement shown by the heat balance components obtained by different methods, indicates the reliability of the predicted values. In the above case the overall difference between values of the heat balance components did not exceed 10%. It should be noted that the difference between predicted values of net radiation from those measured was 2 to 6%; the measured heat fluxes and those predicted by Tseitin's formula were within 7 to 10% of each other; evaporation predicted from the heat balance formula and the heat balance equation and by turbulent transfer theory differed from the measured values by 10% and 2 to 10% depending on weather conditions and forest types, respectively.

BRIEF DESCRIPTION OF THE RESEARCH SITES

Measurements relating to the water and heat balances were made in 18 year old high density pine stands of bonitet II and in 38 to 45 year old high density pine and birch stands. Combinations of forest and understorey stypes were:

18 year old pine-heather (Water Table (WT) at 940 to 1840 mm);

18 year old pine-bilberry (WT at 150 to 1120 mm)
45 year old pine-lichen, bonitet III (WT at 1900 to 2790 mm);
45 year old pine-bilberry, bonitet II (WT at 410 to 1140 mm);
38 year old pine-sphagnum, bonitet V (WT at 200 to 860 mm);
38 year old birch-cowberry, bonitet II (WT at 410 to 1490 mm).

It should be noted that the species studied and their condition of growth are typical of the forests of the Polessie region. Some 63% of the Polessie woodlands are coniferous forest, of which 60.1% is pine. The birch is the most widespread deciduous tree.

A field, of which half is under grass-cover, 50 mm in height, was also placed under observation. The field is on a flat site similar to that of the pine-lichen. The water table is situated at a depth varying between 1390 and 2700 mm.

RADIATION BALANCE

The radiation regime of the forest is extremely complicated and differs from that of fields on other landuses. For example, in clear, cloudless weather the mean diurnal net value of radiation in the open field is 40% less than in the forest (see Fig 59). The difference between the net radiation in the forest and in the field varies between 15% and 30% depending on weather conditions and type of forest, because albedo back radiation varies with forest type. The vertical profile of net radiation varies during a day, but in general the shapes of the distribution curves of solar energy at various levels in the forest are similar to those of the crown. Vertical values of net radiation under the crown are identical and its temporal distribution at a given height is almost uniform with only a small amplitude variation (0.2 to 0.3 cal/cm^2min).

The highest transmission coefficient of incident radiation through the crown is observed in pine-sphagnum (0.35), while the lowest is found in birch forests (0.13). It was also found that for an increase of 0.1 in the stand density, all other things being equal, the transmission coefficient (K_t) is reduced by a factor of 1.6. The K_t coefficient is inversely proportional to the proportion of incoming radiation retained by the crowns. This proportion has a direct qualitative correlation with stand productivity. Thus, the largest volume increment is found for birch (9.5 m^3/ha) and pine-bilberry (7.74 m^3/ha). The lowest increment corresponds to pine-sphagnum (3.0 m^3/ha). The highest retention of incident radiation by crowns is observed in birch forests and pine-bilberry (about 88 to 87%) and the smallest in pine-sphagnum (66%). The same relationship holds for active photosynthesic radiation (AFR) since it is directly proportional to the net radiation of the active forest layer and amounts to 37% of the net radiation for each type of forest in May, June and July and 45% in September and October.

HEAT AND MOISTURE TRANSFER IN THE ACTIVE FOREST LAYER

Because 66 to 68% of incident radiation is absorbed by the crown only a small portion of solar energy is available for heat and moisture exchange with the atmosphere in the active forest layer.

The heat flux into the soil (Q_h) is directly proportional to K_t. In open fields Q_h is 3 to 5 times higher than in dense pine and birch stands and 1.5 to 3 times higher than on moorland. The heat flux resulting from heat accumulation in the forest biomass (Q_{bm}) is on average only 2 to 5% greater than Q_h. There is a time lag of 1 to 2 hours between heat exchange in the biomass and heat exchange in the soil.

The optimum temperature regime in the root layer is observed in medium-age pine-bilberry and birch forests, i.e. in the types of woods where the best growth is observed. The most unfavourable temperature conditions, such as sharp fluctuations and overheating of the upper layers by as much as 19 to 25°C, are found in pine-sphagnum. The temperature regime of the root system of pine-lichen and pine-heather is also well below the optimal conditions, especially in July when the moisture content of the top one metre of soil approaches wilting point.

EVAPOTRANSPIRATION, EVAPORATION AND TRANSPIRATION

The present author's observations have shown that irrespective of forest type and plant cover the diurnal behaviour of evaporation and transpiration follows the net radiation variation of the day. However, the rates of moisture consumption vary with types of forest.

Table 40. Rates of moisture consumption for different types of forest (mm/day).

Type of forest	Months					
	May	June	July	Aug	Sept	Oct
Pine-lichen	3.2	3.8	3.7	2.4	1.5	1.0
Pine-heather (young)	2.7	3.3	3.2	2.2	1.4	0.9
Pine-bilberry (young)	2.7	3.3	3.0	2.0	1.3	0.7
Pine-bilberry	3.6	4.5	4.3	3.2	2.1	1.1
Pine-sphagnum	3.3	4.0	3.8	2.5	1.6	1.0
Birch-cowberry	3.4	4.3	4.1	3.0	1.9	0.9

Pine-bilberry, birch, and pine-sphagnum evaporate the most water, and the high evaporation from stunted pine-sphagnum is probably due to very wet soils and significant evaporation from the saturated moss cover (the evaporation from moss represents 29 to 30% of the total evaporation). Many factors contribute to the evapotranspiration of the forest, including transpiration from the forest, evaporation of precipitation, evaporation from the understorey and transpiration of the understorey and of grass and moss cover. It should be noted that the relative importance of these variables changes with type of forest. Thus, birch forest appears to have the smallest evaporation (E_{ph}), with a mean daily value of about 6% net radiation, whilst the understorey of pine-bilberry has the highest value, (see Fig 60).

It follows that the highest transpiration rates occur in cowberry-birch forest (461 mm) and in pine-bilberry (427 mm). The lowest transpiration rates are those of pine-sphagnum (264 mm). The values given here apply to the entire growing season. The understorey of forest stands of both bilberry and pine-heather transpire about 275 mm during the growing season, which is slightly more than medium-aged pine-sphagnum. Evapotranspiration from open field surfaces is less than that from forest, the difference depending on the type of forest and on weather conditions. Medium aged and young pines transpire 15 to 30% and 5 to 10% more respectively. A direct relationship between transpiration and forest stand growth has been found which was similar to that between growth and the radiation absorption of crowns. Evapotranspiration has been found to be independent of grounwater levels, although it is often indirectly affected by these through the moisture content of the upper root layers. Evapotranspiration has been found to occur from the upper layer (0 to 100 mm); below this depth condensation dominates. Evaporation amounts to 20% of evapotranspiration at most. The transformation in water phase in the root layer hardly affect the total moisture content but are extremely significant with regard to heat and mass transfer in the soil.

The movement of moisture by evaporation consumes energy, (Belotserkovskaya and Romanov, 1969; Romanov, 1961; Shebeko, 1970 a and b). Figure 62 shows the net radiation plotted against evapotranspiration for various types of forest. The relationship is linear and defined by the equation

$$E_s = \alpha R_b \pm C \tag{3}$$

Values of coefficient α were determined by the method of least squares. The root-mean-square deviation is ± 0.0064 to ± 0.0081. The value of α varies for different types of forest or for forests of the same type but of different ages.

HEAT AND MOISTURE BALANCE FOR VARIOUS TYPES OF FOREST DURING THE GROWING PERIOD

The energy needed for evapotranspiration accounts for between 64 and 97% of net radiation; of this transpiration accounts for between 49 and 85%, while heating of forest soils accounts for 0.3 to 4%.

For heat and mass transfer in the active forest layer between 0 and 40% net radiation is used, the turbulent heat transfer decreasing as the rate of evaporation increases. The most active heat and mass exchange was observed at the beginning and at the end of the growing season and amounted to 46% of the net radiation in deciduous forest, and to 34% in coniferous. In the middle of the growing season less than 10 to 20% of the available energy was used on heating the air in the forest.

Evaporation from the forest also appeared to be a major output component of the water balance: between 82 and 100% of precipitation was lost to evaporation and only 8.6% became runoff. Evaporation in June and July always exceeded the precipitation and in June it was from 1.3 to 2 times greater and in July from 1.5 to 2.5 times greater than the precipitation. The variation depended on the type and species of forest, as well as its age, stand, density and the weather conditions during these months. Evaporation loss is accompanied by a general decrease in the moisture storage of the most active root layer, the depth of which varied from 50 to 100 mm according to the level of the water table. In those types of forest where the water table fell below 1500 mm (pine-heather and lichen), the soil moisture content almost reached wilting point.

During the growing season the forest acts as a powerful 'pump' which removes moisture from the soil. The moisture deficit observed in the upper root layers in the period of the highest evaporation affected the productivity of the stand less than the excessive moisture in these root layers. This is illustrated by the fact that the volume increment of pine-lichen is twice as great as that of pine-sphagnum.

REFERENCES

Belotserkovskaya, O. A. and Romanov, V. V. 1967. In: Priroda bolot i metody ikh issledovaniya. (Collected Papers). Leningrad, Nauka.

Belotserkovskaya, O. A. and Romanov, V. V. 1969. Issledovanie poverkhnostnogo i vnutrizalezhnogo ispareniya na verkhovykh bolotakh. Study of surface and bulk evaporation from raised bogs. Trudy GGI, issue 177, Leningrad, Gidrometeoizdat.

Romanov, V. V. 1961. Gidrofisika bolot. Hydrophysics of swamps. Leningrad, Gidrometeoizdat.

Shebeko, V. F. 1970a. Isparenie s bolot i balans pochvennoi vlagi. Evaporation from swamps and soil moisture balance. Minsk, Urozhai.

Shebeko, V. F. 1970b. Gidrologicheskii rezhim osushennykh territorii. The hydroligical regime of areas under drainage. Minsk, Urozhai.

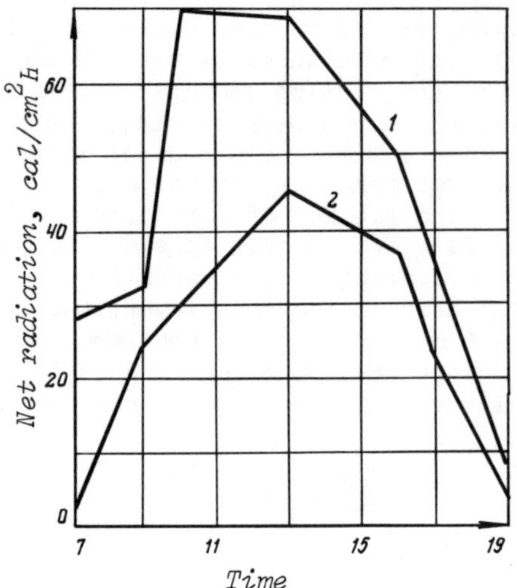

Fig 59. Diurnal variation of net radiation in July 1970.

1. forest(pine-lichen) 2. field.

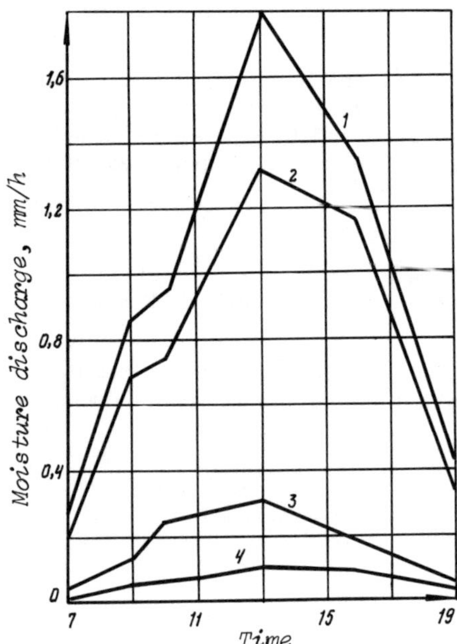

Fig 60. Daily moisture consumption (pine-bilberry), July 1970.

1. evapotranspiration 2. transpiration
3. transpiration by lowbush 4. evaporation

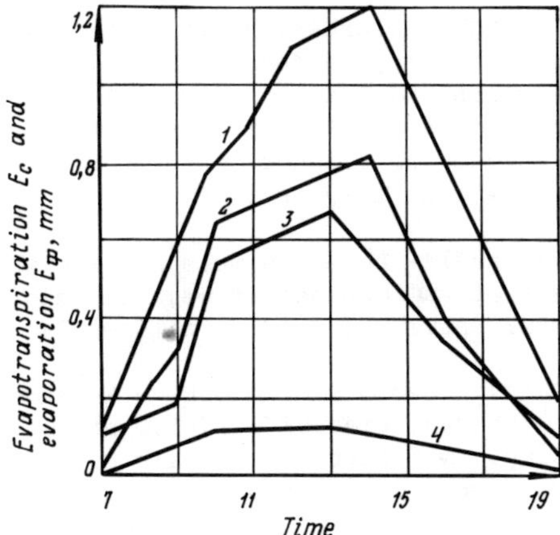

Fig 61. Evapotranspiration and evaporation curves for forest and forest and field, July 1970.

1. evapotranspiration, forest 2. evapotranspiration, field.
3. evaporation, field 4. evaporation, forest

DISCUSSION

Professor L. Wartena (The Netherlands) - Please explain how you found the turbulent transfer coefficient in an active layer of soil?

O. A. Belotserkovskaya (Byelorussian SSR) - The turbulent transfer coefficient in soil has not been found. It is practically impossible to find the turbulent transfer coefficient in soil bearing trees because of the tightly twisted root systems.

Professor L. Wartena (The Netherlands) - In your paper you reported a close correlation between net radiation and evapotranspiration. Do you take average values? Was the correlation always close?

O. A. Belotserkovskaya (Byelorussian SSR) - Yes, we took average values. The direct relation between net radiation and evaporation is quite reliable especially on days of continuous sunny weather. This may be partially attributed to the very accurate values of net radiation measured by Kozyrev-type instruments which are now the most precise in the world. Because evaporation is energy consuming, close correlation between net radiation and evapotranspiration is to be expected.

L. Heikurainen (Finland) - Have you measured interception of water by tree crowns? If so, what method has been used?

O. A. Belotserkovskaya (Byelorussian SSR) - We have measured interception in dense conifers and deciduous trees. But observations were made for few years and data is not absolutely reliable. Therefore additional data of Tkachenko, and general data of Forestry Laboratory, Academy of Science of the USSR (by Malchanov) has been used for the past 10 years.

L. Heikurainen (Finland) - Have you measured the water and heat balances in forests of various densities, i.e. high, medium and low density forests? Have you used fertilizers?

O. A. Belotserkovskaya (Byelorussian SSR) - We made our observations in forests with densities of 1.1, 1.0 and 0.9, i.e. in forests of high density. No, fertilizers have not been used. The soils remained in their natural state.

F. R. Zaidelman (USSR) - On what types of soils were the experiments conducted? What method was used to determine moisture condensation in soils?

O. A. Belotserkovskaya (Byelorussian SSR) - The experiments were carried out on peat and sandy-loam-podzol soils. Moisture condensation in soils was determined by the heat-balance method explained in the paper.

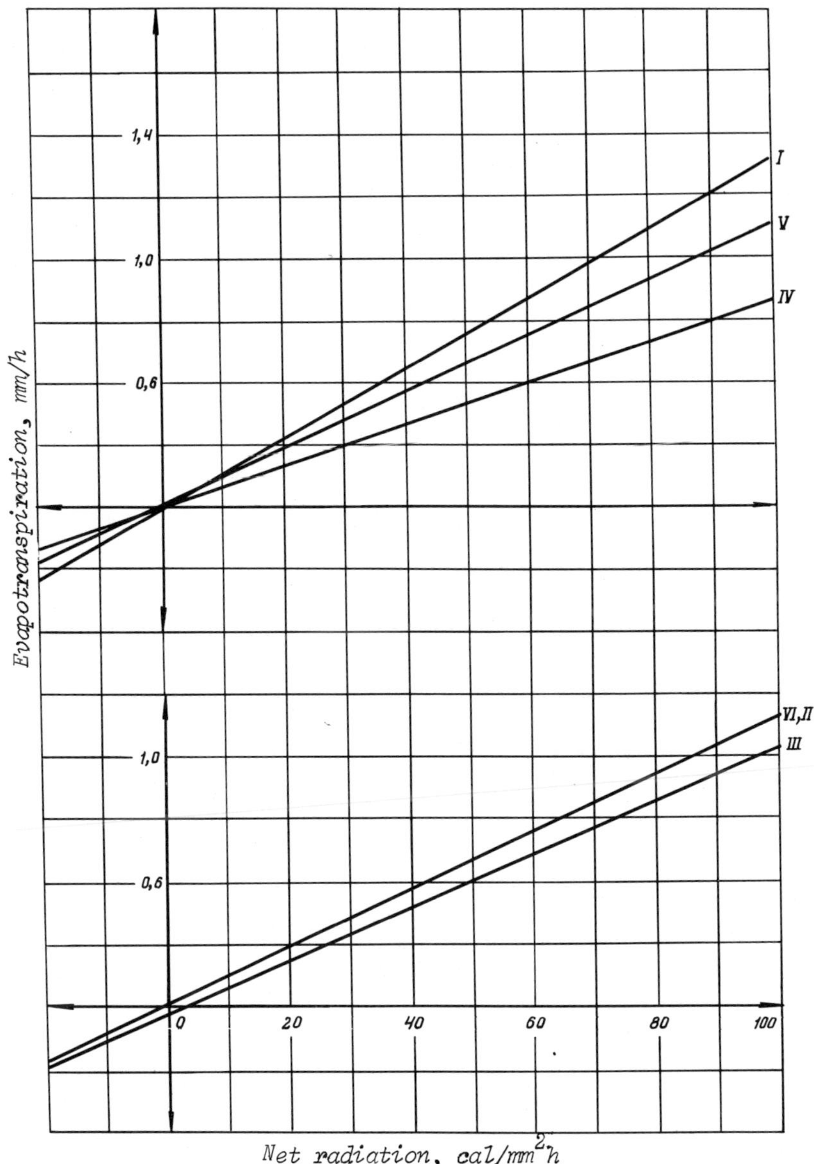

Fig 62. Correlation between net radiation and evapotranspiration.

I. pine-heather	II. pine-lichen
III. pine-bilberry	IV. pinebilberry
V. pine-sphagnum	VI. birch-cowberry

MATHEMATICAL MODELS OF SUBSURFACE DRAINAGE IN HEAVY SOILS

F. Fehér

Research Institute of Water Resources, Budapest, Hungary.

SUMMARY: Subsurface drainage is one of the potential methods of improving heavy soils with poor hydrodynamic characteristics in the hilly regions of Hungary. After exploring hydrological conditions, a mathematical model was devised to assist in the study of a complex soil drainage system. The model consists of a mathematical system comprising a ploughed top layer, subsurface mole drains and a filtered drainage ditch, each of which is described by hydraulic parameters. Mathematical models were used to study the individual elements, their conveying capacity and the flow they have to carry. In this way it was possible to determine the capacity of the complete model and to check the spacing of drainpipes. From the equations describing the individual elements of the system various parameters can be calculated. The determination of the permeability coefficients for the loose ploughed layer and the filter in the drainage ditch will be of special interest.

MODELE HYDRAULIQUE DE DRAINAGE SOUTERRAIN DES SOLS LOURDS

RESUME : Le drainage souterrain constitue un des moyens principaux d'amélioration des sols lourds humides dans les régions hautes de la Hongrie. Après étude des conditions hydrologiques, on élabore un modèle hydraulique pour mettre au point un système complexe de drainage. Le modèle comporte les paramètres hydrauliques suivants : couche de labour, drains tubulaires de filtrage, drainage de "taupe". Pour étudier ces divers facteurs, on étudie la capacité de conductivité et d'écoulement d'un élément particulier à l'aide de modèles mathématiques. C'est ainsi qu'on arrive à déterminer la capacité d'écoulement de tout le système et à contrôler par la méthode hydraulique la distance entre les drains. A partir des équations caractérisant les divers éléments du système, on calcule les paramètres de celui-ci. Il est particulièrement important de déterminer le coefficient de perméabilité pour la couche de labour et pour les drains tubulaires.

INTRODUCTION

Heavy soils having poor hydrodynamic characteristics cover more than 100 000 ha of the western part of Hungary. The annual average precipitation in this area ranges from 800 to 1000 mm. To make agricultural production possible on these clay soils, which contain considerable amounts of colloidal iron compounds, and to improve the reliability of crop yields, measures aimed at the aeration of soils and the removal of excess water were necessary.

Considerations involving the demands of agriculture, engineering, hydrology and hydraulics, as well as economics resulted in the development of a complex subsurface soil drainage system.

The primary purpose of the subsurface drainage system is the removal of undesirable amounts of precipitation water. Studies have shown that it was necessary to design the system for a specific yield of 1.5 to 3.0 l/s ha, depending on the topography of the area.

HYDRAULICS OF SUBSURFACE DRAINS

The soil profile in areas where subsurface drainage is required, usually consists of different layers and is thus neither uniform, nor isotropic. The seepage is completely confined from below, the subsoil being practically impervious. Water supply is intermittent. In mildly sloping areas the seepage pattern is symmetrical, on steeper slopes it is asymmetrical. Difficulties of computation due to discontinuity have been eliminated by defining exactly the parts of the seepage field that may be regarded as uniform and by taking into consideration changes in the hydraulic pressure gradient. The disturbing effect of anisotropy has been avoided by determining accurately the direction of flow velocities occurring in different parts of the seepage field. The surface water cover has been assumed to be balanced over the day.

Three-phase (soil, water, air) soil conditions are favourable to agriculture. The possibility of devising a feasible system, in which three-phase conditions can be maintained, has been studied. A nongravitational three-phase flow occurs under the effect of differential tension between soil layers having different moisture contents. The relation between moisture and tension has been determined for the top soil layer. Hereafter, the following equation has been applied to compute the mean velocity of unsaturated seepage.

$$V_{sz} = Ak/E \qquad (1)$$

where $A = A(S_o, S, \psi)$. At optimum moisture the order of magnitude of A is 10^{-5} to 10^{-7} cm^2/s.

At the seepage distance of interest in subsurface drainage, the mean seepage velocity obtained from equation (1) is in the range of 10^{-8} to 10^{-9} cm/s. It was found practically impossible to develop an economical subsurface drainage system, where the overwhelming portion of seepage would occur under unsaturated conditions. The occurrence of two-phase (soil, water) conditions for a certain period of time cannot be prevented and must therefore be tolerated. At the same time, efforts must be made to minimize the duration of saturation. This requirement was observed when developing the complex subsurface drainage system.

In the parts of the seepage field, where two-phase conditions were assumed to prevail, the validity of Darcy's law was accepted. In this way it became possible to characterize each part by a fictitious seepage velocity. The flow was assumed to be two-dimensional and steady. The Dupuit theory was checked for its validity. Introducing a mapping function of the form

$$\xi = f|Z| = F \ln sh(\pi z/h) \qquad (2)$$

seepage distances were determined in the real and imaginary planes, the ratio of which was used in computing the pressure gradient at the depth to width ratios of interest, for the seepage fields under consideration. The results obtained have indicated the possibility of taking the pressure gradient in the drainage ditch into consideration with a unit value, i.e. the curvature of the streams could be disregarded.

DESCRIPTION OF THE MATHEMATICAL MODEL

The complex subsurface drainage system described with the help of hydraulic parameters (Fig 63) is regarded as the mathematical model. Water from the surface infiltrates into the ploughed and loosened layer, whence it enters the mole drains placed under this layer. In the mole drains the water flows towards the constructed drains, and filters through the drainage ditch into the constructed drainpipes. The elements of the mathematical model are thus the top soil above the mole drains, the mole drains and the filtered drainage ditch. The constructed drainpipe system (fireclay, plastic, etc.) is not regarded as a part of the model. These elements are designed according to conventional methods of pipe hydraulics.

<u>Seepage in the loosened layer</u>: The result of the above investigations demonstrates that the space above the mole drains is capable of passing the water falling on the surface only if saturation conditions are tolerated. Consequently, in this layer two-phase seepage conditions prevail.

The ploughed soil layer is characterized by the seepage coefficient K_1 and the pressure gradient I_1 across the layer. If the mole drain were also in the ploughed layer, the simplification $I_1 = 1$ would be acceptable. However, since the mole drain is in a subsoil of lower permeability (characterized by K_t where $K_t < K_1$) underlying the loosened soil, this approximation is not acceptable. The thickness of the soil layer of permeability K_t is c and in calculations this thickness must be increased in the proportion of permeabilities to obtain a uniform layer of permeability K_t.

The mole drains are spaced distance a apart and the flow to them per unit length is also the conveying capacity of the loosened soil layer:

$$q_1 = K_1 I_1 a \qquad (3)$$

The inflow resulting from precipitation falling on the surface must be conveyed by the deep-ploughed, loosened layer. The inflow per unit length of the mole drain, due to the surface water cover (rainfall depth) becomes:

$$\bar{q}_1 = ai \qquad (4)$$

The loosened layer should, of course, be made capable of passing the rain that falls on its surface, i.e. $q_1 > \bar{q}_1$. Consequently, equation (3) will yield the design values of inflow to the mole drains only in the case of equality. In the case of inequality, equation (4) must be taken into consideration. From equations (3) and (4) the gradient-permeability coefficient -rainfall depth relation is obtained for the loosened layer:

$$i = K_1 I_1 \qquad (5)$$

From equation (5), and the knowledge of precipitation depth on the surface, one of the K_1 and I_1 values is calculated by adopting an arbitrary value for the other. For a particular drainage pattern the spacing of mole drains is fixed so that the value of I_1 can be determined and thus the formula yields the permeability required for the operation of the system. Conversely, where the permeability, or its variations and extreme values are known, it is possible to compute the spacing of mole drains in the critical condition.

Flow in the mole drains: Water enters the mole drains continuously from the loosened layer and is conveyed through them to the drainage ditch. The spacing of the constructed drainpipes is L. Depending on whether the model is symmetrical or asymmetrical, water in the mole drains is carried over a distance of L/2 or L. The equations written for the distance L will be presented subsequently.

The flow to the mole drains from the loosened soil layer is q_1, so that

$$\bar{q}_d = q_1 L \qquad (6)$$

Computing the flow from equation (4), the flow to the mole drain becomes

$$q'_d = iLa \qquad (7)$$

Assuming the mole drains to be complete pipes, their discharge capacity at full-flow is

$$qd = Gd^{5/2} e^{1/2} \qquad (8)$$

where $G = G(g, \lambda, \pi)$.

With the conveying capacity of the mole drains (d = 4 cm) fully employed, the equation q'd = qd would be satisfied. This, however, is not the case even at the minimum slope and the carrying capacity is greater by several orders of magnitude than the flow infiltrating across the loosened layer.

Any attempt to satisfy the equation would involve either the reduction of qd and in turn, that of d which is impractical for constructional reasons, or the increase of q'd by increasing the value of a, and of L. In view of the low reliability of current mole drain operation, L must not be extended at present beyond a certain limit. Nevertheless, this is one of the trends for future development. An increase of a, or any modification thereof is either impossible, or a problem of design, for the reasons mentioned in connection with equation (5).

Flow in the drain ditch: The water discharged by the mole drain into the drainage ditch, which percolates through the filter layer placed therein, enters the constructed drainpipes. To characterize the drainage ditch, the permeability coefficient K_D has been introduced. The number of mole drains discharging into the drainage ditch per unit length is 1/a, consequently the inflow to the ditch is

$$q_D = 1/a\ qd = 1/aiLa = Li \tag{9}$$

The conveying capacity of the drainage ditch is

$$q_D = K_D I_D h \tag{10}$$

Adopting the simplifying assumption $I_D = 1$, from equations (9) and (10), equating inflow and conveying capacity:

$$K_D h = Li \tag{11}$$

equation (11) lends itself to the determination of the permeability coefficient of the layer in the drainage ditch and consequently its design becomes possible. As suggested by Magyar (1968), it is further possible to estimate the drain spacing as well, but before tackling this question it is advisable to synthesize the equations describing the individual elements.

CHECKING THE SPACING OF CONSTRUCTED DRAINS

As will be perceived from the description of the elements of the hydrological model, on its way to the constructed drain water moves by seepage in two elements and by flowing in one. The velocity of flow is, of course, appreciably higher than that of seepage, so that up to a certain limit (controlled by both engineering measures and soil conditions), the role of flow in the mole drains is not decisive. The simplified assumption of the loosened layer discharging directly into the drainage ditch seems thus permissible. Equating the carrying capacities of the two elements and writing equations (3) and (10) for the lengths L and a, respectively yield

$$K_1 I_1 aL = K_D I_D ha \tag{12}$$

Equation (12) solved for L, yields

$$L = \frac{K_D I_D}{K_1 I_1} h \qquad (13)$$

indicating that the ratio of seepage cross-section characterized by the values L and h depends on the ratios of permeability coefficients and pressure gradients.

The pressure gradient-permeability coefficient -rainfall depth relation has been described by equation (5). Thus rainfall depth is combined with equation (13). The resulting expression becomes

$$L = \frac{K_D I_D}{i} h \qquad (14)$$

From this expression it will be appreciated that once the rain falling on the surface is capable of infiltrating through the loosened layer and finding access through the mole drain to the drainage ditch, the spacing of the latter is controlled exclusively by its carrying capacity. The above statement should be qualified by adding that flow in the mole drains is restricted by factors for which no allowance can be made in hydraulic calculations. For this reason the expression is unsuited for estimating the drain spacing. With the assumption $I_D = 1$, it is possible to use the equation in the form

$$L = \frac{K_D h}{i} \qquad (15)$$

which would follow also from equation (11).

APPLICATION OF THE MATHEMATICAL MODEL

The model provides answers to the problems arising in connection with the complex method of subsurface drainage. As regards the spacing of drains, the design of the drainage ditch was demonstrated to be of decisive importance. Methods are available by which the construction of the filter in the drainage ditch, i.e. the provision of a suitable permeability can be solved permanently. Greater difficulties are, however, encountered in ensuring the permeability of the ploughed soil layer, in that no accurate information is available on the changes in the permeability of this layer with respect to both time and location. The formation of mole drains at the required slope has not been solved completely as yet.

Cautioned by the problems listed above it is suggested that equation (15) be used for checking the spacing of drains, rather than for its calculation, careful consideration being also given to the carrying capacity of various elements of the model. The relevant equations lend themselves readily for estimating the parameters of the component elements, commencing with the permeability that it is desirable to attain.

With reference to Figure 64, the parameters involved in the equations are explained below:

$A (cm^2/s)$	proportional coefficient in three-phase seepage;
$E (cm)$	seepage distance in three-phase condition;
F	proportionality coefficient of the mapping function;
$G (cm^{\frac{1}{2}}/s)$	proportionality coefficient of discharge in mole drain;
I_l	pressure gradient in the loosened layer;
I_D	pressure gradient in the drainage ditch;
$L (cm)$	spacing of constructed drains;
S	variable moisture content;
S_o	bonded moisture content;
$a (cm)$	spacing of mole drains;
$c (cm)$	distance of mole drains from loosened layer;
$d (cm)$	diameter of mole drain;
e	slope of mole drain;
$f (cm)$	depth of loosening below surface of soil;
$g (cm/s^2)$	acceleration of gravity;
$h (cm)$	width of drainage ditch;
$i (mm/day)$	specific rainfall depth;
$K (cm/s)$	Darcy's permeability;
$K_l (cm/s)$	permeability of the loosened soil;
$K_t (cm/s)$	permeability of the subsoil;
$K_D (cm/s)$	permeability of the drainage ditch;
$m (cm)$	depth of the drainage ditch;
$q_l (l/s)$	carrying capacity of loosened layer;
$\bar{q}_l (l/s)$	inflow to the mole drain from the loosened layer;
$\bar{q}_d (l/s)$	inflow to the mole drain;
$q'_d (l/s)$	inflow to mole drain from surface rain;
$q_d (l/s)$	conveying capacity of mole drain;
$\bar{q}_D (l/s)$	inflow to drainage ditch;
$q_D (l/s)$	conveying capacity of drainage ditch;
$v_{sz} (cm/s)$	mean velocity of three-phase seepage;
z	characteristic of real plane;
ψ (cm)	tension pertaining to S;
λ	roughness of mole drain;
ξ	characteristic of imaginary plane.

REFERENCE

Magyar, P. 1968. Amelioration of cohesive soils with poor hydrodynamic characteristics. Report by the Research Institute for Water Resources Development.

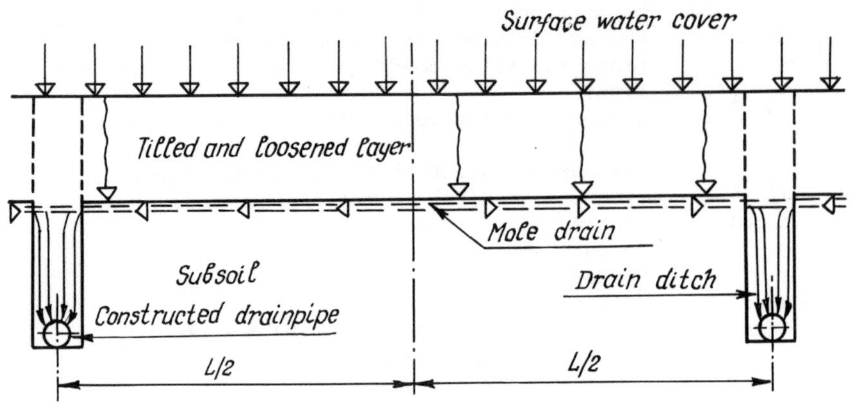

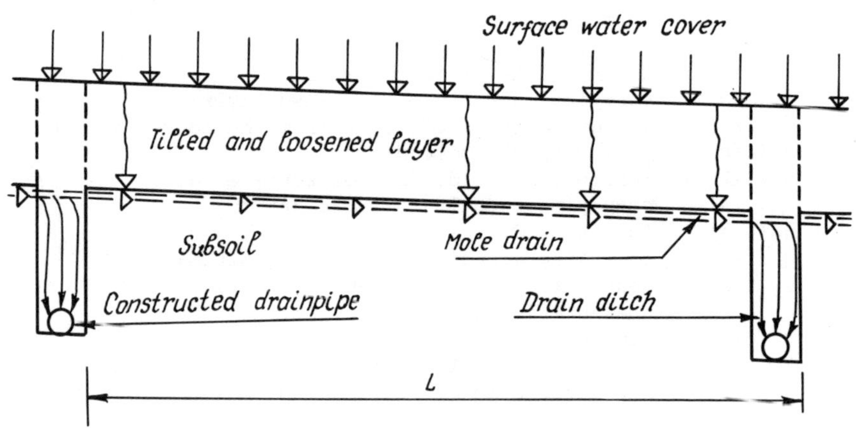

Fig 63. Elements of the mathematical model.

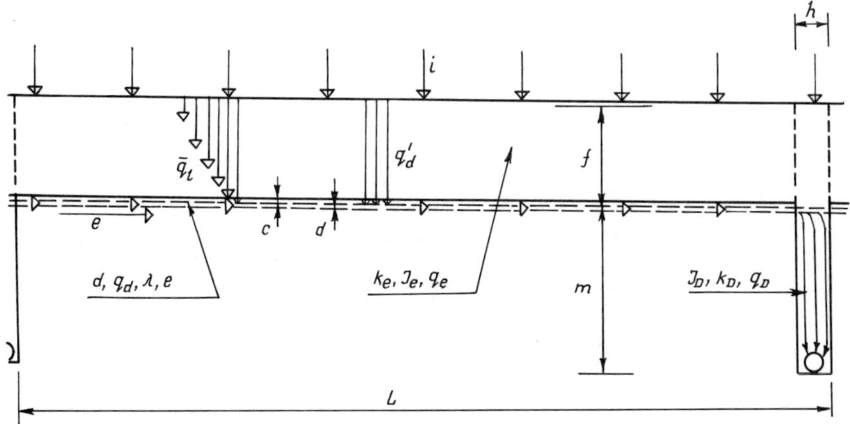

Fig 64. Parameters of the mathematical model.

DISCUSSION

W. H. van der Molen (The Netherlands) - Calculation of infiltration in the ploughed and loosened layer using a conceptual model is very difficult as measurement of their permeability parameters is difficult. How did you find these values?

F. Fehér (Hungary) - The permeability values are measured in the field as in other countries.

Yu. Nikolsky (USSR) - Please explain equation (1) of your paper.

F. Fehér (Hungary) - This is an integral equation. Its parameters depend on the soil moisture content and capillary tension. It implies downward percolation in three-phase conditions.

HYDROLOGICAL STABILITY CRITERIA AND RECONSTRUCTION OF BOGS AND BOG/LAKE SYSTEMS

K. E. Ivanov

*State University,
Leningrad,
USSR.*

SUMMARY: A brief consideration is given of the hydrological conditions associated with bog/lake geological systems that result in a specific type of the natural landscape subject to excessive and unsteady moistening. Some water balance relations are presented that define the stability of central bog lakes and the structure of bog/lake systems as a function of water supply. Predicted relations are presented for steady states of bog/lake and bog systems when the climatic factors and the morphological coefficient of natural or forced drainage are changed. These relations define the limit of drainage or irrigation of the system without their degradation or decay. An example is given of the calculation of the steady-state region and location of real undrained bog/lake and bog systems.

CONDITIONS HYDROLOGIQUES DETERMINANT LA STABILITE ET LA RECONSTRUCTION DES SYSTEMES MARECAGEUX

RESUME : L'existence et la stabilité des systèmes géologiques à formations marécageuses qui donnent un aspect typique aux régions soumises à une humidité excessive et variable dépendent des conditions hydrologiques et de certains traits du bilan hydrique. L'auteur établit des relations entre les facteurs du bilan hydrique selon lesquelles la stabilité des lacs organiques marécageux et la structure des sytèmes à formations marécageuses sont fonction de leur alimentation en eau. Il présente les relations obtenues par le calcul pour les divers états de stabilité lorsque varient les facteurs climatiques et le coefficient morphologique du drainage naturel ou artificiel. Ces relations déterminent les limites dans lesquelles doit se maintenir le drainage ou l'irrigation pour ne pas dégrader ou détruire le système. L'auteur donne un exemple du calcul portant sur la stabilité du système et sur la façon dont sont distribuées les formations marécageuses naturelles non drainées.

The equilibrium that is established in natural geological systems, (Anon., 1971) after the alteration of the hydrological regime results from the construction of irrigation or drainage works, may serve as a basis for forecasting the future consequences of these developments. Any change in the natural hydrological pattern causes inevitable alterations in the balance between the environmental components; these changes in turn affect the surface water level and its variations. The magnitude of the effect and the reversibility or irreversibility of the changes depends on the resistance of the geological system to changes and disturbances in the natural equilibrium. In the present paper the only part of this vast subject to be considered is that relating to transformations of the biology, geology and hydrological regime of bog and bog/lake systems. Stability criteria are very important in connection with the self-regulation of these systems. To study the general laws governing the development of natural bog/lake systems and the diversity of their shape and structure, these systems should be considered both as regular combinations of physiographic lacustrine facies and as interrelated biological and geological systems limited in space and time and interacting with the environment. Study of bog systems has revealed that vast plains on the Earth's surface serve as storage for organic matter which is inevitably accumulated because of excessive moisture. The process of accumulation of organic matter gives rise to a specific physiographic landscape found only in regions of organic soils.

In these bog landscapes it is possible to distinguish a direct interaction between the rate and type of water exchange and the distribution and composition of the vegetation. The interaction occurs at the surface of organic deposits and results in specific forms of co-existance between free water, peat deposits and the vegetation found in the landscapes. These forms of co-existance are one of the principal specific features of bog systems.

All the numerous combinations of bogs and lakes and associated peat deposites which may be encountered in nature will here be referred to as bog/lake systems, classified as follows:

1. A bog system with a lake located at the centre of the bog (Fig 65a);
2. A bog system composed of a single large lake located on the slopes of the bog (Fig 65b, c, d);
3. Various forms of orientated ridge small lake groups, or so-called ridge bog-lake facies (microlandscapes) (Fig 65e);
4. Non-orientated ridge lake groups (Fig 65f);
5. Non-orientated or slightly orientated bog-lake groups (Fig 66).

For convenience in this general consideration, bog systems without lakes will be taken as a particular case of bog-lake systems with a zero free water surface at certain conditions.

The steady state in any physical system implies the possibility of retaining the original structure (or arrangement) and main functions if a part of these functions or properties have changed with respect to the normal (mean) state, (Hilmi, 1966).

Any irreversible changes in the state of the system may be considered as a sequence of unsteady states which follow one another and will never lead the system back to the initial state (infinite series of states). Any reversible process may be considered as a steady dynamic state of the system. If we consider all the main elements of the environment which cause the formation of bogs, it is necessary to bear in mind that the resulting bog and bog-lake systems with more or less complex structures and adequate peat deposits, have an inverse effect on the environment since they begin to change and transform the latter.

The general pattern of interaction between bog systems and the environment (Figure 67) may be presented as a series of direct and inverse relations and mathematically described by a series of hydromorphological equations (3-5). The latter describes the interaction process for a quasi-steady state of the bog system as a series of processes finite in time for fairly short intervals. For these time intervals the surface configuration of peat deposits and climatic conditions may be assumed to be constant and taken as boundary conditions.

In the schenatic diagram of Figure 67 the relations shown by double solid lines are described by a system of geomorphological equations; here relations Nos 1-5 and 16 at a quasi-equilibrium state express the environmental conditions. Direct and inverse relations shown by single solid lines are only realized for very long time intervals (hundreds and thousands of years). During this time the relief, groundwater and surface flows of a bog system may change due to peat accumulation.

The dynamic stability of bog-lake systems is known to depend on the range of structural variations in the interval without any changes in the external boundaries and the ranges of possible inverse effects on the environmental components. Therefore, the dynamic quasi-steady state of various types of bog phytocenoses depends on whether their structure, as well as their physical and physiological properties (distribution of water diffusivity and capillary properties within the active layer, sensitivity of plants to the chemical composition of the water, etc.) are consistent with the main hydrological components of the environment such as water flow, composition and intensity of water and mineral supply, aeration of the rooting layer and water level in the active layer. Within the range of possible variations in the environmental condition for vegetation which do not interfere with the main functions and do not cause decay of the whole system, bog systems may usefully be considered as self-regulating biological and geological systems. Self-regulation is exhibited by changes of the corresponding structural forms of the vegetation and of the corresponding properties of the active layer, (Ivanov, 1970).

For the calculation of the structural characteristics of bog-lake systems and the definition of the ranges of dynamic stability and ecological amplitude of the variation of the main components of the environment, the following criteria may be found to be useful

1. A water balance criterion for individual central lakes may be expressed in the forms

$$\oint_{L_1} q\,dl + pw = \oint_{L_2} q\,dl \qquad (1)$$

where:
- w — the total water area of the lake;
- p — internal feed to the lake which is the algebraic sum of precipitation, evaporation and vertical exchange between the groundwater and the lake basin;
- q — horizontal water exchange with the lake basin per unit length of the projected water line of the lake;

- dl the element of the projected water line in the direction normal to the streamlines in bog microlandscapes adjacent to the lake shore;
- L_1 the length of the lake water line with a horizontal inflow;
- L_2 the same but with a horizontal outflow.

$$\left.\begin{aligned} w &= \frac{\Delta(ql)}{P} \\ w_* &= \frac{\Delta(ql)}{P_*} \end{aligned}\right\} \tag{2}$$

$$P_{rp} = \left(\frac{w_*}{w} - 1\right)P_* \tag{3}$$

where: $\Delta(ql)$ the increment of horizontal water exchange of the lake;

- P_* internal water feed;
- w_* the lake area with no vertical groundwater exchange;
- P_{rp} vertical groundwater exchange.

$$\left.\begin{aligned} \Delta l &= l_1(e^{a_0 s_0} - 1) \\ w &= l_1(e^{a_0 s_0} - 1)\frac{1}{a_0} \end{aligned}\right\} \tag{4}$$

where: Δl the difference between the projected lines of the outflow l_2 and inflow l_1 due to horizontal water exchange;

- $a_0 = \frac{P}{q}$ water feed coefficient which is the ratio of the internal feed P to the horizontal outflow of the surrounding bog micro-landscape;
- q, s_0 linear dimension of the lake along the mean streamline on the inflow and outflow lines.

2. The water balance criterion for a lake-bog system as a whole is expressed by

$$w^*_\Sigma + w^*_{\Sigma M} = 1 \tag{5}$$

$$\frac{w^*_\Sigma}{w^*_{\Sigma M}} + \frac{P_M}{P_{03}} = \frac{q_\smile}{w^*_{\Sigma M}P_{03}} \cdot \frac{l_\smile}{w} + \frac{q_\frown}{W_{\Sigma M}P_{03}} \cdot \frac{l_n}{w} \tag{6}$$

where: w^*_Σ and $w^*_{\Sigma M}$ relative total areas of the lakes and bog micro-landscape;

$P'_M = P_0 - P'_н$ internal water feed of bog micro-landscape;

$P_{03} = P_\theta - P_{н03}$ internal water feed to the lakes;

w area of the whole lake-bog system;

$q_\smile$ reduced flow based on the internal projected outflow line;

$q_\frown$ the same, based on the external projected outflow line;

P_0 normal mean precipitation.

$$P_н = \frac{\sum_1^n w_i P_i}{w_{\Sigma M}}; \quad q_\smile = \frac{\sum_1^{n_1} q_i l_i}{l_{\Sigma 1}}; \quad q_\frown = \frac{\sum_1^{n_2} q_i l_i}{l_{\Sigma 2}} \tag{7}$$

where: w_i area of the micro-landscape type considered;

P_i evaporation from the j-th micro-landscape;

n the number of different micro-landscapes in the system;

q_i and l_i flow and length of the projected outflow line of the j-th micro-landscape;

$l_{\Sigma 2}$, $l_{\Sigma 1}$ total lengths of internal and external projected outflow lines;

n_1 and n_2 the numbers of bog micro-landscapes adjacent to internal and external drainage lines.

3. A biophysical criterion which controls the internal structure of lake-bog systems at a steady rate can be expressed in the form

$$S_M = \frac{SM_M i}{q} = \frac{SK_M(Z_o - Z)i}{q} = \frac{SM_M}{M} \tag{8}$$

$$\left.\begin{aligned} S_M &= nL_M \\ n &= \frac{Si}{\Delta Z\theta} \end{aligned}\right\} \tag{9}$$

$$L_M = \frac{S_M}{n} = \frac{\Delta Z_\theta K_M(Z_o - Z)}{q} = \frac{M_M \Delta Z_\theta}{q} \tag{10}$$

where: S_M the total area of bog micro-landscape along the maximum slope line;

S the length of the whole area considered along the streamline between the divide and the external drainage line;

n the number of lakes along the streamline within the length S;

L_M the statistic mean width of bog micro-landscapes

separating the lakes;

ΔZ_θ the ecological amplitude of the mean bog water levels which does not change the composition of the phytocenoses considered;

M_M specific mean outflow rate of the bog micro-landscapes;

K_M the horizontal permeability coefficient of the active layer in the bog micro-landscapes at the mean groundwater level;

M specific mean outflow rate of the bog-lake system along the streamline;

q mean horizontal outflow rate at a certain point of the bog system;

i the bog slope along the streamline.

4. The criterion of mechanical stability (strength) of the peat matter in the bog phytocenoses is expressed by the resistance of various vegetation associations or of their individual units, to mechanical pressures engendered by inflow and wave oscillations in central lakes, (Ivanov, 1969).

If the structure of a bog-lake system is to be stable and remain unchanged, all the above criteria must be fulfilled. If the biophysical criterion and that of mechanical stability are not fulfilled, then these will only be a change in the inner structure of the system, the external boundaries will remain the same and the system as a whole will be retained. This is the case of a system in dynamic equilibrium. However, if either of the water balance criteria are not fulfilled apart from possible internal changes (adaptations), this will inevitably result in the decay of the whole system due to excessive drainage or to overmoistening, the system eventually being replaced by other natural units.

The range of variation of the morphological drainage coefficient in the water balance criterion for the whole system which does not lead to decay depends on possible internal transformations of the structure of the system and is defined by the relation:

$$\left[\frac{P'_M W^*_{\Sigma M} + P_{03} W^*_{\Sigma}}{q}\right]_{min} < \frac{1 + 1}{W} < \left[\frac{P'_M W^*_{\Sigma M} + P_{03} W^*_{\Sigma}}{q}\right]_{max}$$

When this criterion was applied to the climatic conditions of the West-Siberia Plain, it appeared that bog-lake systems along north latitudes 63-65° are most stable, the range of steady states decreasing towards the south.

In the areas to the south of 56-57° N.L., bog-lake systems are unstable and require external catchments. On 59-60° N.L. water feed to bog micro-landscapes is equal to precipitation feed deficit in central bog lakes. In these latitudes, bog-lake systems located at water divides disappear and more bog systems without central lakes at water divides are encountered.

Calculations using the water balance criteria for systems as a whole allow us to define the stability and instability regions of bog-lake and lake systems on plains for various latitudes and longitudes and to draw graphs for them, to define the location of stability regions for individual bog systems in various areas and to predict possible transformations by drainage without disturbance of the natural equilibrium of the landscape, decay of the system and degeneration of peat deposits.

REFERENCES

Anon., 1971. Topologiya geosistem - 71. Topology of geosystems - 71. Decisions of the Symposium. Irkuts, Academy of Sci. of the USSR. Siberia Dept.

Hilmi, G. F. 1966. Osnovy fiziki biosfery. Fundamentals of biosphere physics. Leningrad, Gidrometeoizdat.

Ivanov, K. E. 1969. Erozionnye yavleniya na nolotakh i ikh rol v formirovanii ozerno-bolotnykh landshavtov Zapadnoi Sibiri. Erosion effects on bogs and their contribution to formation of bog-lake landscapes. Leningrad, Gidrometeoizdat. (Trudy GGI, vol. 157).

Ivanov, K. E. 1970. Nekotorye voprosy issledovaniya vzaimosvyazei rastitelnyks soobshchestv i gidrologicheskogo rezhima zabolochennykh zemel. Some aspects of investigations into relations between vegetal associations and hydrology of marsh-ridden areas, Riga, Zinatne.

Fig 66. Fragment of non-orientated and slightly orientated bog/lake units within very large bogs at excess moisture zone.

1. central bog lines.
2. pine-sphagnum bogs.
3. sphagnum, carex, cotton-grass bog microlandscapes.

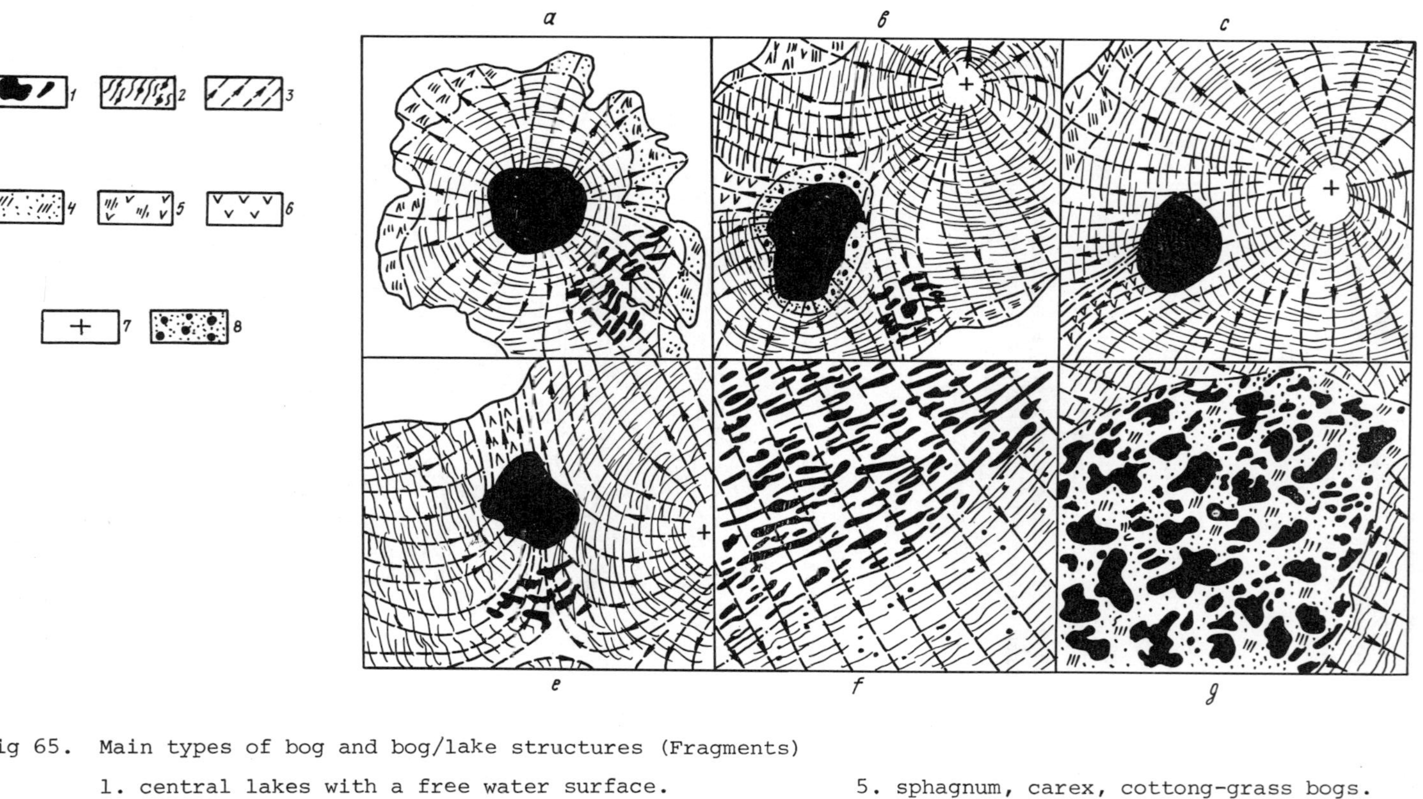

Fig 65. Main types of bog and bog/lake structures (Fragments)

1. central lakes with a free water surface.
2. water-hole and ridge bog facies.
3. directions of seepage and seepage surface flow lines.
4. sphagnum, cotton-grass, lowbush bog, microlandscapes.
5. sphagnum, carex, cottong-grass bogs.
6. carex bog microlandscapes.
7. epicentres of bog bulges.
8. pine, sphagnum, lowbush microlandscapes.

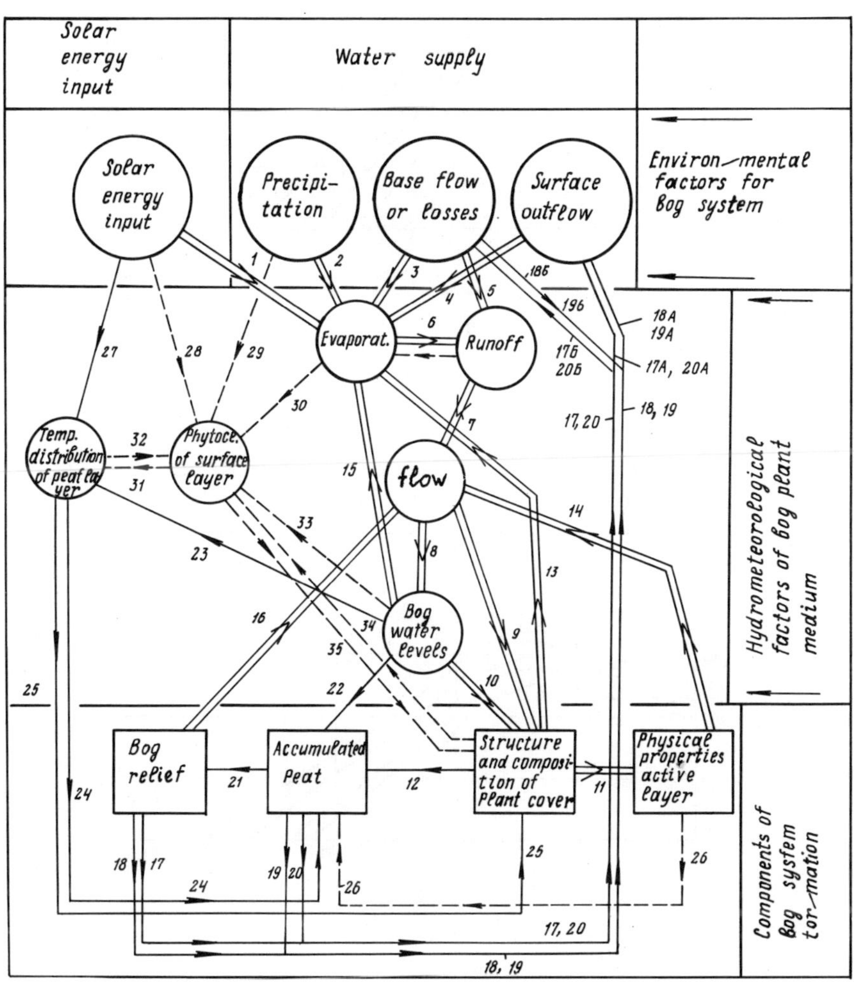

Fig 67. Schematic diagram of interaction (direct and inverse relations) between hydrometeorological elements and elements of the development process and states of bog systems.

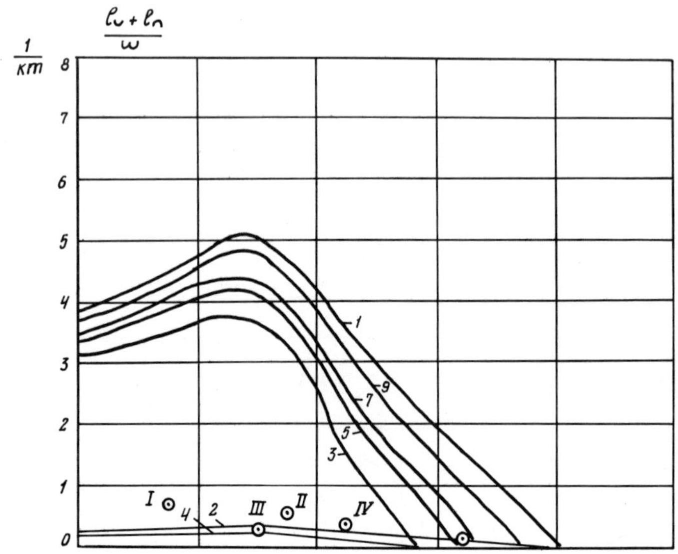

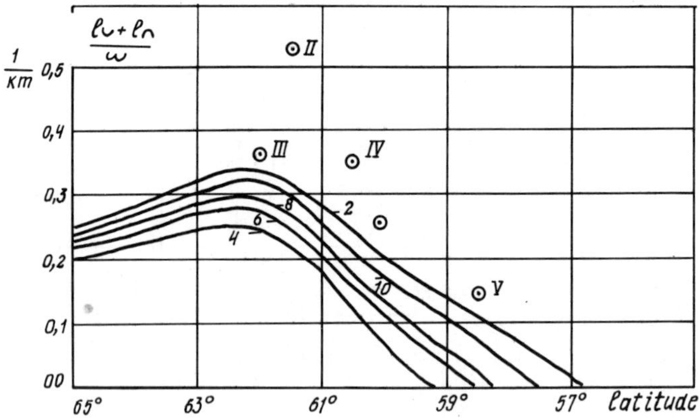

Fig 68. Plot of limiting morphological coefficients of internal and external drainage of bog/lake system versus the latitude fro system stability in plains of the excessive moisture zone.

DISCUSSION

G. Kovacz (Hungary) - My question is concerned with the equation presented in your paper. Does it give a quantitative or a qualitative estimate? Is it possible to use a computer for your calculations?

K. E. Ivanov (USSR) - This equation makes use of the values obtained from hydrogeological studies of bogs. These are precipitation, evaporation and horizontal flow in bog formations. A computer may be used although the present calculations were made without computers.

P. A. Kiselev (USSR) - What is the range of the problems that can be solved by the suggested relations? Can your method be applied to the whole area or only to particular plots?

K. E. Ivanov (USSR) - It may be applied to the whole area as far as the stability is concerned. The present method also includes the rate of drainage of the area.

V. F. Shebeko (BSSR) - The quantities in your equation are stochastic. What particular quantities did you use? What is the rate of disturbance of stability criteria in bog drainage?

K. E. Ivanov (USSR) - We first used evaporation from large areas and precipitation. Later, we also included mean values for different types of grounds. As we know the reserve of a bog system, we can obtain the water exchange rate. The equations are valid for the deposits resulting from drainage, but if these criteria are adopted, it should be mentioned that within a steady range this will be the most favourable rate of drainage with respect to peat burning and with respect to its mechanical stability.

THE WATER BALANCE OF LOWLAND AREAS IN NORTH-WESTERN REGIONS OF THE FRG

R. Eggelsmann

Peat Soil Research Institute,
Bremen, FRG

SUMMARY: The paper begins with an examination of the climate and soils of the lowlands of the north-western regions of FRG. After a discussion on peat hydrology, relationships are established between soil moisture, groundwater, runoff morphology, evaporation and microclimate in catchments having different plant associations. Lastly the water balance and water storage in peat soils and their influence on the hydrography of the landscape before and after drainage and cultivation are examined.

BILAN HYDRIQUE DES BASSES PLAINES DU NORD-OUEST DE LA REPUBLIQUE FEDERALE D'ALLEMAGNE

RESUME : L'auteur décrit d'abord les caractéristiques des sols et du climat de la région étudiée. Après étude de l'hydrologie des tourbières, l'auteur examine les relations qui existent entre l'humidité du sol, les eaux souterraines, le régime de l'écoulement, l'évaporation et le microclimat des bassins versants à végétation variée. Enfin, il traite du bilan hydrique et du stockage en eau des régions de tourbières et de leur influence sur l'hydrographie des sites avant et après drainage et mise en culture.

INTRODUCTION

The north-west of the FRG belongs to the humid climate zone of Europe. Prevailing strong winds from the west and south-west create a maritime climate with a mild winter and a relatively cool summer, as shown by the climatic data presented.

Table 41. Climatic data of north-west FRG. Raised bog.

Mean or sum	Air temperature °C	Rel. air humidity %	Rainfall mm
Year	8.2	83	742
January	0.2	89	364 (winter total)
July	16.7	78	378 (summer total)

A high proportion of soils in the north-west of FRG is influenced by high groundwater levels and by rainfall. The proportion of peat soils is considerable, (Table 42).

Table 42. Soil types in the north-west of the FRG.

Soil types	Area	
	mill.ha	% of total
Glei, river and marshy soils	1.00	15.6
Peat soils (peaty, fen, and raised bog)	1.25	19.5
Sum	2.25	35.1
Total area	6.40	100.0
Agricultural area	4.30	67.2

All modern uses to which these soils may be put require that they be adequately drained. On account of the different opinions which prevail regarding the influence of drainage upon the water balance of the environment, it was necessary to carry out some research on the subject. For the past two decades our Institute has been undertaking hydrological investigations on the influence of drainage, cultivation and vegetation on runoff, groundwater, soil moisture, evaporation and microclimate*. Most of the research was devoted to peatlands, especially the raised bogs near our Koenigsmoor Experimental Farm between Hamburg and Bremen. We compared the results with those of similar watersheds.

* These investigations were conducted in co-operation with the Board of Agricultural Engineering of the FRG (Kuratorium für Kulturbauwesen) and with financial support of the German Commom Research (Deutsche Forschungs gemeinschaft).

THE HYDROLOGICAL BEHAVIOUR OF PEAT

From the hydrological point of view the following phases may be distinguished:

a) natural, even-growing peat;
b) predrained peat bot used for agriculture; includes peat for industrial extraction;
c) drained and cultivated peat, in the north-west of FRG, mainly under grass;
d) deep ploughed peat (sand mixed with cultivation peat), mainly arable land;
e) afforested peat.

There is no virgin peat in the north-west of the FRG. During the course of the last century the whole country was drained, sometimes for the purpose of road or railway construction. Only some very small peat areas are still untouched and we are trying to preserve these peats for future scientific research.

In this area we have only small areas of peatland under forest but it is possible that might increase in the future. Investigations are underway to determine the influence of afforestation on the water balance.

PEAT SOIL MOISTURE CONTENT

Most of our hydrological and climatic data pertain to raised bogs of the lowlands, which are the bogs under discussion; the hydrology of fens is reputedly somewhat similar.

Soil measurements in both virgin and predrained peats have shown that they are filled with water throughout the year, except for the surface peat layers which lose some soil moisture by evaporation. On the other hand, the drained and cultivated peats have a soil air content of 10 to 25% by volume, especially in summer time (Fig 69) and a moisture content of 60 to 85% by volume.

RUNOFF PATTERNS

Simultaneous measurements of rainfall, groundwater and surface runoff have shown that after heavy or prolonged rain (Fig 70) and after snow melt (Fig 71) the groundwater and surface runoff patterns in virgin or predrained peats are more extreme than those of drained peat grassland. In the latter case, the flood peaks are much more delayed and flatter than in the first case. Observed high-water peaks, in predrained uncultivated peat are frequently two to four times higher than those of well drained peat grassland.

The different runoff patterns are attributable to the predominance of surface runoff in the predrained peat area whereas in grassland areas most of the runoff percolates through the soil to the tile drainage.

Apart from this, it is also considered that the contrasting soil structures are very important (Fig 69), since the considerable available soil pore volume in the soil strata above the water table in the peat grassland can absorb large quantities of water.

EVAPORATION

The evaporation from peats is generally higher than that from mineral soils. The type of vegetation, however, is known to have a considerable influence upon evapotranspiration (Table 43).

Table 43. Mean evaporation for different plant associations in the Atlantic region catchments (annual rainfall 740mm).

Year	Plant association	Evaporation (mm)		
		Winter	Summer	Annual
Peat	Sphagnetum	136	396	532
	Callenetum	136	370	506
	Grassland	95	396	491
Sand	Arable land, forest	91	385	476
Loamy sand	Grassland			

The annual distribution of measured evaporation is shown in Fig 72. It can be seen that evaporation from peat grassland is greater than that from predrained peat in autumn and winter only. This explains clearly the different rates of growth of grassland and of evergreen Ericaceae and Sphagna. However, although the total annual evaporation from Callunetum and from grassland is nearly equal, in swampy raised bogs containing much Sphagnum moss, the evaporation is higher (Table 43).

INFLUENCE OF WATER CIRCULATION ON THE MICROCLIMATES OF PEATS

It is generally known that the soil temperature and the microclimate near the surface are much more extreme in peats than in mineral soils, (Romen, 1894; Geiger, 1950).

During the last decades investigators in the FRG have tried to obtain more information about the influence of drainage and agricultural engineering on microclimate, (Baden and Egglesman, 1958; Schmeidl, 1960, 1962, 1964 and 1965; Mies, 1968).

Fig 73 shows the differences between air temperature at the soil surface and at 2 m above the surface (mean monthly values for 1958). The greatest differences occur with Sphagnetum which is compared here to other similar plant associations.

Fig 73 shows the net night radiation above different peats. Again the Sphagnetum can be seen to have the lowest radiation.

Other measurements have confirmed that drainage and reclamation have improved the microclimatic conditions of peat (Table 44).

WATER BALANCE OF CATCHMENTS WITH AND WITHOUT PEATS

Such comparisons are only possible for catchments which have similar exposure and climate, and differ only by soil and vegetation. Fig 74 shows the main components of the hydrological budget for three catchments. The diagrams represent the rainfall, actual evaporation, runoff, change of water storage and groundwater level. With similar rainfall in summer, the evaporation from the peat catchments is slightly higher and the runoff less than is the case for catchments with mineral soils. The annual variation

Table 44. Mean monthly absolute and relative air humidity at a height of 5 cm above the plant-covered surface of a raised bog (measured at noon).

1951	Callunetum		Grassland	
	Vapour pressure mm	Relative humidity %	Vapour pressure mm	Relative humidity %
May (24 days)	8.4	65	10.3	79
June	12.8	63	14.2	77
July	14.8	71	15.3	84
Mean	12.0	66	13.2	80

of the water storage and groundwater level of catchments with peat grassland and with mineral soils (upper Ems) are somewhat similar, only the absolute depths of the water table being different.

THE ROLE OF PEATS IN WATER STORAGE

Unlike mineral soils, the peats of raised bogs and fens are generally pretty moist. As with mineral soils, only the gravity water drains off (about 3 to 10% by volume of peat soil pore space). This is attributable to the higher evaporation from peat soils and the corresponding reduction of runoff with respect to mineral soils.

Several creeks have their source in peats or peaty soils. Our investigations have shown that the peats in question are valley peats, spring peats or slope peats (Talmoor, Quellmoor, Hangmoor) or that there is a swampy peat (Versumpfungsmoor) on which a raised bog is developing. The occurrence of swampy peat is due to springs which are fed by groundwater from gravel and coarse sand. The real springs are hidden under peat layers. The runoff in these creeks and drains originates in groundwater of the swampy mineral subsoil. There are no causal relations between the steady runoff and the water storage of peat.

Several hydrological investigations of peats in Europe and North American have shown that in general they have no water retention capacity, since the peat soil pore space is almost always filled with water. Such results have been observed for both lowland peats (Baden and Eggelsmann, 1964; Bay, 1960; Uhden, 1967) and for peat in hills or mountains (Ferda and Pasak, 1969; Gottlich, 1961; Nys, 1958; Schuch, Schmeidl and Wanke, 1970).

CONCLUSIONS

Several hydrological investigations have shown that the water circulation in virgin, drained and cultivated peats is rather similar in lowland peats. In predrained or in industrial peat cutting areas or directly after drainage works, however, the high-water peaks increase and the runoff pattern becomes extreme. About ten years after draining when the surplus water has gone and the catchment has acquired agricultural vegetation, a new hydrological balance arises.

REFERENCES

Baden, W. and Egglesmann, R. 1958. Uber das Bodenklima verschiedener Hochmoorkulturen und sein Einfluss auf den Pflanzenwuchs. Z.f. Acker- und Pflanzenbau, Berlin-Hamburg, 106. 127-152.

Baden, W. and Egglesmann, R. 1964. Wasserkreislauf eines nordwestdeutschen Hochmorres. KfK-Heft 12. Verlag Wasser und Boden, Hamburg.

Bay, R. R. 1960. Runoff from small peatland watersheds. J. of Hydrology, Amsterdam, 9. 90-102.

Egglesmann, R. 1960. Uber den unterirdischen Abfluss aus Mooren. Wasserwirtschaft, 50. 149-154, Stuttgart.

Egglesmann, R. 1964. Die Verdunstung der Hochmoore und deren hydrographischen Einfluss. Deutsche Gewässerkundliche Mitt. 8. 138-147. Koblenz.

Egglesmann, R. 1967. Oberflächengefalle und Abflussregime der Hochmoore. Wasser und Boden, Hamburg, 19. 247-252.

Egglesmann, R. 1971. Uber den hydrologischen Einfluss der Moore. Telma, Rannover. 1. 37-48.

Ferda, J. and Pasak, V. 1969. Hydrological and climatic function of Czechoslovac peat bogs. Zbraslav.

Romen, Th. 1894. Bodenphysikalische und meteorologische Beobachtungen, mit besonderer Berückichtigung des Nachtfrostphonomens, Berlin, Mayer & Müller.

Geiger, R. 1950. Das Klima der bodennahen Luftschicht, 3 Ed. Braunschweig, Vieweg & Sohn.

Gottlich, Kh. 1961. Achtjährige Untersuchungen über die Wasserbilanz und die Strukturverhältnisse eines gedränten, kultivierten Niedermoores im südwestdeutschen Alpenvorlandes Bayr. Landw. Jb. 38, 718-739.

Kuntze, H. 1967. Wasserhaushalts - und Klimaforschung bei Grundwasserboden im humiden Klima. Z.F. Kulturtechnik und Flurbereinigung, 8. 142-162.

Nys, L. 1958. Nilan des eaux internes et externes dans les tourbieres hautes. VI Intern. Congr. Univ. Moorforschung. Brussels. 48-59.

Mies, M. 1968. Vergleichende Darstellung von meteorologischen Me ergebnissen und Wärmehausuntersuchungen an drei unterschiedlichen Standorten in Norddeutschland. Ber. Inst. Meteorologie und Klimatologie. Techn. Univ. Hannover, Nr. 2.

Schroeder, G. 1952. Die Wasserreserven des Oberen Emsgebietes. Bes. Mitt. Dt. Gewasserkundl. Jb. 5. Koblenz.

Sperling, W. 1955. Zusammenhange zwischen Niederschlag, Abfluss, Verdunstung und Schwankungen des Wasservorrates im Oberen Emsgebiet. Gas- und Wasserfach. 96, Heft 6/12.

Schmeidl, H. 1960. Vergleichende Wasserhaushalts- und Klimabeobachtungen auf kultivierten und unkultivierten Hochmooren und Südbayern (III. Teil) Mitt. für Landeskultur, Moor-und Torfwirtschaft, 8. 108-128.

Schmeidl, H. 1962. Klimatische Vergleiche in Moorgebisten. Wetter und Leben, 14. 77-82.

Schmeidl, H. 1964. Bodentemperaturen in Hochmoorböden. Bayr. Landw. Jb. 41. 115-122.

Schmeidl, H. 1965. Oberflächentemperaturen in Hochmooren. Wetter und Leben, 17. 87-97.

Uhden, O. 1967. Niederschlags- und Abflussbeobachtungen auf underuhrten, vorentwässerten und kultivierten Teilen eines nordwestdeutschen Hochmoores der Esterweger Dose am Küstenkanal bei Papenburg, KfK-Heft 15. Verlag Wasser und Boden, Hamburg.

Schuch, M; Schmeidl, H. and Wanke, R. 1970. Wasserhaushalt und Klima einer kultivierten und unberuhrten Hochmoorflache am Alpenrand. KfK-Heft 19. Hamburg, Verlag Wasser und Boden.

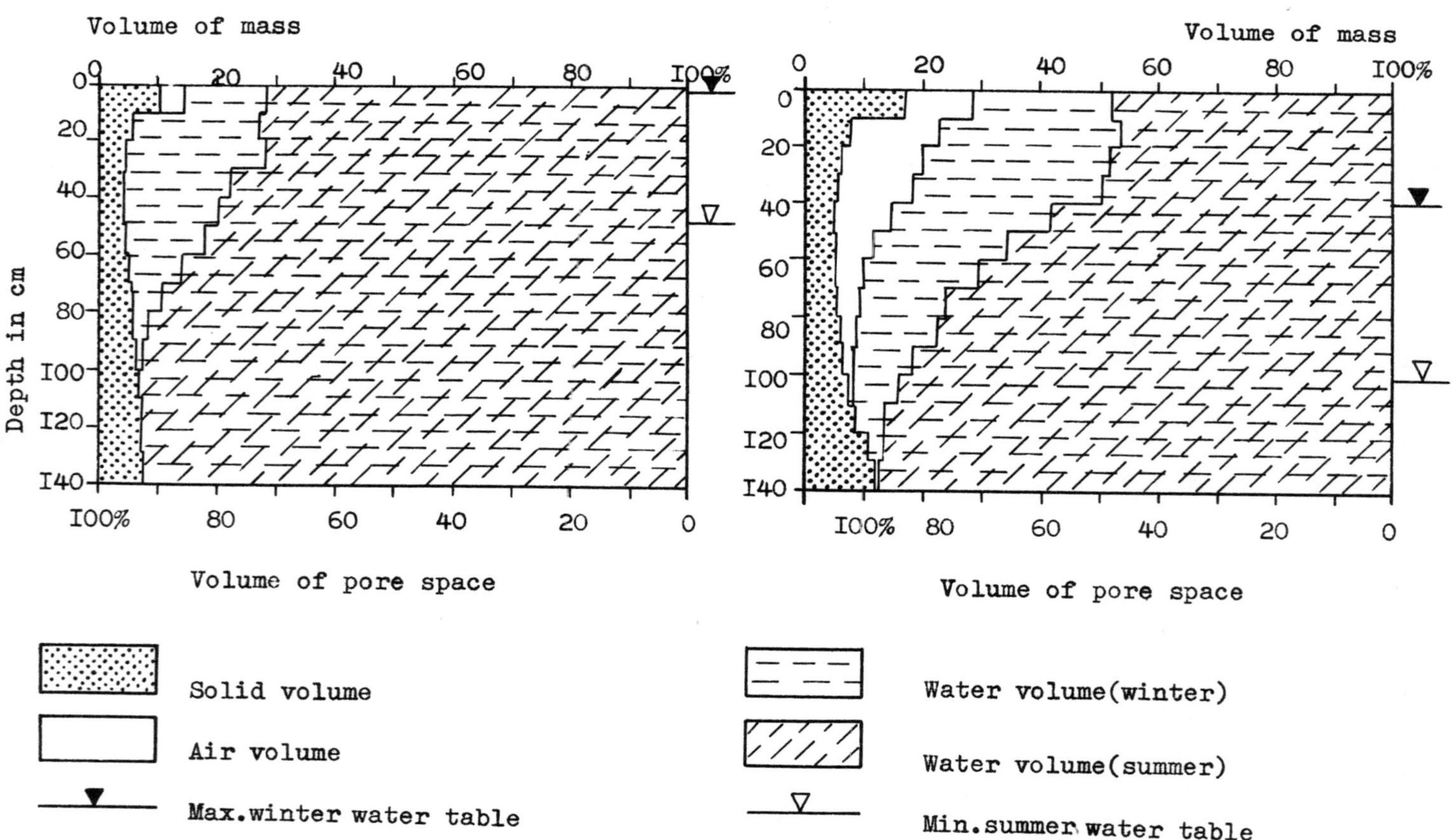

Fig 69. Soil structure diagram of a predrained raised bog (left-hand side) and for a cultivated raised bog (right-hand sied) near the Koenigsmoor Experimental Farm.

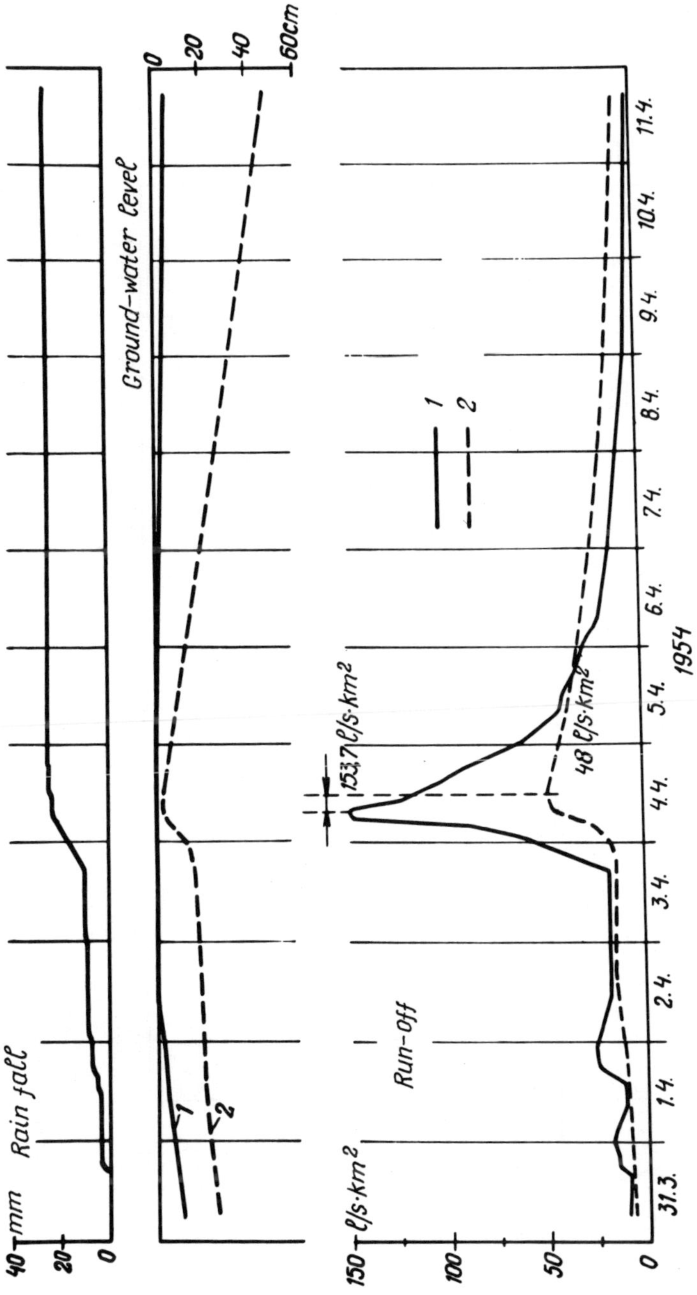

Fig 70. Diagram of rainfall, groundwater and surface runoff for watersheds with different plant associations.

1. Callunetum. 2. Raised bog grassland.

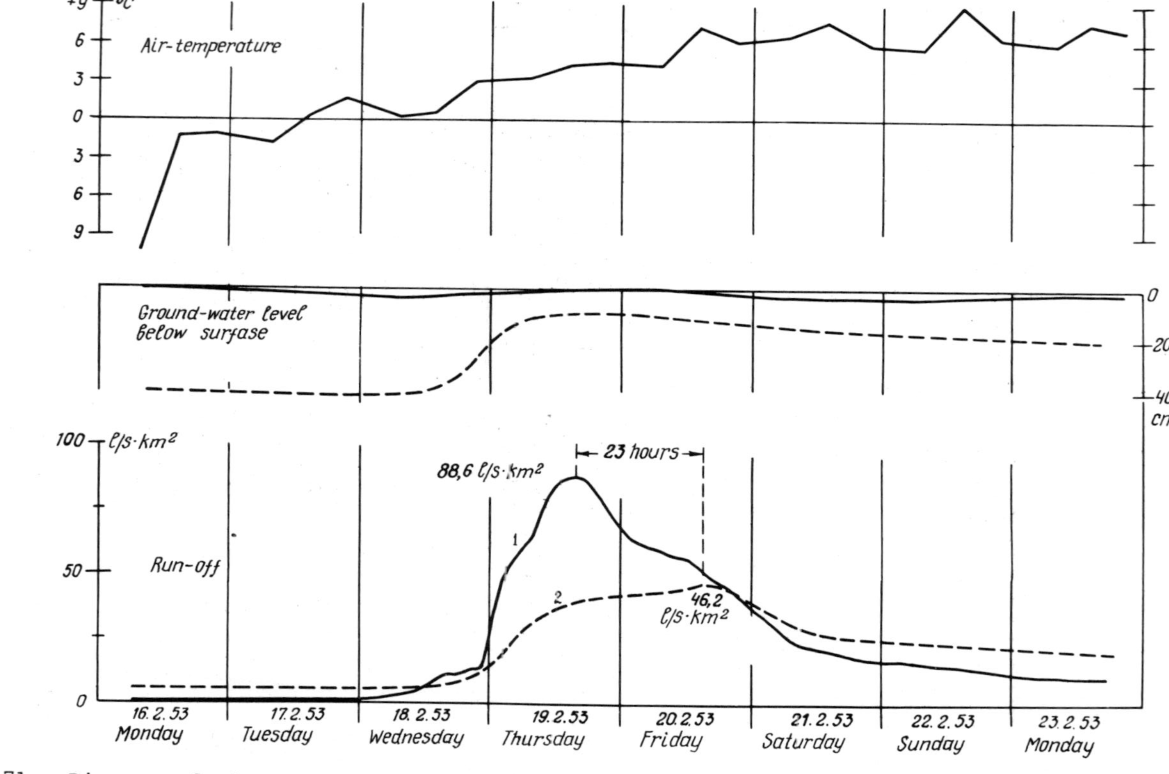

Fig 71. Diagram of air temperature, groundwater and surface runoff at the time of snow melting for watersheds with different plant associations.

1. undrained raised bog. 2. reclaimed raised bog.

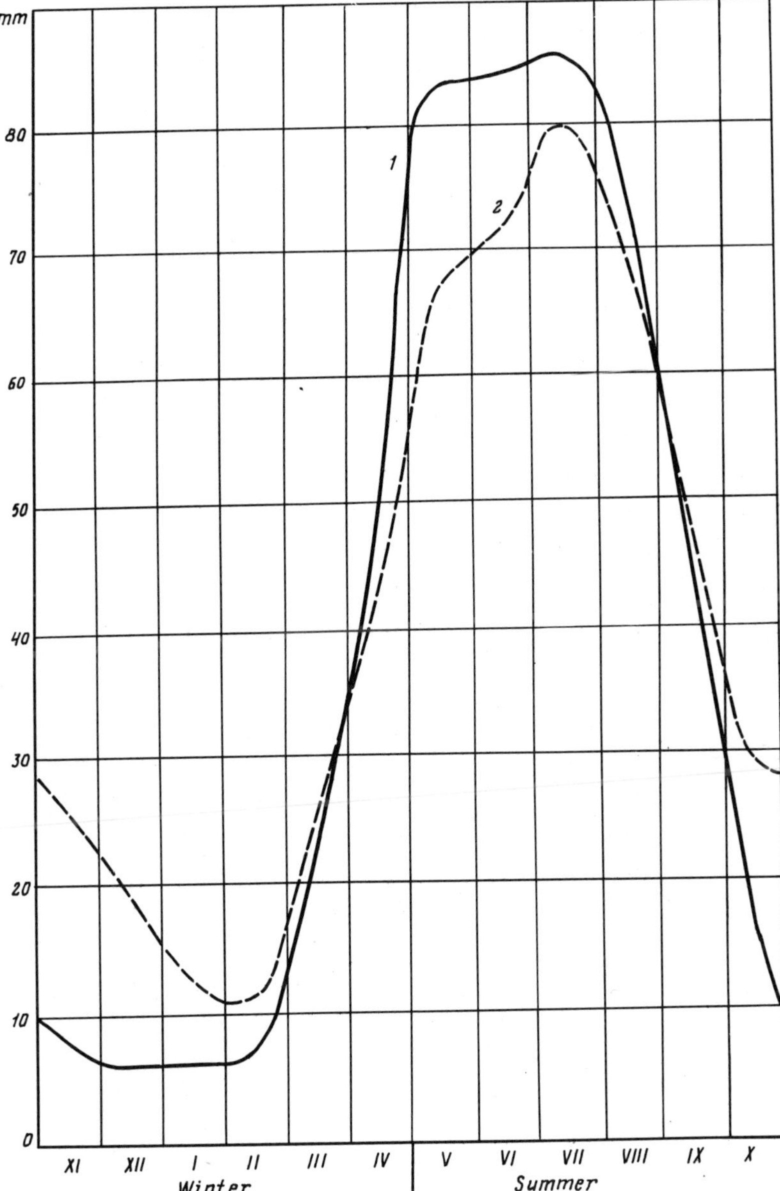

Fig 72. Annual distribution of evaporation for different plant associations in raised bogs.

1. Raised bog Callunetum
2. Raised bog grassland

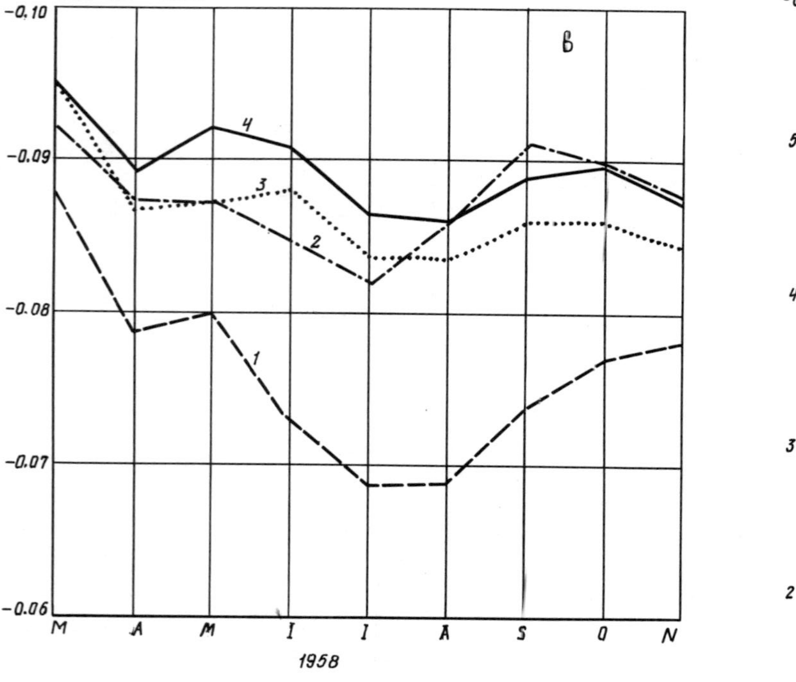

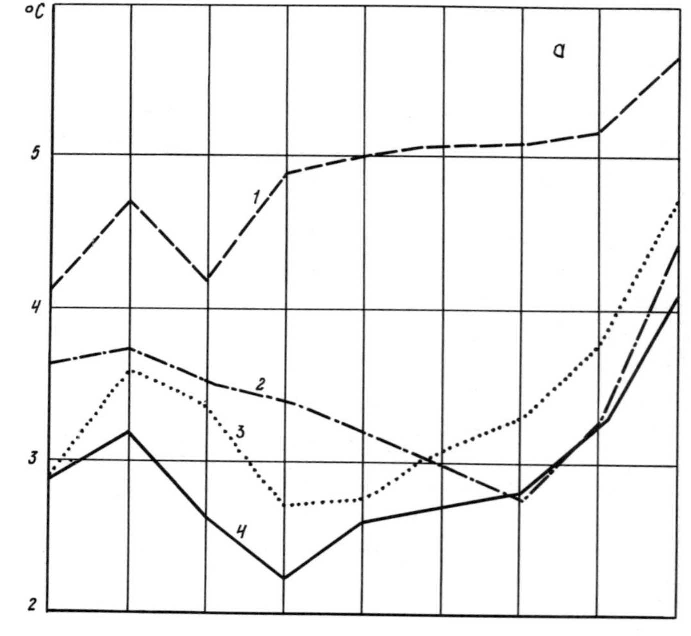

Fig 73. Difference between air temperature at 2 m and at 0.05 m above soil and net radiation balance for different plant associations of high bogs.

1. Sphagnetum
2. Callunetum
3. Raised bog grassland
4. Fern

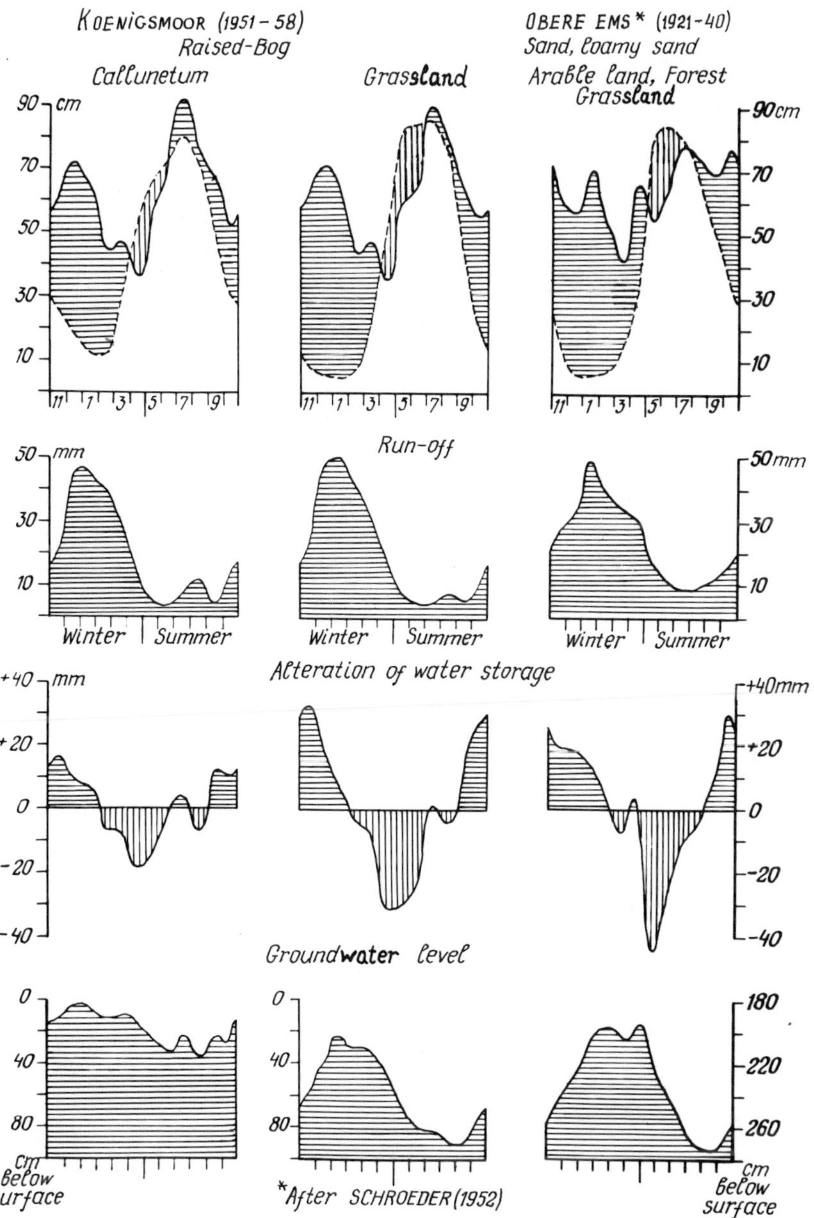

Fig 74. Hydrological budget for different watersheds with and without peats in the north-west of the FRG (annual rainfall about 740 mm).

MATHEMATICAL SIMULATION OF RUNOFF FROM SMALL PLOTS OF UNDRAINED AND DRAINED PEAT AT GLENAMOY, IRELAND

J. Dooge and R. Keane

University College Dublin,
Dublin, Ireland.

SUMMARY: An attempt is made to examine the feasibility of the use of mathematical models to predict the effect of the drainage of peat areas by the adjustment of the parameters of such models. Rainfall and runoff data for an undrained plot and a drained plot of peatland in Glenamoy in north-west Ireland are analysed. Two models are used to fit the records for the month of January 1968. The computer program developed for this purpose is briefly described. The preliminary results indicate that the approach is feasible and that further work along these lines would be worthwhile.

UTILISATION DES MODELES MATHEMATIQUES POUR CONNAITRE L'ECOULEMENT PROVENANT DE TOURBIERES DRAINEES ET NON DRAINEES

RESUME : L'auteur étudie la possibilité d'utiliser des modèles mathématiques pour prévoir les effets du drainage des terrains tourbeux en ajustant les paramètres de ces modèles. Il analyse les données concernant les précipitations et l'écoulement pour une parcelle soumise au drainage et une parcelle sans drainage à Glenamoy dans le nord-ouest de l'Irlande. Il établit deux modèles dans lesquels il utilise les données obtenues pour le mois de janvier 1968. Il fournit une brève description du programme d'ordinateur élaboré à cet effet. Les premiers résultats indiquent que le procédé est valable et que l'expérience mérite d'être poursuivie.

NATURE OF STUDY

The study reported in this paper was concerned with the development of a methodology for investigating the effect of drainage on the hydrology of a small area of peat. The study was exploratory and the emphasis was on devising methodology rather than on the actual results obtained.

The general properties of the drainage of blanket peat at Glenamoy have been described by Burke (1963) and the method of measuring runoff from the two experimental areas studied are described in an earlier paper at this Symposium, (Burke, 1972). The two plots were each about 0.35 ha in area. The undrained plot is surrounded by two channels, each about 150 mm deep. The outer channel effectively isolates the plot from the influence of the surrounding area. The inner channel takes the complete surface runoff from the plot and directs it to the outlet where it passes over a V-notch, the head of which is continuously recorded. The drained plot is surrounded by a drain approximately 1 m deep. The plot is drained internally by field drains spaced at about 3.5 m and at a depth of about 0.8 m below the surface. The run-off from the drained plot is recorded in a manner similar to that from the undrained plot. As described by Burke, the readings from the recorders are processed by computer to give daily runoff totals.

DATA USED IN PRELIMINARY STUDY

The processed data for January 1968 were used in the preliminary study of the possibility of simulating runoff and the effect of drainage by a mathematical model. A preliminary study of the run-off charts showed that their accuracy was not as good as had been hoped. It was thought that this was partly due to the effect of vegetation clogging the V-notch or the recording mechanism.

It was necessary to adjust the data so as to allow for the effect on the outflow of rainfall prior to January 1968 and also for the fact that some rain falling during January 1968 would not have appeared as outflow during that month. It was also desir-able to take a period in which the input and output would be app-roximately equal. Therefore, a period was chosen where the stor-age at the beginning and the end were similar so that only small corrections for storage would be required.

The attempt to adjust the rainfall and runoff data was not entirely satisfactory but ultimately the period from 10th to 26th January was chosen for study. The rainfall for this period was adjusted to allow for evaporation and also to correct for the change in storage between the beginning and the end of the period. The values of the rainfall adjusted in this way are shown in the first column of the table in Appendix A. The adjusted values of the measured runoff from the undrained and drained plots are sh-own in the second and third columns of the same table.

CONCEPTUAL MODELS USED

Two conceptual models were used during this preliminary study to simulate the runoff from Glenamoy peat. Both of them assumed that the peat areas acted as linear time-invariant systems. The operation of such systems is completely determined by the impulse response or instantaneous unit hydrograph of the system, (Kraij-enhoff van de Leur, 1966; Dooge, 1968; Becker and Glos, 1969).

The first model used was that represented by an impulse response in the form of a gamma distribution

$$h(t) = \frac{\exp(-{}^{t}/_{k})\,({}^{t}/_{k})^{n-1}}{n-1} \qquad (1)$$

where n and k are parameters of the response and (n-1) is interpreted as a generalised non-integral factorial, i.e. as the gamma function of n. This type of impulse response was used by Nash to represent the direct flood response of a catchment to precipitation excess and was used by Kalinin and Milyukov (1958) to represent the response of an open channel to an unsteady inflow at the upstream end.

The second model used to simulate runoff was that of two storage elements in parallel, with the inflow divided between them in a given ratio (a). This is a three parameter model since we can vary the time constants of the elements (k_1 and k_2) and the value of (a). The physical concept underlying this model is that of a fast response component and a slow response component acting in parallel.

DETERMINATION OF THE MODEL PARAMETERS

The optimum parameter values for simulating the given data were determined by matching the moments of the impulse responses of the models to the moments of the impulse response indicated by the input and output data. Since the systems are linear and time-invariant, it was possible to use the theorem which indicates that the first three moments of the impulse response can be obtained from the corresponding moments of the output and input by the relationship

$$U_R(h) = U_R(y) - U_R(x) \quad (2)$$

where $U_R(\ )$ represents the R-th moment of the function in brackets and h, y and x represent the impulse response fuction, the output function and the input function, respectively. Equation (2) above is valid whether the moments are taken about the origin or about the centres of area of the individual functions.

In the case of the gamma distribution model, it is most convenient to work in terms of the first moment about the origin and the second moment about the centre. Then:

$$nK = U_1'(h) = U_1'(y) - U_1'(x) \quad (3a)$$

where $U_1'(\ )$ is the first moment about the origin; and

$$nK^2 = U_2(h) = U_2(y) - U_2(x) \quad (3b)$$

where $U_2(\ )$ is the second moment about the centre. Since the values on the right hand sides of equation (3a) and (3b) can be determined from the input and output data we can solve the equations for the values of n and K.

In the case of the second model, it is more convenient to work in terms of the first three moments about the origin. These are

$$aK_1 + (1 - a)K_2 = U_1'(h) = U_1'(y) - U_1'(x) \tag{4a}$$

$$2a(K)^2 + 2(1 - a)(K)^2 = U_2'(h) = U_2'(y) - U_2'(x) \tag{4b}$$

$$6a(K_1)^3 + 6(1 - a)(K_2)^3 = U_3'(h) = U_3'(y) - U_3'(x) \tag{4c}$$

which can be solved for the three parameters a, K_1 and K_2.

A program was written in Fortran IV for the IBM 1130 computer to find the first three moments of the unit hydrograph from the input and output data. The data needed for the program are the daily values of rainfall, evaporation and runoff during the period and, if storage is to be corrected for, the estimated storages at the beginning and end of the selected period and the estimated recession constant for the catchment.

PRELIMINARY RESULTS

The techniques outlined above were applied to the data for the undrained and drained plots at Glenamoy for the period from 10th to 26th January 1968 shown in Appendix A. When the data were entered into the computer program, it was found that the determination of the moments was very sensitive to the manner in which the data had been adjusted from the raw values of rainfall and runoff. This was particularly marked in the case of the undrained plot. The results indicated that it would be necessary to use shorter time intervals than a day for the satisfactory analysis of the data. However, a set of results was obtained from the data in Appendix A which seemed to be fairly consistent. The moments which were obtained from the data are given in Table 45. Units in this table are days and powers of days.

By matching moments as indicated in equation (3a) and (3b), the parameters for the gamma distribution model are found (Table 46). As can be seen from Table 46 the increased delay time from 1.43 days to 5.69 days in the drained case is simulated largely

Table 45.

Central moment	Undrained	Drained
First	1.43	5.69
Second	7.55	97.85
Third	76.10	2565.60

Table 46.

Parameter	Undrained	Drained
n (dimensionless)	0.25	0.33
K (days)	5.81	17.20

by the increase in the values of K and only to a slight extent by the increase in the value of n.

For the second model the following values were found for the parameters by using equations (4a) and (4c). As can be seen from Table 47, almost all the effect of drainage in the case of the

Table 47.

Parameter	Undrained	Drained
a (dimensionless)	0.33	0.37
K_1 days	0.44	0.46
K_2 days	3.36	8.80

second model is reflected in the parameter K_2. These results indicate that the models might be useful for predicting the effect of drainage on small areas of peat. Existing records could be analysed in order to obtain the values of a and K (or of a, K_1 and K_2) for the undrained condition. A comparatively short record would be sufficient in order to gain a reasonable estimate of the first moment or lag under post-drainage conditions. This value could be used in equation (3a) or (4a), together with the pre-drainage values of n (or of a and K_1) to determine the new value of K (or of K_2) and these three values used to predict flows under post-drainage conditions.

The values of the parameters given in Table 46 and Table 47 above were used to simulate the daily runoff of the two plots on the basis of the effective rainfall. The actual runoff for the undrained and drained areas from 10th to 26th January and the simulated daily runoffs for the 3-parameter model for the same period are shown in Appendix A.

As can be seen from the table, the model simulates the recession reasonably well but seriously underestimates the peak runoff. This is probably due to the fact that the simulation is on a daily basis. Better results might be obtained with hourly data.

While the results of the study were somewhat disappointing, enough was achieved to indicate that the approach may be useful in predicting the effect of drainage on the runoff from the peat areas. The results are certainly encouraging enough to suggest that further studies be made on the basis of more recent data which are known to be more reliable and which cover all periods of the year.

REFERENCES

Becker, A. and Glos, E. 1969. Grundlagen der system hydrologie Mitteilungen des Institutes fur Wasserwirtschaft. Heft 32, Berlin GDR.

Burke, W. 1963. Drainage of blanket peat at Glenamoy. International Peat Congress, Leningrad, vol. 2, pp. 809-817.

Burke, W. 1972. Aspects of the hydrology of blanket peat in Ireland. Hydrol. Marsh-ridden areas, Minsk.

Dooge, J. C. I. 1968. The hydrological cycle as a closed system. Bull. Inter. Ass. Sci. Hydro. vol. 13. 1. p. 58-68.

Kalinin, G. P. and Milyukov, P. I. 1958. Priblizhennyi raschet neustanovivshegosya dvizheniya vodnykh mass. Trudy TsIP, no. 66.

Kraijenoff van de Leur, D. A. 1966. Runoff models with linear elements recent trends in hydrograph synthesis proceedings and information. TNO, The Hague, no. 13.

Nash, J. E. The form of the instantaneous unit hydrograph. Int. Ass. Sci. Hydro. Pub. vol. 3, no. 45, p. 114-121.

APPENDIX A

Values of Daily rainfall and runoff.

Effective rainfall	Measured runoff (mm)		Simulated runoff (mm)	
	Undrained	Drained	Undrained	Drained
0.00	0.54	3.41	0.00	0.00
6.98	5.20	4.59	2.27	3.84
3.89	2.46	2.20	3.37	4.10
9.04	5.91	5.60	4.80	5.11
9.75	11.61	11.20	6.52	6.28
9.37	10.11	6.98	7.34	6.71
6.88	6.91	9.40	7.11	6.33
8.83	4.84	6.07	7.41	6.59
0.13	2.56	6.56	5.38	4.81
0.00	0.38	3.98	3.27	3.27
0.48	0.30	3.46	2.48	2.91
0.00	0.35	3.10	1.86	2.57
0.00	0.21	2.70	1.34	2.24
0.00	0.09	2.40	0.99	1.99
0.00	0.02	2.26	0.74	1.78
0.37	0.04	1.99	0.65	1.68
0.22	0.02	1.73	0.57	1.55

TOPIC III

TRANSFORMATION OF THE HYDROLOGICAL REGIME OF MARSH-RIDDEN AREAS OF THE TEMPERATE ZONE BY MODERN RECLAMATION TECHNIQUES: THE PREDICTION OF THE HYDROMETEOROLOGICAL EFFECTS

TRANSFORMATION OF THE HYDROLOGICAL REGIME OF MARSH-RIDDEN AREAS IN THE TEMPERATE ZONE BY MODERN RECLAMATION TECHNIQUES AND THE PREDICTION OF THEIR HYDROMETEOROLOGICAL EFFECT

(Review Report)*

V M Zubets and M G Murashko

Byelorussian Research Institute of Reclamation and Water Economy, Minsk, BSSR.
and
Central Research Institute of Complex Use of Water Resources, Minsk, BSSR.

Introduction

The aim of drainage reclamation is to provide the optimum moisture regime for marsh-ridden areas which suffer from excess water and insufficient aeration. Removal of excess water improves not only the moisture regime but also air, heat and nutrient regimes. The moisture regime of the soil should be regulated within a given range depending on the moisture of plants and on economic aspects of the use of the area to be reclaimed.

The pattern of individual inputs to outputs from the water budget of reclaimed lands depends on specific physiographic and hydrogeological conditions. General conclusions can be drawn from investigations of changes in the water balance under various conditions which should be taken into account in the planning and implementation of land reclamation measures, not only to make the most effective use of reclaimed lands but also to improve natural conditions.

Effect of reclamation on the water balance

The scale of modern water reclamation construction requires scientific investigations of the changes that occur in the water balance of a large area.

In the initial phases of reclamation, assumptions were made based on a small amount of experimental data. However, in the past 10 to 15 years, attention has been given to a more thorough study of the problem. During this period, a number of experimental and theoretical investigations have been made concerning

---

* Some of the material in compiling this report has been provided by Drs A Bulavko, M Golberg, A Lavrov and N Smolsky of Byelorussia

the effect of marsh-drainage works on the water balance and water resources of reclaimed areas.

Reclamation has two major hydrological effects: first, the water balance of the reclaimed area itself is changed, and second, hydrological changes occur on the land adjacent to reclaimed areas and on entire river basins.

These investigations show that the groundwater characteristics of reclaimed lands depend on many factors, of which the most important are the quantity and temporal distribution of precipitation falling on the swamp, the inflow of groundwater to the area, the rate of evaporation, transpiration, and the removal of water by natural and artificial means. These factors depend in their turn on the climatic conditions and on the source of supply to the waterlogged area. These sources include the stream discharge rate and changes in soil water storage, the extent of canalisation, the physical properties of the underlying strata, the vegetation, etc. Because the input and output components of the water budget are not in equilibrium, the groundwater level must vary with time. Therefore, when concerned with rates of fluctuation of the groundwater level in reclaimed lands, the type of water supply, the method of reclamation and the climatic conditions must be kept in mind.

The main features of the water balance of large areas mainly result from the climate, primarily precipitation, which is the main source of water supply. Reclamation measures cannot have any significant effect on such global processes as moisture exchange in the earth's atmosphere.

At the same time, reclamation may affect the discharge rate of rivers, depending on the climatic, pedological and physiographic characteristics of the catchment area, the extent and type of the swamp, and the reclamation works. In some cases this influence is slight, in others it is quite appreciable. In the first years after reclamation, the annual and seasonal runoff increases due to a decrease in evapotranspiration and to the consumption of excess in underground water reserves which accumulated during the previous water regime. Later, with increased agricultural use of reclaimed lands, the streamflow rate becomes more uniform and annual runoff approaches its original value.

In the humid zone of Europe, one of the largest areas where draining of waterlogged land is carried out on a large scale is the Polessie Lowland basin in the Byelorrusian and Ukrainian SSR. The total catchment area of the Pripyat river is more than 12 000 000 ha, of which 4 500 000 ha are swamps or marsh-ridden lands. Using this area as an example, the prediction of possible effects of drainage reclamation on the water balance of the territory will now be considered.

From the large number of reports on the hydrological conditions of reclaimed catchment areas in Byelorussian Polessie, and on investigations of runoff and the water balance of catchment areas of large reclamation systems, it has been possible to trace the main trends in the effect of reclamation on streamflow.

As a result of draining and developing swamps in Polessie, a decrease in the surface runoff and an increase in the groundwater flow may be expected in the reclaimed area. The pattern changes in total runoff will vary depending both on the ratio of surface to groundwater components, and on the rate of discharge by drainage.

Changes in the total river flow of individual catchments depend on the proportion of the area which is drained swamps, the rate of groundwater supply, and the type of drainage. A very considerable increase in groundwater flow is observed after reclamation of swamps which have high pressure artesian supply. When large systems of swamp with a small groundwater supply are drained, the change in groundwater flow is practically negligible. Further, after reclamation of swamps with a low groundwater supply the annual streamflow hardly changes. Drainage of swamps with a plentiful supply of groundwater decreases the annual volume of runoff by as much as 30% for some catchments, depending on the proportion of the catchment reclaimed.

Because of reclamation, the peak discharge and the volume of the spring flood runoff in Polessie may increase in some years and decrease in others. This depends on the hydrometeorological conditions in the spring and in the preceding winter. Over a period of many years the average values of individual peak discharges and the volumes of the natural spring flood discharges may increase by 20% in the northern parts of the left bank tributaries of the Pripyat river. There is however practically no change in catchments in the southern part of Polessie. In other cases, the drainage of swamps may decrease spring flood peaks as a result of the high storage capacity that is created in the unsaturated zone of the developed swamps. The volume of summer and autumn runoff increases after drainage and development of swamps with high groundwater supply but changes where the groundwater supply is low.

The prediction of groundwater levels is also important. From observations made over more than 15-20 years it is known that natural groundwater levels are extremely variable. This is not only true for the variety of hydrogeological conditions which in Byelorussian Polessie depend on the geomorphology of the region but also from one year to another over periods in a particular area. Many of the factors which affect the hydrological regime, particularly the climatic and hydrogeological characteristics, vary considerably both in time and in space.

In spite of the large number of variables which cause changes in the water level, the general patter of natural groundwater behaviour has been fairly well-defined. A rise of level in spring, a long recession in summer and autumn, a rise in the autumn-winter period, and a short winter recession can be distinguished. As a rule annual peaks are observed in spring (March to May) and sometimes in summer (June). The lowest levels are found in the summer or autumn (July to November) and in winter (December to February) during periods without precipitation or when precipitation ceases to infiltrate. Groundwater resources are mainly recharged in spring, when infiltration of snowmelt and precipitation are maximum.

Long term variations in level are connected with climatic changes over periods of 3 to 5, 11 to 13, 20 to 22, 35 and even 100 years. Of these, the 11-year cycle of variations in groundwater level are the most distinct. A comparison of groundwater levels and solar activity since 1951 clearly shows the dependence of the mean annual groundwater level on the Wolf number.

Observations of the groundwater level in a number of areas subsequently reclaimed over a period of more than 45 years have indicated that, under natural conditions, the level of groundwater in the central zone of the Soviet Union has two annual maxima, (spring and autumn) and two annual minima (summer and winter). The mean amplitude of the variation in groundwater level is 400-1200 mm) the storage capacity of the saturated zone is 100 to 150 mm of precipitation without raising the level of the water table.

It is known that during the growing season a dynamic balance is established between precipitation and soil moisture storage in a reclaimed area. The soil moisture regime established after initial drainage remains stable for long periods and is subject only to seasonal fluctuations which result from moisture consumption by plants, and the subsequent recharge of the soil moisture during the year. The state of soil moisture storage also affects the groundwater level.

In dry years there is insufficient moisture in the root layer, especially of seeded pastures, in the non-chernozem zone of the European part of the USSR. To provide plants with the necessary amount of water when large areas are reclaimed, channel irrigation systems have to be constructed.

The drainage of large areas and their subsequent use for agriculture give rise to a redistribution of the principal components of the groundwater balance. In dry areas immediately adjacent to drained bog systems, the groundwater level may fall by 0.3 to 0.7 m, the fall decreasing with distance from the drained area. The effect of drainage on the groundwater level of the adjacent areas can be detected at distances of up to 1.0 to 1.5 km. In the first years after the beginning of drainage, the level drops rapidly. In addition, data now available indicate a local decrease in level.

To study long-term slower changes in the groundwater regime, different approaches should be used for confined and unconfined aquifers.

According to available data it is possible to conclude that land reclamation affects the movement of phreatic and confined aquifers of Polessie in different ways. Groundwater levels fall considerably on areas adjacent to drained marshland during the subsequent 3 to 5 years, but this fall is to some extent compensated for by an increase in groundwater flow. During this initial period water levels in confined aquifers change less, but because of the slow rate of water exchange in Polessie (200 to 400 years) their depletion, together with increased recharge of unconfined groundwater resources, may cause a more prolonged and constant disturbance of the regime.

Although some progress has been made in the study of the effect of drainage on the water resources of the area, a number of problems have not yet been clarified and require further study. The main aims in this respect should not be just the accumulation of new experimental data; a theoretical examination of the available information should be made to explain the mechanism of the effects of drainage in various hydrological, soil, climatic and hydrogeological conditions. The basic tenets of the investigation methodology must be worked out, keeping in mind the specific features of the problem.

In hydrological investigations mathematical and physical simulation methods should be widely applied when making long-term predictions of changes in the underground and surface water regime of the areas under drainage.

Further studies of changes in the quality of surface and underground waters due to drainage as well as investigations of the thermal characteristics of the soil are also required.

Effect of reclamation on hydrometeorological conditions

Measures for the reclamation of marsh-ridden areas change the thermal radiation, physical, and moisture properties of the soil, which in turn affect the microclimate of the drained areas. The change in the microclimate occurs largely because the moisture content of the soil decreases with a resulting decrease in the heat capacity and thermal conductivity. In the warm period of the year the daily mean temperature of the upper 200 mms of the soil is 3 to 4°C lower in a natural swamp than in a drained one. However, the temperature gradient in a drained swamp is 2 to 3 times higher than in an undrained swamp resulting in almost equal temperatures at 1.5 metre depth in both drained and undrained swamps. The spring temperature of the upper 200 mm layer of peat soil rises above 10°C 10 to 15 days earlier when the land is drained; in autumn the temperature remains above 10°C longer. During the growing period the mean daily air temperature at a height of 2 m over a drained peat land is somewhat higher than that over the undrained peat area.

From records of temperatures of drained and undrained peat lands it is concluded that the air and soil temperatures should be expected to rise after reclamation.

Peat lands are much more prone to freezing than mineral soils. Observations show that the minimum surface temperature of a drained swamp is on average 3 to 5°C lower than that of a loamy mineral soil (on some clear nights the difference between minimum temperatures can be as much as 10°C). On the peat soils of the Byelorussian Polessie, frosts are possible at any time during the summer. The freezing temperatures on peat soils are normally 4 to 6°C, and sometimes 7 to 10°C, lower than on mineral soils.

Because of the low albedo of peat soils and the comparatively low heat conductivity and heat capacity of the surface layer, the maximum temperatures at the surface of drained peat bogs are on average 5°C (on particularly clear days 10 to 15°C) higher than those on loamy soil during the warm part of the year. The heat loss from peat beds is however considerably more rapid, hence, from May to September the mean daily temperature of the upper 200 mm of peat soil is 3 to 5°C lower than that of dry areas. During the growing period the cumulated temperatures above 10°C at a depth of 100 mm in drained peat bogs is 400-500°C less than for loams. Correspondingly, the frost-free period is

30 to 60 days shorter than on loamy soils and in some years there is practically no frost-free period in peat beds. Considerable differences in the temperature of the soil cause differences in air temperatures. In the warm months (May to September) the mean monthly temperature at a height of 2 metres over drained bogs is 1 or 1.5°C lower than over dry lands. On certain clear days the difference in mean daily temperatures may reach 2 to 3°C. The frost-free period in the air over drained swamps is on average 45 days shorter.

In the warm part of the year the absolute daytime humidity over drained bogs is several millibars higher than over dry soil; at night the difference decreases and may even be reversed. However, the relative humidity over peat lands is always higher, usually by 5 to 15%. When the relative humidity in the warm period is high, even these small differences cause frequent mists and dew falls on drained bog systems.

Summary of the papers presented

The papers on this topic presented to the Symposium deal with many aspects of investigations made into the effect of reclamation on the hydrological regime of areas with different conditions.

The paper 'Bog reclamation and its effects on the water balance of river basins' by Bulavko and Drozd (Byelorussian SSR) presents results of investigations into changes in the components of the water balance of river basins of Byelorussian Polessie where reclamation measures cover from 6 to 25% of the basin area.

Reclamation increases the density of the stream network, changes the microrelief, decreases evaporation and time of inundation of the floodplain, thereby changing the distribution of the runoff within the year.

The authors state that the most difficult problem is to estimate the effect of climatic fluctuations on changes in the runoff. After reclamation, conditions promote an increase in the groundwater component of the river flow and consequently an increase in low flows. It has not yet been established whether the magnitude of spring runoff tends to increase or decrease.

An estimate is given of the disturbance of the statistical properties of hydrological series as a result of reclamation work on watersheds.

Glazacheva's (USSR) paper 'The effect of reclamation on rivers and lakes in Latvia' gives an analysis of observations in the Latvian SSR and comments on the increase of maximum and decrease of minimum flows resulting from drainage measures. In describing the decrease observed in groundwater levels on adjacent areas, the author notes the combined effect of hydrometeorological factors and reclamation measures.

Heikurainen's (Finland) paper 'Hydrological changes caused by forest drainage' gives data on the change in the components of the water balance (retention of precipitation, evaporation, transpiration, surface and groundwater runoff) for virgin and bog swamp forests which have been either drained recently or drained long ago. An increase in low flows for the recently drained peat bogs was found with a decrease in evaporation. No adverse effects have been observed on the hydrological regime of forest areas adjacent to the drained area.

Changes in the humic compounds content of waters draining from reclaimed peat bogs are analysed, and it is stated that further investigation is necessary.

In the paper 'Effect of fen bog drainage on annual runoff in the Estonian SSR' Hommik and Madissoon (USSR) examine the effect of drainage of bogs and waterlogged mineral lands on the annual runoff, by comparing annual runoff determined by meteorological factors with that measured at gauging stations. Differences between the actual and the calculated runoff are correlated with percentage of the catchment which comprises bog, waterlogged mineral, drained and other land.

The quantity of the annual runoff and other components of the water balance of fen bogs are studied on experimental plots which have different drainage rates. From these investigations, it has been possible to determine the main hydrological factors which influence the design of drainage systems for the fen bogs of the Estonian SSR. These factors are the effect of reclamation on annual runoff, exchange between surface and groundwater month by month during the growing period, and the normal annual runoff from drained fen bogs.

In the paper 'The effect of land reclamation by drainage on the regime of rivers in Byelorussia', Klueva (Byelorussian SSR) examines changes in the hydrological regime of rivers due to a combination of drainage and reclamation measures. River catchment areas of the Byelorussian SSR where swamps are widespread due to specific physiographical conditions have been investigated. This work involved a comparison between the hydrological data of 16 catchments for the periods before and after reclamation with corresponding data from 30 reference catchments which are in a natural state or where the effect of reclamation is only slight. Minimum summer and winter specific daily discharges, maximum discharge and the magnitude of spring flood runoff as well as seasonal and annual discharges have been analysed.

From these investigations, conclusions have been drawn on the effect of reclamation on the hydrological regime of rivers in the area under consideration.

'The influence of agriculture on water eutrophication in lowland areas' by Kuntze and Eggelsman (FRG) describes an investigation of leaching of the mineral fertilizers used on reclaimed peat soils, and the effect of this process on changes in the quality of water in drains. It is shown that because of their solubility the calcium, potassium and nitrogen content of water increases considerably. To avoid eutrophication of water it is essential that the quantity of spring fertilisers is limited to that needed by plants cultivated on reclaimed areas. The quantity of potassium, phosphorus and nitrogen in water increases during the wet periods of the year.

The paper 'Problems of groundwater regime and its prediction in drained regions of the Byelorrusian Polessie' by Lavrov, Fadeeva, Sachok, Vakhovsky and Vasiliev (USSR) describes the relation between the groundwater level on reclaimed areas and the most important factors controlling the hydrological regime: fluctuation of the water level in drains, precipitation, etc. The authors conclude that fluctuations in groundwater level mainly result from changes in meteorological factors. The effect of the various factors is found by regression equations derived from a large quantity of observed data. The equations serve as a basis for making both long-term and short-term predictions.

In the paper 'The effect of drainage on stream flow' by Moklyak, Kubyshkin, and Karkutsiev (Ukranian SSR) a long-term comparison of the discharge in rivers draining reclaimed land and boglands of the Ukranian SSR indicates a change in the streamflow after reclamation. It has been found that after drainage, the annual, spring and maximum discharges of many rivers decreased; there are however situations where the runoff did not change for a number of reasons including the small size of the drained area. A brief list is given of the main factors which may affect the runoff in rivers draining reclaimed land.

The results of long-term observations of changes in the runoff of open and forested peat bogs as a result of drainage are given in the paper 'Influence of forest drainage on the hydrology of an open bog in Finland' by Mustonen and Seuna (Finland). Observations show an increase in the winter and summer minimum runoff due to a decrease in evaporation. Drainage lowered groundwater levels in peat bogs by an average of 300mm, and caused an increase in the spring maximum runoff of 131%.

The paper by Shebeko (Byelorussian SSR) deals with the hydrological regime and its prediction in catchments comprising reclaimed land. The paper describes variations of the components of the water balance of swamps due to reclamation. Over a period of many years the average evapotranspiration for individual catchments increased by 13%, the mean annual surface runoff by 12 to 14%, while the groundwater flow also increased.

To quantify the effect of reclamation on the distribution of runoff, coefficients relating to the rate of drainage and to the stream network are used. Relations are obtained which define the magnitude of the specific daily mean discharges of floods due to both rainfall and snowmelt. It is shown that maximum discharges of rain floods tend to decrease while the summer-autumn runoff from reclaimed swamps which have large groundwater flows tend to increase, after reclamation.

The paper by Verry and Boelter (USA) entitled 'The influence of bogs on the distribution of streamflow from small bog upland catchments' shows that the seasonal distribution of streamflow from small bog upland watersheds depends mainly on hydrological conditions. Organic soils in the bog part of the watershed have only a minor effect. Contrary to general opinion, neither upland or lowland bogs regulate the streamflow.

The seasonal distribution of flow from lowland bogs is fairly uniform due to a stable base flow. Flow from upland bogs in spring is due mainly to snowmelt and spring rainfall. The annual maximum discharge from both types of watersheds is due to snowmelt, and is a function of the watershed size, water content of the snowpack, and of the influence of rain falling on snow rather than of the hydrogeological conditions of the bog. Both types of bogs produce a short-term regulation of storm flows by delaying the release of storm waters thereby decreasing the maximum.

Conclusions

A review of the papers presented to the Symposium shows that a large number of investigations have been carried out on the effect of drainage measures on the hydrological regime of waterlogged areas under various physiographic conditions. In addition, those aspects of the problem which require further attention can now be defined more clearly. The limited data available on hydrological and hydrogeological conditions do not enable changes in the hydrological regime caused by drainage reclamation to be quantified.

The data available on individual swamp and reclamation systems do not permit temporal changes in the hydrological regime of larger regions to be accurately predicted.

Separate investigation of the surface water balance and the groundwater balance is inadequate. This deficiency can be overcome by (1) the creation of experimental research stations equipped with modern instrumentation to measure all the components of both surface water and groundwater balances, (2) further study of all aspects of their interaction, (3) simulation of the processes and analysis of data using electronic computers.

It is essential that all forms of communication which promote the exchange of results of investigations carried out in different countries are improved so as to contribute in every possible way to the development of methods for the design and construction of reclamation systems.

THE TRANSFORMATION OF THE HYDROLOGICAL REGIME OF MARSH-RIDDEN AREAS BY LAND RECLAMATION, AND FORECASTING OF ITS INFLUENCE ON HYDROMETEOROLOGICAL CONDITIONS WITH REFERENCE TO THE NETHERLANDS. (Review Report).

F. C. Zuidema

Ijsselmeerpolders Development Authority, Lelystad, the Netherlands.

The amount of water contained within an area at a given time will change, when water either flows to or from the area. Depending on land use, the amount of water stored may be considered too great or too small; in a marsh-ridden area the amount of water is too large.

In Chapter 2 of the IHD study 'Status and trends of research in hydrology' the terms 'swamp' or 'marsh' have been described 'as applied to land which is either seasonally or permanently wet and whose condition arises from a wide range of geographical causes' (UNESCO, 1972). An excellent description of the terms 'bogs' and 'fens' has been given earlier by Dooge (1972). Thus, to improve these areas the runoff must be increased as the inflow to the area (mostly rainfall) can be little influenced by man. Isolation of the area from its surroundings by embankments and stream regulation can prevent uncontrolled surface runoff from entering the area.

Water regulation and water control mainly result in accelerated and/or greater runoff. The depth of the groundwater level below the ground surface is a simple parameter for characterising bad drainage conditions; more attention will be paid later to this in this paper. For many centuries, human activities have influenced and transformed the hydrological regime according to local circumstances and the kind of land use.

In the case of marsh-ridden areas the following questions arise:

I. To what level should the groundwater table be lowered?

II. How should this change of level be obtained and how should it be maintained?

III. What environmental effects can be expected as a result of the hydrological changes in the reclaimed area?

IV. Is it possible to forecast the effects of human activities on hydrometeorological conditions and how can our approach be improved?

Each of these four questions will now be examined.

I. TO WHAT LEVEL SHOULD THE GROUNDWATER TABLE BE LOWERED?

The various activities of man make various demands upon the upper soil layers and the groundwater table. These demands are related either directly or indirectly to the soil moisture content, which for given soil gives some information on its physical behaviour:

1. the air content of the soil;
2. the forces by which the water is retained in the soil;
3. the pore size distribution, which is related to the capillary system.

4. the moisture bearing retention capacity of the soil.

When the air content and pore size distribution of the soil are known, the extent of the unsaturated layer of soil can be estimated. Moreover, the response of soil temperature changes in air temperature can be calculated. The air content and the moisture content of the soil are fundamental to the existence of bacteria, and are therefore an important factor in determining soil fertility.

Soil moisture content is, however, not easily measured. At equilibrium, when the supply of water to the soil is equal to the discharge out of the profile, there is a close relationship between soil moisture and groundwater level. Because the latter can be measured very easily, the level of the water table is often used as a measure of good or bad drainage conditions, and in winter(when most runoff occurs) this approach is adequate. But at equilibrium, different soil types require different groundwater levels to attain the same quantity of available moisture at the same depth in the soil profile. In the summer this quantity of available moisture is extremely important.

As Table 48 shows, the soil moisture requirements for a given land use differ in summer according to soil type. Reference will also be made to the drainage rates for various kinds of bogs, given by Ivitsky (1972).

In addition to soil properties, soil use, climate and economics influence the desired groundwater level. Regarding agricultural uses the information on the effect of the water table depth on crop yields which form the basis for the requirements shown in Table 48 will be given.

Visser (1958) gives the relation between crop yield and the level of the water table during the growing season (Fig. 75). Decrease of yields with high water tables can be ascribed to lack of aeration of the soil, while decreases at lower water table levels are due to a shortage of water. As both conditions may occur depending on climatological conditions, it is evident that the curves may undergo a horizontal shift to the right in wet years and a shift to the left in dry years (Wesseling, 1969).

The effect of wet and dry years on the dry matter yield of grassland on peat soils is shown on Fig. 76. The year 1949 was very dry, while the other years had normal climatic conditions. Still larger differences have been found for grass on a profile of 200 - 400 mm clay on peat (Fig. 77), (Sieben & Smits, 1969).

Feddes (1971) has calculated the influence of climate on yield, first giving the total quantity of available moisture for different soil profiles (Fig. 78). All figures relate to a water table depth of 900 mm and the soil moisture is present in the root layer. During the growing period supplementary moisture will reach the roots by capillary action from the root zone and from groundwater.

The total amount of moisture needed for optimum plant growth can still be supplied by rainfall. Thus an apparent loss of yield due to lack of soil moisture may be totally or partly reduced. With the aid of a rainfall depth-frequency distribution for consecutive days, the magnitude and frequency of occurrence of

Table 48. Drainage requirements and desired groundwater levels during summer in the Netherlands for various soils and land uses

land use	soil type	drainage requirements: highest groundwater level in mm below s.s.	drainage requirements: at discharge of .. mm/day	desired groundwater level during summer in mm below s.s.
arable crops	sand			1000-1400
		500	7	
	sandy loam			1200
	clay	300	10	1000-1400
grasslands	sand			600- 700
	sandy loam			1000
	clay (on peat) > 600 mm			1200-1300
	clay (< 400 mm) on peat	200	10	500 (see fig.3)
	clay (> 400 mm) on peat			700-1000
	dense clay			200- 500 (dry summer)
				1000-1200 (wet summer)
	peat			600- 700 (see fig.2)
horticulture	sand			500- 600
		500	7	
	sandy loam			1000-1200
		300	10	
	clay			900-1100
orchards	clay	600	10	1300-1500
		700	5	
		950	1	
forests		200-500	5 - 7	100->1000
areas with buildings under construction		500	10	-
areas with completed buildings		700 mm below ground level		
sports fields and camping sites		50	15	

this compensating rainfall can be calculated. The results of this calculation are given in Fig. 79.

From Figure 79, it can be seen that for potatoes grown on clay on sandy loam with the water table at a depth of 1.2 m a yield of 31 $ton.ha^{-1}$ or more can be expected with a 1 in 5 year probability. It also appears that there is a 50% probability that at water table depth of 0.90 m, the yield will be 49 $ton.ha^{-1}$ or more.

An attempt to describe the effect of high water tables on crop yields has been given by Sieben (1964). On the basis of considerable data from experimental fields in the new Ijsselmeer-polders Sieben calculated the so-called SOW_{30} values (SOW = Som overschrijdingen wintergrondwaterstand or: Sum of excess winter groundwater level, SEW).

The SOW_{30} value is the sum of the products of days and height of water table in cm above a level of 30 cm below the surface during the winter (November to February inclusive). If for example the water table is at a depth of 20 cm below surface during 2 days, the SOW_{30}value for these two days is 2 x 10 = 20 day cm. The yields obtained from the fields were plotted against the SOW_{30} values (Fig. 80). The results show that continuous high water tables have a detrimental effect on both winter and summer crops.

It is remarkable that, although summer crops are not in the fields during periods of high winter water tables, the detrimental effect on the crop yield is nearly the same as that suffered by the winter crops. Sieben suggests that this phenomenon is due to a deterioration of the soil structure, which causes poor aeration and nitrogen mineralization during the growing season.

Economic aspects

In addition to physical and climatic aspects, the economics of the planned reclamation must be considered. A plan for a total improvement of an area can be optimized by a cost benefit analysis (Van der Molen, 1969).

Costs will increase with the depth to which the unconfined water table has to be lowered to obtain the desired drainage depth.

In general the optimum crop yield is related to the level of the water table, and more or less the same factors apply with respect to management in general.

Figure 81, derived from Van der Molen (1969) gives the relation between the total costs of a project of land improvement and drain spacing. Figure 81 also gives an example of the influence of the cost of construction work on the optimum solution of a subsurface drainage problem.

Curve K shows that wider drain spacing leads to lower yearly drainage costs (interest, amortization and maintenance), but may result in a greater loss due to lower crop yields or departure from optimum farm management. The loss is related not only to the drain spacing, but also to the drainage requirements of the soil, so the five curves 'a' to 'e' apply to five different soil types and to normal drain spacings from 8 to 20 m.

The optimum solution is given when the sum of losses and costs is a minimum (curves A to E). In practice, drain spacings less than 8 m are not applied for economic reasons.

II. METHODS OF LOWERING THE WATER TABLE AND OF MAINTAINING IT AT THE REQUIRED LEVEL

After determining the factors which influence the choice of the required groundwater level, the next step is to consider how this can be accomplished and controlled.

The technical requirements must be met by design criteria. As has been shown, these requirements depend in part on the proposed use of the reclaimed land.

In marsh-ridden areas, man's activities should aim to accomplish optimum conditions from both the technical and economic points of view. Hence, the application of optimization theories, though complex, will lead to a refinement of the criteria.

Table 47 shows a number of drainage requirements, applicable to the Dutch climate and in general based only on hydrological considerations. To this list can be added the wide range of requirements of nature reserves. The direct relation between a given water table level and the discharge for steady state flows applies to all land uses.

Winter requirements differ greatly depending on whether the land is to be used for urban or rural development.

The intensity, frequency and duration of rainfall plays an important role in the choice of the discharge criteria. In this case the frequency distributions of the total amount of rainfall in periods of minutes up to several days are of great value. In the Dutch new town of Lelystad data are even collected on rainfall intensities over intervals of 30 sec, to improve the design of structures related to peak runoff rates.

Reclamation works increase and accelerate the runoff from an area due to a change of storage. Land use has an influence on the dimensions of open waterways and the capacity of structures such as weirs, culverts and pumping stations.

This general statement is supported by the results of investigations in forest swamps with respect to dry years by Bavina (1972). It was found that in normal years the runoff in forest swamps and in afforested river basins was equal.

Both reclamation and land use must obviously be preceded by adequate investigations and measurements in and around the area considered, in order to ensure good and reliable design; this point is also mentioned by Novikov (1972). It is clear that the phenomenon of subsidence must be fully examined in such investidations and in low lying areas of soft clay and peat the considerable subsidence which will result from land reclamation cannot be neglected. In general, with the help of research, forecasting of subsidence will be possible, provided that the physical behaviour of the soil is understood. In addition to the papers on this subject which are presented at this symposium, the I.H.D. Symposium on land subsidence, held in Tokyo in 1970 should also be mentioned; the proceedings of this Symposium have been published by UNESCO (1969).

If all these factors are taken into account, land reclamation should be successful and produce no detrimental side effects. By studying each project the design criteria can be adjusted to meet the requirements of the local situation.

III. WHAT ENVIRONMENTAL EFFECTS CAN BE EXPECTED AS A RESULT OF THE HYDROLOGICAL CHANGES IN THE RECLAIMED AREA?

This question which arises from the two preceeding questions will be illustrated with some examples from the Netherlands. In most of these aspects which have general validity can be identified.

The constant struggle to control water in the Netherlands in previous ages met with failures as well as successes. Despite considerable endeavour, probably some 500 000 ha of land have been lost since the early Middle Ages, but because from the 16th century onwards land has also been gained, it seems likely that at present these losses have been compensated for (Fig. 82).

The Zuyder Zee Reclamation Project(under progress from 1930 to about 1995) is essentially a continuation of the historical pattern. Its aim is to further reduce the length of the coastline, to improve water management by the creation of storage basins and by the partial reclamation of the enclosed areas. Here, the change is from sea to land which is then to be put to various uses. Hence there is in fact no transformation of an original marsh-ridden area, but, when the seabottom has dried up, its behaviour is similar to that of a marsh.

Three effects which are manifested in this area due to reclamation works will be examined:

1. changes of deep groundwater flows and these effects of levels on the polder environment
2. changes of phreatic levels
3. meteorological changes in the area.

In addition to these effects the influence of open water on the occurrence of night frost will be mentioned in section III.4.

III. 1. Changes of deep groundwater flow and the effects of levels on the polder environment

Below Lake Ijssel (formerly the Zuyder Zee) and its vicinity, there is a very permeable sandy geological formation, about 200 m thick (Figure 83). In this formation, groundwater is flowing from these areas having a higher potential (Veluwe, Utrecht-hills) and from areas where the groundwater is fed by rainwater, to areas with a lower potential (polderland in Holland and the new reclaimed land in Lake Ijssel), where it occurs as upward seepage water. The extent of the seepage decreases when the sandy aquifer is covered by layers of holocene clay and peat, which have a low hydraulic conductivity. At the bottom of the sand medium, clay layers can have a similar retarding influence on the groundwater flow.

As a result of polder development in the Lake Ijssel the potential of the deep groundwater flow will be reduced (Figure 89), while the poor permeable top layers will be penetrated by excavation works eg for canals and ditches.

A change in the existing groundwater flow will result from these reclamation activities, possibly giving a harmful reduction of the water table in the vicinity. This has happened in a peat area near the North East Polder, where no border canal separates the area from the polder. Although predictions of changes in the deep groundwater level agreed well with actual changes, no satisfactory prediction could be made as to the extent of the change in phreatic level. (Veenenbos, 1950).

In the Lake Ijssel area various methods of predicting the hydrological consequences of given layouts of the polders were tried. This is particularly relevant to the design of the border lake between the "old" land and the new polders. Such methods are generally based on both hydrogeological and geophysical investigations. Moreover, in the absence of natural groundwater flow, pumping tests can give information about the properties of the sandy layers eg KD, product of hydraulic conductivity (K) and thickness (D) of the sandy layers.

In the case of the North East Polder, the KD-value was derived from measurements in two old polders in the locality: one with a deep water table and the other with a higher groundwater level, which resulted in a permanent seepage flow.

In the area of Eastern Flevoland, where a deep groundwater flow exists, the equipotential lines are close together (Figure 84). Here the boundary conditions for the groundwater flow could be computed from measurements of the discharge of small streams, and the best dimensions of the border lakes could therefore be found.

For Southern Flevoland Markerwaard such an approach is impossible, because over many years water will be forced out of the pores in the very thick clay layers. This is due to changes in the load caused by drainage of the soils. The variations in hydraulic conductivity, particularly in low permeability layers, are so large that the changes in the hydrological regime resulting from polder development cannot be calculated.

Simulation with an analogue model (two parallel plates) does not appear to be relevant; only a digital model could give a solution.

The equipotential lines in Figures 84 and 85 show a considerable difference between the value of the potential of the deep groundwater flow in 1960 and in 1970, due to reclamation works. In 1968 in particular, the Southern Flevoland polder was dewatered. These lines can be drawn from regular measurements of the potential in hundreds of small wells. The situation which exists in 1970 is not however the final one, and continuing changes are not easily predictable for the western part of Lake Ijssel and adjacent areas.

Therefore, continuous observations of potentials is desirable in order to obtain indications as to future changes.

III. 2. Changes of phreatic water levels

The changes in the phreatic level will be illustrated by the same polder development in the Lake Ijssel area. As soon as dike building is completed and the water inside of the diked area has been pumped out, a marsh-ridden area has in principle been created. Several measures then become necessary to maintain the required groundwater level.

Soils range from sand to silty loam, and the physical condition of the loamy soils differs considerably from that of comparable older soils. They are characterized by large pore spaces, an extremely low hydraulic conductivity (17×10^{-2} mm/day) and a very low water storage capacity. When the polder is first created all the pores are completely filled with water, making the sediments extremely soft and the area inaccessible.

There is a close relation between the clay content (soil particles of diameter $< 2\ \mu$) and organic matter, and the water content.

When the polder has been drained by pumping, a drainage system is established which consists of main canals, main ditches and temporary field ditches. This sytem can discharge only standing surface water that can flow out under gravity.

Reeds are sown from aircraft in the first year after the polder is created, and as reclamation proceeds these reeds are replaced by arable crops. The loss of water, which is largely irreversible, and the physical, chemical and microbiological processes are called the ripening of the soil and can be characterized by the following effects: decrease of water content, increase in accessibility, lowering of level (subsidence), crack formation (increase in hydraulic conductivity), and aeration of the soil (possibility of growth of crops). The process of soil-ripening depends on the following factors: weather conditions, vegetation, upward seepage, thickness of the clay layer and clay content of soil.

The manner in which the water content decreases after the creation of the polder is shown in Figs 86 and 87.

Figure 12 shows how the speed with which the water content decreases in the topsoil compared with the subsoil, and also in the first years after reclamation compared with a later period.

Due to the drying-out process the soil particles contract, leading to subsidence and crack formation (Fig. 13). The loss of volume shown in this figure is due solely to loss of water and this has a considerable effect on subsidence.

Reference will now be made to the discussion between Zubets (1972) and Van der Molen (1972). The essential point of this discussion was that the process of compaction of the clay soil (1.5 - 2 m thick) is much slower than that of a similar peat soil. Only a small part of the water loss is reflected by an increase of air content (cracks). This part of the water loss however is very important for the hydraulic conductivity of the young sediments. The sediments split up into prisms, mainly hexagonal, with a principal diameter of 250-500 mm. The hydraulic conductivity rises from 17×10^{-2} mm/day at the outset to sometimes over 100 m/day a few years later. Moreover, the water storage capacity of the soil increases from zero to some 10 per cent. This is in contradiction with the figures on the relationship between subsidence and hydraulic conductivity for peat soils given by Eggelsman (1972). When the cracks have reached a depth of 600 to 700 mm below the land surface the open field ditches of depth 500 mm can be replaced by tile drains (3 to 4 years after the installation of the open field ditches). The deep tile drains lower the water table, which further promotes physical ripening e.g. crack formation in the deeper soil layers. These processes end when the cracks have reached the level of the tile drains (Figure 88).

The optimum spacing of tile drains in ripening soils differs from that in normal soils. Sometimes the spacings needed later are twice that needed initially; the most appropriate spacing can be determined by field trials.

Where tile drainage is used with ripening sediments special attention is paid to subsidence. In time, the tile lines may come too near the land surface, and should therefore be placed somewhat deeper (1000-1300 mm) than in normal soils in the Netherlands (900-1000 mm).

III. 3. Meteorological changes in the area and its environs and their effect on crop yields

The reclamation of the Eastern Flevoland polder changed the distribution of the atmospheric isotherms (average daily values)in the Lake Ijssel area, in such a way that the air temperatures over the polder increased. Keeping in mind the extent of the long term reclamation programme an increase of this effect can be expected, and the general climate will become more similar to a land climate. As a result there will be relatively lower morning temperatures (approximately 1 to 2°C) and relatively higher temperatures in the afternoon and evening (approximately 1 to 2°C). The higher temperatures create a greater instability of the atmosphere in summer, which will reduce the number of hours of sunshine per day. This is a very important point in relation to the drying of cereal kernels (Figure 89). Van Kampen (1969) investigated the relationship between cumulative daily incoming radiation and the drying of the kernels of colza (A), barley (B), oats (C) and wheat (D). On cloudless days the incoming radiation is about 250 calories, and at harvest time every extra hour of sunshine is of great value.

Feddes (1971) mentions research on tulip bulbs, which shows that an increase in the number of days with higher air temperatures will quicken the death of leaves, shorten the period of growth and, as a result, the yield of bulbs will be lower, especially for larger bulbs.

III. 4. The influence of open water on the occurrence of night frost

In addition to the above changes in the hydrological regime, the effects of reducing the area of open water in new areas should be mentioned. In areas with peat soils the original transport system for land produce (lumps of peat, crops) was shipping. The change into transport by motor vehicles and an improvement in the drainage systems has led to a reduction of the area of open water. This results in a higher probability of night frosts in springtime.

The value of the product of the thermal conductivity λ and the heat capacity per unit volume C defines the sensitivity to night frost (Rethmeier, 1966):

λ C:		
	wet sand	1.75
	water	1.4
	wet peat	0.38
	dry sand	0.16
	dry peat	0.05

The figures above show that dry peat has the highest sensitivity and that the frost danger in wet sand is lower than in the direct vicinity of open water, provided that the suction of the wet sand is equal to field capacity right up to the soil surface; this is the case in marsh-ridden areas.

IV. IS IT POSSIBLE TO PREDICT THE EFFECTS OF HUMAN ACTIVITIES ON HYDROMETEOROLOGICAL CONDITIONS AND WHAT IS NEEDED FOR AN IMPROVED APPROACH

The examples demonstrate that it is difficult to answer this question by a simple 'yes' or 'no'.

In general, research into the principal components of the mechanisms of change, supplemented by local investigations, measurements and experience, are needed. In general, quantification of all the factors is not made.

In all cases, an intensive preliminary investigation of the works, supported by scientific study, will provide the more accurate prediction of the effects.

Wallace (1971) concludes from a study on the effects of land-use change on the hydrology of an urban watershed, that "the study was based on the premise of a relationship between land-use pattern and hydrology. A feedback mechanism was postulated; land-use patterns were assumed to have been influenced by watershed characteristics and the runoff characteristics, in turn, to have been affected by land-use pattern. The effort to analyze and quantify such a mechanism was complicated by many other factors which affect both land-use patterns and hydrology, and by an absence of data required for precise analysis. Nevertheless, certain conclusions may be drawn from the given analysis". Many people would reiterate these views.

Finally it is hoped that the contributions to this Symposium will increase research on hydrological changes in marsh-ridden areas and will lead to more international exchange of experience and results.

References

Bavina, L.G. 1972. The water balance of marshes of the forest zone of the European part of the USSR. Symp. Hydrol. Marsh-Ridden Areas, Minsk. Paper 8 on theme II.

Dooge, J.C.I. 1972. Water balance in marshlands. Symp. Hydrol. Marsh-Ridden Areas, Minsk. Review paper on theme II.

Eggelsmann, R. 1972. Physical effects of drainage in peat soils of the temperate zone and their forecasting. Symp. Hydrol. Marsh-Ridden Areas, Minsk. Paper 7 on theme I.

Feddes, R.A. 1971. Water, heat and crop growth. Wageningen (English with a Dutch summary).

Ivitsky, A.I. 1972. Hydrological foundations of drainage of swamps. Symp. Hydrol. Marsh-Ridden Areas, Minsk. Paper 2 on theme I.

Kampen, J.H. van 1969. Optimizing harvesting operations on a large-scale grain farm. Van Zee tot Land no. 46, Zwolle. (English with summaries in Russian and Dutch).

Novikov, S.M. 1972. Methodological basis of stationary and exceptional hydrological investigations of marshes. Symp. Hydrol. Marsh-Ridden Areas, Minsk. Paper 7 on theme II.

Rethmeier, B.C. 1966. Grondtemperaturen en de groe van mais. (Temperatures of soils and the growth of maize). Cursus "Meteorologie in de landbouw". (Dutch).

Sieben, W.H. 1964. Het verband tussen ontwatering en opbrengst bij de jonge zavelgronden in de Noordoostpolder (Relations between conditions and crop yield for young light clay soils in the Noordoostpolder). Van Zee tot Land nr. 40, Zwolle. (Dutch with an English summary).

Sieben, W.B. and Smits, H. 1969. The Veluwe lake, Part III. Geo-hydrological investigations, soil survey and soil-water-plant relationships in the border area of this lake. Reports about the Zuiderzeereclamation project no. 7, The Hague (Dutch with English summary).

UNESCO, 1969. Proc. Symp. on land subsidence. Studies and reports in hydrology 8, Unesco Paris (English and French).

UNESCO, 1972. Status and trends of research in hydrology 1965-'74. A contribution to the International Hydrological Decade studies and reports in hydrology 10, Unesco Paris p.62 (English and French).

Van der Molen, W.H. 1969. Ontwateringsnormen voor de landbouw. Cultuurtechnische verhanderlingen, p.277-296, The Hague (Dutch).

Van der Molen, W.H. 1972. Subsidence of peat soils after drainage. Symp. Hydrol. Marsh-Ridden Areas, Minsk. Paper 16 on theme II.

Veenenbos, J.S. 1950. De bodemgesteldheid van het gebied tussen Lemmer en Blokzijl in het randgebied van de Noordoostpolder (Soil conditions of the area between Lemmer and Blokzijl in the border-area of the Noordoostpolder). Soil survey of the Netherlands, part V, The Hague (Dutch with an English summary).

Visser, W.C. 1958. De landbouwwaterhuishouding van Nederland. Agricultural water management in the Netherlands). Comm. Onderz. Landbouwwaterhuish. Ned. TNO-Rapp. 1 (Dutch with English summary).

Wallace, J.R. 1971. The effects of land use change on the hydrology of an urban watershed. Rep. Georgia Institute of Technology, Atlanta, Georgia, U.S.A. (English).

Wesseling, J. 1969. Soil-Water-Plant Chapter I, Committee for Hydrological Research TNO, Proceedings and informations no. 15, The Hague, p.9-27.

Zubets, V.M., Murachko, M.G. 1972. Symp. Hydrol. Marsh-Ridden Areas, Minsk. Review paper on theme III.

DISCUSSION

I. F. Rusinov (USSR) - 1. Please give figures of cereal crop yields on drained lands in your country? 2. How much fertilizer is used? 3. Are drained lands irrigated?

F. C. Zuidema (The Netherlands) - 1. I don't remember the figures exactly. Everything depends on the soil type, say, 5000 kg/ha for winter crops; 3000 to 4000 kg/ha for summer barley.
2. The government is planning to use soils for different applications. Farmers should have fertile land. Various fertilizers are used, about 10-30 kg of phosphorus per hectare. 3. Yes, we use additional irrigation on sandy soils.

L. Heikutainen (Finland): Can you find any relation between the groundwater levels and (P_F) values in your correlations of the runoff and underground levels?

F. C. Zuidema (The Netherlands) - We haven't made such investigations.

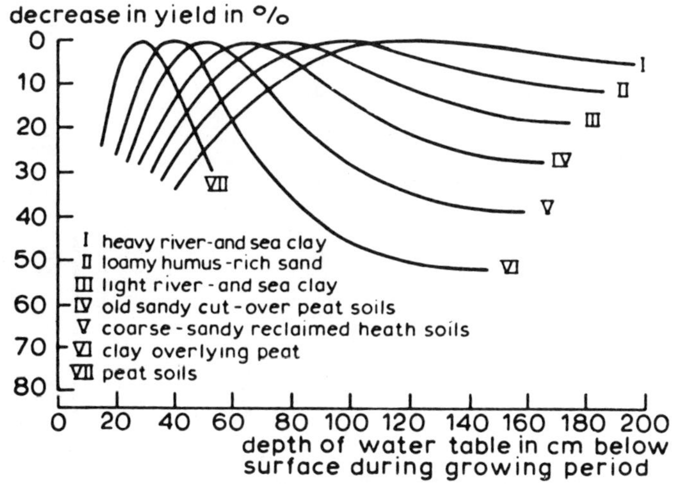

Fig. 75. The effect of the mean depth of the water table during the growing season on final crop yield for seven groups of soils.

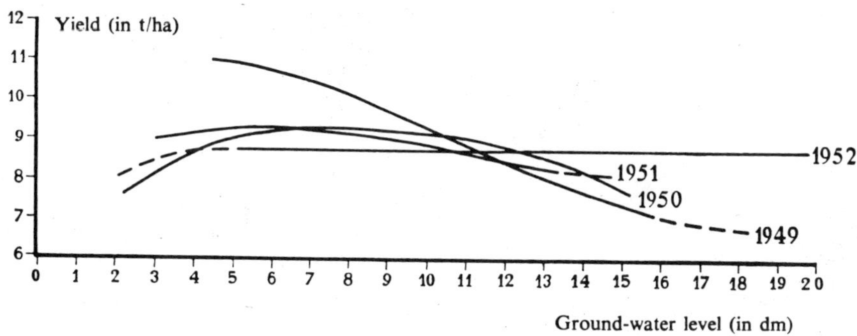

Fig. 76. Relation between the annual yield of dry matter of grassland on peat soils and the average groundwater level in the period 15 June-23 August, in two regions (1949-1952).

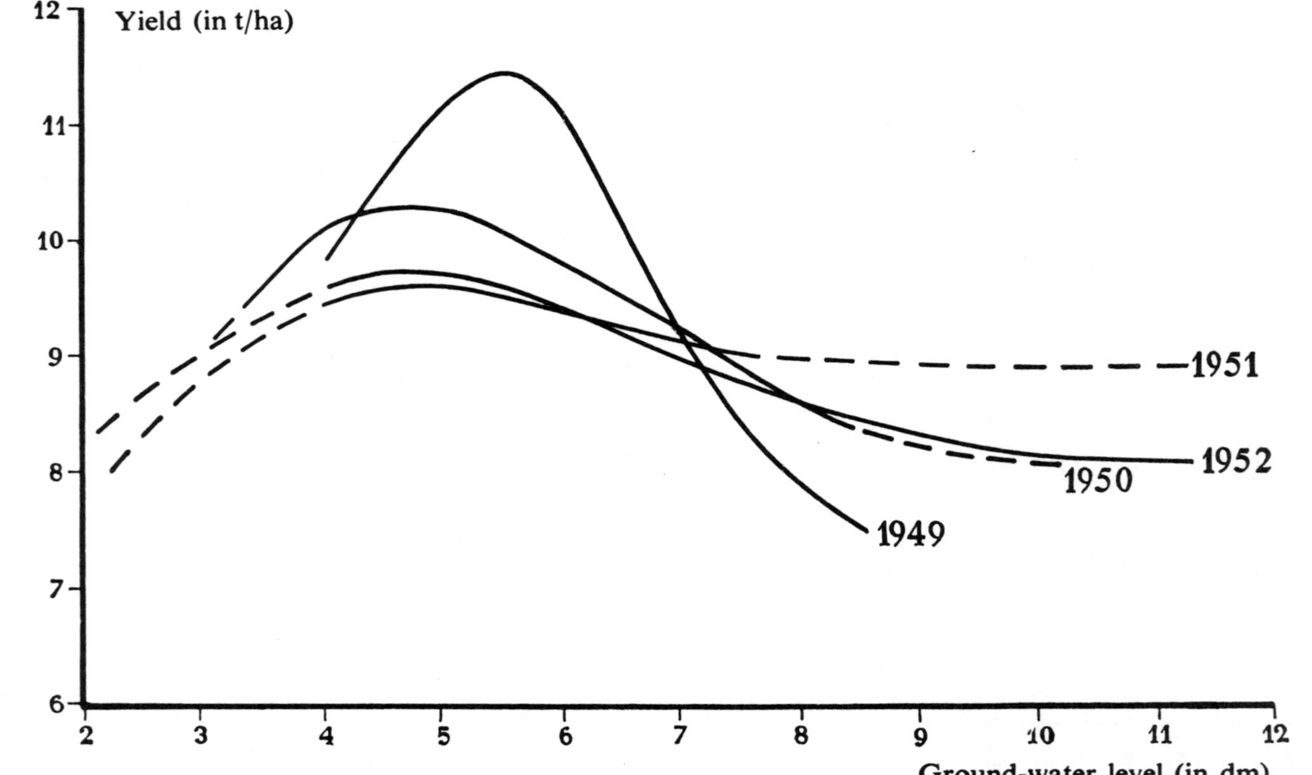

Fig. 77. Relation between the annual yield of dry matter of grassland on clay-on-peat soils (clay layer 2-4 dm) and the average groundwater level in the period 15 June-23 August (1949-1952)

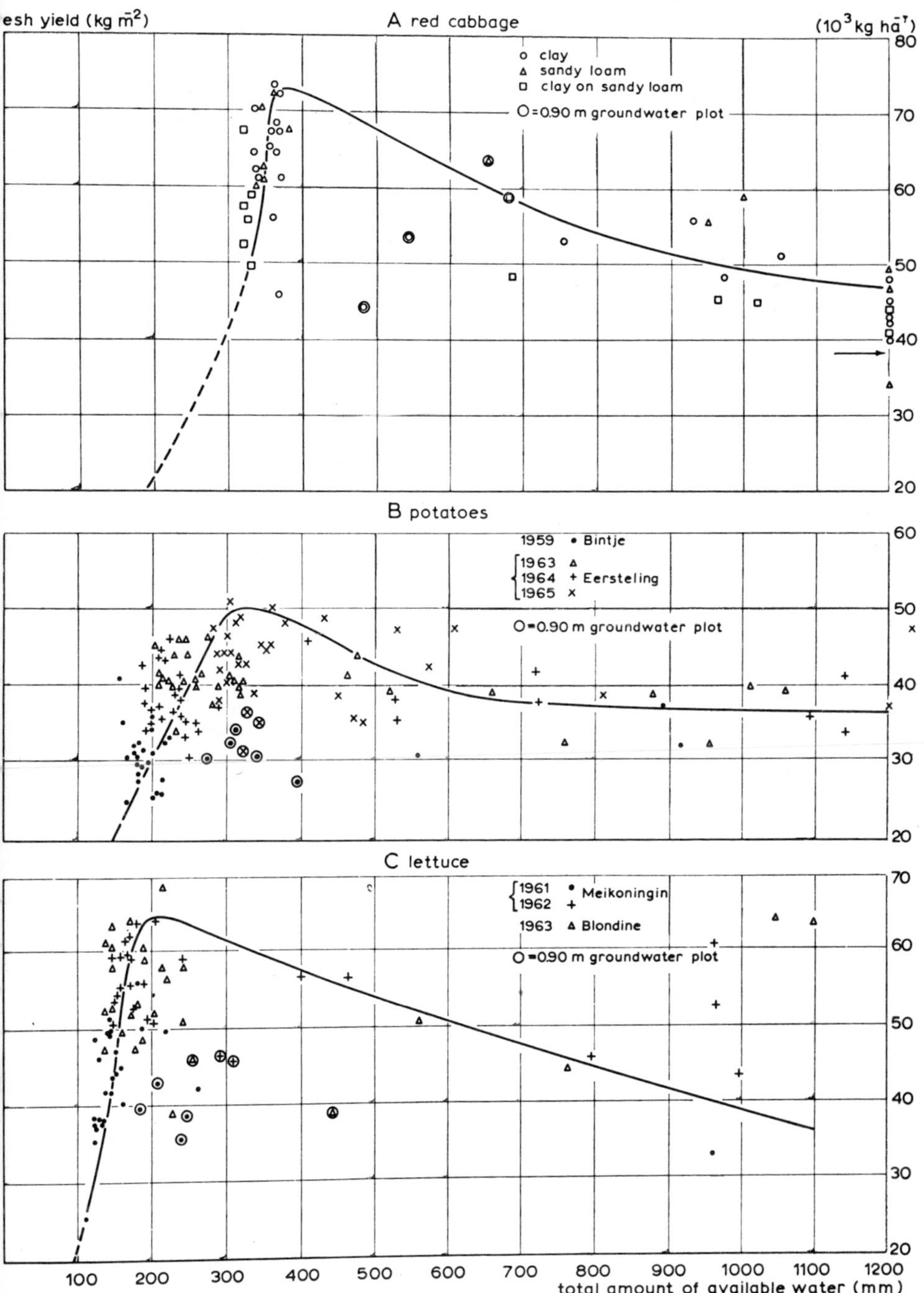

Fig. 78. Dependence of fresh yield of red cabbage, potatoes and lettuce, as measured on the three profiles of all groundwater plots, on total amount of available water

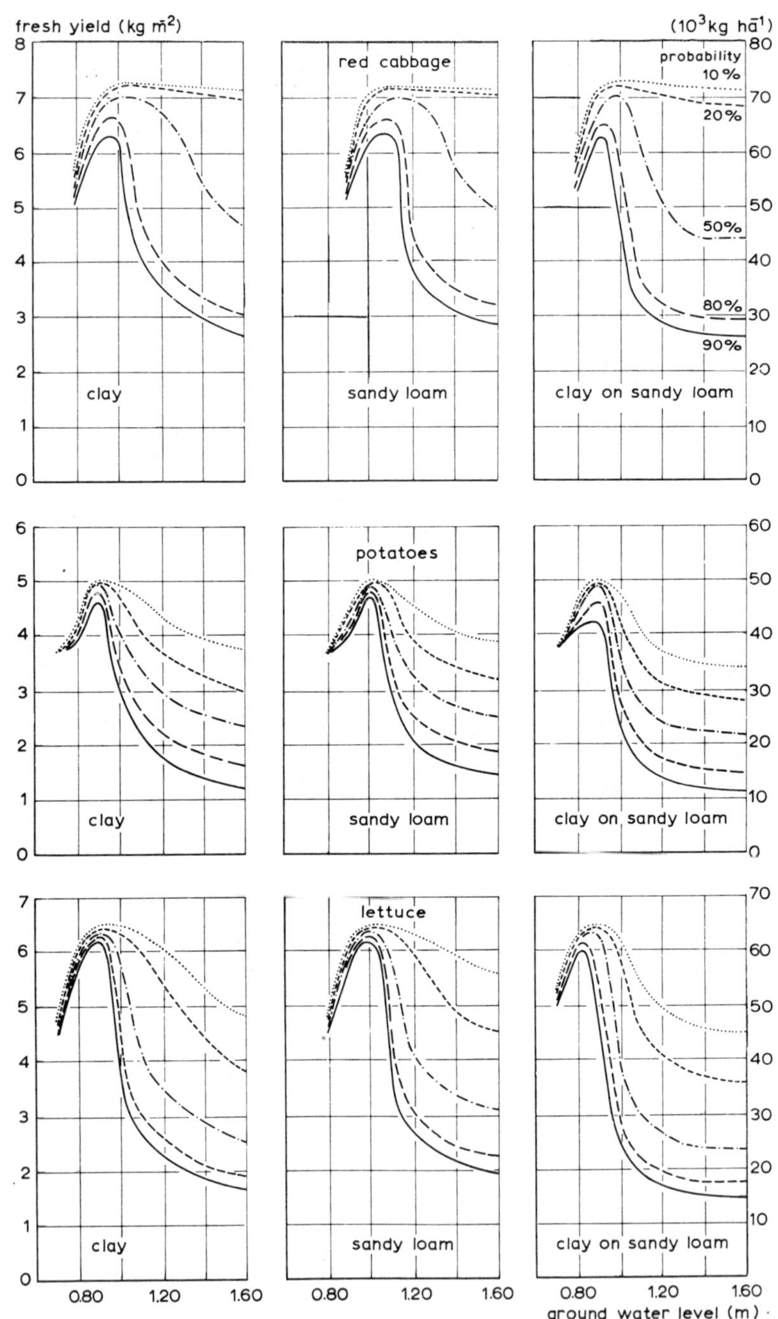

Fig. 79. Dependence of fresh yield of red cabbage, potatoes and lettuce grown on clay, sandy loam and clay on sandy loam respectively, on groundwater table depth over the growing season at five probability levels of exceedance

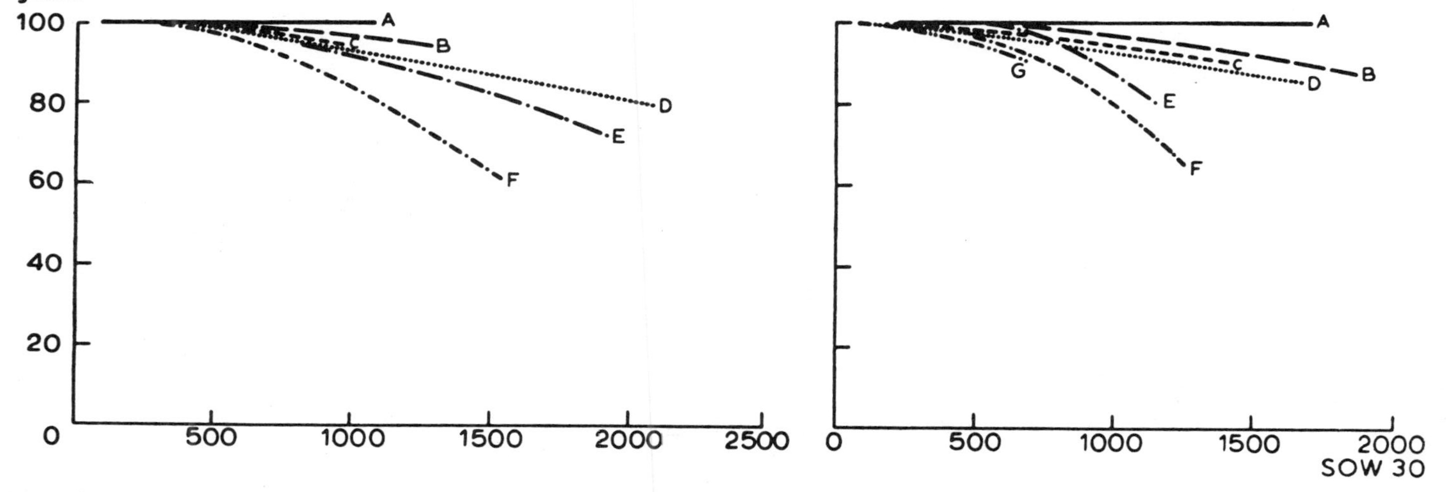

Fig. 80. The yields of various winter grain crops (left) and summer crops (right) as influenced by high water tables. The occurrence of high water tables is expressed in SOW-30 values, i.e. the sum of the products of days and cm height of water table above the level of 30 cm below surface (Nov 1-March 1)

winter grain crops:

100: optimum yield
A: winter wheat 1948
B: " " 1950
C: " " 1947
D: colza 1950
E: winter wheat 1951
F: winter barley 1946

summer crops:

100: optimum yield
A: oats 1947
B: spring barley 1952
C: spring wheat 1954
D: peas 1953
E: oats 1952
F: spring barley 1951
G: spring wheat 1951

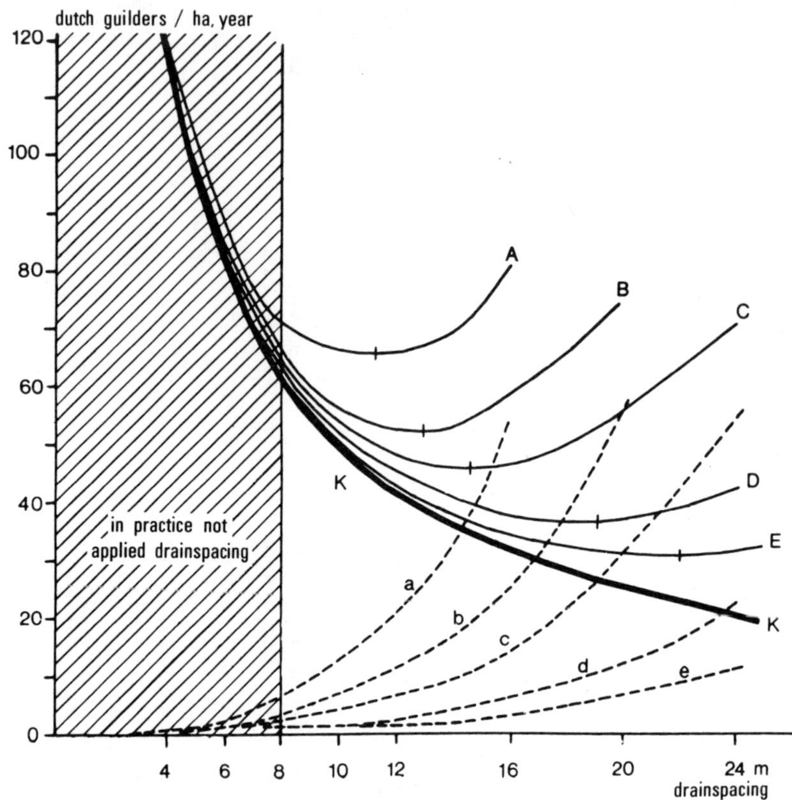

Fig.81. Relation between total costs (Dutch guilders/ha,year) for a project of land improvement and the drainspacing (metres) as a function of intensity of drainage, which depends on the type of soil

K : costs of subsurface drainage

a up to e : curves, showing loss due to lower yields or to below optimal farm management, on five various soil types

A up to E : curves, showing sum of costs of drainage and loss due to lower yields for the various soil types

Drainspacing a, A: 8 m
b, B: 10 m
c, C: 12 m
d, D: 16 m
e, E: 20 m

I in the curves A to E: place of minimal total costs and economically most profitable drainspacing

range of drain spacing not applied in practice

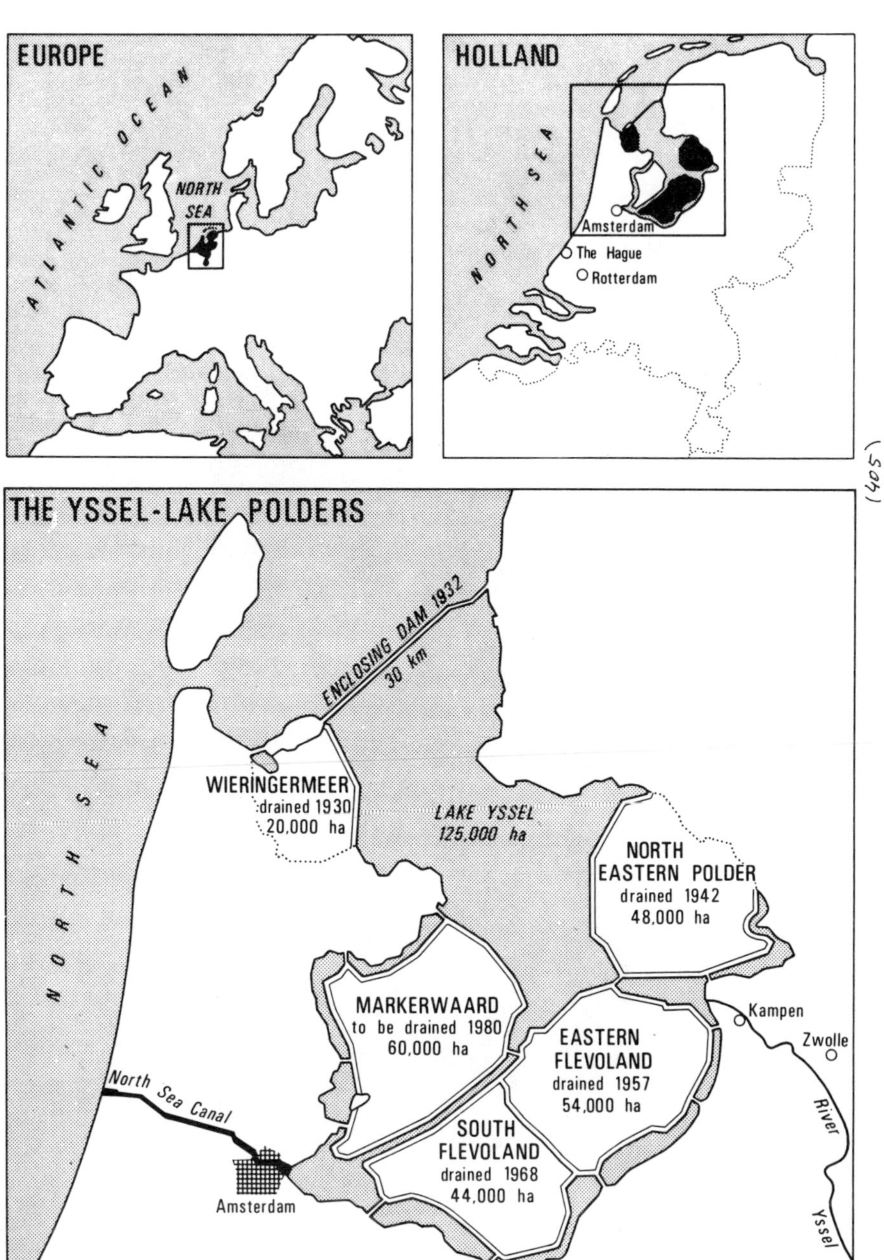

(405)

Fig. 82. Map of the Netherlands and the Zuyder Zee Project (Ijsselmeerpolders)

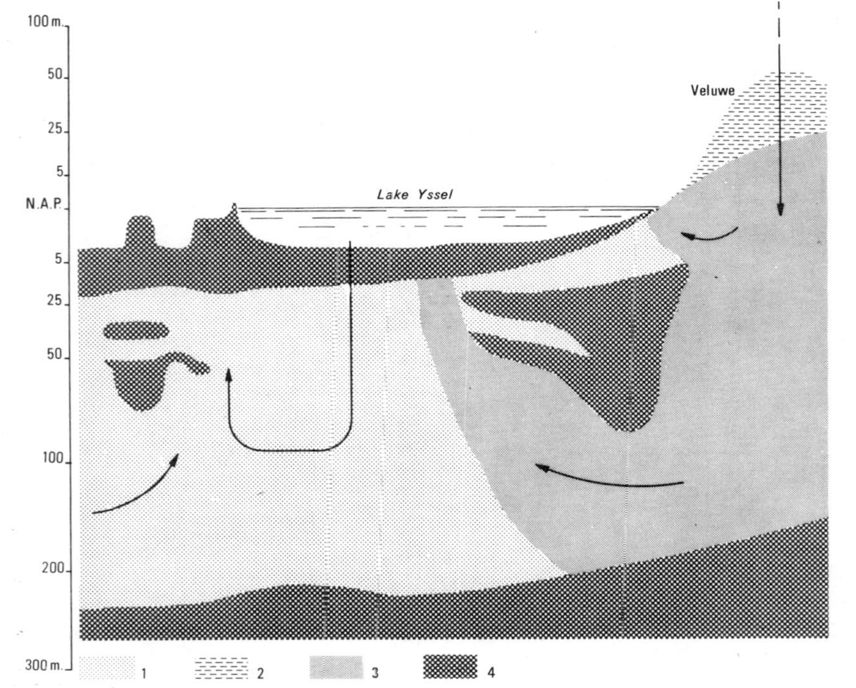

Fig. 83. Geological profile of the province North Holland, the Lake Ijssel and the Veluwe. The location of this cross-section is given in figure 84.

1. sand, filled with brackish to salt water
2. aerated sand
3. sand, filled with fresh water
4. low permeability layers (peat and clay)

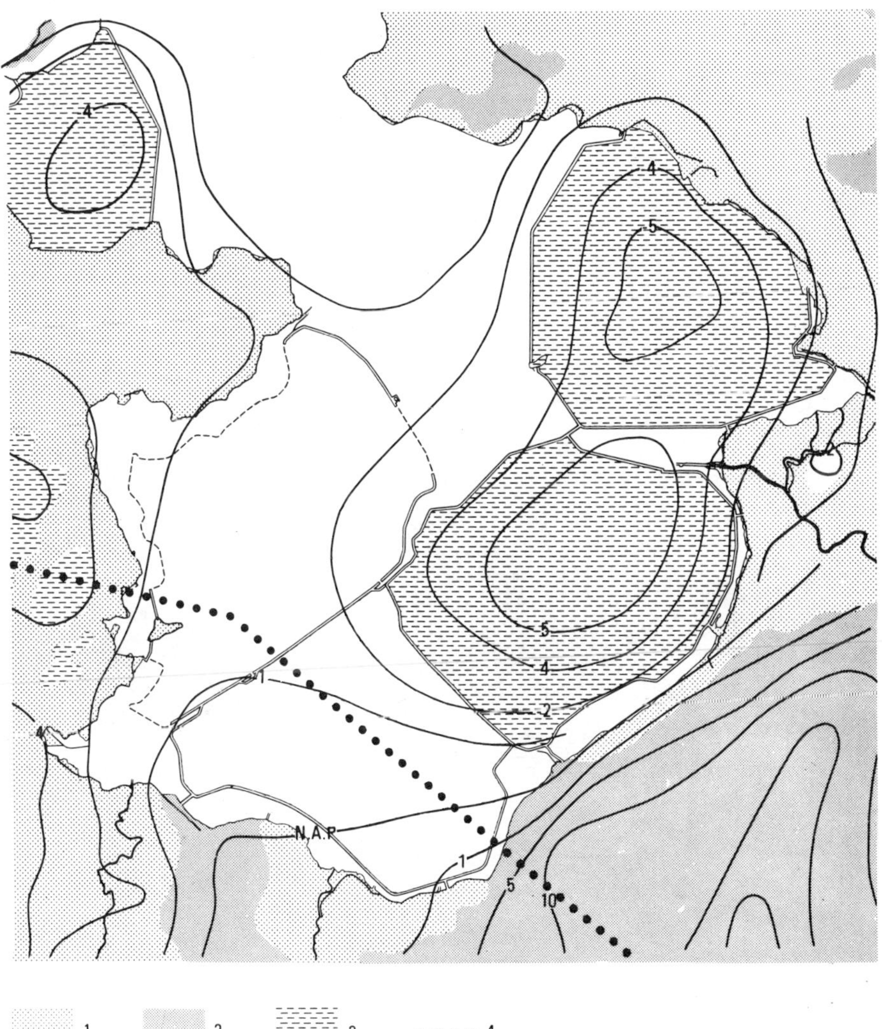

Fig. 84. Equipotential lines for an aquifer at a depth of 200 metres in 1960 in a part of the Netherlands
1. area with low permeability top layer
2. high sandy soils
3. area with polder level of 4 m below mean sea level
4. location of the profile of figure 83

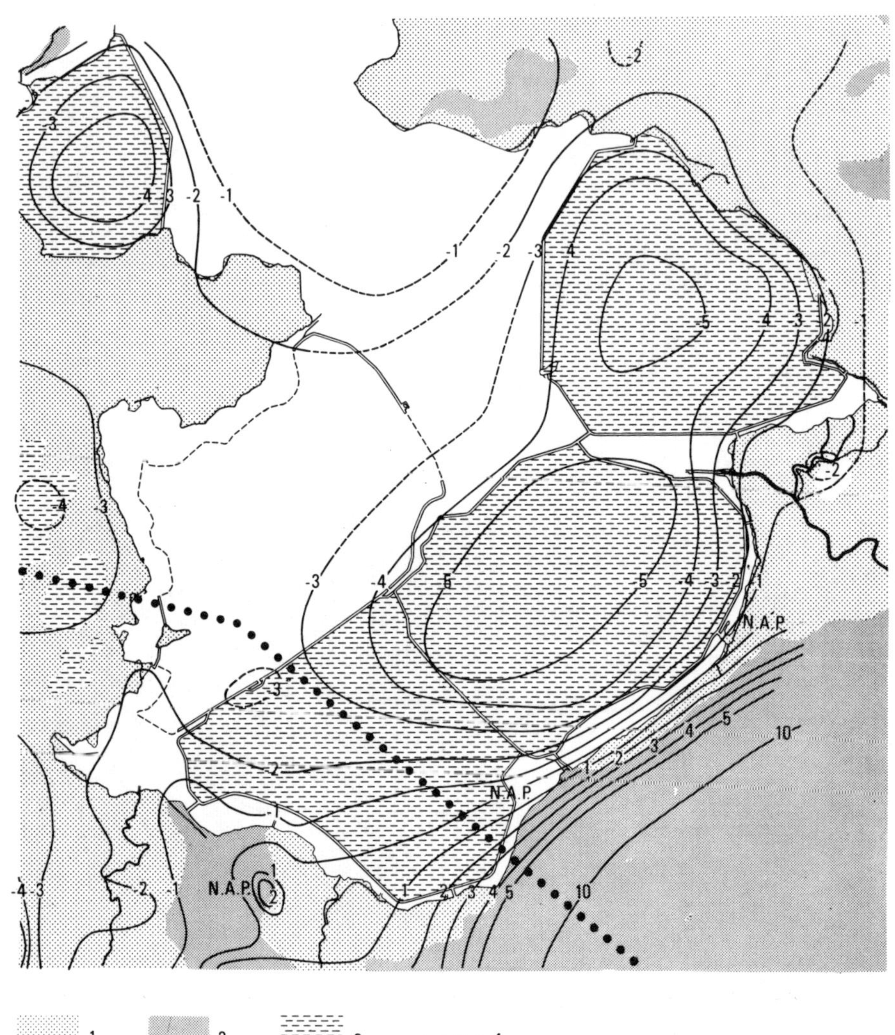

Fig. 85. Equipotential lines for an aquifer at a depth between about 20 and 200 metres in 1970 in a part of the Netherlands

1. area with low permeability top layer
2. high sandy soils
3. area with polderlevel of 4 m below mean sea level
4. location of the profile of figure 83

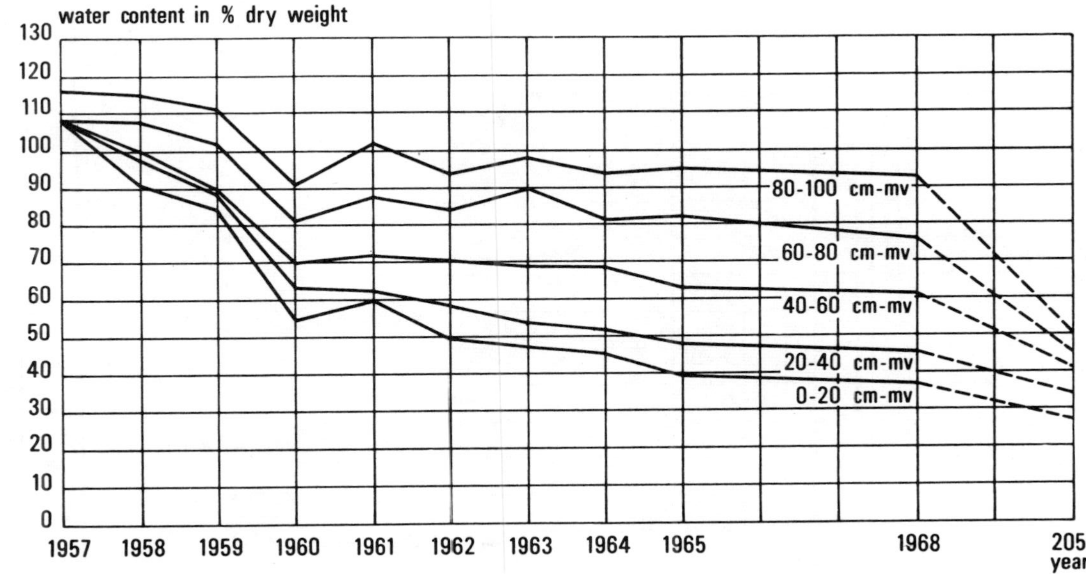

Fig. 86. Decrease in the course of time of the water content (A) per 100 gram dry soil (clay content 30 per cent) of different soil layers, after emergence of the subaqueous sediments in the Lake Ijssel (polder Eastern Flevoland)

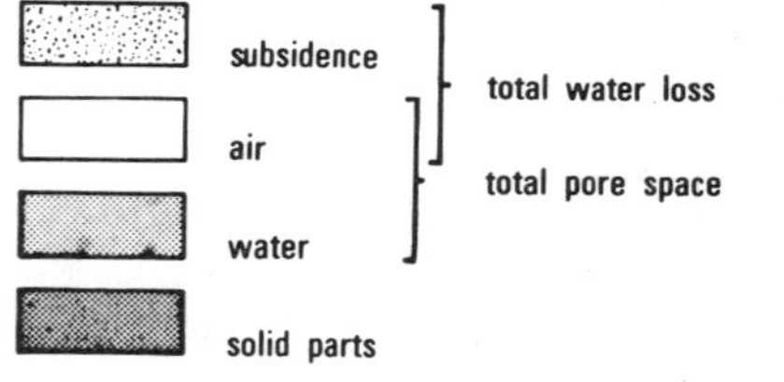

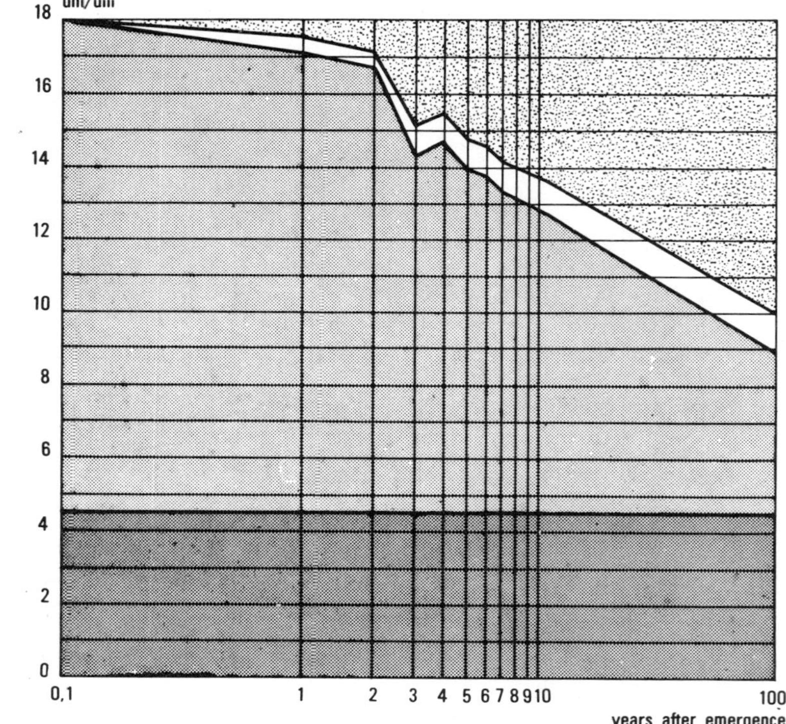

Fig. 87. Decrease of the total volume and the volume of water and the increase of the volume of air of subaqueous sediments of Lake Ijssel during the first century after emergence

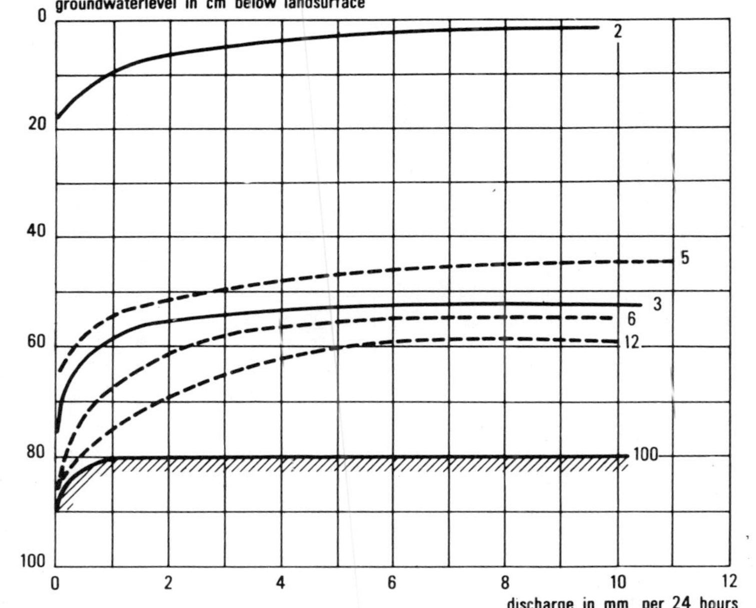

1. drainage system: open field drains
2. drainage system: tile drains
3. relation is estimated (influence of subsidence
(number at ends of curves indicate years after emergence)

Fig.88. Relation between groundwater level and discharge of detail drainage system at various moments after emergence of a subaqueous sediment in the Lake Ijssel (polder Eastern Flevoland) with about 30 per cent of clay

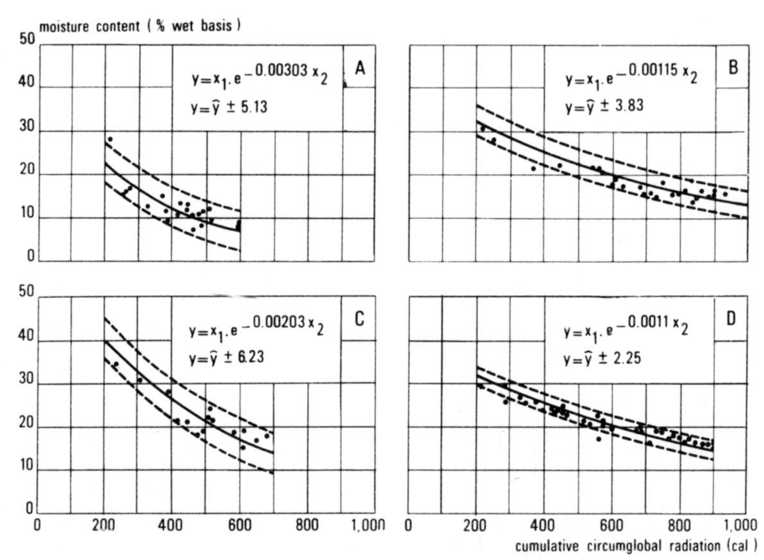

Percentage time discharge was equalled or exceeded

Extraterrestrial solar radiation (gm cal/cm2 day)

Fig.89. Relationship between cumulative daily circumglobal radiation (at top of atmosphere) of the kernels of:
A: colza B: barley C: oats D: wheat

ENVIRONMENTAL CONSERVATION IN THE GERMAN DEMOCRATIC REPUBLIC WITH SPECIAL CONSIDERATION OF THE PRESERVATION AND ENHANCEMENT OF AREAS OF WATER, MOORLAND AND WETLANDS

L. Bauer

Institute of Agriculture,
Halle,
German Democratic Republic

SUMMARY: Planned management of the environment and nature conservation based on scientific and ecological principles is carried out in the German Democratic Republic (GDR) under the Law for Environmental Conservation enacted on 14th May, 1970. The systematic selection of nature reserves on a scientific basis required by the IUCN has been achieved.

The GDR participated in the MAR and AQUA projects for the examination and systematic conservation of internationally important areas of water and wetlands under the auspices of IBP, IUCN and other international organizations. A complex of hydrologically and limno-biologically valuable reserves in the GDR has been placed under protection and is being investigated systematically. This report deals with the types of water areas involved and with the practical application of nature conservation research.

PROTECTION DU MILIEU ET DEFENSE DE LA NATURE DANS LA REPUBLIQUE DEMOCRATIQUE ALLEMANDE. CONSIDERATIONS PARTICULIERES SUR LA PRESERVATION DES BASSINS, DES MARAIS ET DES TERRAINS MARECAGEUX

RESUME : La République démocratique allemande se livre à un travail systématique de mise en valeur des ressources naturelles et de défense de la nature en s'appuyant sur les données scientifiques de l'écologie. Cette activité se fonde sur la loi dite de défense des richesses naturelles et culturelles adoptée le 14 mai 1970. Conformément aux recommandations de l'Union internationale pour la conservation la nature et de ses ressources (UICN), on a procédé à une classification des ressources naturelles du pays, d'un point de vue scientifique.

La République démocratique allemande a participé aux projets MAR et AQUA pour l'étude et la conservation des bassins et des marais d'intérêt mondial, dans le cadre des activités de l'UICN et d'autres organisations internationales ainsi que du Programme biologique international (PBI).

L'ensemble des régions ayant une valeur hydrologique et limnobiologique dans le pays fait l'objet de mesures de préservation. L'auteur étudie les types de bassins concernés par les mesures de préservation et traite des mesures prises à cet effet.

The quality of the historical relationship between man and nature has been broken because of the scientific and technical revolution of the present day. The far-reaching, and partly irreversible effects of man's exploitation of natural resources, including the water balance and the form of the landscape, have now taken on almost global dimensions.

Deliberate environmental conservation and planned environmental management based on scientific and ecological principles are vital to the prevention or reduction of harmful influences, which result from exploitation of natural resources. Conservation of the biosphere and the control of any changes imposed upon it, are the duty of every responsible nation. Utilization and conservation of the natural bases for human existence complement each other and are inseparable. The deliberate conservation of nature is a relatively new aspect of the relation between man and nature.

This fact has been taken into account in the German Democratic Republic. Since May 14, 1970 there has been a 'Law for the planned management of socialist environmental conservation' (Landeskulturgesetz). Its enforcement is the responsibility of the Ministry for Environmental Conservation and Water Management. The aims of the law are effective nature conservation and planned development and shaping of the natural and artificial environment. This will facilitate the preservation, improvement and effective utilization of the natural resources vital to society, ie soil, water, air, flora and fauna, and improve the aesthetics of the country. This law closely controls the following activities:

- the shaping and cultivation of the landscape and the conservation of nature;
- the cultivation and conservation of the soil (with regard to both the most economical and rational utilization of the ground, the maintenance and increase in soil fertility;
- the utilization and conservation of forests and woods;
- the utilization and conservation of areas of water;
- the prevention of air pollution;
- the utilization and harmless elimination of waste products;
- the reduction of noise;
- the cultivation and development of the landscape for recreation.

The aim is to create a landscape which combines high productivity, makes for efficient organisation, and provide perfectly harmonious scenery. Nature reserves for specific purposes have been located in what is a relatively densely populated and highly industrialized country which has utilised the soil for many centuries.

Natural biogeocoenoses have increasing importance as ecological reference, and experimental areas in which the extent of man made changes on the biosphere can be evaluated. Limits on the concentrations of various types of waste products have to be determined as required by law. Because the cumulative effects of biologically active substances are related to their concentration, it is necessary to intensify research into their toxic effect on ecosystems and into their metabolism in organisms, populations and ecosystems. Clearly identifiable indices of their biological influence upon the environment must be found. Investigations of the stability, control and productivity of ecosystems representative within the GDR should be used to evaluate the effect of pollution on the natural processes of the environment. These ecosystems will assist the recognition of long-term environmental changes. The nature reserves will therefore have a new purpose in addition to their traditional function of protecting species of animals and plants threatened by extinction.

Nature reserves set within the environment but little affected by land use change can form the basis for a systematic investigation of changes in the biosphere. Thus a network of systematically selected nature reserves has various functions, including fundamental nature conservation, ecological research, and regional planning. The nature reserves of each country should be representative of the most important natural region and environmental units, and of their biocoenoses. In these reserves ecological long term investigations can be made. The nature reserves are also generally typical of a particular region and provide shelter to those animal and plant communities which are threatened with extinction. As nature reserves are scientifically characteristic of different forms of the earth's surface they also serve an educational and research role from which the geological development of the landscape can be understood.

A regional network of nature reserves is important in every country having a long history of development of mineral resources, forestry, water and agricultural areas. Exact knowledge of the interactionof the different physiographic factors on the various aspects of agriculture and of the changes in metabolic processes are essential if the natural resources of a country are to be intensively developed using modern techniques. This knowledge cannot be obtained in the laboratory. Because individual processes are controlled by the overallenvironmental situation, it is necessary to investigate the laws governing these processes primarily in situ. Furthermore an understanding of the processes is only possible when they can be observed in an undisturbed state. Protected ecosystems are therefore the most suitable bases for ecological research, and can be thought of as open air laboratories.

The International Union for Conservation of Nature and Natural Resources (IUCN) recommended that a system of national parks and nature reserves be selected on a scientific basis and established in individual countries throughout the world.

To facilitate the investigation of internationally important areas of water and wetlands and hence determine measures for their protection, two important activities were initiated by IUCN in the 1960s:

(a) The project MAR: 'List of European and North African wetlands of international importance'

(b) The project AQUA: 'Source book of inland waters of the world proposed for conservation'

These projects come within the scope of the International Biological Programme (IBP) run by UNESCO, and required international scientific cooperation. The areas of water and wetlands in question have been investigated by an international study team, and representative types have been conserved or proposed as nature reserves. The results of the projects MAR and AQUA have now been published (Olney, 1965; Luther and Rzoska, 1971).

The two handbooks classify the areas of water and wetland

reservations on a limno-biological basis, and give valuable information on their characteristics, related to natural science, genesis, degree of utilization, state of conservation etc.

The systematic selection of nature reserves on a scientific basis requested by the IUCN has been made in the GDR over the past two decades and is now complete.

A network of nature reserves have been selected which represent all types of landscape, soils, biocoenoses and ecosystems in the GDR. Areas of water and wetland reserves are a special category of nature reserves. The reduction of water pollution and the rational utilization of water in an industrial country like the GDR must be emphasised. Water in all its forms of occurrence on the Earth is a vital factor in maintaining the ecological balance. Thus for practical water management and environmental conservation, water must be carefully managed both quantitatively and qualitatively. In addition to civil engineering works, it is essential that all measures which assist the natural storage of water and the water retaining capacity of the environment be promoted. Improved conservation of natural water storage and retention areas such as marshland, abandoned stream channels, river meanders, lakes and ponds is therefore generally necessary. Natural or artificial regulation of runoff is in the best interests of water management and environmental conservation. The water law of 1963 and the law for environmental conservation of 1970 control these factors in the GDR.

A number of examples of the most valuable areas of water and wetlands which remain in a near-natural condition have been conserved as nature reserves. These include:

- lengths of streams and river including a strip of land along the banks;
- abandoned stream channels in various delta formations;
- all types of moorland and marshes;
- various types of lakes including those of glacial origin, those formed by erosion, karst lakes etc;
- the environments of ground moraines;
- coast lagoons, shallow bays on the Baltic Sea, lagoons on the Baltic Sea (Haff);
- cataracts;
- limestone caves, some with lakes;
- various springs and their environment.

The nature reserves representative of areas of water and wetlands serve primarily as the basis for education on and research into the natural interrelationships between water and environmental conservation. The results of these investigations provide the fundamental knowledge needed for a wide range of applied engineering and biological problems in all branches of water conservation.

Research into ecological aspects of surface, ground and soil water can often only be carried out in nature reserves which are not influenced by man's activities. The results of this research are of increasing practical importance in water management, environmental and nature conservation, agriculture and forestry(especially with respect to reclamation). The conservation of such areas for native animal and plant species often saves them from extinction due to the increasing industrial use of the environment. In regions of the GDR which have a dry climate, for example in the Magdeburger Borge and the Thüringer Becken, areas of water are rare and nature reserves and conserved areas of water are of particular value. Because conserved areas of water and wetlands provide resting places for migrating birds, they are also of great importance in international nature conservation.

Specific long term plans and tasks have been worked out for each nature reserve. Modern environmental conservation and nature conservation in the GDR has a scientific basis and is therefore closely related to the fundamental science of hydrology and limnology, as well as to practical work concerning drainage and land use.

References

Bauer, L. 1972. Handbuch der Naturschutzgebiete der Deutschen Demokratischen Republik. (Handbook of nature reserves of the German Democratic Republic). 5 volumes, Urania-Verlag, Leipzig.

Luther, H., Rzoska, J. 1971. Project AQUA, a source book of inland water proposed for conservation. IBP Handbook, No.21, London.

Olney, P.J.S. 1965. Project MAR. List of European and North African of international importance. IUCN Publications, new series, No.5.

THE EFFECT OF LAND RECLAMATION BY DRAINAGE ON THE REGIME OF RIVERS IN BYELORUSSIA

K. A. Klueva

Hydrological Research Laboratory for Byelorussian Polessie, Minsk, BSSR.

SUMMARY: Consideration is given to estimation of the regime of rivers affected by a range of reclamation measures based on investigations in the river basins of Byelorussia, where marsh covers a considerable area because of physiographical conditions.

Hydrological data from 16 basins for the periods before and after reclamation have been compared with corresponding data from 30 control basins, either in their natural state or only slightly influenced by reclamation. The minimum summer and winter daily moduli of runoff, maximum discharge and runoff of spring floods, seasonal and annual runoff have been analysed leading to an understanding of the influence of reclamation on the hydrological regime of these rivers.

INFLUENCE DE L'ASSECHEMENT PAR DRAINAGE SUR LE REGIME HYDROLOGIQUE DES RIVIERES DE BIELORUSSIE

RESUME : L'auteur étudie les modifications apportées au régime hydrologique des rivières par les mesures d'assèchement par drainage. Cette étude porte sur les bassins de rivières de la Biélorussie, où les marais sont largement répandus en raison de conditions physico-géographiques spécifiques. Les résultats obtenus proviennent de la comparaison des données hydrologiques de 16 bassins pour les périodes qui ont précédé et suivi l'aménagement avec les données de 30 bassins de contrôle à l'état naturel ou peu influencés par l'aménagement. On a analysé les indices d'écoulement minimal journalier d'été et d'hiver, les débits maximaux et l'indice d'écoulement de la crue de printemps, l'écoulement saisonnier et annuel. Ces recherches ont montré l'influence de l'aménagement par drainage sur le régime hydrologique des rivières du territoire considéré.

The present study evaluates the changes in the regime in Byelorussian rivers attributable to reclamation by drainage. Data have been obtained from 16 river basins, where a range of reclamation measures has been undertaken at various times. In these river basins the proportion of marshland varies from 7 to 49 per cent and drained land covers from 5 to 25 per cent of the area of each basin. The main characteristics of these reclaimed basins are shown in Table 51.

Of the 16 basins, 10 basins (Nos I-X) are situated in Byelorussian Polessie in hydrological region VI or partly in regions IV and V (Anon, 1966). The others are in different hydrological regions as follows: basin XI in the Dnieper lowland lies in region VI; basin XII in the Central-Berezina plain; basins VIII and XV on the north-western slopes of the Minsk Hills and basin XVI on the western slopes of the Lukoml Hills lie in region III; basin XIV in Byelorussian Poozerie lies in region I. The basin of the Oressa river, a tributary of the Ptich, down to its confluence near the village of Andreevka, is largely reclaimed and has been subjected to most study. The total area of the Oressa drainage basin is 3620 km^2, of which 3580 km^2 lies above the village of Andreevka. The basin consists of a flat plain with poorly-defined watersheds. The relief of the upper basin is undulating with low sandy hills and ridges associated with marshy depressions. The central part of the area is lower; the number of marshes increases greatly and the left-bank area is almost entirely marsh-ridden (Mar'ino-Zagalsk marsh). Peats and sandy soils predominate. Before drainage, about 40% of the area of the basin was marsh and 90% of this was lowland marsh. The peat depth in the marshes ranges between 1 and 2.5 m and the underlying formations are permeable sands. Since reclamation (in 1926-28, 1929-35, and in 1936-41), the regimes of the Oressa and Ptich have significantly changed downstream of their confluence.

Between 1948 and 1969 all the drainage systems of the Oressa basin were restored and regulated and additional drainage and river training works were constructed. The total area of drained marshes in the Oressa basin is about 900 km^2. The reclaimed land is used for agriculture and large farms have been established there, such as the State-Farm named after the 10th anniversary of the Byelorussian SSR and Luban State-Farm. A dam and a reservoir for a fishery farm have been erected 86 km from the confluence of the Oressa (9 km upstream from Luban).

Evaluation of the changes in regimes caused by reclamation has been made by comparing the hydrological data of reclaimed (design) basins for the periods before and after reclamation with the corresponding data of control basins, situated within the same hydrological region, and in which conditions are natural or only slightly affected by reclamation. Comparison of the data of the reclaimed and control basins within the same hydrological region reduces considerably the effect of varying precipitation and storage in different years. The year of completion of all the main reclamation works in each basin was usually chosen as the point to divide the complete observational series into the period before and the period after reclamation.

Selection of analogue control basins was a difficult task, since, within homogeneous physiographical regions, the majority of gauged rivers varied in their basin sizes, whilst rivers with similar basin dimensions more often than not, had been subjected to reclamation. Therefore, some of the basins selected differ considerably from the design basins in area and in some other characteristics. The scatter of points on some graphs which compare the data of design with that of control basins may be attributed to a variation in the timing of flow in the rivers under comparison

Table 51. Main characteristics of reclaimed and control basins

Reclaimed (I-XVI) and control (1-17) basins	River slope		Basin area km^2	Mean basin elevation, m	Mean basin slope	Afforestation,%		Total area of marshes %	Reclaimed areas, % of basin area	Channel network density factor
	Mean %	Weighted mean				Forest on marshes	Dry forest			
I.Ptich-Luchitsy	0.47	0.23	8770	155	-	15	34	22	(10)	-
II.Oressa-Verkhutino	0.45	0.23	520	150	5.1	23	42	31	(15)	0.61
III.Oressa-Luban	0.34	0.22	1290	156	-	14	47	22	(17)	0.50
IV.Oressa-Kutinki	0.29	0.17	1830	153	-	13	40	20	-	-
V. Oressa-Andreevka	0.23	0.15	3580	147	-	18	36	25	(25)	0.54
For I-V the control basins are:										
1.Sluch-Novodvortsy	0.38	0.35	910	167	3.5	16	20	31		0.49
2.Ptich-Krinka	0.70	0.39	2010	186	-	15	22	23		0.54
VI.Vit-Borisovshchina	0.37	0.23	782	132	2.8	37	28	49	(25)	0.40
VII.Cherten-Nekrashevka	0.50	0.37	445	145	4.5	16	59	19	-	0.24
For VI, VII the control basins are:										
3.Slovechna-Kuzmichi	2.2	0.74	914	171	10.6	38	35	50		0.38
4.Ubort-Krasnoberezhie	0.34	0.34	5260	156	-	37	28	48		-
VIII.Merechanka-Stavok	1.0	0.78	118	152	8.7	0	17	25	(6)	0.44
For VIII, control basins are:										
5.Yaselda-Bereza	0.31	0.27	916	164	4.7	7	30	42		0.39
6.Tsna-Dyatlovichi	0.58	0.37	969	155	3.6	16	46	62		0.40
7.Lan-Loktyshi	0.71	0.26	909	184	8.4	1	15	17		0.45

(continued)

Table 51. Main characteristics of reclaimed and control basins (continued)

Reclaimed (I-XVI) and control (1-17) basins	River slope		Basin area km^2	Mean basin elevation, m	Mean basin slope	Afforestation, %		Total area of marshes %	Reclaimed areas, % of basin area	Channel network density factor
	Mean %	Weighted mean				Forest on marshes	Dry forest			
IX.Grivda-Ivatsevichi	0.59	0.42	699	168	8.8	1	13	14	-	0.52
For IX control basins are:										
5.Yaselda-Bereza	0.31	0.27	916	164	4.7	7	30	42		0.39
8.Neman-Grodno	0.41	0.18	33600	171	-	5	21	14		-
X.Grebelka-Birchuky	1.4	1.3	67.0	188	3.4	1	17	29	(15)	0.69
For X control basins are:										
9.Oressa-Verkhutino*	0.45	0.23	520	160	5.1	23	42	31		0.61
1.Sluch-Novodvortsy	0.38	0.35	910	167	3.5	16	20	31		0.49
XI.Uza-Pribor	0.32	0.26	760	136	6.5	0	4	14	(10)	0.28
For XI control basins are:										
10.Ut-Pribytki	0.48	0.43	340	151	7.3	0	6	14		0.41
11.Besed-Svetilovichy	0.34	0.23	5010	162	-	2	17	11		-
XII.Ola-Mikhalevo	0.79	0.37	380	159	7.0	3	8	13	(5)	0.33
For XII control basins are:										
12.Resta-Sukhari	1.3	1.2	114	190	11.8	2	11	7		0.44

* data taken before reclamation

(continued)

Table 51. Main characteristics of reclaimed and control basins (continued)

Reclaimed (I-XVI) and control (1-17) basins	River slope		Basin area km^2	Mean basin elevation, m	Mean basin slope	Afforestation,%		Total area of marshes	Reclaimed areas, % of basin area	Channel network density factor
	Mean %	Weighted mean				Forest on marshes	Dry forest			
13.Sushanka-Susha	0.93	0.81	153	166	3.2	38	49	44		0.22
11.Besed-Svetilovichy	0.34	0.23	5010	162	-	2	17	11		-
XIII.Ilia-Vladyki	2.8	1.6	402	233	18.4	5	38	10	(9)	0.56
XIV. Berezovka-Sautky	1.4	1.2	554	170	20.7	3	8	7	(6)	0.54
For XIII,XIV the control basins are:-										
14.Sorvech-Krivichy	0.38	0.27	797	187	6.3	14	16	27		0.44
15.Viliya-Steshitsy	1.3	0.86	1230	204	-	5	41	12		-
XV.Usha-Molodechno	2.8	1.4	552	210	22.2	8	16	21	(19)	0.38
For XV control basins are:										
16.Isloch-Borovikovshchina	3.4	1.6	624	254	26.8	1	14	7		0.41
15.Viliya-Steshitsy	1.3	0.86	1230	204	-	5	41	12		-
XVI.Essa-Gadivlya	0.35	0.32	530	182	9.6	16	24	24	(12)	0.45
For XVI control basins are:										
15.Viliya-Steshitsy	1.3	0.86	1230	204	-	5	41	12		-
17.Ulla-Promysly	0.32	0.30	3330	172	13.8	11	22	17		-

during certain years. Some scatter is also due to the varying effect of reclamation on runoff, depending on precipitation during the year under study and in previous years. However, the general distribution of the points gives some idea of the changes in the river regimes after reclamation.

In the stage of the study that is described here, characteristics of the hydrological regime are derived from actual observations during the periods before and after reclamation, without standardisation of the flow volume of specific years. The Oressa at the village of Andreevka and the Ptich at the village of Luchitsy illustrate the changes that have taken place after reclamation in the normal runoff, as well as in the runoff of various frequencies.

Conclusions were only drawn after comparison of data from each design basin with data from several control basins (2 to 5 basins), some of which serve for several design basins. The main characteristics of the control basins are presented in Table 51 (to save space no more than three control basins are presented for each design basin).

Minimum daily mean discharge for summer and winter periods

Analysis of discharge data showed that after reclamation the minimum summer daily mean discharge increased on average by 30 to 150% in all the design basins. The Ilia river at the village of Vladyka is the only case for which no considerable changes in minimum summer daily mean discharge were found. Figure 90 illustrates the relationship between the minimum summer specific discharges of reclaimed and control basins.

Changes of long-term mean discharge and minimum summer daily mean discharge of different frequencies after reclamation in the Oressa and Ptich basins are summarized in Table 52. To obtain long-term data for the Oressa at Andreevka, a semi-graphical method of reduction was used, as well as graphs of relations between the long-term mean values of the parameters of frequency curves for the control basin of the Ptich at the village of Luchitsy. The change in long-term values of the minimum winter daily mean discharge, and its values for various discharge frequencies, for the Oressa at the village of Andreevka and the Ptich at the village of Luchitsy are presented in Table 52. The parameters of frequency curves of winter daily minimum values for the Oressa at Andreevka before and after reclamation are determined in the same way as the parameters of summer daily mean minimum discharges.

The minimum winter daily mean discharge in 8 basins increased after reclamation by an average of 20 to 130%. In other basins under study, no significant change of the values was found.

Figure 91 shows the graphs of the relation between several reclaimed basins and the reference basins. The figure shows that for the Oressa at the villages of Luban and Andreevka, as well as for the Merechanka at the village of Stavok, minimum winter values have increased, while for the Ptich at the village of Luchitsy, for the Uza at the village of Pribor and for the Essa at the village of Gadivlya no significant variations are observed.

Table 52. Changes of mean long-term minimum daily mean discharges and minimum daily mean discharges of different frequencies in summer and winter for the Oressa and Ptich after reclamation expressed on a % basis

Characteristic	Long term mean			Minimum specific discharge of different frequency, %				
	M 1/sec.km^2	C_v	C_s	50	75	80	95	97
	Oressa at Andreevka							
M_s	+ 49	- 7	0	+ 49	+ 54	+ 55	+ 62	+ 63
M_w	+ 36	- 11	0	+ 38	+ 44	+ 47	+ 61	+ 69
	Ptich at Luchitsy							
M_s	+ 40	- 5	0	+ 41	+ 42	+ 44	+ 46	+ 49
M_w	0	0	0	0	0	0	0	0

Spring snowmelt flood characteristics

Maximum discharges decreased after reclamation by an average of 17 to 30% (Table 53) on 7 basins; on the other basins studied, no significant changes occurred. Figure 92 shows two graphs of the relations between the maximum discharge of the spring snowmelt flood in the reclaimed basins and in the control basins. The first graph indicates a decrease in maximum discharge of the Oressa after reclamation. The second graph shows that the data before and after reclamation are similar, and it is impossible to trace any changes in the maximum discharge of the design basin of the Uza at the village of Pribor due to reclamation.

The height of the spring snowmelt flood, like the maximum discharges, decreased after reclamation by 10 to 30% on 6 basins, whilst on the other reclaimed basins no significant changes were observed. Figure 93 shows the graphs of the relationship between the flood levels of two reclaimed basins and two control basins.

Table 54 furnishes data on changes after reclamation in the maximum discharge and gauge height of the snowmelt flood of the Oressa at Andreevka for both the long-term mean and various frequencies. Mean long-term characteristics of the maximum discharge for the Oressa before reclamation are obtained by applying the data from the control basin (the Ptich at Luchitsy), and after reclamation by using the series of actual observations. Long-term data on stage for the periods before and after reclamation are obtained by analysing the data on the Ptich at the village of Luchitsy.

Spring snowmelt flood

No significant changes were observed in the timing of the beginning of spring snowmelt flood **or** in the date of the peak discharge after reclamation.

The end of the spring flood occurred on average 8 to 24 days earlier on 7 basins. The duration of the flood decreased accordingly. On the other basins under study, no significant changes were observed in the date of ending or of the total duration of the spring snowmelt flood after reclamation.

Table 54. Changes of mean annual maximum discharges of different frequencies and stage of spring flood of the Oressa at Andreevka after reclamation (%)

Characteristic	For long-term period			Maximum discharges and stage of flood of different frequencies,%				
	normal discharge	C_v	C_s	1	3	5	10	25
Q-discharge	- 30	- 4	- 6	- 33	- 32	- 32	- 31	- 30
h-stage	- 20	- 6	- 11	- 24	- 23	- 23	- 22	- 21

Figure 94 shows graphs of the relationship between the dates of the beginning, end, duration and peak discharge occurrence of the spring snowmelt flood of the Oressa at the village of Andreevka and the Ptich at the village of Krinka.

Annual distribution of streamflow

The characteristics of annual streamflow distribution have been examined for conventional years starting from a high flood period. Data on seasonal streamflow were arranged within strict seasonal boundaries, common for all the years and all the basins, and with approximation to the nearest month. Only for the Ptich and the Oressa rivers was seasonal streamflow considered both within constant and variable seasonal boundaries.

The analysis of seasonal streamflow showed that 6 basins were characterized by a decrease of spring streamflow after reclamation of 10 to 30% on average, while summer and autumn streamflow increased by 20 to 60% and winter streamflow by 15 to 75%. On 9 basins, there were no significant changes in spring streamflow after reclamation; however, streamflow during low water minimum period increased with an increase of streamflow during either both the summer-autumn and winter periods or in the summer-autumn period alone (Table 53).

In the case of the Ilia at the village of Vladyka, the streamflow did not change in any season after reclamation.

Figure 95 shows the relationship between the seasonal streamflow of the Oressa reclaimed basin at the village of Andreevka and that of the control basin of the Ptich at the village of Krinka.

Annual runoff

The annual runoff of 9 basins increased after reclamation by an average of from 10 to 20%. On the other 7 basins there was no

Table 53. Approximate information on percentage changes in the streamflow parameters after reclamation

Reclaimed Basin (Gauging Station)	Before Reclamation	After Reclamation	Daily mean Minimum Discharge		Maximum discharge	Seasonal Streamflow			Annual runoff
			summer	winter		spring	summer autumn	winter	
I.Ptich-Luchitsy	1895-1917 1926-34	1935-40 1947-59	+ 40	0	- 17	0	+ 25	0	+ 10
II.Oressa-Verkhutino	1926-39 1945-70	1962-70	+ 115	+ 130	- 25	- 10	+ 60	+ 75	+ 20
III.Oressa-Luban	1950-62	1963-70	+ 65	+ 70	- 30	- 15	+ 50	+ 70	+ 20
IV.Oressa-Kutinki	1926-28	1936-39	+ 50	+ 80	- 30	- 25	+ 15	+ 70	0
V.Oressa-Andreevka	1926-34	1935-41 1945-70	+ 50	+ 40	- 30	- 20	+ 25	+ 40	+ 15
VI.Vit-Borisovshchina	1936-41 1945-63	1964-70	+ 50	-	-	-	+ 30	-	0
VII.Cherten-Nekrashevka	1944-60	1962-70	+ 150	0	- 30	- 30	+ 50	0	0
VIII.Merechanka-Stavok	1951-61	1962-70	+ 145	+ 125	0	0	+ 45	0	+ 20
IX.Grivda-Ivatsevichi	1945-54	1955-67	+ 30	+ 20	- 20	- 10	+ 20	+ 15	0

continued

Table 53. Approximate information on percentage changes in the streamflow parameters after reclamation (continued)

Reclaimed Basin (Gauging Station)	Before Reclamation	After Reclamation	Daily mean Minimum Discharge		Maximum discharge	Seasonal Streamflow			Annual runoff
			summer	winter		spring	summer autumn	winter	
X.Grebelka-Birchuki	1927-39	1949-63	+ 50	+ 90	O	-	+ 20	-	O
XI.Uza-Pribor	1931-40 1944-57	1958-70	+ 60	O	O	O	-	O	O
XII.Ola-Mikhalevo	1956-63	1964-70	+ 50	O	O	O	+ 25	+ 50	+ 15
XIII.Ilia-Vladyki	1956-59	1960-70	O	O	O	O	O	O	O
XIV.Berezovka-Sautkii	1950-62	1963-70	+ 35	+ 20	O	O	+ 25	+ 20	+ 10
XV.Usha-Molodechno	1958-61	1962-70	+ 30	O	O	O	+ 15	+ 8	+ 10
XVI.Essa-Gadivlya	1952-62	1963-70	+ 130	O	O	O	+ 35	+ 25	+ 15

Note: O indicates that there is no noticeable change of the streamflow;
(-) shows that it was impossible to find any change of the streamflow because the available observation series was too short for the period after reclamation and due to the absence of a close correlation with the data from the reference basins.

appreciable change (Table 53, Figure 96).

The annual runoff of given return periods of the Ptich at Luchitsy and the Oressa at Andreevka increased by not more than 15% after reclamation. The variation coefficient of the annual runoff changed little after reclamation.

It should be noted that for the last five years (1965-1970), the annual runoff of the Oressa at Andreevka is peculiar on account of its high value in comparison with the annual runoff of neighbouring rivers. This affect was not observed during the first and subsequent years after reclamation with the exception of individual years. This phenomenon is still not explained.

Conclusions

Reclamation in all the basins under study resulted in an increase of streamflow in summer and autumn period; the streamflow distribution within the year became more uniform because of redistribution: the spring flow component of the annual runoff decreased while the summer-autumn and winter flows increased as well as the values for the whole specified period.

Annual runoff may either increase or remain the same depending on the behaviour of the spring flood discharge and the rate of change of the low-stage discharge.

Reference

Anon. 1966. Resursy poverkhnostnykh vod SSSR (Surface water resources in the USSR), vol.5, Belorussiya i Verkhnee Podneprovie (Byelorussian and Upper Podneprovie). Pt.1. Ed. by V V Kupriyanov. Leningrad, Gidrometeoizdat.

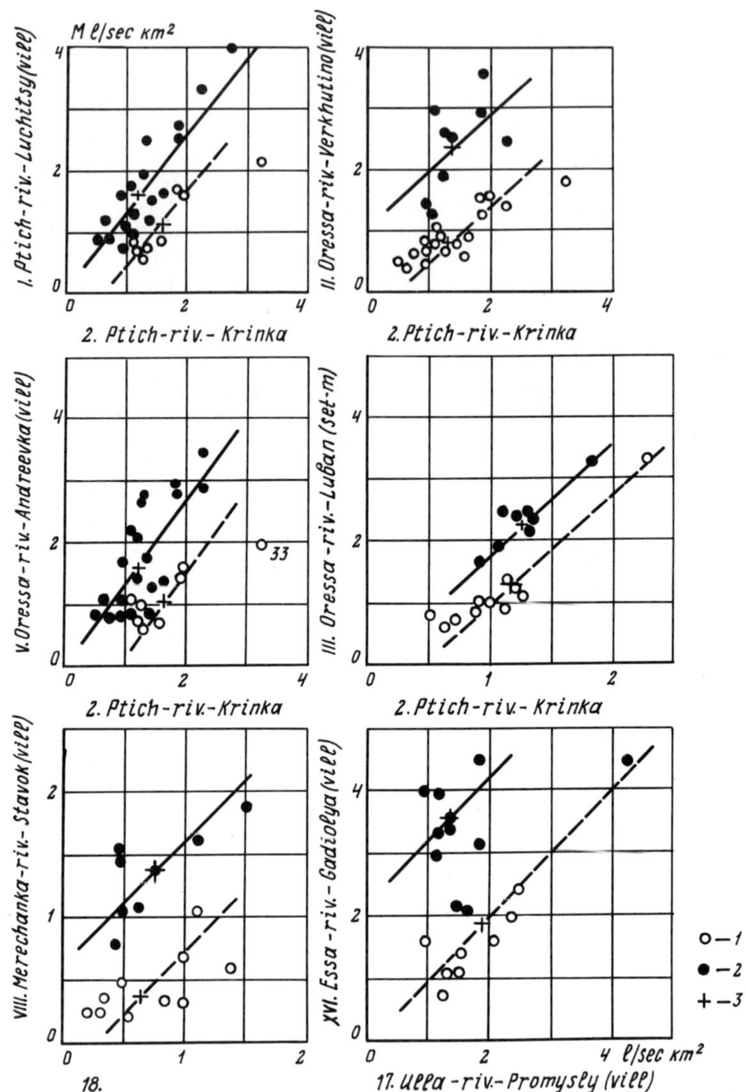

Figure 90. The relationship between minimum summer daily mean discharge on reclaimed (Nos. I through III, V, VIII, XVI) and control (Nos. 2, 17, 18) basins.

1. before reclamation;
2. after reclamation;
3. mean value for the period of simultaneous observations.

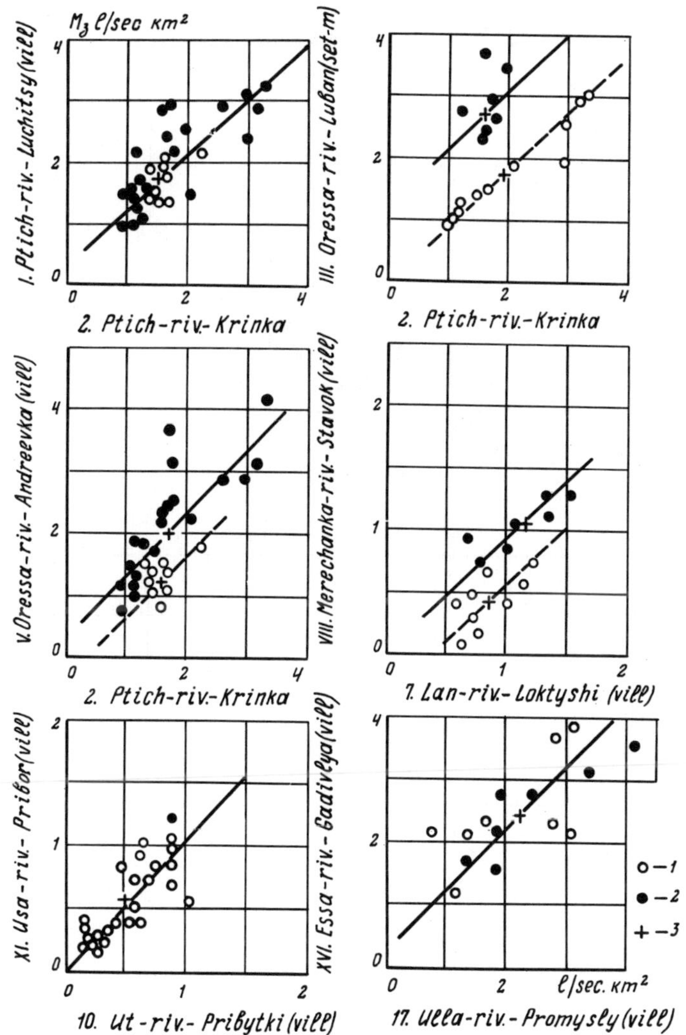

Figure 91. The relationship between minimum winter daily mean discharge of reclaimed (Nos. I, III, V, VIII, XI, XVI) and control (Nos. 2, 7, 10, 17) basins. 1. before reclamation, 2. after reclamation, 3. mean value for simultaneous observations.

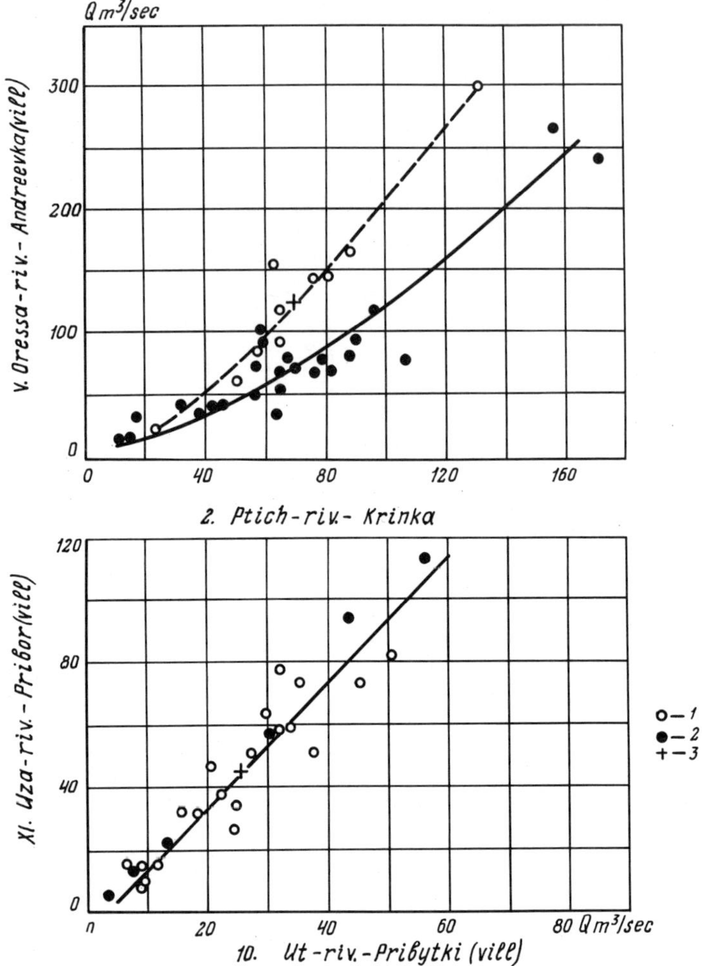

Figure 92. The relationship between maximum discharge of spring snowmelt flood in reclaimed (the Oressa and Uza) and control (the Ptich and Ut) basins.
1. before reclamation; 2. after reclamation;
3. mean value for simultaneous observations.

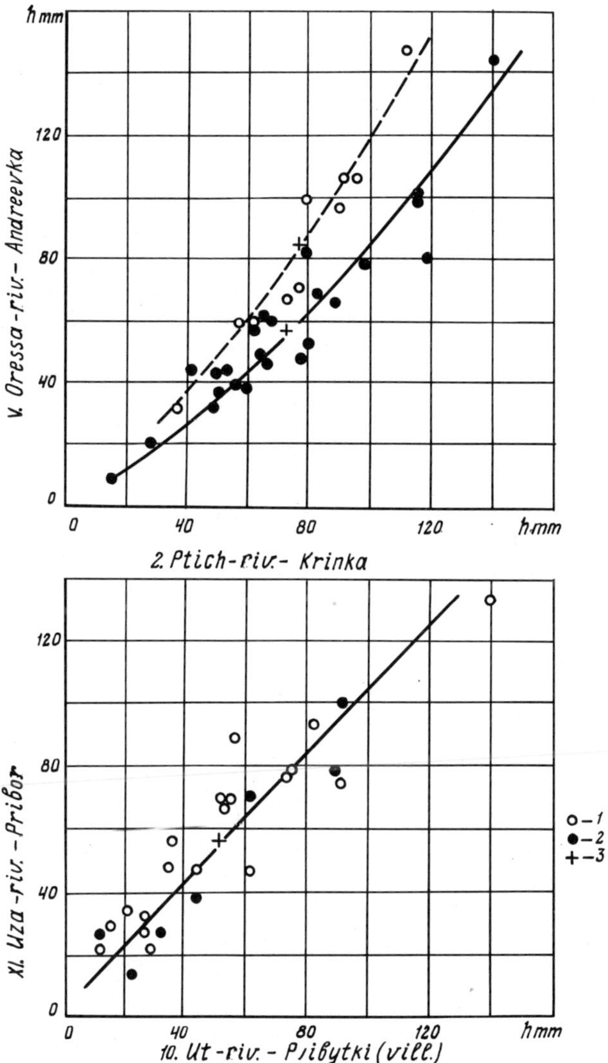

Figure 93. The relationship between spring runoff depth of reclaimed (the Oressa and Uza rivers) and control (the Ptich and Ut) basins.
1. before reclamation; 2. after reclamation;
3. mean value for simultaneous observations.

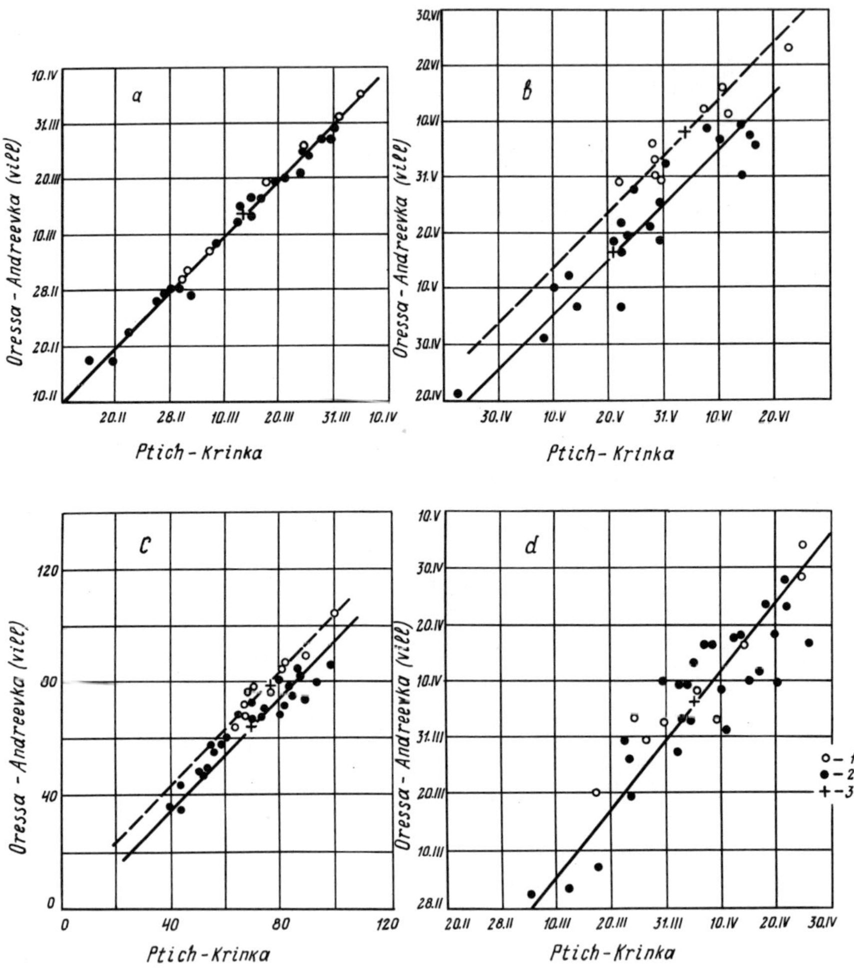

Figure 94. The relationship between (a) beginning and (b) end, (c) duration and (d) maximum discharge of spring snowmelt floods of the Oressa at the village of Andreevka and of the Ptich at the Krinka Sanatorium
1. before reclamation in the Oressa basin;
2. after reclamation;
3. mean value for simultaneous observations.

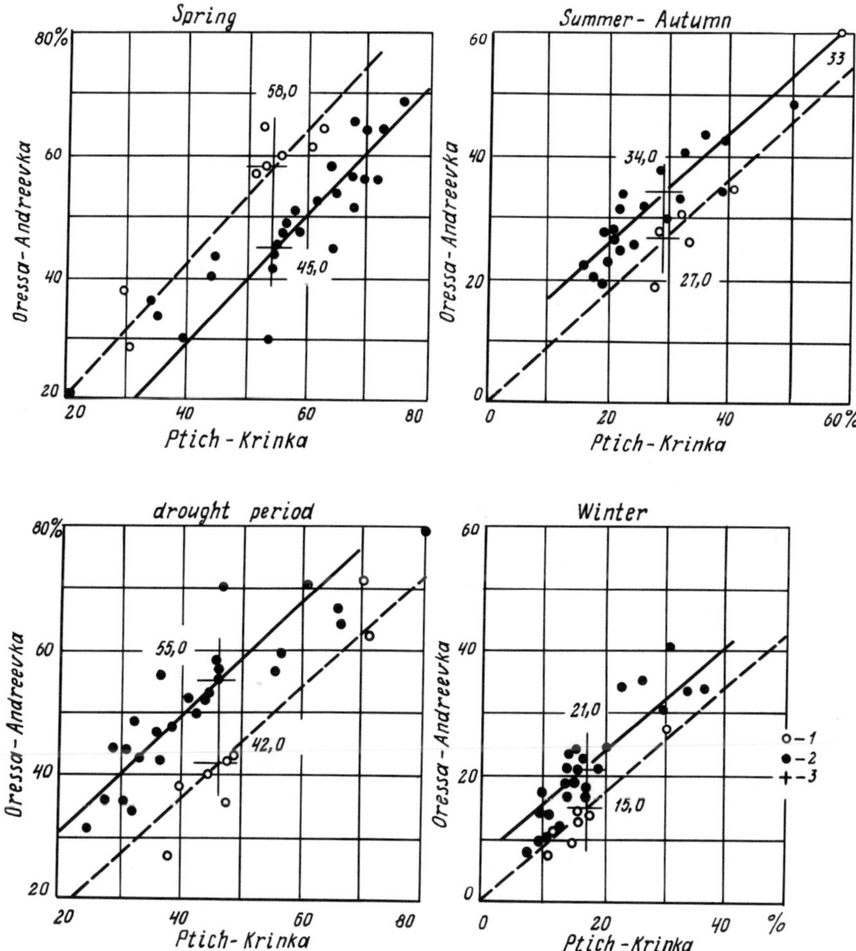

Figure 95. The relationship between seasonal (% of the annual value) streamflow of the Oressa at the village of Andreevka and of the Ptich at the Krinka Sanatorium.

1. before reclamation in the Oressa basin;
2. after reclamation;
3. mean value for simultaneous observations.

Figures on graphs show the seasonal streamflow as a (%) of the annual runoff of the Oressa river.

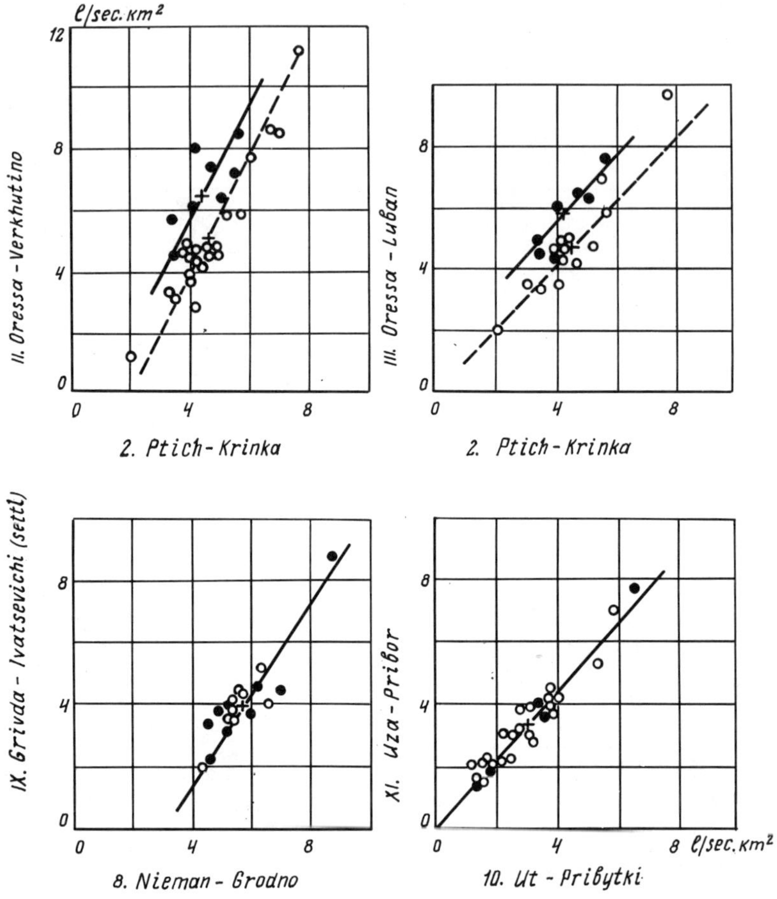

Fig 96. The relationship between mean discharges of specific reclaimed (Nos II, III, IX, XI) and reference (Nos 2, 8 10) basins.
1. before reclamation; 2. after reclamation;
3. mean values for simultaneous observations.

DISCUSSION

F. Zuidema *(The Netherlands)* - Was streamflow measured or estimated?

K. A. Klueva *(USSR)* - Real measurements were used.

G. V. Bogomolov *(BSSR)* - Your figures for changes in the streamflow should be related to the natural conditions in each basin. How far do the control basins differ from the natural ones?

K. A. Klueva *(USSR)* - Before starting the analysis of the data we took several catchments as controls. These were chosen carefully from locations in the same regions as the altered basins. The preliminary conclusions show differences which may be attributed to the degree of swampiness, the nature of the drainage and the type of catchment. We are making more sophisticated comparative analyses to understand these effects.

V. F. Shebeko *(USSR)* - 1. In what way did you account for any other aspects of human activity in a basin in your estimation of the effect of reclamation on the streamflow? 2. In what way did you account for hydrometeorological conditions in the periods of comparison of the streamflows before and after drainage of swamps?

K. A. Klueva *(USSR)* - 1. We only took into account the combined effect of reclamation measures. 2. The relationships between the corresponding streamflow values were compared in order to eliminate the effect of variations in precipitation and storage.

S. M. Perekhrest *(Ukrainian SSR)*: - To what period may the analysis be referred if reclamation took place over many years?

K. A. Klueva *(USSR)* - In the fifties, we looked at the effect of drainage on the Oressa considered in individual stages. The conclusion was drawn that changes did not occur simultaneously with the reclamation works but only after these had been completed (1935). With additional reclamation works now in progress no changes have been detected.

THE EFFECT OF DRAINAGE WORKS ON STREAMFLOW

V. I. Moklyak, G. P. Kubyshkin,
G. N. Karkutsiev

*Ukraininan Research Institute of
Water Engineering and Reclamation,
Kiev, Ukr. SSR.*

SUMMARY: Comparison of characteristic discharges of rivers draining reclaimed marsh areas of the Ukraininan SSR has revealed flow modifications due to drainage. Relative flow values (annual, spring and maximum discharges) in the post-drainage period are generally found to decrease, although in some cases the flow does not change partly because of the small size of the drained areas. The main factors affecting the flow of a river draining reclaimed land are summarised.

EFFET DU DRAINAGE SUR LE DEBIT DES RIVIERES

RESUME : La comparaison, sur plusieurs années, des débits caractéristiques de rivières marécageuses ukrainiennes en des régions soumises au drainage a mis en évidence les modifications que l'assèchement a produites dans l'écoulement. On a observé une diminution des valeurs relatives de l'écoulement (débit annuel, débit printanier et débit maximal) après assèchement, mais, en certains cas, l'écoulement n'a pas été modifié, ce qui est dû aux dimensions réduites des terrains d'assèchement et à d'autres raisons. Les facteurs essentiels qui peuvent influencer l'écoulement des rivières sont brièvement énumérés.

Analysis of the plotted curves has revealed that the annual flow in rivers, such as the Irpen, which have drained valley bogs has decreased, as can be seen in Figure 97. Double mass curves of the specific annual yield (l/s km^2) at two points is shown in the Figure 97; the Irpen basin at Mostishche has been drained, whilst the Irsha river at Ukrainka acts as the control basin. The discontinuous curve of the graph starts in the period prior to 1948 when both the watersheds were under natural marsh-ridden conditions. The second portion of the curve with a lower slope is based on data from the period of continuous observations from 1958 when 8 200 ha in the flood plain of the Irpen were drained. The flow decrease factor can be estimated by taking the ratio of the slopes of the two lines. The drained area represents 33% of the total watershed area, and it is situated in a flood plain covering about 100 km^2, the flow from the upstream basin is measured at Mostishche. In this case, there appears to be a significant effect of drainage on flow.

In contrast, there are cases when drainage has no effect on the flow. For example, the drained river basin of the Nedry at the Berezan settlement when compared with the control basin of the Udai at Priluki (Figure 98). The conclusions for other rivers are given in Table 55.

Table 55. Estimation of the effect of drainage on the annual flow characteristics

Drained basins	Control basins	Annual runoff	Volume spring flood	Maximum snow-melt discharge
Tnya, (at Broniki)	Smolka, (at Suslov) Zherev, (at Babinichi)	+	+	+
Grezlya (at Davydki)	Uzh, (at Korosten) Zdvizh (at Garvonsh-china)	-	-	-
Irpen, (at Mostishche,	Irsha, (at Ukrainka) Teterev (at Zhitomir)	-	-	-
Oster, (at Krivitskoe)	Udai, (at Priluki) Term, (at Budki)	-	-	-
Nedra, (at Berzan)	Udai, (at Priluki)	O	O	O

Key: + indicates that the discharge increased during the post-drainage period;

- indicates discharge decreased;

O indicates that there was no change.

In the Ukrainian SSR, there are 4 200 000 ha of marsh-ridden land. Drainage systems are now operating on 1 400 000 ha of this area. During the current Five-Year Plan, reclamation works will be completed on an area of 600 000 ha, bringing the total drained area to be 2 000 000 ha.

The behaviour of the levels and discharges of swamped rivers in the Ukraine has been observed since the 1920s. In 1958, a special bog-discharge station was established on the Trubezh river near the village of Baryshevka in connection with the Trubezh drainage works. The operational program of this station includes study of the hydrology, topography, climate, vegetation, soil and other characteristics. The information accumulated since 1961 enables preliminary conclusions to be drawn regarding changes that have taken place in some of the water balance components, but unfortunately the data are not strictly comparable because of the effect on the natural flow regime of introducing water from the Desna catchment for irrigation of the drained lands in dry summer periods.

Available data on marsh land river flows in the northern Ukraine are not adequate for extensive analysis of the effect of drainage on the flow nor for comparison of water balance components under natural conditions with those affected by drainage which is usually carried out only on a portion of the watershed area. Data from fixed observation stations allow the flow values in pre-drainage periods to be compared, but as the observation period was not very long, the results obtained should not be considered to be representative of most of the drained watersheds in the Ukraine.

Investigations were based on the following research technique. A river with a well known flow and whose basin contains marsh-ridden areas was chosen. During the years before construction of drainage works the gauge recorded the natural flow of the basin which had not yet been subjected to human activity. After construction of drainage works, flow was affected by the new conditions which included development of the drained lands. Adjacent to the drained basin under study, a control basin was chosen with similar hydrographic, climatic, edaphic and other conditions but without, or with negligible, drained areas. By comparing the flow characteristics of both basins, the change of flow in the drained basin was estimated. When data were available for comparison, two or three control basins were used to improve the validity of the conclusions.

Practical applications of the above method of defining the main flow characteristics, such as mean annual discharge, maximum snowmelt discharge and volume of the spring flow are considered below.

ANNUAL FLOW

Determination of the effect of drainage on the annual flow is based on comparisons with the flow data of control rivers, of which there are generally more than one; the data for five such rivers are shown in Table 55.

A decrease of flow after drainage of bogs implies a loss, which sometimes may be large, of a portion of the annual flow. Such losses may be caused by additional evaporation from the drained lands or by groundwater discharge of a portion of the surface flow.

When, in the process of development of drained bogs for arable cultivation, irrigation is introduced evaporation may be about 20% higher than that from undrained bogs. These losses by evaporation, however, only cover a part of the total losses of flow under the influence of drainage. No change in evaporation in the areas adjacent to the drained regions is to be expected. Groundwater levels during the pre-drainage period in these areas are always higher than those during the post-drainage period but it is not yet clear how the losses of surface flow may be apportioned.

Double mass curves showing the relationship between the annual specific yield of adjacent rivers indicate possible changes in the flow from the drained watersheds. Most commonly these tend to decrease. Further investigations are needed to explain both the annual flow decrease in the rivers of the Ukraine and the annual flow increase in the rivers of Byelorussia during the post-drainage period.

Spring flow volume

Double mass curves have been plotted to show the relationship between the specific yield of spring flows in two sections of marsh land rivers: one section lies in a basin where a drainage system was constructed in the year T + t and the other with undrained bogs acts as a control. A decrease in the spring flows is generally found but cases are also encountered where the flow is unchanged or tends to increase (Table 55).

Maximum snowmelt flows

The maximum discharge is known to depend on the volume of the spring flow and on the rapidity of melt. Thus, maximum discharges in rivers with drained flood plains should not decrease in the years of low spring floods; rather they should increase because of higher flow velocities. During the years of high spring floods, a decrease of maximum discharge can be expected under the influence of a non-uniform velocity distribution in the cross-section of flow and, in part of a reduction of spring flow volumes due to storage by absorption by peat and by the increased network of canals with sluices and reservoirs. Drainage does not always affect the maximum discharge, although after drainage, in some rivers the maximum discharge may either decrease or increase (Figure 99). The maximum discharge is stable when there is a negligible change in flow conditions after drainage of the flood plain.

In Ukrainian Polessie and Middle Prydneprovie the maximum river discharges are produced by either long duration rain of low intensity or short duration storms. The maximum storm flows of two adjacent rivers with similar basin characteristics vary but as the meteorological influence is more pronounced than the effect of drainage, the comparative methods described above failed to reveal the influence of drainage on the runoff of rainfall.

The investigations completed up to now are only preliminary and need verification by other techniques but suggest that the effect of drainage on the flow depends on a number of factors which are listed in Table 55.

The effect of drainage on the streamflow depends also on the dimensions and location of the swamp in the basin and on the drainage techniques adopted. For control of the soil moisture regime by drainage only, a simple system of canals is constructed. This drainage method is still in use but is being progressively replaced by canals used for both drainage and irrigation which render crop yields largely independent of the weather.

With control by irrigation and drainage, the quantity of water lost to evaporation certainly increases and the regime of streams in the drained areas during the period of vegetation growth will differ from that of areas having drainage only.

Table 56. List of additional effects of drainage on runoff

Causes of decreases in flow	Causes of ncreases in flow
Reduction in the hydraulic conductivity of peat beds.	Increased discharge of water accumulated in bogs by ditches and drains.
Loss of surface runoff by storage in the active layer of the drained peat bed.	Temporary flow increase following straightening, deepening and clearance of vegetation from streams and ditches.
Flow loss by storage in soil on slopes. Flow losses by storage in closed depressions created by uneven peat subsidence. Increase of evapotranspiration from drained and intensively used agricultural lands. Use of sluices and regulators in canals.	Decrease in evaporation from drained, but undeveloped areas. Increase of surface and groundwater slopes. Additional outflow of unconfined and artesian waters from aquifers exposed by canals. Draining of closed marshy areas by means of the canal system.

It is at present impossible to explain all the above relationships. Accumulation of field observations takes time; such data together with adequate theories of the processes involved in stream flow should provide the answers to many such questions in the future.

DISCUSSION

V. F. Shebeko (Byelorussian SSR): In what way did you account for bilateral control, by drainage and irrigation, in your estimation of the effect of drainage on the streamflow.

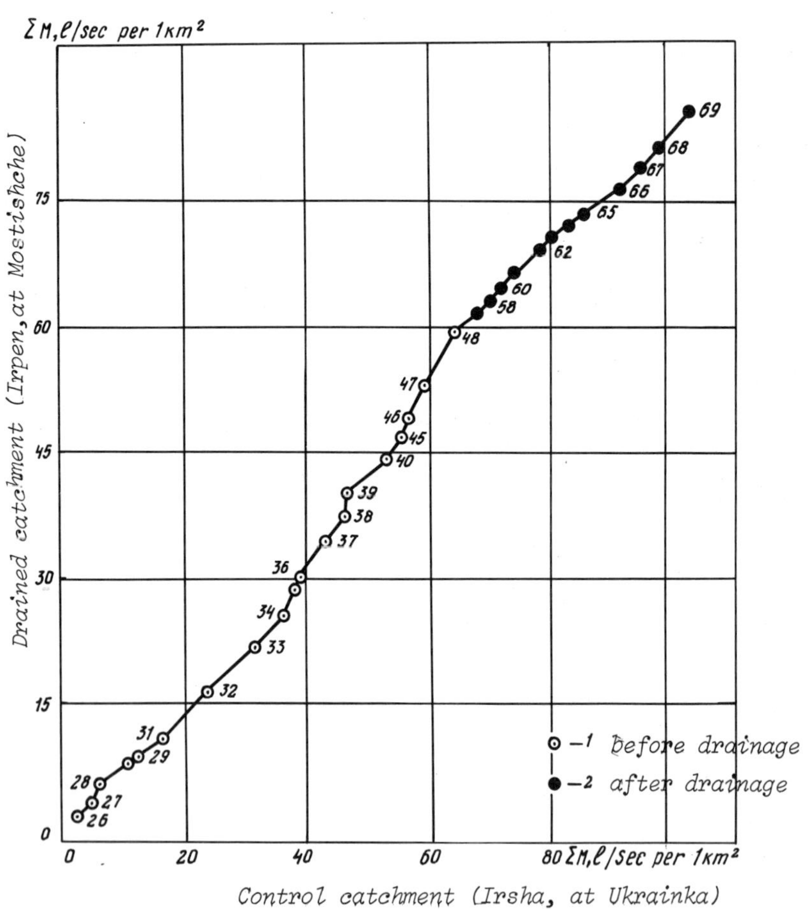

Figure 97. Double mass curves of mean annual specific yield of the drained catchment of the Irpen at Mostishche related to the control Irsha catchment at Ukrainka

V. I. Moklyak (Ukrainian SSR): Bilateral control cannot have any effect on the snowmelt flood, although it causes sharp changes in the regime. We made special studies to compare the situations of bilateral and unilateral control. We worked mainly on rivers of Byelorussia and found that unilateral control results in a sharp decrease of the minimum discharge.

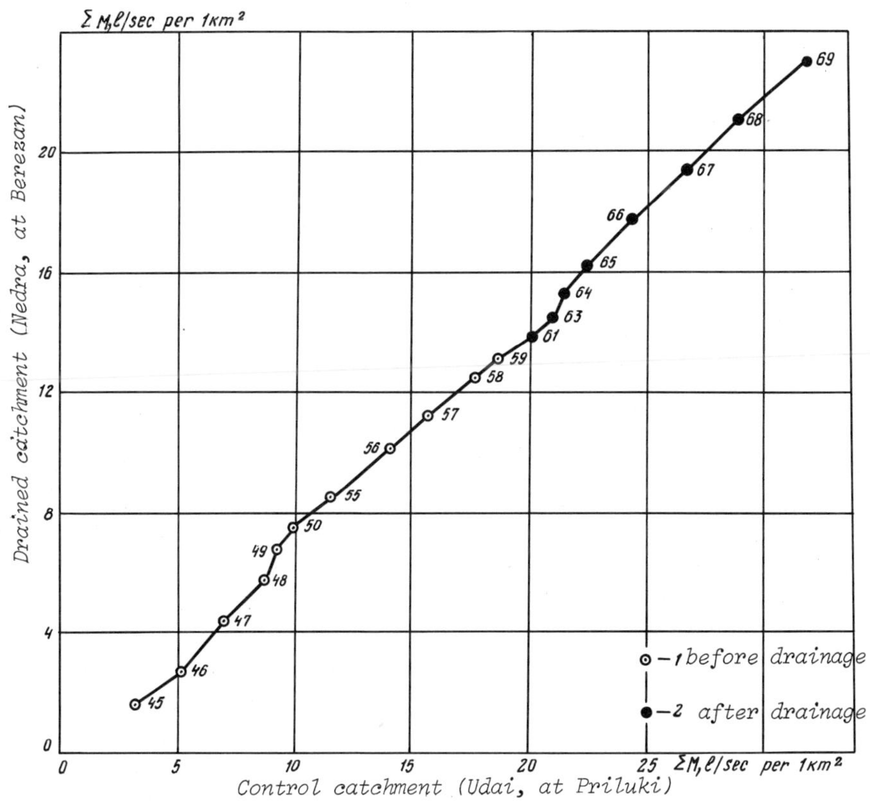

Figure 98. Double mass curves of mean annual specific yield of the drained catchment of the Nedra near the village of Berezan related to the control Udai catchment near the village of Priluki

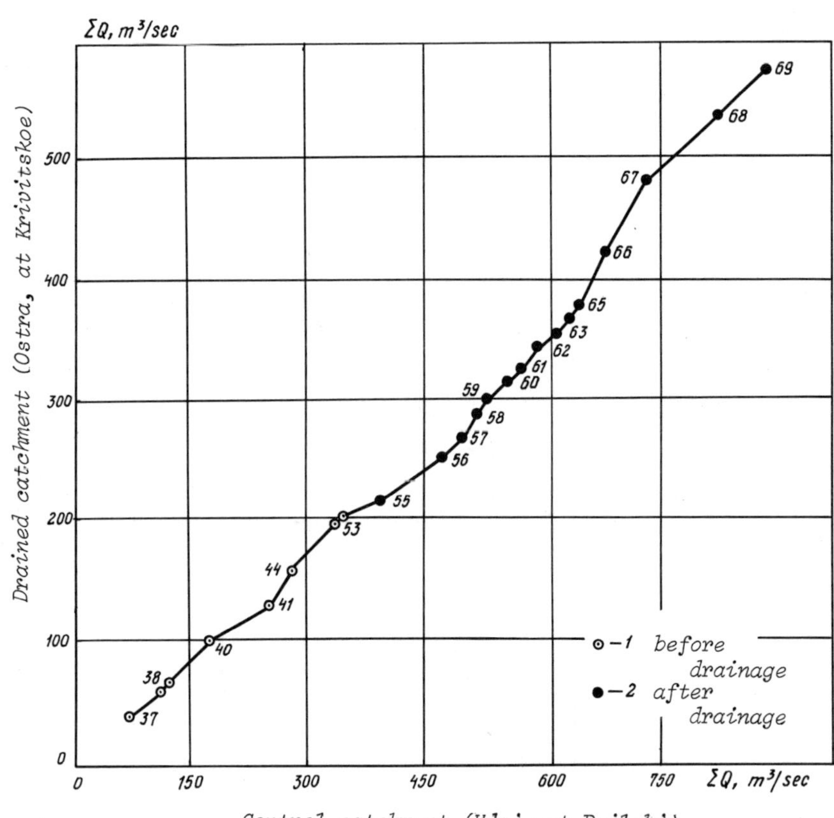

Figure 99. Double mass curves of maximum spring flood discharges from the drained Osta catchment near the village of Krivitskoe related to the control Udai catchment near the village of Priluki.

PROBLEMS OF GROUNDWATER REGIME AND ITS PREDICTION IN DRAINED REGIONS OF THE BYELORUSSIAN POLESSIE

A. P. Lavrov, M. V. Fadeyeva, G. I. Sachok, A. P. Vahkovsky and V. F. Vasilyev

Byelorussian Geological Research Institute, Minsk, BSSR.

SUMMARY: This investigation of groundwater regime has been carried out in Byelorussian Polessie where there are numerous marshes which are now intensively drained. The analysis of annual hydrographs of groundwater level has led to the solution of the problems under examination. The existing conditions in certain regions are simulated, and the observational data for a period of 15 to 20 years are statistically treated.

The relations between the groundwater level and the most important factors which affect its regime (water level fluctuations in the river, precipitation, air temperature) are described, as is the dependence of the natural groundwater regime on particular geomorphological elements. The possibility of regime prediction is also discussed. Changes in the regime of groundwater caused by drainage are also analysed.

PREVISION DU REGIME DES EAUX SOUTERRAINES DANS LES TERRITOIRES SOUMIS AU DRAINAGE EN POLESIE BIELORUSSE

RESUME : Les auteurs présentent les résultats de leur étude des eaux souterraines, régime et prévisions, et des problèmes s'y rapportant.

Cette étude a été effectuée en Polésie biélorusse, région comportant de nombreux marais, qui, à l'heure actuelle, sont drainés intensivement. Pour résoudre les problèmes posés, on a procédé à une analyse des hydrogrammes annuels de fluctuation du niveau des eaux souterraines. Les conditions existantes, dans divers secteurs, ont été simulées et les données des observations du régime sur une durée de 15-20 ans ont été traitées par les méthodes de la statistique mathématique.

Les auteurs établissent les relations entre le niveau des eaux de sous-sol et les principaux facteurs qui affectent le régime (fluctuation du niveau de la rivière, précipitations atmosphériques, température de l'air) ; ils montrent aussi l'influence de certains éléments géomorphologiques particuliers et les possibilités d'effectuer des prévisions. Ils analysent également les modifications que le drainage apporte au régime des eaux souterraines.

Drainage of swamplands and marsh-ridden areas changes the hydrogeology of the regions under reclamation. The groundwater regime needs careful study both before and after drainage to estimate the qualitative and quantitative changes.

The groundwater regime of Byelorussian Polessie has been systematically studied by the Byelorussian Geological Research

Institute, the Byelorussian Hydrogeological Expedition and by other institutes using a wide network of observation wells characterizing the groundwater regime of those aquifers which play a significant role in the national economy. Particular zones have been under constant observation for 15 to 20 years. This has made it possible to establish the main pattern of the regional groundwater regime, to find the relationship between unconfined and artesian waters and the effect of rivers and climate on this relationship. These observations have also helped to evaluate to what degree reclamation changes the groundwater regime. Such problems can be solved by analysing groundwater hydrographs and by statistical methods such as correlation analysis and the theory of random processes and sequences (autocorrelation and cross-correlation functions and spectral densities).

Fluctuations in the groundwater level are affected by meteorological conditions and by water level variations in nearby rivers. In the annual cycle of the groundwater levels, three peaks separated by troughs may be distinguished: a maximum spring rise (April - May), a summer rise (July - August), and an autumn rise (October - November) (Figure 100). These fluctuations are reflected in three periodic changes in the water level on the spectral density charts at one year, 6 month, and 4 month intervals. These charts also show variations with a periodicity of several years (Figure 101).

The closest and most stable relation existing between river and groundwater levels throughout a year is observed in flood-plain terraces. The highest correlation coefficients, varying between 0.74 and 0.96, are recorded from January through May, indicating a close hydraulic interrelation and synchronism of changes between river and groundwater levels over this period. In other months, the correlation coefficient is somewhat smaller. The level of low water in the river in September, October and, to a lesser degree, in November, shows a correlation with the groundwater level until April in the succeeding year. (Figure 102, Well No. 149, Mozyr).

The value of the correlation coefficient of the relationship between the groundwater level and precipitation does not usually exceed 0.44 to 0.52 and only in June and August does it reach 0.61 to 0.78; the effect of precipitation on the groundwater level is still indicated in October. The rainfall in July and September affects the groundwater level one month later (Figure 102, Well No. 149).

The correlation between the groundwater level and air temperature is still weaker. Correlation coefficients of 0.45 to 0.71 obtain in February, March, July and August; July and August temperatures are likely to have some influence on the water table until April.

When the water table is close to the surface, fluctuations in the groundwater level differ somewhat from the case discussed. Isolines of the spring maximum levels of various degrees of probability are compressed, and the peaks are smoothed (Figure 100, Pinsk, Well No. 30). The groundwater level in October, November and December affects the river discharge until May, while the inverse relation is hardly ever observed during this period.

The correlation of groundwater level with precipitation is weak and noticeable only a month or two later (Figure 102).

In the interfluvial areas where the water table is at about 5 metres depth, annual variations in the levels are very small, and no direct relationship is observed between the water level and the river stage or with rainfall (Well No. 48, Pinsk).

Standard deviations of the average monthly groundwater levels from April to September are smaller than in the cold period of the year, ie levels for these months can be predicted with a smaller absolute error. The values of levels for adjacent months, except March - April, are closely correlated (Table 56). This fact may be used to predict the mean levels of the following month using the linear regression equation of the type $y = ax + B$ with a probability better than 50%. The correlation coefficient of the groundwater level at 1.5 to 2.0 m (Bereza and Alexandrov Stations) attains a value of 0.9 and sometimes more; at depths over 4 m between rivers the coefficient is 0.99 to 0.97 (Well No. 48, Pinsk).

Variations in the artesian water table level are in general similar to those of the phreatic levels. This may be attributed to a close hydraulic relationship between aquifers in the zone of active water exchange. Interrelation between phreatic and artesian water should be carefully studied and taken into account in the solution of problems concerning sources of replenishment of swamplands. This interrelation should not be neglected when drawing up plans for land reclamation and when evaluating its effect on the groundwater regime of adjacent areas. The quantity of artesian water reaching the area under reclamation must be determined as must any change that might result from a change in the relationship between the phreatic and artesian water levels in the process of land reclamation. Calculations made at several points show that the swamplands receive 30 to 35 mm of groundwater every year from sand deposits underlying the peat.

It is not only the value of groundwater recharge of swamplands that is changed as a result of drainage; the proportions of all the water balance components are also changed to a different extent in each area.

It is established that a drop in the groundwater level decreases the rate of evaporation and rainfall infiltration, and that at depths below 0.75 to 1.0 m a positive water balance or steady groundwater feed is preserved at all times (Figure 103).

Observations of the groundwater regime of drained swamplands and adjacent areas reveal a drop in the groundwater level due to drainage. Where the groundwater level of the swamp area before drainage was on the surface of the ground or close to it, it is now at a depth of about 1.0 m. As a result of this decrease in the groundwater level during the initial period of drainage, groundwater flow in the zone affected by drainage systems is lessened; some time later, new conditions of intensive water exchange over the whole area appear, the lateral influx of water from the adjacent areas increases, and the infiltration recharge and evaporation also increase. The amplitudes of groundwater level fluctuations on drained areas are increased by as much as 2 or 3 times due to greater thickness of the non-saturated zone after drainage.

We can draw certain conclusions and make generalizations on the basis of the regime observations. To this end, graphs of groundwater level fluctuations in wells located at different distances from the swamplands being drained were analysed; the observations were statistically treated, and the conditions existing in various sectors were simulated on a Lukianov-type hydraulic integrator.

For this purpose, observations started long before the drainage of the Luninets swamplands and adjoining mineral lands are of great value. This bog system is associated with the flood-lands of the Bobrik river, a tributary of the Pripyat. Peat, 1 to 2 m thick, rests on sands of varying grain size.

As a result of drainage, the mean annual groundwater level in the area has fallen by 10 to 80 or 90 cm, depending on the distance of the wells from the drainage ditches. The influence of the drainage system on the water table is felt mainly at the summer-autumn low water period; in April to May, the groundwater level after drainage approaches the normal level before drainage. A certain rise in average monthly variations of the groundwater table is also observed in the winter (January - March) and summer (June - October) low-water periods due to the formation of an unsaturated zone (Table 57).

The mean annual rate of groundwater decrease was determined from observations in 25 wells situated at various distances from the drainage ditches. The water levels in the years before drainage (5 to 9 years) were compared with those after drainage (7 years). An equation was obtained relating the drop in the groundwater level (ΔH) to the distance from the drainage ditches (ℓ);

$$\Delta H = 36.62\ell^{-0.6515} \qquad (1)$$

The correlation coefficient of the relationship between decrease in groundwater level and the distance from the ditches is 0.71.

Equation (1) has been used to calculate possible decreases of the groundwater levels at various distances from the drainage system, for areas that have conditions similar to those of the Luninets marsh.

Distance from drainage system, ℓ, m	500	1000	1500	2000	3000	4000	5000	7000
Groundwater level decrease ΔH, m	0.65	0.41	0.31	0.26	0.20	0.16	0.14	0.11

Simulation on the hydraulic integrator at the Byelorussian Geological Research Institute by T. D. Krivetskaya, revealed

Table 57. Correlation coefficients of average monthly groundwater levels

Observation Points	Number of years	Months											
		1-2	2-3	3-4	4-5	5-6	6-7	7-8	8-9	9-10	10-11	11-12	12-1
1. Well No. 48 (Pinsk Station)	15	0.98	0.92	0.81	0.85	0.99	0.99	0.99	0.99	0.99	0.99	0.97	0.96
2. Well No. 30 (Pinsk Station)	19	0.68	0.86	0.66	0.81	0.85	0.92	0.86	0.80	0.62	0.84	0.89	0.87
3. Well No. 5 (Bereza Station)	18	0.89	0.91	0.73	0.87	0.76	0.89	0.89	0.93	0.90	0.96	0.92	0.85
4. Well No. 111 (Stolin Station)	16	0.98	0.83	0.67	0.84	0.87	0.91	0.84	0.97	0.93	0.92	0.86	0.63
5. Well No. 149 (Mozyr Station)	21	0.91	0.76	0.42	0.70	0.73	0.81	0.88	0.93	0.92	0.84	0.93	0.91
6. Well No. 28 (Alexandrov Station)	18	0.86	0.84	0.50	0.86	0.79	0.86	0.71	0.72	0.80	0.86	0.90	0.90

Table 58. Average monthly values and mean square deviations of groundwater levels in the Luninets Swampland, cm

	Years	Para-meters	Months												Mean annual
			I	II	III	IV	V	VI	VII	VIII	IX	X	XI	XII	
		$\bar{x}$	-20	-20	-14	-10	-12	-16	-25	-27	-32	-30	-24	-19	-21
Well No. 17	1954-60	S_x	24	28	23	21	18	18	20	20	23	21	23	20	21
		$\bar{x}$	-64	-67	-49	-15	-15	-33	-59	-79	-92	-90	-74	-64	-58
	1961-69	S_x	31	34	40	18	16	24	31	29	27	28	20	14	22
		$\bar{x}$	-18	-18	-10	- 6	-14	-26	-40	-41	-44	-38	-27	-20	-25
Well No.8	1951-61	S_x	22	23	16	7	8	9	16	15	21	26	22	18	11
		$\bar{x}$	-10	- 9	-69	-36	-58	-81		-110	-122	-115	-97	-91	-89
	1962-69	S_x	24	28	34	16	24	32	36	33	25	23	14	13	20

that five years after drainage the drop of the groundwater level at a distance of 25 m from the drainage system was 0.70 m, and 0.09 m at 1250 m. Ten years later, the level had dropped by 0.77 m and 0.18 m, respectively. It has been established that the effect of drainage is felt 2 years later at a distance of 1250 m from the drain and 6 years later at 2250 m. No effect was found after 10 years at a distance of 3000 m.

At the Glubonets swamp, located between two rivers, the decrease of pre-flood groundwater levels associated with bog, soil, till and morainic deposits was estimated from the data of 17 wells. The dependence of the groundwater level decrease on the distance from the main drainage canal, is as follows:

$$\Delta H = 236.2\ell^{-0.7795} \qquad (2)$$

The correlation coefficient of the relationship between decrease in groundwater level and the distance from the ditches is 0.81.

Equation (2) has been used to calculate the possible value of the pre-flood drop in the groundwater level at various distances from the main drainage ditch for areas similar to the Glubonets sector.

Distance from drainage ditch ℓ,m	500	1000	1500	2000	3000	4000	5000	7000
Level decrease ΔH, m	1.86	1.08	0.79	0.63	0.46	0.37	0.31	0.24

In accordance with the "Drainage and Development of Polessie Lowland Project", worked out in 1968, a further large extension of drainage has been planned which will embrace the areas of the Byelorussian and Ukrainian Polessie. Land reclamation, as envisaged by the Project, will complicate the prediction of changes in the natural conditions of a large geographical region. Hence the use of analog and digital computers for processing of accumulated data becomes of the greatest importance.

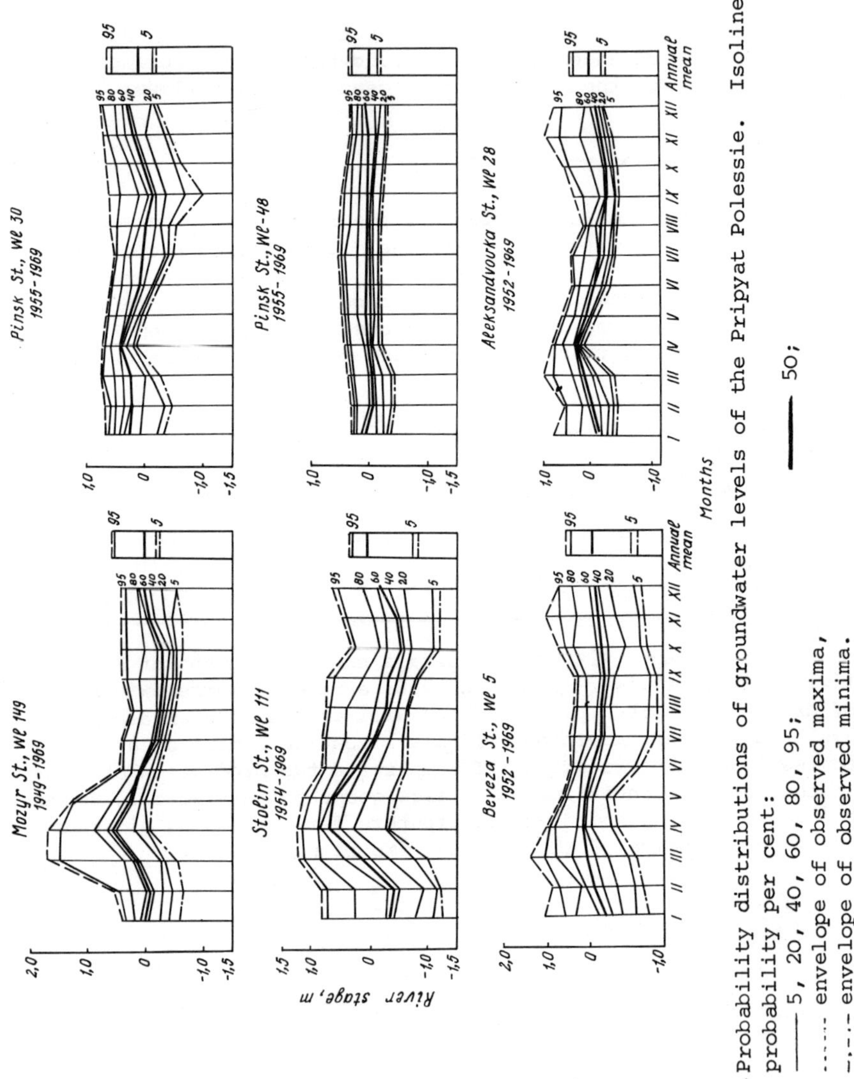

Figure 100. Probability distributions of groundwater levels of the Pripyat Polessie. Isolines of various probability per cent:
—— 5, 20, 40, 60, 80, 95; ▬▬ 50;
----- envelope of observed maxima,
-.-.- envelope of observed minima.

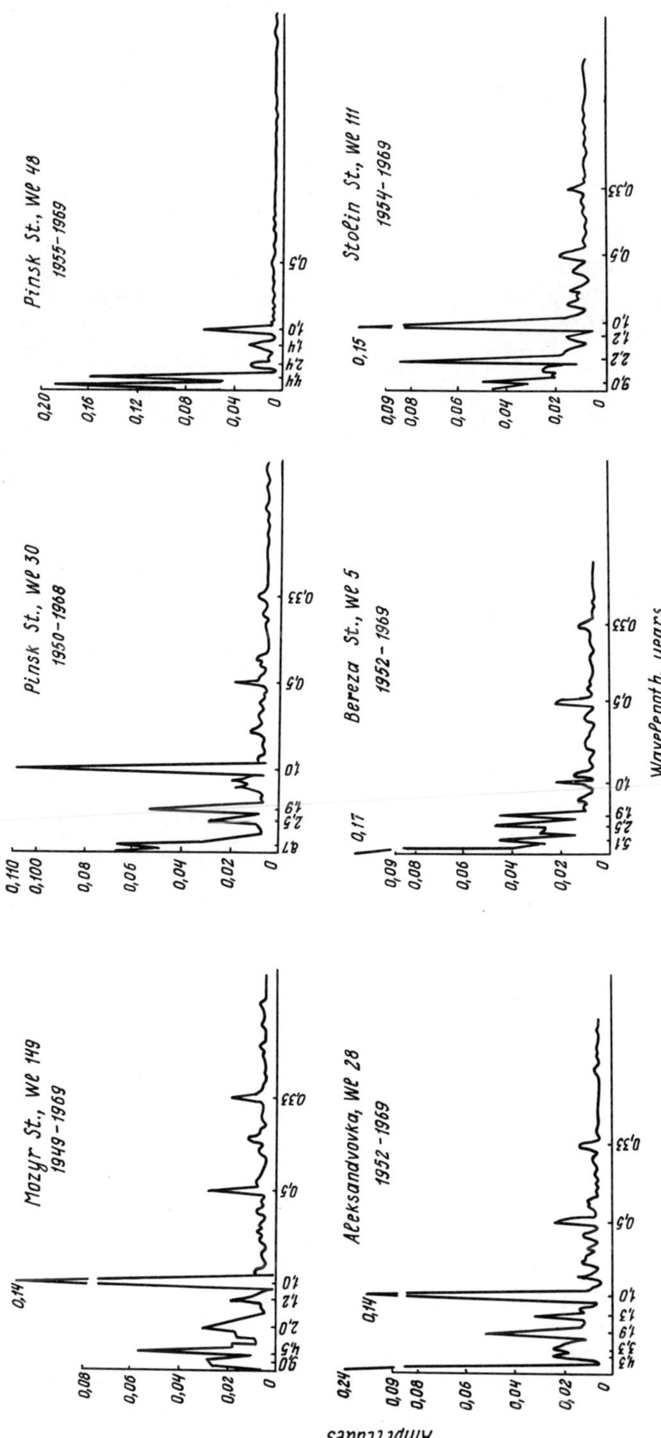

Figure 101. Spectral densities of autocorrelation functions of groundwater levels

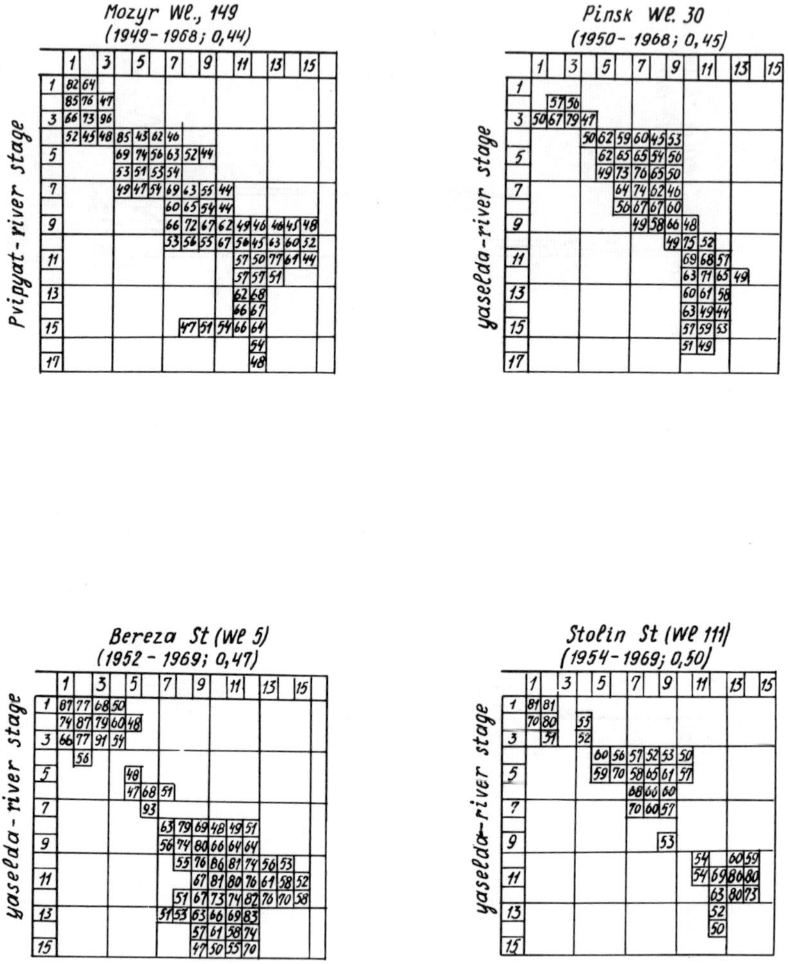

Mozyr Wl., 149
(1949 - 1968; 0,44)

Pripyat-river stage

Pinsk Wl. 30
(1950 - 1968; 0,45)

yaselda-river stage

Bereza St (Wl 5)
(1952 - 1969; 0,47)

yaselda-river stage

Stolin St (Wl 111)
(1954 - 1969; 0,50)

yaselda-river stage

Figure 102. Correlation matrices of mean monthly values of groundwater levels and regime factors at different stations. Years of observation and the tabulated values of correlation coefficient at 95 % significance level are given in brackets. The abscissa shows the levels of the water table. The matrices contain two decimal digits in the correlation coefficients.

Mozyr (We 149)
(1949 - 1968; 0,44)

precipitation	1		3		5		7		9		11		13		15		17	
1			52															
			51															
3																		
5																		
						51	78	71	71	61								
7								45			47		61	32	45			
								62	69	54	48			51	52			
9										47						60	58	49
											45		57					
11																		

Pinsk St (We 30)
(1955-1968; 0,51)

precipitation	1		3		5		7		9		11		13		15
1			52	55											
3															
				53					52		69	72	58		
5							53								
								52	59						
7															
										73		51			
9															
										53	68	53			
11															

Mozyr (We 149)
(1949-1968; 0,44)

Air t°	1		3		5		7		9		11		13		15	
1																
		72	54													
3			51													
									-44	-50						
5																
						-50	-50									
7							-48	-47	-42	-45	-47	-50	-58	-56	-56	
								-65	-60	-49	-60	-71	-62	-60	-56	
9													-50	-47		
11															-44	
												43	42			

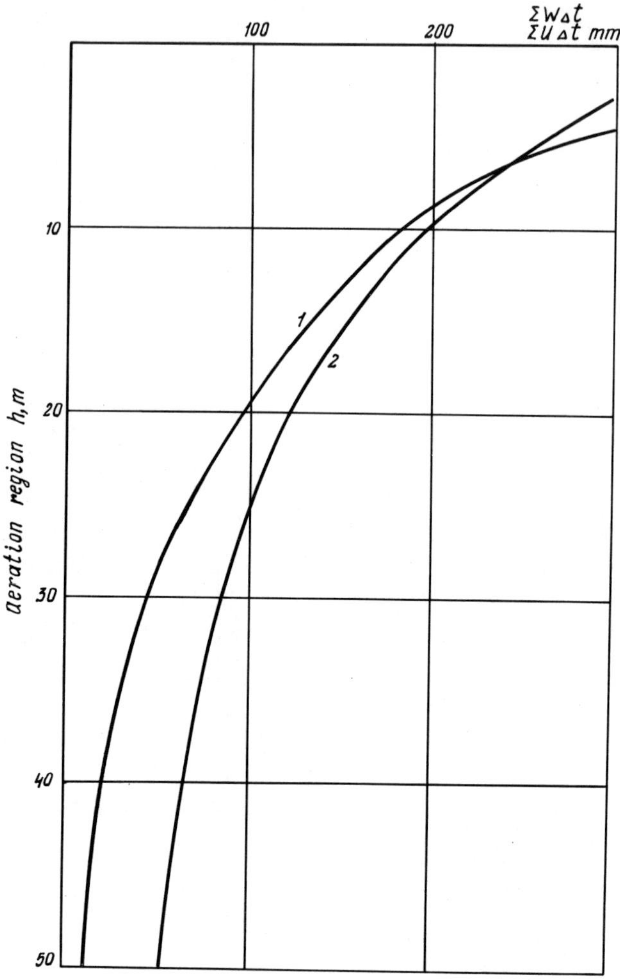

Figure 103. Plot of normal seepage and evaporation from the table versus the depth of the unsaturated zone.

DISCUSSION

G. Kovacz (Hungary): - Have you used measured values to plot Figure 4 of your paper?

A. P. Lavrov (Byelorussian SSR): - These curves are based upon the groundwater balance in Polessie. What is most interesting and valuable is that at a depth of 3 to 4 m a positive balance still remains. The figure is plotted from actual measurements.

P. A. Kiselev (Byelorussian SSR): - 1. What is the influence of hydrological factors on the surface flow from areas under reclamation? 2. Is it necessary to consider the geology when analyzing changes in runoff? 3. Should the deeper groundwater horizons be taken into consideration in simulating the conditions in bog systems?

A. P. Lavrov (Byelorussian SSR): - 1. To solve this problem we have divided Polessie into hydrogeological regions and singled out the areas where groundwater flow is of different significance and proportion. 2. In hydrological investigations of an area, hydrogeological conditions should be taken into account as well for correct prediction of possible changes due to reclamation. 3. Phreatic and artesian water should be considered together.

BOG RECLAMATION AND ITS EFFECT ON THE WATER BALANCE OF RIVER BASINS

A. G. Bulavko and V. V. Drozd

Central Research Institute of Complex Use of Water Resources, Minsk, USSR.

SUMMARY: Some modern concepts of the title subject are given together with the results from investigations of the effect of drainage systems on the hydrological properties of river basins. Changes in some components of the water balance, including changes of total discharge and groundwater flow resulting from drainage systems are studied. The distortion of the statistical pattern of hydrological variables due to reclamation works in watersheds is estimated using non-parametric criteria.

L'AMENAGEMENT DES MARAIS ET SON INFLUENCE SUR LE BILAN HYDRIQUE DES BASSINS DE RECEPTION DES RIVIERES

RESUME : Cette étude expose les conceptions qui ont cours aujourd'hui sur le sujet traité et donne les résultats de l'étude des particularités hydrologiques des bassins de réception des rivières qui ont bénéficié de systèmes d'aménagement par drainage. On y expose les modifications apportées à certains éléments du bilan hydrique des bassins versants, tels qu'écoulement total et écoulement souterrain, par l'établissement de systèmes de drainage. A l'aide de critères non paramétriques, on évalue les écarts qui se présentent dans l'homogénéité statistique des séries hydrologiques à la suite des travaux d'aménagement effectués dans les bassins versants.

Reclamation of bogs and marsh-ridden lands by a combination of engineering and agrotechnical measures controls the water regime of these areas. The objective is to create the optimum conditions needed to obtain continuously high crop yields on reclaimed lands.

Land reclamation measures result in predictable changes to the water balance of the area. These changes are accompanied by changes to water balance of lands adjacent to the reclaimed areas and in the whole catchment. It is necessary, therefore, to make a special study of the hydrological consequences of land reclamation.

The effect of reclamation on the water balance of areas first attracted the attention of scientists in this country at the end of the 19th century when the Western Bog Draining Expedition led by I. I. Zhilinsky, made tentative estimates of the influence of drainage on the hydrological regime of the Pripyat basin. However, because of the small scale of the land reclamation and the limited data, it was not possible to make satisfactory estimates.

Later, in the Soviet Union and in other countries, a considerable research effort was devoted to the subject, (Bulavko, 1961, 1968 and 1970); no common view developed, however.

It has been suggested that drainage for land reclamation, particularly in areas with sandy soils, results in:

1. a deterioration in the water balance for the catchments;
2. reduces the water resources of the surrounding areas;
3. causes small rivers, areas of water and wells to dry up;
4. reduces the condition of water supplied to agricultural land and forests in areas adjacent to the reclaimed swamps;
5. promotes wind erosion of soils.

A different viewpoint suggests that the groundwater levels lowered by 1 to 1.5 m in reclaimed lands cannot extend far beyond the reclaimed areas and will not damage the water balance of adjacent lands.

The views of those who believe that land reclamation has a deleterious effect on the water balance have been disproved many times in the literature, the most convincing arguments being presented by Maslov (1971).

The effect of land reclamation on the water balance is far from being of pure scientific interest. It has important practical implications and the type and rate of further reclamation in the Soviet Union depend on their understanding.

After reclamation stream density increases 2 to 5 times, which is considered to increase runoff. Because peat settles near drainage ditches, reclaimed lands acquire a specific mesorelief, which consists of transverse slopes and regularly connected depressions, which encourage runoff. Reduced water levels at water intakes on regulated rivers result in increased slopes at the confluence of tributaries and mainstream; this also facilitates runoff. A fall in the groundwater level on reclaimed lands leads to a considerable decrease in evaporation which is not completely compensated for by intensive agricultural utilization of reclaimed lands. Evaporation losses from reclaimed and cultivated bogs are about 15% less than before drainage, and this factor again encourages runoff. The reduced time of flood land inundation which follows river regulation decreases evaporation and increases the storage capacity of the unsaturated zone.

At present no data are available from which the individual effects of these factors can be estimated. It is clear, however, that most of these promote increased runoff. This is proved by observations made on six catchments of the left bank Pripyat tributaries, where from 6 to 25% of the area was reclaimed.

Any experimental catchments must meet the following requirements:

(a) considerable influence of land reclamation;

(b) fairly reliable hydrological observations over a period

of several years before and after land reclamation;

(c) the existence of one or two similar catchment with the same period of hydrological observations as the experimental catchment, but with natural conditions, (Bulavko, 1970).

The main analysis comprised a comparison of the runoff characteristics of the experimental catchment before and after land reclamation, with corresponding runoff values from the reference catchment. The period of hydrological observations for each of the three catchments (experimental catchment, main and auxiliary references) were assumed equal, and were divided into two periods, before and after land reclamation.

When comparing the characteristics of runoff for different periods, it was necessary to eliminate the effect of climatic variations during these periods. This was achieved by reducing the runoff during these periods to a standard means, using reduction coefficients. First, the runoff from the catchment after drainage was reduced to the same standard as for initial conditions, and the result obtained was compared with the actual runoff.

If the actual runoff of the catchment prior to and after land reclamation is denoted by subscripts 1 and 2, and the runoff from the reference catchment is shown by subscripts 1a and 2a, then the reduction coefficient will be:

$$K_r = \frac{R_{2a}}{R_{1a}} \tag{1}$$

and the reduced runoff of the experimental catchment for the period after reclamation will be:

$$R_{r2} = K_r R_1 \tag{2}$$

A change in the runoff agter reclamation may be written as:

$$\pm\Delta R = R_2 - R_{r2} \tag{3}$$

or

$$\pm\Delta R = R_2 - K_r R_1 \tag{3a}$$

The rate of change of runoff from the entire river catchment after reclamation may be expressed as:

$$K_{ch} = \frac{R_2}{R_{r2}} = \frac{R_2}{K_r R_1} \tag{4}$$

Results of the analysis are summarized in Table 59.

As can be seen from Table 59 the total annual river runoff increases after reclamation in five catchments, especially during the summer-autumn (June-November) and winter (December- February) flow periods. This increase can be attributed to a decrease in evapotranspiration during the summer-autumn low flow period due to the drop in the groundwater level, and to a decrease in groundwater storage in the winter low flow period.

It was also shown that after land reclamation, the spring runoff may either increase or decrease, depending on the combined effect of a number of counteracting factors, such as the storage available in the nonsaturated zone before spring floods and the existance of a developed artificial stream network. In some cases, such as the catchment of the Bobrik river, the decrease in spring runoff can exceed the augmentation of runoff which occurs during the remainder of the year.

Results from the main and auxiliary reference catchments are similar to those from all six larger catchments of Mukhavets, Merechanka, Bobrik in west Byelorussia, Vit in the east and Grebelka and Oressa in the north. This verified the pattern of changes in runoff found by the present and other authors.

Changes in the pattern of river runoff after land reclamation have been estimated from analyses of changes in the groundwater component in three catchments; variations in surface runoff were obtained as the difference between total and groundwater runoff.

Groundwater runoff is determined by analyzing the total runoff hydrographs using the techniques described by Drozd (1966); its changes are found from equations (1) to (4). The suitability of using reference catchments for this purpose has been examined by correlating groundwater runoff for the large catchment with that for the reference one, for the period before land reclamation.

Data on changes in the total runoff (Table 59) and its groundwater component gave the changes in surface runoff shown in Table 60.

Examation of Table 60 shows an increase in the groundwater supply to all three catchments after drainage. In one case the surface runoff has increased, while in the other cases it has fallen. The decrease in the total runoff of the Bobrik catchment is caused by a decrease in the surface runoff, which is similar to the decrease in the Oressa river. The augmented groundwater component due to reclamation improves the distribution of the river runoff. Similar changes in the groundwater component of runoff after reclamation can be demonstrated for all catchments by relating runoff to total precipitation, (Drozd, 1970).

The analysis shows that the total runoff in rivers is increased after drainage of the area due to a decreased evaporation loss. Changes in the surface runoff which normally occur in spring may be positive or negative.

Statistical methods have been used to evaluate the significance of disturbances to the uniformity of hydrological series using the non parametric criteria of Van der Varden and Willcockson (1968). When examining possible significant changes in the river runoff and its components, the null hypothesis is rejected with only a few samples, and the hydrological series are considered to be statistically non-uniform as a result of reclamation. In cases where changes are less marked, the null hypothesis is accepted and series is considered to be uniform.

Table 59. Change in the runoff after reclamation.

River, Gauging station	Absolute changes, mm			Rate of change (K_{ch})		
	Annual	Low flow period		Annual	Low flow period	
		Summer-autumn	Winter		Summer-autumn	Winter
Grebelka, Birchuki	1.9	4.5	2.6	1.02	1.25	1.22
Mikhovets, Pruzhany	31.9	13.8	15.5	1.35	1.87	1.11
Merechanka, Stavok	27.1	16.1	4.3	1.27	1.73	1.35
Oressa, Verkhutino	20.8	18.6	10.6	1.14	1.50	1.50
Vit, Borisovshchina	12.4	6.5	3.6	1.15	1.42	1.26
Bobrik, Parokhonsk	-7.1	3.5	4.2	0.93	1.16	1.31

Table 60. Changes in the runoff pattern.

Runoff characteristics	Runoff changes for the year after reclamation, mm		
	Oressa, Verkhutino	Vit, Borisovshchina	Bobrik, Parokhonsk
Total	20.8	12.4	-7.1
Groundwater	34.5	9.4	5.3
Surface	-13.7	3.0	-12.4

It has been shown that the total runoff series of the Grebelka, Merechanka and Bobrik rivers are uniform at the 1- and 5-% significance levels. Groundwater runoff is generally more likely to change, and its series is therefore non-uniform at both significance levels. Hydrological series of the surface runoff can be taken as uniform at a 1% significance level; at a 5% significance level the uniformity hypothesis is rejected for the Mukhavets and Vit catchments.

The present investigation demonstrates the following trends in the hydrological regime after land reclamation;

(a) during the first years after drainage annual runoff is increased due to a decrease in evapotranspiration and a reduction of centuries-old stores of surface waters; subsequently intensive agricultural use of reclaimed lands, leads to a less variable runoff pattern, evaporation increases and the annual runoff approaches the pre-drainage value and may even decrease slightly;

(b) spring flows either increase or decrease;

(c) the minimum runoff and summer low flows (the most critical period for the utilization of water resources of rivers) increases considerably;

(d) the proportion of groundwater component in the total runoff increases; this improves the distribution of river runoff and is a beneficial result of land reclamation.

Overall it is concluded that the reclamation of bogs and marsh-ridden lands when carried out on a scientific basis and strictly in accordance with engineering requirements has a favourable effect on river runoff, particularly on the important characteristic low flow which increases considerably. It should also be stated that the effect of drainage on the components of the water balance is often overestimated. Changes in these components sufficient to radically change the water balance of drained catchment are not likely.

REFERENCES

Bulavko, A. G. 1961. Vliyanie osysheniya bolot na elementy vodnogo balansa rek Belorusskogo Polesiya. Bog drainage effect on water budget components of rivers in Byelorussian Polessie. Moscow.

Bulavko, A. G. 1968. Novye dannye o vliyanii gidromelioratsii na elementy vodnogo balansa rechnykh vodosborov. New data on water reclamation effect on water budget components of river catchments. In: Kompleksnoye ispolzovanie i okhrana vodnykh resursov. Minsk.

Bulavko, A. G. 1970. Vliyanie osushitelnoi melioratsii na rechnoi stok v Belorusskom Polesie. Effect of drainage on river discharge in Belorussian Polessie. In: Vodnye resursy i ikh ispol'zovanie. Minsk.

Drozd, V. V. 1966. Opyt vydelenya podzemnogo stoka po gidrografu. Experience of individual estimation of underground runoff from a hydrograph. In: Ispolzovanie i okhrana vodnykh resursov Belorussii, pt 1, Minsk.

Drozd, V. V. 1970. O nekotorykh voprosakh statisticheskoi obrabotki podzemnogo stoka. Some aspects of statistical treatment of underground runoff. In: Vodnye resursy i ikh ispolzovanie, Minsk.

Maslov, B. A. 1971. O nekotorykh posledstviyakh osyshitlenykh melioratsii. On particular consequences of drainage reclamation. Gidrotekhnika i Melioratsiya, No 4.

Van de Varden and Willcockson, 1968. Tables of Mathematical Statistics, Vol. 46.

DISCUSSION

S. M. Perekhrest (Ukrainian SSR) - Your data show an increase in the runoff from drained and developed basins. Water balance calculations by water engineers and hydrologists imply that runoff and precipitation are not sufficient for water consumption, and that reservoirs need to be constructed. If you are right, the water engineers are wrong. Is this correct?

A. G. Bulavko (Byelorussian SSR) - I do not think the water engineers have made a mistake. A moisture deficit was also found in Polessie before reclamation. In this case precipitation is more important than runoff. Precipitation is harmless before drainage and important after drainage.

S. M. Perekhrest (Ukrainian SSR) - You said that deficits were also found before drainage. But at that time, bogs were of no importance in the economy of Byelorussia and production of sedge grasses was also of no importance. Drained bogs are of great significance to the national economy and hence smaller crop yields due to an insufficient supply of moisture may be harmful to collective farms. Is this acceptable?

A. G. Bulavko (Byelorussian SSR) - This cannot be allowed. Water engineers therefore design double control systems.

S. M. Perekhrest (Ukrainian SSR) - It can be seen from the paper by B. G. Shtepa, Assistant Minister of Reclamation and Water Economy of the USSR, that all future drainage systems will involve double operation. Is it possible to use your data for designing new systems?

A. G. Bulavko (Byelorussian SSR) - We have only studied the existing situation. If some changes occur, we shall start new research to provide the answer.

R. Eggelsman (FRG) - I am pleased to note that your results agree with the data of our hydrological work on marshes in the northwestern region of the FRG before and after drainage.

THE INFLUENCE OF BOGS ON THE DISTRIBUTION OF STREAMFLOW FROM SMALL BOG-UPLAND CATCHMENTS

E. S. Verry and Don H. Boelter

Forest Service, United States Department of Agriculture, Grand Rapids, Minnesota, USA

SUMMARY: The distribution of flow from small bog-upland catchments is governed primarily by the hydrogeological relationships; organic soils in the bog part of the catchment have only a minor effect. The seasonal distribution of flow from groundwater bogs is fairly uniform due to base flow from the groundwater basin. Flow from perched bogs is primarily concentrated in the spring and results from snowmelt and early spring rains. Maximum annual stream flow peaks from both types of catchment usually result from snowmelt and are a function of the water content of the snowpack and rain on snow rather than the hydrogeological relationship of the bog. Both types of bogs provide short-term regulation of storm-flows resulting from rainfall by reducing peaks and delaying the release of storm flow volumes.

INFLUENCE DES MARECAGES SUR LA DISTRIBUTION DE L'ECOULEMENT PROVENANT DE BASSINS DE RECEPTION DE PETITS MARECAGES DE PLATEAU

RESUME : La distribution de l'écoulement provenant des bassins versants des petits plateaux marécageux est principalement gouvernée par des rapports hydrogéologiques ; les sols organiques dans la partie marécageuse du bassin n'ont qu'un effet restreint. La distribution saisonnière de l'écoulement provenant des marécages qui sont au niveau de la nappe phréatique environnante est assez uniforme à cause de l'écoulement de base du bassin de la nappe phréatique. L'écoulement des marécages qui sont situés au-dessus de la nappe phréatique est principalement concentré au printemps ; c'est le résultat de la fonte des neiges et des premières pluies du printemps. Les crues maximales annuelles des cours d'eau des deux types de bassin résultent d'ordinaire de la fonte des neiges et sont fonction de la quantité totale de neige ou de pluie sur la neige plutôt que de la fonction hydrogéologique du marécage. Les deux types de marécages provoquent une régulation à court terme des chutes de pluie d'orage en réduisant les crues et en retardant la sortie des eaux.

The 15 to 16 million acres of organic soils in the Lake States of north-central United States range from relatively small lake-filled bogs of the Laurentian Shield and morainic areas to large 'built-up' bogs in former glacial lake basins such as Lake Agassiz (Heinselman, 1963). Although the 'built-up' bogs account for a large portion of all peatlands catchment studies have not been carried out in these areas because the watersheds are difficult to define.

In contrast, hydrological data are available from lake-filled bogs. These bogs are completely surrounded by uplands and usually make up less than half of their respective catchment area. The larger the bog, the larger the proportion of a catchment it usually occupies. This discussion is based on hydrological research at the Marcell Experimental Forest in northern Minnesota where lake-filled bogs range in size from 3.2 to 21.0 hectares and make up 12 to 33 per cent of their respective catchment areas, (Bay, 1966).

Lake-filled bog-upland catchments

Chebotarev (1962) and Bay (1970) have shown the importance of understanding the local hydrogeological systems of individual peat-lan in order to correctly interpret their hydrological value. Bay (1966) has described two hydrogeological types of lake-filled bogs: perched bogs and groundwater bogs. These terms are analogous to current ecological classifications based on the water chemistry of bogs (Gorham, 1966) (Heinselman, 1970).

Perched bogs are analogous to ombrotrophic bogs, meaning that their water is derived mostly from ion-poor precipitation. Perched bogs have developed in basins (ice block depressions) perched above and relatively independent of the regional groundwater system. Seepage to the groundwater system is negligible because of the nearly impervious layer of fine textured glacial till or lacustrine clay, and the low hydraulic conductivity of the peat itself (Boelter, 1965). Romanov (1968) also suggested that organic colloids can fill the pores of permeable mineral substratum resulting in a water-impervious layer.

Groundwater bogs are analogous to minerotrophic forested fens meaning that their water is derived mostly from mineral-enriched groundwater. In groundwater bogs the water table is simply an expression of the regional groundwater table and the bog is a discharge area for the groundwater system. Both bog types usually have outlets to surface streams which exhibit striking differences in their flow regimes due to the hydrogeological relationships of the two kinds of catchments.

Forested bogs in northern Minnesota are usually dominated by a ground cover of sphagnum moss (Sphagnum spp.Dillenius, Hedwig) and a stand of black spruce (Picea mariana Mill, B.S.P.), though northern white-cedar (Thuja occidentalis L.) and tamarack larix Laricina Du Roi, K. Koch) are not uncommon. Non-forested peatlands may have sphagnum moss with a cover of Ericaceae shrubs of lowland brush or be a reed-sedge meadow.

Seasonal distribution of streamflow

Much of the annual streamflow from catchments with perched bogs occurs prior to June 15; the distribution of flow from catchments containing groundwater bogs is much more uniform. These differences can be explained primarily in terms of the hydrogeological conditions. The bog itself has only a minor effect on the distribution of annual flow from either of the catchment types.

High spring runoff from perched bogs is the result of the melting snowpack and early spring rains. Even though the greatest amount of precipitation occurs during late spring, summer and early autumn, runoff is quite low at this time. This rainfall is quickly channelled back to the atmosphere by the process of evapotranspiration at the expense of runoff and deep seepage. Evapotranspiration increases with increased energy input from solar radiation and completion of the conductive link between the soil and air by the flush of annual leaves in mid-May and their persistence to early October (Figure 104). Average rainfall on the Marcell Experimental Forest for the period May 1 to November 1 is 546 mm and evapotranspiration estimates average 505 mm (Bay, 1967). Assuming no changes in storage, average runoff for the period is only 41 mm less than 8 per cent of the rainfall. The reduction of this very large evaporative loss of summer rainfall offers the possibility of increasing low streamflow through land management practices.

Long term runoff characteristics of a larger river basin in northern Minnesota containing extensive 'built-up' peatlands exhibit similar variations in flow (Winter, Maclay & Pike, 1967). Chebotarev (1962) and Vidal (1960) report long periods without flow from bogs in the USSR, GDR and FRG.

The seasonal distribution of flow from the groundwater bog is much more uniform, ranging from 14 per cent in the first quarter (January, February, and March) to 31 per cent in July, August, and September. The principle source of water is the groundwater basin and changes in flow can be related to groundwater levels illustrated by a typical upland well hydrograph (Figure 105).During winter there is a gradual decline in the level of the water table. With the beginning of snowmelt and disappearance of frost, there is a very quick, sustained high rate of recharge for the period April 8-24. The water table in this well is 1.2 to 1.5 metres below the surface in an area of sandy soils. The high recharge rate extended about five days beyond the disappearance of snow and reflects the time necessary for gravity flow to pass through the sands. From late April until mid-June there is another stable, though lesser rate of recharge. This reflects unsaturated drainage of the soil and parent material and may include some short periods of saturated recharge. By mid-June, however, the continuity of water flow to deep seepage is broken by the high evapotranspiration demands made on summer rains. This, coupled with discharge to surface areas, such as the bog, accounts for the steady decline of the water table until the next snowmelt spring rain recharge cycle.

There has been speculation that bogs, because of their high water storage capacity and flat slope, would regulate the distribution of streamflow. This does not appear to be the case with small lake-filled bogs in bog-upland catchments. Streamflow from catchments with perched bogs is not well distributed (Bay, 1969). They exhibit flow-duration curves (Figure 106) with steep slopes because streamflow and evapotranspiration loss deplete basins that have very little perennial storage. Winter, et al (1967) and Vidal (1960) also report that peat areas do not significantly affect streamflow during dry periods.

Streamflow from groundwater bogs is more uniformly distributed because the bog is a surface discharge point for the regional groundwater system, which provides a more constant supply of water. The flow duration curve in Figure 106 has very little slope, showing a nearly constant rate of flow for 70 per cent of the time, because of the large amount of perennial storage(Verry, 1970). The characteristic of the flow is again due to the hydrogeological situation; the bog has only a minor effect. Instead of regulating flow, the groundwater bog may do the opposite by releasing excess water more quickly than mineral aquifers during periods of high precipitation and using more water by high evapotranspiration during dry periods (Sander, -).

Stormflow

Although bogs generally have only minor effects on seasonal flow, small bogs do appear to modify storm flows. Typical storm hydrographs from perched bogs exhibit long drawn out recession curves (Figure 107, (Bay, 1969)). The recession leg of the dormant season hydrograph (bottom) approaches a straight line on semilogarithmic paper rather quickly and remains well above the pre-storm discharge level. This configuration indicates a temporary storage and slow release of stormflows due primarily to the nearly level bog topography and the large detention storage of surface peats. The recession leg of the growing season hydrograph (top) is broken by daily evapotranspiration losses that reduce streamflow. However, the storage and slow release of stormflow is still evident from the small decay rate during periods of low evapotranspiration.

No data are available for direct comparison of stormflow peaks to those from other catchments which do not have a bog component. However, the relatively high level and low decay rate of unbroken streamflow recession curves indicate that peak flows must be reduced in comparison with those that would occur without the temporary storage effect of the bog.

Peak flows are related to climatic and physiographic variables including the short term storage capacity of the catchment. In bogs, storage capacity is greatly dependent on the position of the water table in the peat profile. Therefore, storm peaks are directly related to the water table position (Bay, 1969). Maximum runoff occurs when water tables are high because there is little available storage capacity and water moves directly to the bog outlets. Surface peats also have high hydraulic conductivities and drain quickly; deeper peats are more decomposed, retain more water, and drain very slowly (Boelter, 1970).

Conclusions

The seasonal distribution of streamflow from small bog-upland catchments is governed primarily by the hydrogeological relationships. The organic soils in the bog portion of the watershed have only a minor effect. Contrary to popular belief, neither perched nor groundwater bogs have a regulating effect.

The flat topography and physical detention of flow by the organic soils do provide short term regulation of individual storm flows by reducing peaks and delaying the release of stormflow volumes. Maximum annual peaks do not appear to be a function of bog type, but rather of watershed size and water content of the snowpack and rain-on-snow.

References

Bay, R.R. 1966. Factors influencing soil-moisture relationships in undrained forested bogs. Int. Symp Forest Hydrol.p. 335-343. Pergamon Press, Oxford.

Bay, R.R. 1967. Evapotranspiration from two peatland watersheds. 14th Gen. Assembly. Inst. Union of Geod. & Geophys. Switzerland, p. 300-307.

Bay, R.R. 1969. Runoff from small peatland watersheds. J. Hydrol. 9(1): 90-102.

Bay, R.R. 1970. The hydrology of several peat deposits in northern Minn., U.S.A. Third Int. Peat Cong., Quebec, Canada. 1968, p.212-218.

Boelter, D.H. 1965. Hydraulic conductivity of peats. Soil Sci. 100 (4):227-231.

Boelter, D.H. 1970. Important physical properties of peat materials. Third Int. Peat Cong., Quebec, Canada. 1968.p.150-154.

Chebotarev, N.P. 1962. Theory of stream runoff. Moscow Univ.p.464 (Russian transl. TT65-50027).

Gorham, E. 1966. Some chemical aspects of wetland ecology. Proc. of the 12th Muskeg Research Conf., Tech. Memorandum No.90, 20-38.

Heinselman, M.L. 1963. Forest sites, bog processes, and peatland types in the Glacial Lake Agassiz Region, Minnesota, Ecol. Monogr. 33: 327-374.

Heinselman, M.L. 1970. Landscape evolution, peatland types, and the environment in the Lake Agassiz Peatlands Natural Area, Minn. Ecol. Monogr. 40 (2): 235-261.

Romanov, V.V. 1968. Hydrophysics of bogs. Gidrometeorologicheskoe izdatel'stvo, Leningrad. 299 p. (Russian transl. TT67-512289).

Sander, J.E. Bog-watershed relationships utilizing electric analog modelling. Mich. State Univ., Unpubl. Ph.D. Dissertation, 225 pp.

Verry, E.S. 1970. Water quantity and quality differs greatly between perched and groundwater bogs. Presented at Lake Superior Biological Conf., Superior, Wisconsin, Sept. 25-27.

Vidal, H. 1960. Vergleichende. Wasserhaushalts-und Klimabeobachtungen auf unkultivierten und kultivierten Hochmooren in Südbayern. Mitteilungen Für Landkulture, Moorund Torfwirtschaft 8: 50-107.

Winter, T.C., Maclay, R.W., Pike, G.M. 1967. Water resources of the Roseau River watershed, northwestern Minnesota. Hydrol.Invest. At las HA-241, Dept. Interior, U.S. Geol. Survey.

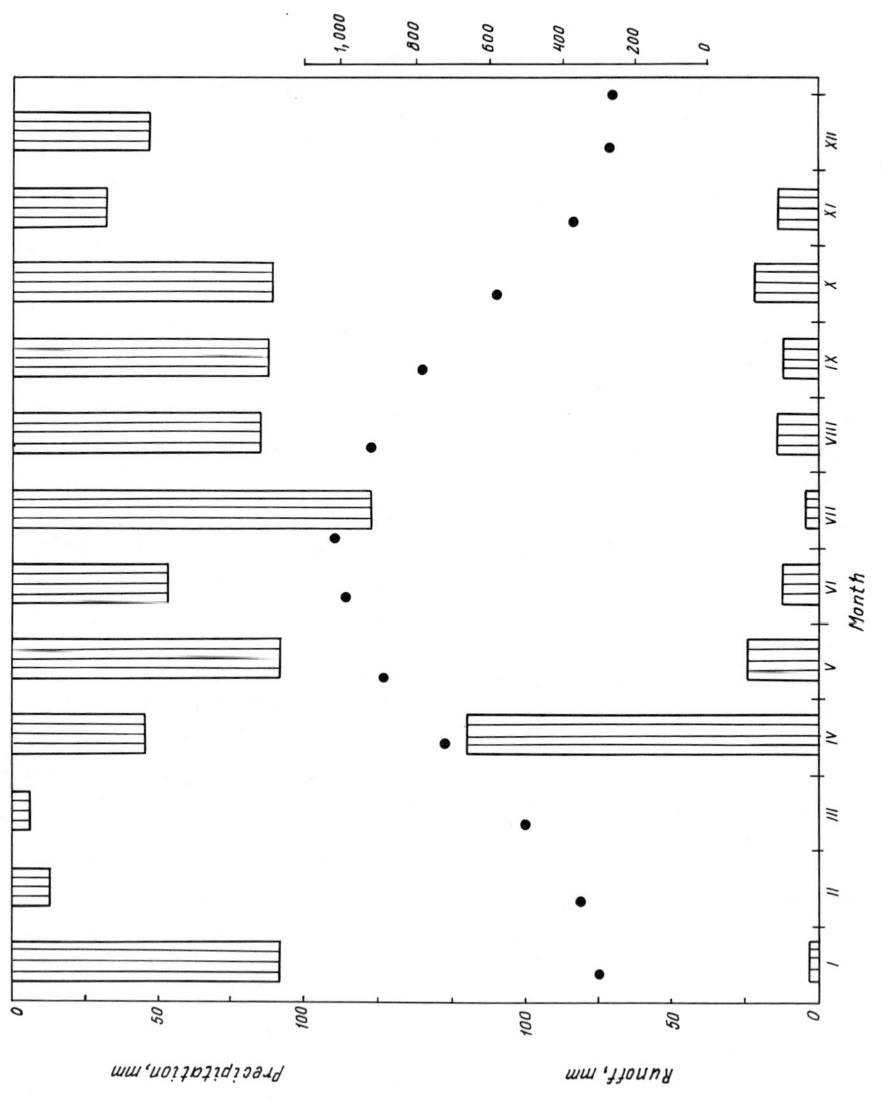

Figure 104. Monthly precipitation-radiation-runoff relationships for a perched bog for 1969.

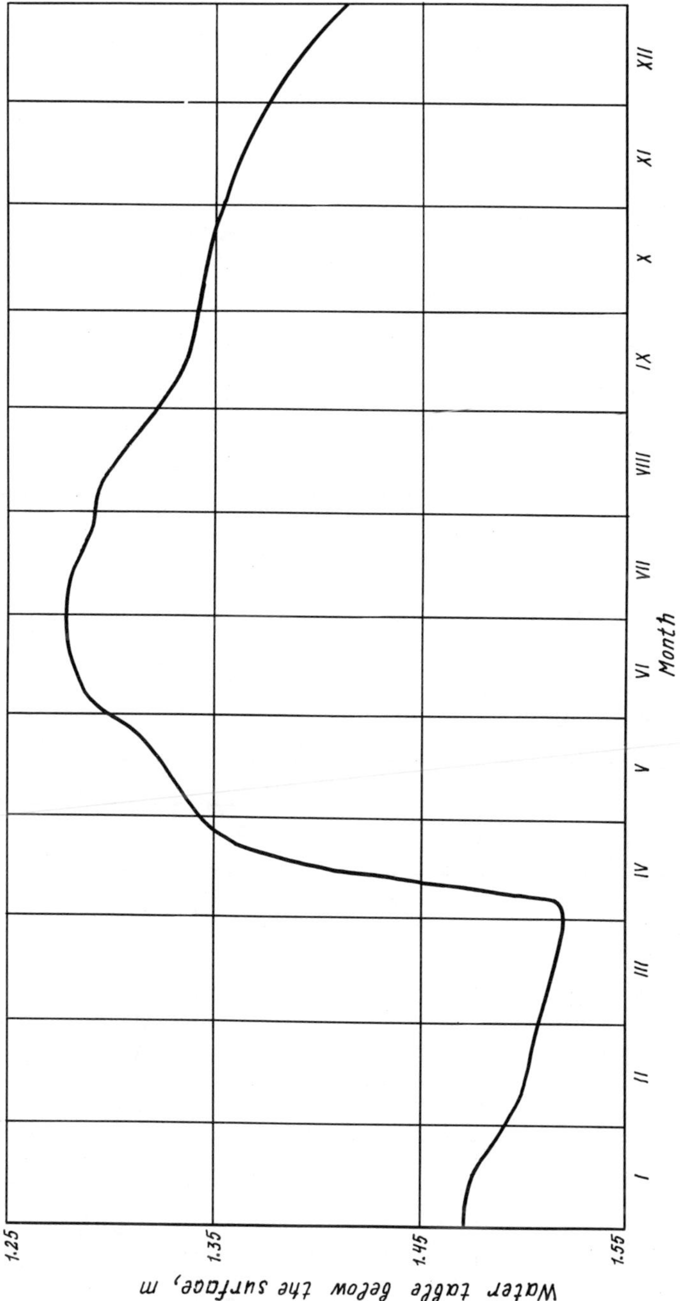

Figure 105. The 1969 hydrograph of the regional groundwater table.

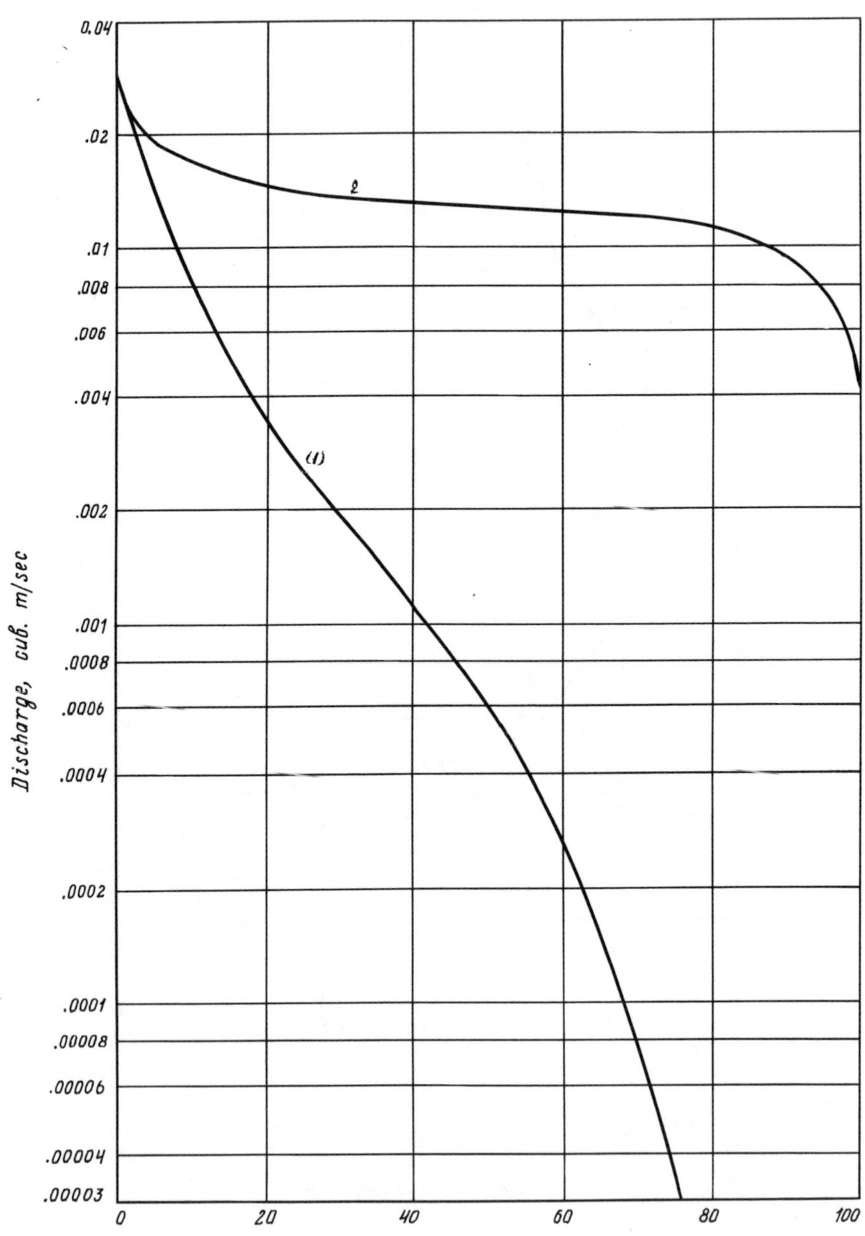

Figure 106. Flow duration curves for perched and groundwater bog catchments of similar size.

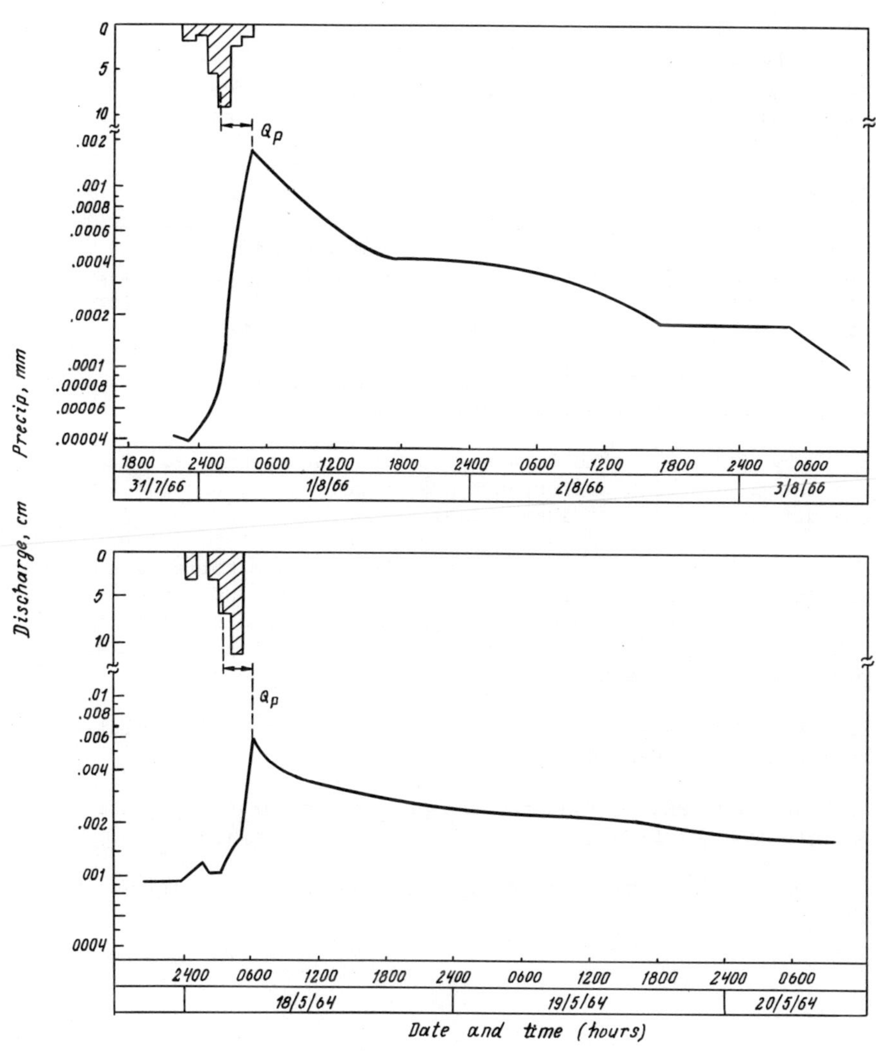

Figure 107. Hydrographs and accompanying hyetographs for two storms from a perched bog catchment.

DISCUSSION

L. G. Bavina (USSR) - In your paper you have reported evapotranspiration of 505 mm. In what way did you obtain this value? Did you make calculations or did you make measurements?

E. Verry (USA) - The value is calculated from the water-balance equation. We tried to measure the evapotranspiration and designed and installed a special container in peat soil. The results obtained were close to the predicted values, but these measurements are too labour-consuming.

E. D. Vvedenskaya (USSR) - 1. Does the groundwater level begin to increase after or before the snowmelt? 2. Have you made any experiments on snowmelt water infiltration into frozen grounds and if so, can you give any results?

E. Verry (USA) - 1. The hydrograph was determined for a mineral portion of the catchment. The groundwater levels are found to increase after snowmelt. 2. No experiments have been made on snowmelt water infiltration into frozen soils. Soil becomes frozen with no snow cover. Infiltration seems to be smaller in frozen soil, but we have not made any measurements.

S. M. Novikov (USSR) - 1. How many catchments have you studied? 2. How do swamps affect the streamflow?

E. Verry (USA) - 1. The data are for one catchment. 2. On six catchments with different types of soils we could not find any differences due to swampiness.

U. Schendel (FRG) - You have stated that organic soils have little effect on the streamflow rate. I cannot understand this. Peat soils are known to have a high moisture-holding capacity, thus the streamflow may be retarded, as water fills the numerous pores, especially after dry period.

E. Verry (USA) - I suppose that the seasonal streamflow distribution is not affected by bogs.

R. Eggelsman (FRG) - What are the proportions of the surface and groundwater flow?

E. Verry (USA) - The quantity of seepage water is hardly measurable.

THE INFLUENCE OF AGRICULTURE ON EUTROPHICATION IN LOWLAND AREAS

H. Kuntze and R. Eggelsmann

Institute for Peat Research and Applied Soil Science, Bremen, FRG.

SUMMARY: Hydrological and chemical investigations were made in four lowland catchments with sand and peat soils. Results over the last ten years show a relatively high content of lime, which is accompanied by a rising yield. Potassium, phosphorus and nitrogen seem rather mobile. Their contents in water are higher in wet periods than in dry periods, but are not paralleled by a rise in yield.

INFLUENCE DE LA MISE EN VALEUR AGRICOLE SUR L'EUTROPHICATION DE L'EAU PAR LA LEXIVIATION DES SUBSTANCES NUTRITIVES DANS LES PLAINES BASSES

RESUME : Des recherches hydrologiques et chimiques ont été effectuées dans quatre bassins versants au sol tourbeux et sableux. Les résultats obtenus pour les dix dernières années montrent que la teneur en chaux devient relativement élevée avec l'accroissement des récoltes. Les quantités de potassium, de phosphore, d'azote sont plutôt variables. L'eau en contient des proportions plus élevées dans les périodes de pluies que dans les périodes sèches, mais ces proportions ne s'accroissent pas avec les récoltes.

Introduction

In the last decades pollution of creeks and rivers by industry and towns has increased. During this time the application of fertiliser has led to an increase in yield. The important question therefore is whether nutrient leaching increases with fertilising.

Previous investigations of nutrient leaching were usually made in lysimeters. These results however gave little information about eutrophication of river and groundwaters, because the hydrological conditions within a lysimeter are not comparable.

For more than ten years the authors have made chemical water analyses as part of hydrological investigations.

Geography

Investigations were made in four catchments which lie between Hamburg and Bremen in the Lueneburger Heide on Pleistocene sands about 40 m above sea level. Table 61 gives their area and soil type. Polders are catchments which have insufficient slope for natural drainage and are therefore drained by pumping.

About 45% of the Wuemme watershed belongs to the Natural Reservation Park at Wilsede. There is no industry in any of the four watersheds. The density of population is about 15 to 20 inhabitants per km^2. Agriculture is predominantly arable land, grassland and forest.

Table 61. Area and soils of catchments

Watershed	Area km^2	Predominant soils
Wuemme	96	Podzol and hydropodzol
North Creek	3	Podzol and peats
Polder "Sand"	0.4	Sand mixed peat cultivation
Polder "Peat"	0.4	Highbog cultivation

Application of fertiliser

The average rate of fertiliser application for the period 1960-70 on arable land and grasslands is given in Table 62. The crop rotation in the lowland area consists of grain crops, potato, winter grain crops, ley farming.

Table 62. Average rate of fertiliser application (1960/70) for Northwestern Germany Lowland

Crop	Rate of application of fertiliser, kg/ha				Farmyard manure*
	CaO	P_2O_5	K_2O	N	
Arable land	275	55	120	80-120	30 000
Grassland	215	45	130	100	10 000

* every third year

It is suggested that the rate of fertiliser application will be equal to the losses of nutrient due to harvesting and leaching. Ca^{++} is mostly contained as a compound in P- and N- fertilisers.

Methods of investigation

In the catchments rainfall, runoff, evaporation, groundwater levels and water storage were measured. The values of these hydrological variables were determined on a monthly, half-year and yearly basis.

Monthly or quarterly water samples were taken from every catchment. These were analysed to determine the ash and organic matter content, pH, and the content of Ca,P,K and N, and sometimes of Mg, Mn, Cu and B. Then the quantities of nutrients lost by leaching were calculated from the chemical analysis and the rate of runoff.

Discussion of results

The waters of the four watersheds are of different acidity (Table 63) depending on the proportion of peat and peaty soils. The greater the proportion of peat, the larger is the organic content and the lower is the acidity.

Table 63. Content of organic matter and pH-value in waters of lowland watersheds

Watershed	Predominant soils	Organic matter mg/l	pH
Wuemme	Podzol, hydropodzol	42	7.2
North Creek	Podzol and peats	61	4.9
Polder "Sand"	Sand mixed peat cultivation	83	4.7
Polder "Peat"	Highbog cultivation	99	4.3

During the ten years after the land was improved the yields from arable land and grassland have increased due to reclamation, better tillage, seed and fertiliser application (Figure 108).

In the same period the contents of P,K and N in the waters did not show a parallel increase (Figure 109).

The waters which are rich in humic acid have a noticeable N content, although there is almost no other source of N pollution than agriculture. N comes into the waters both from peat and peaty layers and from fertilisers, but their relative proportions are as yet unknown. It appears that N contents are higher in wet periods than in dry periods (Figs. 2,3,4). K contents are relatively high, because adsorption of K from fertilisers in the sand and peat soils is small. In summer (Figure 110) and in dry periods the K content of water is smaller than in winter or in wet periods (Figure 111).

In summer P contents were less than 0.5 ppm and range in winter between 1 and 2 ppm. Adsorption of P in sand and peat is less than in other mineral soils. P in peats is very mobile, and a very small quantity of phosphorus is required to promote immediate algal growth. The growth of blue algae is greatest in May and June, provided the sun's radiation is strong enough.

On average the ratio of N : P : K in water samples is of the order of 1: 0.1 : 1, as the mass production of weed is limited.

In general the greatest nutrient contents occurred in rainy periods with high runoff (Figure 111), while the nutrient contents in water in dry periods was correspondingly less.

The relative quantities of nutrients leached, decrease in the order Ca - K - N - P.

Conclusions

It is obvious that in the agricultural areas investigated the content of Ca, K, N and especially P in water is relatively high. If these nutrients are very mobile and easily leached in soils with a high content of organic material, particularly in acid peat soils, the type and quantity of fertilisers should be limited to the amount of these substances used by plants, in order to avoid eutrophication of waters.

References

Baden, W. 1963. Sorption and Wanderung von Ca,P,K und N in Moorböden. Int.Ass.Symp.Hydrology Berkeley, Publ.Nr.65,pp.80-92

Baden, W. 1965. Die Dalkung und Düngung von Moor und Anmoor. In Scharrer-Linser:Handbuch der Pflanzenernährung und Düngung. Bd. III.Springer-Verlag, Wien-New York.

Eggelsmann, R.; Kuntze, H. 1972. Vergleichende chemische Untersuchungen zur Frage der Gewässereutrophierung auf landwirtschaftlich genutzten Moor - und Sandböden. Landw.Forschung. Frankfurt (in press).

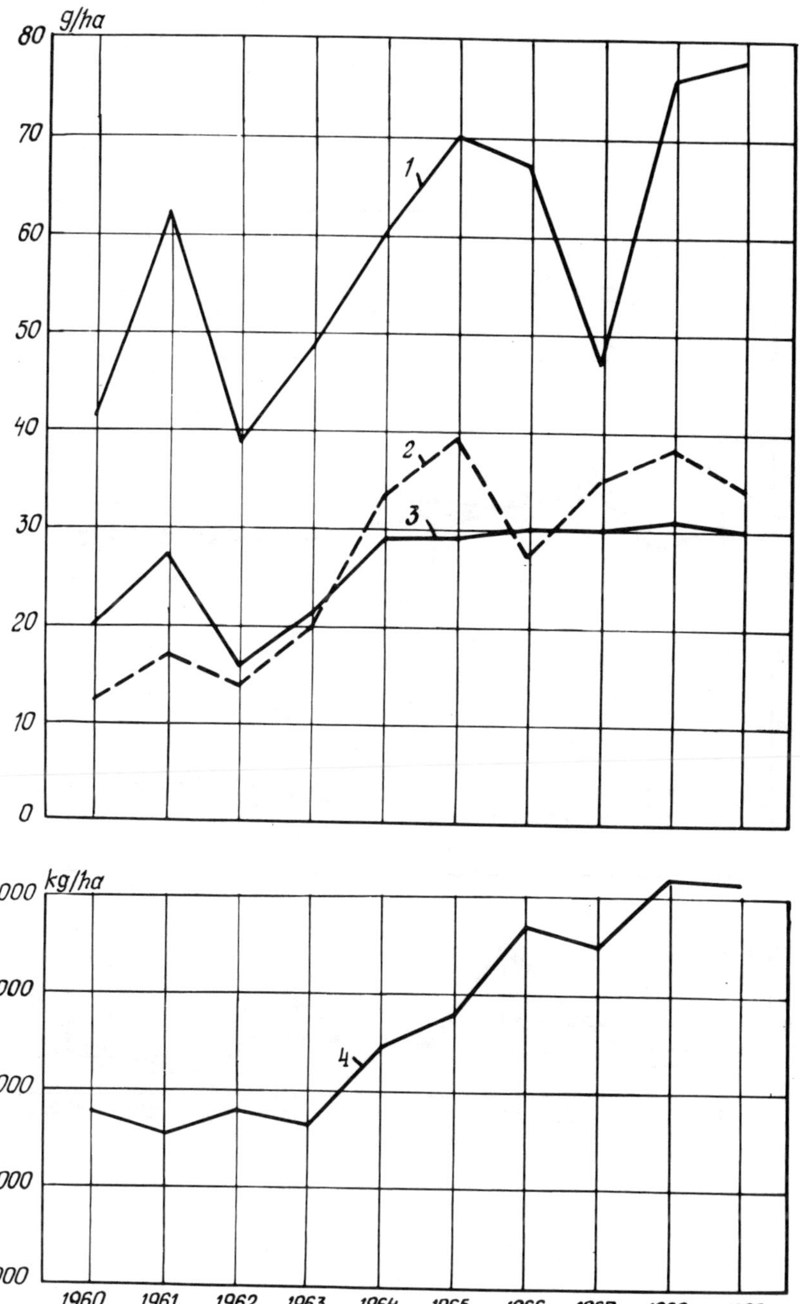

Figure 108. Yield of arable land and production of milk (1960/69) in the Koenigsmoor Experimental Farm.

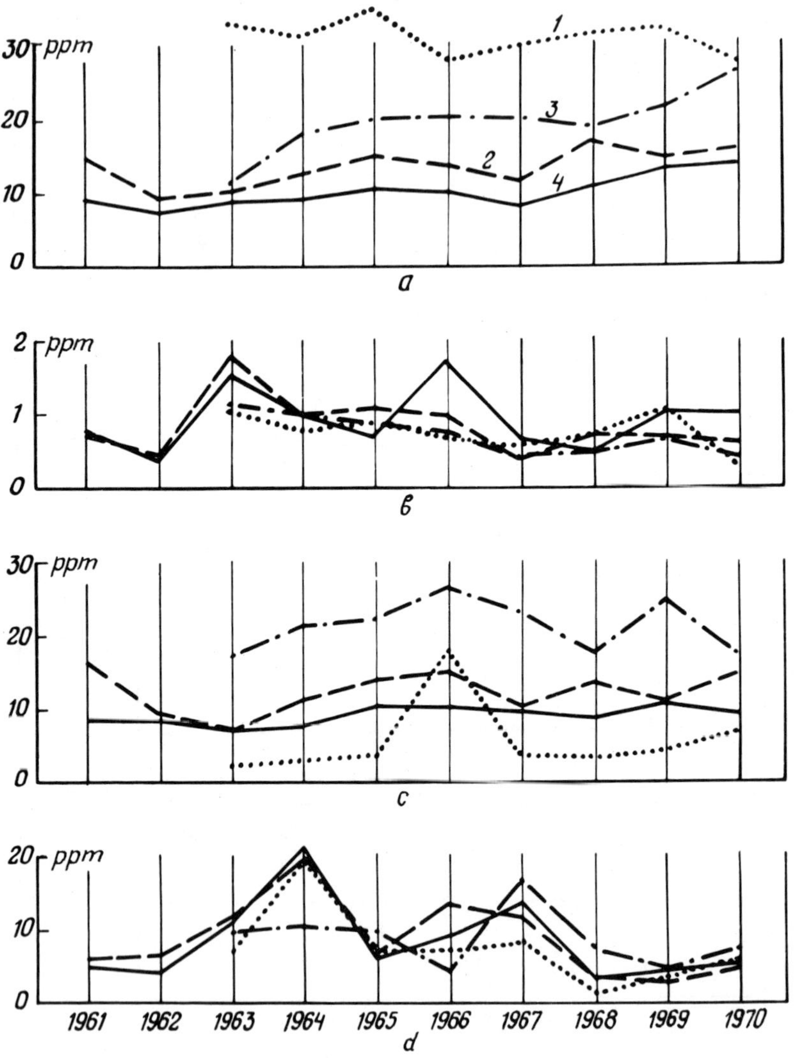

Figure 109. Mean annual contents of Ca, P, K and N in the water of four catchments in the lowlands of north-west FRG.

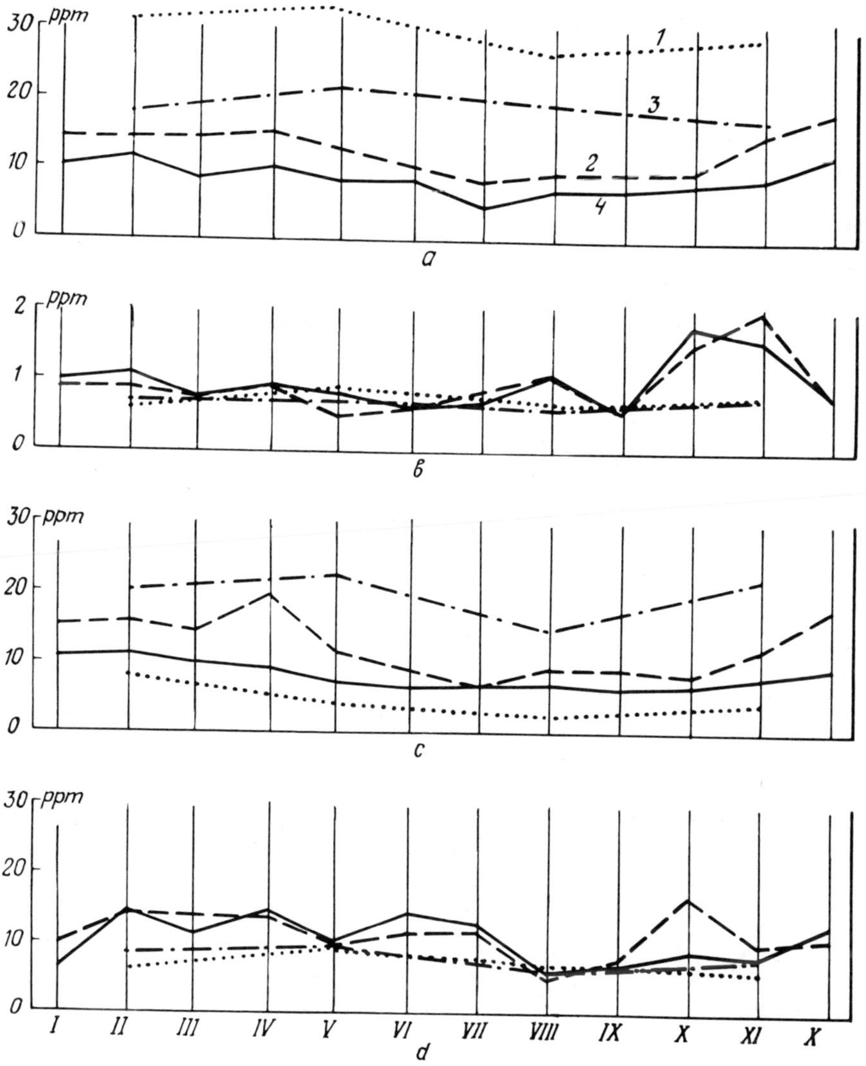

Figure 110. Mean monthly content (1961/67) of Ca,P,K and N in the water of four catchments in the lowlands of northwest FRG.

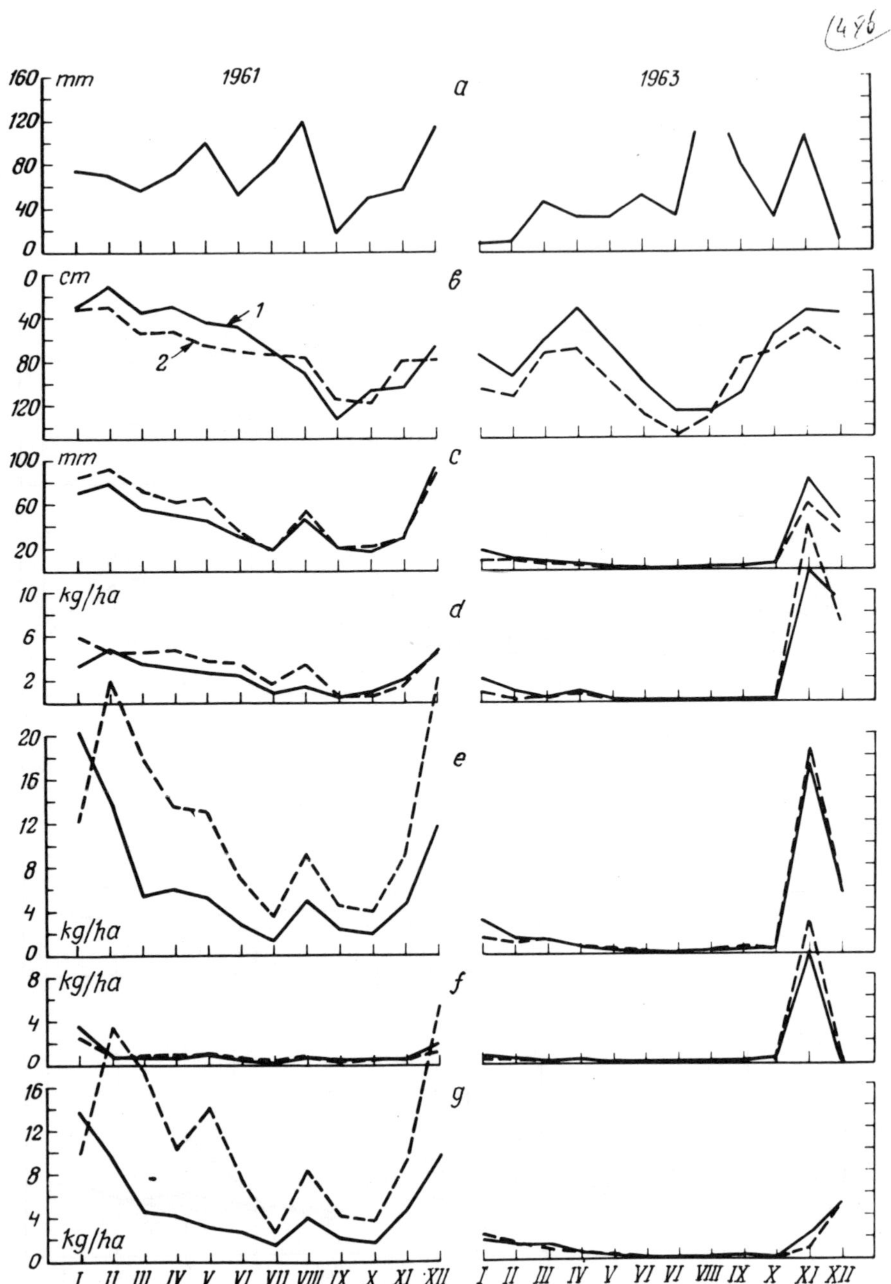

Figure 111. Water storage and leaching of nutrients in peat and sand soils.

THE INFLUENCE OF FEN BOG DRAINAGE ON ANNUAL RUNOFF IN THE ESTONIAN SSR

K. T. Hommik and G. I. Madissoon

Estonian Research Institute of Agriculture and Land Improvement, Saku, USSR.

SUMMARY: The effects of draining bog and excessively wet mineral soil on the quantity of annual runoff are studied by comparing the runoff calculated from meteorological factors with measured runoff. Differences between the actual and the calculated runoff are correlated with the percentages of the watershed which are bog, excessively wet mineral soil, drained soil and other soils.

The quantities of annual runoff and of the other components making up the water balance of fen bogs are studied on experimental areas subject to different drainage rates. From the research, the effects of drainage on annual runoff, on the exchange of water with the underlying ground in particular months of the growing season, and on the normal annual runoff from drained fen bogs are determined; the last results are the basis of hydrological calculations used for designing drainage systems in the Estonian SSR.

INFLUENCE DE L'ASSECHEMENT DES MARAIS SUR L'ECOULEMENT ANNUEL EN ESTONIE

RESUME : Les auteurs étudient l'influence de l'assèchement des marais et des sols minéraux excessivement humides sur les valeurs de l'écoulement annuel ; pour ce faire ils comparent le taux d'écoulement mesuré d'après les facteurs météorologiques et les valeurs jaugées au limnimètre. Les différences entre l'écoulement effectif et l'écoulement calculé sont en corrélation avec l'étendue des surfaces marécageuses, l'importance des sols minéraux excessivement humides, les sols drainés et autres dans les bassins hydrographiques.

La valeur de l'écoulement annuel et les autres facteurs du bilan hydrique des marais ont été étudiés sur des terrains expérimentaux soumis à diverses intensités de drainage. Les résultats obtenus ont permis de déterminer l'influence de l'assèchement sur l'écoulement annuel, sur les échanges des eaux de surface avec les eaux souterraines durant chaque mois de la période de végétation et l'écoulement moyen annuel des marais drainés, ce dernier facteur servant de base aux calculs hydrologiques pour l'élaboration des projets de systèmes de drainage dans la RSS d'Estonie.

Investigations of the effects of draining fen bogs and excessively wet mineral soils on the annual runoff have indicated the need to determine the effects of other physiographical factors on runoff. For this purpose measured values of annual runoff Y_f were compared with the runoff Y_1 determined from meteorological data.

The difference $\Delta Y = Y_f - Y_1$ was correlated with parameters relating to the physiograph of the watershed

$$Y_1 = X - Z$$

Where: x annual precipitation;
Z annual evaporation determined from meteorological data.

Z was determined by the formula (1) (attributed to Oldekop), to calculate the annual runoff and water balances of Estonian rivers, (Hommick, 1958):

$$Z = Z_{max} . \tan h \frac{X}{Z_{max}} \qquad (1)$$

where: $Z = \alpha d$

d mean annual saturation deficit of the air;
α coefficient which is constant for the given period.

The coefficient α was determined from equation (1) by an iterative method for watersheds having data on streamflow (Y), precipitation (X), evaporation (Z = X - Y) and air saturation deficit (d), and with no significant inflow or outflow of groundwater. The mean calculated value of the coefficient α was used in equation (1) to determine evaporation from meteorological data.

By correlating the differences between measured mean annual runoff and that determined from meteorological factors (ΔY_m) with parameters relating to the physiography of watersheds, the following regression equation for conditions in the Estonian SSR was obtained:

$$\Delta Y_m = 0.11A_b + 0.43A_h + 0.56A_u + 0.83A_{bd} + 0.05B_d + + 0.97B_u - 0.31C + 9.6_{q95\%} - 31.6 \qquad (2)$$

which has a multiple correlation coefficient, R = 0.90

Where: A_b percentage area of fen bog in the watershed;
A_h percentage area of high bogs;
A_u percentage area of excessively wet soils with forest bush cover;
A_{db} percentage of intensively drained fen bogs (for agricultural use);
B_d percentage of woods on soils of normal wetness;
B_u percentage of woods on excessively wet soils;
C percentage of open soils of normal wetness and intensively drained mineral soils;
$q_{95\%}$ mean minimum runoff, i.e. value of daily runoffs which lie below the 95% chance of exceedence level.

For a watershed wholly covered by fen bog the difference is negative and equal to - 20.6 mm. In watersheds which comprise high bogs or excessively wet mineral soils covered with brushwood, ΔY_m is also small, but positive. A considerable increase of ΔY_m is found in watersheds with stunted forest cover on excessively wet soils and under fen bog which is intensively drained. After drainage of fen bogs for agricultural use, the annual runoff is expected to increase under the climatic conditions of the Estonian SSR by an average 72 mm. Intensive drainage of mineral soils covered with brushwood, however, gives an average decrease of annual runoff of 87 mm.

Equation (2) relates to the average effect of each parameter on the quantity of runoff, and the values given may vary widely. For example, the factor for the percentage of forest on excessively wet soils, B_u, depends on the volume of wood per hectare. When this volume is equal to or greater than 100 m^3/ha, B_u tends to zero, whereas for a volume of about 30 m^3/ha or less, it approaches unity. In the case of fen bogs of small surface slopes (<0.002), the factor A_b tends to zero or can even be negative, whereas for slopes >0.005, it increases to 0.5.

The difference ΔY also depends on the climate of the year in question. For example, when the annual runoff has only a 5% chance of being exceeded (5% frequency), the equation for the difference is:

$$\Delta Y_{5\%} = 0.46A_h + 1.57A_u + 1.69A_{db} + 1.46B_u - 0.51C - 13.6_{q95\%} - 24.1 \qquad (3)$$

(R = 0.81)

whereas when the annual runoff has a 90% chance of being exceeded (90% frequency) the equation is:

$$\Delta Y_{95\%} = - 0.61A_b - 0.05A_h - 0.24A_u + 0.31A_{db} + 0.33B_d + 0.61B_u + 25.8_{q95\%} - 25.5 \qquad (4)$$

(R = 0.82)

Equations (2) to (4) show that excessively moist soils affect the quantity of annual runoff in different ways depending on the climate of that year. Whereas for a moderately wet or wet year, an increasing percentage of wet soils increase the annual runoff in a dry year an increasing percentage of these soils tend to reduce runoff and increase evaporation.

Equation (4) shows that after fen bogs are drained the annual runoff of a dry year would increase by 92 mm which is 20 mm more than the mean annual runoff. When draining mineral soils covered with brushwood, an increase in the annual runoff of a dry year and a corresponding decrease in the evaporation of 24 mm is to be expected. For a year with average climate a decrease in runoff of 87 mm is to be expected.

The evaporation values determined by the radiation balance method have shown that for a dry year (90% frequency), evaporation from drained soddy gley soils and from drained peaty gley soils during the growing period was 19 mm and 35 mm less than from undrained gley soils with brushwood cover, respectively, (Hommik, 1966). Evaporation measurements made by the radiation balance method on drained and undrained fen peaty soils for a wet year (10% frequency) have shown that during the growing season evaporation from a drained meadow area is 115 mm higher than from an undrained area. According to equation (2) runoff from drained fen bogs is 169 mm greater than that from undrained fen bogs. Thus, in a wet year, drained fen bogs must receive a groundwater inflow of about 284 mm.

To make a detailed study of the water regime of drained fen bogs and the effects of the rate of drainage on particular components of the water balance, experimental areas have been set up on two bogs 'Pikavere' and 'Kärde'. In one experimental plot, drain spacings were varied, but the same depth was maintained throughout. In the other area the depths of the drains were varied and spacings were equal. In the areas with different drain depths, all components of the water balance were measured, and in areas with different drain spacings runoff and groundwater level were measured. Measurements at 'Kärde' were made by the staff of the Bog Station 'Tooma' of the Hydrometeorological Service Board of the Estonian SSR, and on 'Pikavere' by scientific workers of the Estonian Research Institute of Agriculture and Land Improvement.

On all plots the peat layer was over 2 m thick before draining, but hydrometeorological conditions on the two areas was different.

The following equation was used to determine the water balance for the experimental areas:

$$X - Y - Z \pm W \pm S = 0 \qquad (5)$$

where: W change of water storage in a 1 m layer;
S groundwater flow.

The average flows during the period of observation were obtained as the residual term in equation (5) and are given in table 64.

Analysis of the results reveals the following patterns:

1. The groundwater inflow is larger in the case of more intensive drainage. In the Pikareve experimental area the annual inflow to the plot with less intensive drainage was 35% of the inflow to the plot with more intensive drainage and in the Kärde area 66% of the inflow to less well drained areas.

2. The largest inflow during the growing season is in June.

3. The net inflow during the growing season (June to October) is considerably larger than the inflow in the winter period, and represents from 66 to 122% of the mean annual inflow. As the run-

off from June to October is relatively small (on average 9% of the mean annual runoff), the entire inflow is lost to evaporation. In winter the runoff increases due to groundwater inflow.

4. The groundwater inflows given in the table agree with the average value obtained from equation (3) and with measured evaporation values on drained and undrained bogs.

To clarify the relation between the runoff volume and drainage intensity, the data from all experimental drainage areas have been analyzed. Average curves have been plotted of specific discharge versus the groundwater level within the drains. Comparing these curves with the so-called normal curve (a discharge curve

Table 64.

Experimental area	Drain depth (m)	Inflow Outflow	Months							Total
			Nov-May	June	July	Aug	Sept	Oct	June-Oct	
Pikavere	0.90	Inflow		78		28		6	88	72
		Outflow	16		14		10			
	1.20	Inflow	54	119	20	26		5	150	204
		Outflow					20			
Kärde	1.10	Inflow	52	90	54	21	34		180	232
		Outflow						19		
	1.25	Inflow	165	90	43	36	34		188	353
		Outflow						15		

that corresponds to the normal drainage intensity established for the Estonian SSR), drainage rate factors were found:

$$f = \sqrt{\frac{q}{q_n}}$$

where: q measured (actual) specific discharge;
q_n specific discharge for the normal curve, which corresponds to the groundwater level at which q was measured.

The factor f represents the difference between the actual drainage rate and the standard one.

The mean annual discharges of the experimental areas were expressed as a function of f by the equation

$$q_m = 0.097f + 0.011$$

for which the correlation coefficient was 0.76.

On average, the normal drainage f = 1 is equivalent to a normal annual discharge of 0.108l/s ha or 340 mm per year. Depending on the local factors, groundwater flow may vary.

The average normal annual depth of runoff Y_ℓ found from meteorological data for all catchments in the Estonian SSR is 285 mm.

The difference ΔY_m for drained fen peat soil is 51 mm (from equation 2). As $Y = Y_\ell + \Delta Y_m$, the normal annual depth of runoff from drained fen bogs has an agerage value of 336 mm. Thus, the results of more detailed investigations of runoff from drained bogs confirm the conclusions drawn from equation (2). As a result, a normal discharge of 0.108 l/s ha has been taken as the basis for planning drainage systems for fen bogs in the Estonian SSR. (A number of discharges used in hydrological design are usually expressed in terms of the normal annual runoff). A considerable (300 mm and more) annual groundwater inflow to drained fen bogs is observed. Its value depends on hydrogeological conditions and drainage rate, and is lost by evaporation.

Analysis of the differences between the actual annual depth of runoff and that determined by meteorological factors enables the effects of catchment characteristics and drainage on the annual runoff to be determined.

REFERENCES

Hommik, K. I. 1958. Vliyanie osusheniya na rezhim stoka v usloviyakh Estonskoi SSR. Effects of drainage on runoff regime under conditions of the Estionian SSR. Tallin.

Hommik, K. I. 1966. Osnovy rascheta osushitelnykh sistem. Elementary designing of drainage systems. Tallin.

DISCUSSION

G. P. Kubyshkin (USSR) - What method have you used to obtain the runoff from climatic factors?

K. T. Hommik (USSR) - The runoff has been found from measured precipitation and evaporation using Oldekop's formula.

G. P. Kubushkin (USSR) - What is the difference between the calculated runoff and the actual value as a percentage?

K. T. Hommik (USSR) - No considerable differences have been found. Deviations are only observed for particular catchment areas.

HYDROLOGICAL CHANGES CAUSED BY FOREST DRAINAGE

L. Heikurainen

Academy of Sciences,
Tapida, Finland.

SUMMARY: The drainage of peatland for timber production is being performed on an extremely large scale in Finland; by 1972 a total area of about four million hectares had been drained, and the annual area drained in the late 1960's was between 200 000 and 300 000 ha.

This paper deals with possible post-drainage changes in the magnitude and variations of the runoff, in the humus content of the drained water, in the moisture characteristics of the mineral soils surrounding the drainage areas, and in the groundwater resources. It is concluded that the hydrological changes caused by drainage are a function of the time that has elapsed since draining occurred.

MODIFICATIONS HYDROLOGIQUES CAUSEES PAR LE DRAINAGE FORESTIER

RESUME : En Finlande, le drainage des tourbières est effectué à grande échelle en vue de l'exploitation des forêts ; jusqu'en 1972, près de 4 millions d'hectares au total ont été drainés, la moyenne annuelle des surfaces drainées pendant les années soixante étant de 200 000 à 300 000 hectares.

Cette étude traite des modifications qui en résultent pour le volume d'écoulement et ses variations périodiques, pour la teneur en humus de l'eau d'écoulement, pour la teneur en eau des sols minéraux environnants aussi bien que pour les ressources en eaux souterraines. On en conclut que les modifications hydrologiques causées par le drainage sont à étudier en fonction du temps écoulé depuis que le drainage a été effectué.

Extent of forest drainage in Finland

Before World War II peatlands and paludified forests in Finland were drained exclusively by hand. There was sufficient labour available, and as sufficient funds were reserved for this work, the total area drained in the best years reached a figure as high as 80 000 hectares. According to statistics, a total area of 700 000 hectares was drained during the period 1928-39.

During the war, forest drainage was, of course, not practised, and it was only in the 1950's that activities were resumed. New methods were developed: first, the digging of ditches, and in the mid 1950's, ploughing. The funds reserved for forest drainage were also increased, and the following table shows the development of forest drainage activities during the 1950's and 1960's. The figures refer to the areas drained.

1950 ca	10 000 ha	1956 ca	62 000 ha
1952 "	15 000 "	1958 "	81 000 "
1954 "	40 000 "	1960 "	115 000 "
1962 "	145 000 "	1968 "	270 000 "
1964 "	183 000 "	1970 "	295 000 "
1966 "	233 000 "		

By the beginning of 1972 a total area of 4.0 million ha of peatlands and land containing excess water had been drained. There is considerable divergence of opinion as to areas which should be drained in the future. Some people consider that the area of still drainable peatlands and paludified mineral soils in Finland is 4.0 million ha, and stress that this area should be drained as quickly as possible. Other people maintain that the area in which drainage will be economical is considerably smaller, for example, slightly less that 3.0 million hectares, and that a relatively large part of this area should be set aside for nature conservation. It is evident that drainage activity will continue during the 1970's and even in the 1980's at a rate of 200 000 to 300 000 ha annually. Consequently, in the late 1980's, the area drained for forestry purposes in Finland will be between six and eight million hectares.

Introduction

Due to the high rate at which forest drainage is now being carried out, the bulk of the Finish peatlands will be drained during the next 15 to 20 years. As, originally, the peatland area of the country covered about one third of the total land area, it is clear that the hydrological consequences of such an activity are of the greatest practical importance. Deep concern for the consequences of forest drainage has arisen and this can be readily understood when previous concepts of the hydrological role of peatlands are considered. It was thought that peatlands retain the water produced by melting snow, and from heavy rains in the autumn and then gradually discharge it into brooks and rivers. Despite the fact that recent research has shown the assumed regulating effect of peatlands on the runoff to be erroneous, (Boden and Eggelsmann, 1964; Bay, 1967 and 1970; Boelter, 1966; Ferda, 1966), the concept still persists and influences opinion.

Although natural peatlands do not regulate runoff to any great extent, large-scale drainage might lead to harmful hydrological changes. The following is a brief survey of the most important changes in the hydrology following drainage. The examination is based primarily on studies already carried out or current in Finland and deals with the following topics:

1. the amount of runoff and its seasonal variation;
2. the humus content of the discharged water;

3. the water relationships of the mineral soils surrounding the drainage area;
4. the groundwater resources.

Effect of drainage on runoff

On the basis of a long-term study of the influence of forest drainage on runoff, Mustonen and Seuna (1971), indicated that forest drainage increased the annual runoff by 95 mm, or 43%. In absolute terms, the increase was greatest during the snowmelt period, and in terms of relative values, in the winter and during other periods of low runoff. According to Multamäki (1962), the annual runoff increases due to drainage, whereas the maximum runoff decreases.

Both of these studies dealt with recently drained areas (within the last 10 years), and they lead to the conclusion that drainage increases the runoff during the first years after drainage. The increase in the runoff is particularly marked during low runoff periods, but the maximum runoff may also increase. The effect of drainage on runoff 20 to 50 years later, which may be regarded as the permanent influence of drainage on runoff, can only be estimated on a theoretical basis. From various sources of information, it has, however, been concluded that the distribution of the annual runoff after drainage will be as in Table 64.

Interception shows a strong increase only when the forest has clearly revived after drainage. Evaporation decreases immediately after the groundwater level falls due to drainage and this decrease continues because the developing tree stand decreases the amount of energy available for evaporation. Transpiration increases as the tree stand developes. Evapotranspiration decreases sharply immediately after drainage but later it may even exceed the pre-drainage values.

Variations in the temporal distribution of the runoff as a function of the time that has elapsed since draining cannot be predicted without more detailed investigation. However, the greater water storage capacity of peatlands after drainage and the delayed snowmelt as the tree stand develops, probably produce lower flood peaks of longer duration. The influence of the ditches themselves is, however, probably the reverse.

Current investigations of the factors which influence the water balance of virgin and recently drained peatlands and of peatlands that have been drained decades ago, will during the next two or three years provide further knowldege for the solution of the above problems.

The humus content of drained water

The water in most of the lakes in Finland contains humus which comes from the peatlands. In general, the occurrence of humus in water is considered to be ecologically unfavourable. Humus easily carries phosphorus which is the most important factor in

Table 65. Distribution of annual runoff after drainage.

		Virgin peat-lands	Recently drained areas	Old drainage areas
		% of precipitation		
Interception		5	6	20
Evaporation		35	20	15
Transpiration		10	11	20
	Total	50	37	55
Surface runoff		35	35	25
Subsurface runoff		15	28	45
	Total	50	63	45

eutrophication, and also becomes easily involved in oxygen-consuming biological processes, etc. On the other hand, the presence of humus in water leads to a decrease in biomass production because the penetration of solar radiation is small. It is probable that the importance of the humus content of water - and of the fluctuations in this content - in respect of the ecology of lakes and watercourses is as yet unknown. Furthermore, it must be remembered that there are several kinds of humus, including both small and large molecule humus compounds. Studies have indicated that different kinds of humus vary greatly in importance in respect of the ecology of the water.

It is quite evident that the humus content of the water discharging from a drainage area is increased during the installation of drainage, (Ferda, 1966). Experience has shown, however, that the water becomes 'clean' again soon after the work has been completed. Many people concerned with forest drainage have voiced the opinion that the water discharging from old drainage areas containes less humus than that coming from virgin peatlands.

Research on this topic is still in its infancy, though an attempt has been made to assess, by frequent sampling and comparison of samples taken from virgin, recently drained and earlier drained peatlands, the changes taking place in the humus content of the discharged water as a function of the time that has elapsed since drainage. By means of fractionisation, the humus substances are separated into two parts: a small molecule fulvous acid fraction and a large molecule, colloidal humus acid fraction.

The water relationships of mineral soils surrounding drainage areas

Many people who oppose forest drainage for reasons of nature conservation have stressed the possibility that drainage of peatlands impairs the water balance of the mineral soils surrounding the drained areas, even to such an extent that the increase in tree growth which is produced in the drained area is lost through a decrease in the growth on surrounding sites.

Small scale studies directed at assessing the validity of this view have indicated that drainage has only affected tree growth along the border of peatlands hampered by excess water, and in this case, the influence has certainly been beneficial. No harmful effects have been observed under any conditions, when tree growth has been used as an indicator.

A research team working under the author's supervision started this summer, a large-scale investigation of this problem. The radical growth of trees is used as an indicator of the water relationships. In this way it will be possible to obtain, in a relatively short time, information on the problem. The final solution of the problem, however, requires long-term invesgtigations including the use of calibration periods, control areas and numerous groundwater holes.

On occasions it is claimed that forest drainage decreases groundwater resources. Such a statement is based, however, on mere opinion or isolated observations. Several studies have indicated that the increase in groundwater resources from peatlands or the groundwater runoff, is extremely small, (Virta, 1966; Baden and Eggelsmann, 1964; Boelter, 1966). Forest drainage only lowers the water table of peatlands by 200 to 300 mm, and change of this magnitude probably being of negligible importance for groundwater runoff.

The studies dealt with in this paper have been considered necessary, although it is evident that no catastrophic hydrological consequences are to be anticipated from forest drainage in Finland. If a risk of this existed, it is likely that there would have been some visible signs of it by now.

REFERENCES

Baden, W. and Eggelsmann, R. 1964. Der Wasserkreislauf eines nordwest-deutschen Hochmoores. Hamburg. Verlag Wasser und Boden, 156 pp.

Bay, R. R. 1967. The hydrology of several peat deposits in Northern Minnesota, USA. Third International Peat Congress, p. 212-128.

Bay, R. R. 1970. Water table relationships on experimental basins containing peat bogs. IASH-Unesco. Sympos. on the Results of Res. on Reprs. and Exper. Basins, Wellington, NZ., December 1970.

Boelter, H. 1966. Hydrological characteristics of organic soils in Lake States Watersheds. Journal of Soil and Water Conservation, 21, 2.

Ferda, J. 1966. Dàrskà raselinisté z hlediska jejich hydrologicke funkce. 2. sedeleni: Odtokove pomery. Summary: The Darko peat bogs from the point of view of their hydrological function. 2. Communications: The runoff conditions. Vedecké prace. VUM, Praha.

Ferda, J. 1966. Dàrskà raselinisté z hlediska jejich hydrologicke funce. 3. sedeleni: Chemismus vod. Summary: The Darko peat bogs from the point of view of their hydrological function. 3. Communication: Chemism of waters. Vedecké prace VUM, Praha.

Mustonen, S. and Seuna, P. 1971. Metsäojituksen vaikutuksesta suon hydrologiaan. Summary: Influence of forest draining on the hydrology of peatlands. Publ. of the Water Res. Inst. 2.

Multamäki, S. E. 1962. Die Wirkung von Waldentwässerung auf die Ablaufverhältnisse von Torfboden. Comm. Inst. Forest. Fenn. 55.23.

Salmela, K. and Salonen, K. 1959. Suojeluojan vaikutus soistuvan kankaan mäntyjen sädekasveen. Manuscript in the Dept. of Silviculture, Univ. of Helsinki.

Virta, J. 1966. Measurement of evapotranspiration and computation of water budget in treeless peatland in the natural state. Comm. Physico-Math. Societ. Scient. Fenn., 32, 11.

DISCUSSION

I. I. Kulikov (Ukrainian SSR) - How large is the forest area drained in your country?

L. Heikurainen (Finland) - In my paper you can find a table which furnishes the figures of drainage for the recent decade. In 1970 an area of about 295000 ha was drained.

S. K. Vompersky (USSR) - Are there any cases of forest overdrainage in Finland and what are the criteria of overdrainage?

L. Heikurainen (Finland) - There are two ways to determine the optimal drainage level, namely, measurements and experiments. We have found experimentally that under the conditions found in Finland the optimum water level should be 600 mm below the bog surface. If the level is lower, it means overdrainage. From the theoretical review which I made you can see that the optimum drainage should be at PF = 2.2, and the water level at 200 mm to 1000 mm, depending on the type of peat and other conditions. However, no overdrainage occurres in the case considered, as the drain spacings are adhered to carefully and PF = 2.0.

V. V. Kupriyanov (USSR) - Do you think evapotranspiration decreases just after reclamation and increases after bog cultivation? Have you any data to illustrate this conclusion?

Are there any projects to preserve some bogs (especially high bogs) in their natural state in your country?

L. Heikurainen (Finland) - Yes, such data are available.

In our country 3 000 000 ha are to be preserved in their natural state. There is a large area under preservation (400 000 ha) where peat will not be touched. These areas are reserved for scientific investigations.

G. Goffeng (Norway) - Do you often use your method? Does if affect soil regeneration?

L. Heikurainen (Finland) - Your question is concerned with the rotational device, isn't it? The device is not used very often. It is only used for 5% of all the ditches. The method is good. All organic materials are preserved at the site, no nitrogen fertilizers are required.

SEDIMENT DRAINAGE AND CONSOLIDATION AS A RENEWAL TECHNIQUE IN A SMALL BOG LAKE

S. A. Smith

Department of Geology,
University of Wisconsin, Madison, USA

SUMMARY: The draining of Jyme Lake, Wisconsin, to test the efficacy of bottom sediment consolidation as a lake renewal technique, provided the opportunity to study the groundwater and surface water relationships in a small bog lake and the effects of dewatering on the soil and water system. Although located in a topographic depression, Jyme Lake and the surrounding bog act as a groundwater recharge area. The consolidation of the bog during the draining of Jyme Lake resulted in an initial increase in the hydrostatic head in the sand and gravel beneath the lake and bog sediments. This pattern was reversed when a widespread slumping of the bog occurred in the near-shore areas. Because the lake was not fully drained, only limited lake sediment consolidation took place. Lake drainage and bottom sediment consolidation is still being evaluated as a lake renewal technique.

DRAINAGE ET CONSOLIDATION DES SEDIMENTS : TECHNIQUES DE RENOVATION D'UN PETIT MARECAGE

RESUME : Le drainage du lac Jyme (Wisconsin) en vue de tester l'efficacité de la consolidation du sédiment de fond comme technique de rénovation d'un lac a donné l'occasion d'examiner les corrélations entre eaux souterraines et eaux de surface d'un petit lac marécageux ainsi que les effets de l'assèchement sur le sol et sur le système hydraulique. Le lac Jyme et les marécages environnants, tout en étant localisés dans une dépression topographique, agissent comme source d'alimentation des eaux souterraines. La consolidation du marais au cours du drainage du lac Jyme a eu comme résultat initial l'augmentation de la pression hydrostatique dans le sable et le gravier sous les sédiments du marais et du lac. Cette situation s'est renversée lorsque s'est produit un glissement des terrains dans la région environnante. Le lac n'ayant pu être complètement drainé, la consolidation des sédiments de fond n'a été que partielle. Le drainage des lacs et la consolidation des sédiments de fond constituent néanmoins une méthode valable de rénovation des lacs.

Introduction

This paper is based on an in-progress study by the Inland Lake Demonstration Project, a joint venture of the University of Wisconsin and the Wisconsin Department of Natural Resources. Project objectives include the demonstration of techniques to restore, maintain, and protect a high quality environment within and adjacent to inland lakes in the Upper Great Lakes Region.

The Project is currently involved with renewal techniques applicable to small, shallow lakes which are subject to fish winterkill and/or extensive rooted aquatic plant growth. Lake renovation by drainage of the lake and sediment consolidation is a possible alternative to dredging in lakes which have a thick accumulation of highly organic, low density bottom sediments.

The distribution of pressure within a saturated soil mass is governed by the well-known Terzaghi (1943) equation,

$$p = \bar{p} + u$$

where p is the total pressure on any plane within the soil, $\bar{p}$ is the effective, or grain-to-grain pressure due to the weight of the solids, and u is the neutral, or pore water pressure due to the weight of the water. Decreasing the pore water pressure then, results in an increase in the grain-to-grain pressure if the total pressure remains unchanged. In highly compressible soils such as peat or lake muck, an increase in the grain-to-grain pressure can produce a decrease in the void volume and result in a significant reduction of the sediment volume. Without removing any solids, sediment consolidation can deepen a lake, limit aquatic plant growth and provide a sufficient volume of oxygenated water to enable fish to survive winter conditions.

Consolidation of organic lake bottom sediments had been observed during the draining of Snake Lake, a small eutrophic lake lake in northern Wisconsin (Born & Stephenson, 1972). Although the goal of this earlier Inland Lake Demonstration Project study was to achieve a nutrient reduction by the draining and dilution of lake water, about 1 metre of consolidation occurred in the near-shore areas where lake sediments had been drained. Similar consolidation of organic soils has often been observed during the draining and reclamation of bogs and marshes (Thomas, 1965;Stevens and Speir, 1969).

Jyme Lake in north-central Wisconsin at the Kemp Biological Station, University of Wisconsin, was used to test the feasibility of draining to achieve sediment consolidation and lake deepening. In addition, it presented a valuable opportunity to study the groundwater and surface water relationship in a small bog lake and the effects of lake and bog dewatering on the soil and water system. This paper is presented in two parts: first, a description of the area and the physical characteristics of the lake and bog, and second, the response to draining, including the development and extent of bog slumping.

Description of the area

Several hundred lakes of various sizes occupy the poorly-drained, pitted outwash plain of north-central Wisconsin. In many of the smaller depressions, commonly referred to as 'potholes', bog development has progressed to such a degree that peat accumulations in excess of 3 metres are not uncommon, and open water may be non-existent.

Jyme Lake is typical of the small bog lakes of the area. As

shown in Figure 112, it occupies one end of an elongated wetland area. The surface area of the lake is 0.45 hectares, and it is surrounded by a 3.7 hectare bog. Peat soundings revealed the asymmetrical cross section of the pothole as shown in Figure 113, with the lake occupying the deepest part. Peat and muck accumulations exceed 6 metres beneath the lake, but are less than 3 metres throughout most of the bog. Bog vegetation consists of stunted tamarack (Larix laricina) and black spruce (Picea mariana) with a ground cover of leatherleaf (Chamaedaphne calyculata) and Sphagnum mosses. The lake supports only a small minnow population.

The bog is of the quaking variety within 9 to 15 metres of the lake shore, and although standing water is present on the more distant parts, it is generally capable of supporting the weight of a man.

Geology

The lake basin is surrounded on all sides by low hills made up of poorly-stratified sand and gravel outwash. Soundings disclosed the presence of similar material beneath the lake and bog. With the exception of a thin clay or marl layer directly overlying the sand and gravel beneath the lake (Figure 113), virtually no inorganic sediment has been deposited in the lake basin due to the high permeability and coarse grain size of the surrounding outwash material.

Cores were obtained from the lake and bog to determine the physical characteristics of the organic sediments. Combined with the information obtained by soundings, they indicate that most of Jyme Lake is underlain by a highly-decomposed, dark brown to black, gelatinous ooze or muck containing only a few plant fibres. The upper 2 metres of the lake muck is quite fluid. Percent solids range from 1.6% to 5.1% by weight in the upper 5.4 metres of the muck, and porosity averages about 94% below the fluid layer. Green algae proliferates on the surface of the muck in the shallow parts of the lake, and in the deeper water the muck is covered with a thick mat of moss, which appears to be a major contributor to the volume of sediments.

The peat of the surrounding bog is brown or yellowish brown and consists mainly of moss fibres with smaller amounts of wood fragments and leaves. Samples of the peat were not taken below 1.7 metres, but similar deep peats of the area are fibrous peats throughout most of their thickness (Hole and Schmude, 1959). In the upper metre, the peat contains about 8% solids by weight, and has a porosity of approximately 91%. Soundings indicated that the peat changes little at depth, and it directly overlies the sand and gravel throughout most of the bog.

Hydrogeology

Jyme Lake is located on a peninsula and is surrounded on three sides by a 1500 hectare lake, Tomahawk Lake. Tomahawk Lake is less than 150 metres away from Jyme Lake at its closest point, and is 2.6 metres lower than the surface of Jyme Lake (Fig.112).

No streams enter the lake basin, although a small intermittent spring flows into the bog at the northeast end. There is no natural surface drainage out of the basin, but a small shallow ditch has been recently dug to drain standing water out of the bog into Tomahawk Lake. The present water level is maintained mainly by direct precipitation on the bog and lake.

A series of sand point wells was driven a short distance into the sand and gravel beneath Jyme Lake and the adjacent bog (Fig. 112). The initial water levels in these piezometers were more than 2 metres lower than the surface of Jyme Lake at some locations (Fig. 113).

The water table in the bog is at the surface and slopes slightly with the surface towards Jyme Lake at a rate of about 0.6%. At the edge of the bog the water table may be perched in the peat. The water table in the sand and gravel at piezometer JL 6 is more than 2 metres lower than the water table in the bog less than 5 metres away.

The vertical permeability of the lake muck, as determined from intact cores, ranges from about 0.2×10^{-4} cm/sec in the upper layers to about 1.1×10^{-3} cm/sec at a 5 metre depth. The permeability of the upper portions of the bog peat is about the same as that of the deeper lake sediments.

Although the level of Jyme Lake and the water table in the bog are considerably higher than the regional groundwater and surface water elevations, the lake and bog are not totally perched in the classical sense. Saturated conditions exist in the sand and gravel at all piezometer locatings in the lake and bog. The difference in head between lake level and the groundwater level in the sand and gravel, combined with the small but finite permeability of the lake and bog sediments, indicate a downward flow of water through the peat and muck.

Water chemistry data correlate with the groundwater flow patterns indicated by the piezometer and lake levels. Jyme Lake water shows little influence from the mineralizing effect of movement through the soil and is very soft, with a calcium carbonate hardness of less than 5 mg/l. The specific conductance is 10 micromhos/cm at 20°C compared to about 100 micromhos/cm for water from Tomahawk Lake.

Pumping response

Operational procedure

Water from Jyme Lake was pumped intermittently for a period of 10 days at flow rates as high as 11 000 litres per minute using a high capacity centrifugal pump. Rubber tubing conveyed the water from Jyme Lake to Tomahawk Lake. After about 12 hours of operation, pumping rates decreased as sticks and muck repeatedly blocked the pump intake. Pumping was finally terminated because of pumping difficulties and the imminent freeze-up of the lake.

The lake level and piezometer levels were checked as often as twice a day during the initial phases of pumping and at least once a month after pumping stopped. Bog and lake sediment levels were measured during consolidation from the tops of the piezometer pipes and from several open-ended pipes that were driven into the sand and gravel to serve as additional reference points. A few of the piezometers and pipes were rendered unusable during the operation due to lateral displacement of the peat.

Lake level and bog slumping

The decline of the lake level during pumping is indicated in Figure 114. When the pump was running at full capacity, the lake level dropped at the fairly constant rate of 15 cm per hour. Precipitation during the 10 day pumping period was minimal, and the lake level remained essentially unchanged when the pump was shut down.

At noon on the second day of pumping, after the lake level had been lowered about 1.5 m, a fissure began to develop in the bog in the vicinity of the access road. The bog started slumping and the fault spread rapidly parallel to the lake shoreline and about 15 to 30 m away from the edge of the water. By the end of the day the fault extended around the northern end of the lake and along most of the western side.

The slumping continued even though the pump was shut down for the next five days. Four days after pumping had started and two days after the initial slumping occurred, the fault encompassed about 75% of the lake's perimeter. The bog on the lakeward side of the fault had dropped over 0.9 m vertically and had moved 0.3 to 0.6 m towards the centre of the lake. Twenty-five days after pumping had started, the vertical displacement along the fault was about 1.7 m, somewhat less than the total 2.3 m draw-down of the lake.

The slumping occurred due to the removal of support from beneath the quaking part of the bog as lake water and muck were pumped out of the lake, but actual rupture took place in the more stable peat. The slump was probably triggered by mechanical vibrations from an electrical generator and other machinery on the access road. Figure 115 shows the fault on the third day after pumping started.

Groundwater levels and sediment consolidation

The response of the water levels in the piezometers to the declining lake level is shown in Figure 114. Over 0.3 m of initial rise was recorded in most piezometers, and JL8, in the centre of the lake (Fig. 112), showed a head increase of more than 0.6 m. After about four days had elapsed, all piezometer levels were declining.

The initial rise of the groundwater levels is probably closely connected to the consolidation of the bog. As the water level in the lake dropped, a corresponding decline in the water table in the bog was observed in core holes. The buoyant support of water was removed from the upper layers of the bog, increasing the effective pressure and initiating consolidation.

Even though solids represent only a small per cent of the total weight of the peat, lowering the water table produced a substantial increase in the effective pressure in the upper bog layers. The initial effective pressure was very small because of the low submerged weight of the solid material, mostly moss fibres and wood fragments. When the water table dropped, the peat remained at least partially saturated, and the total unsubmerged weight of the solids plus the weight of the capillary pore water was transferred to the fibres.

Because of the low permeability of the peat, the water table decline did not produce an immediate decrease in the hydrostatic pressure in the lower bog layers. Instead, the increased load due to the peat above the water was initially balanced by an increase in the hydrostatic pressure. As pore water was squeezed out of the void spaces in response to the new hydraulic gradient, consolidation began in accord with the Terzaghi theory of consolidation (1943).

The increase in the hydrostatic pressure in the lower peat layers was transmitted to the sand and gravel and was recorded as a rise in groundwater levels. This head increase can be shown to be the approximate amount expected for the observed water table lowering. If we assume (1) that the water table was lowered instantaneously, and (2) that the peat remained completely saturated above the water table, then, according to the Terzaghi theory, hydrostatic pressure should increase by an amount equal to the water table decline. Even though these conditions were not completely met at Jyme Lake, the recorded piezometer head increases approximately coincided with the average observed water table decline.

Although continuous water level recorders were not installed, it is probable that the peak on each piezometer hydrograph closely coincides with the faulting of the bog near the piezometer. Water levels declined first closest to the initial faulting in JL2, and more slowly in those piezometers further away from the first and subsequent ruptures. Faulting provided an immediate zone for pore pressure relief, and enable the hydrostatic pressure in the lower peat layers and the sand and gravel to decline to equilibrium levels. Two weeks after pumping had started, piezometer levels had dropped to near their original levels.

As much as 25 cm of consolidation was measured in parts of the bog 25 days after the start of pumping. Small amounts of consolidation were recorded at distances of 30 m from the fault, although the water table was still at the bog surface about 15 m away from the fault. This indicates that in some parts of the bog consolidation was keeping pace with the declining water table.

Because of the pumping difficulties, Jyme Lake was not completely drained, and only limited lake muck consolidation took place. It was virtually impossible to separate the effects of consolidation from the effects of muck flow and bog slumping.

Summary and conclusions

Jyme Lake and the surrounding bog act as a groundwater recharge area but because of the low permeability of the peat and muck beneath the bog and lake, the actual volume or recharge to the groundwater system is probably small. The draining of Jyme Lake provided a unique opportunity to observe the effects of dewatering on the soil and water system in a small bog lake. The initial increase of the hydrostatic pressure in the sand and gravel beneath the lake and bog during the first phases of consolidation was reversed when slumping occurred in the near-shore areas of the bog. Although unusual, this pattern of response can be explained in the light of the soil mechanical properties of the peat.

Much of the potential consolidation of the lake bottom sediments did not occur because the lake was not fully drained and the water table was not lowered below the sediment surface. The fluid lake muck repeatedly blocked the pump intake and prevented pumping at full capacity. In spite of pumping difficulties, consolidation was observed in the bog, although laboratory testing indicated that the peat was much less susceptible to consolidation than the lake muck.

The bog consolidation at Jyme Lake and the consolidation that occurred during the draining of Shake Lake indicate that the consolidation of organic sediments is a promising lake renewal technique. The high costs and spoil disposal problems associated with lake deepening by dredging can be avoided, but the effectiveness of the method depends on the complete draining of a lake and maintaining the water table below the sediment surface. In addition to the problems associated with pumping, bog slumping should be recognised as a potential problem during the draining of bog lakes. Sediment consolidation is still being evaluated as a lake renewal technique, and future field testing is likely.

References

Born, S., Stephenson, D.A. 1972. Dilutional pumping of Shake Lake, Wis.Dept. of Natural Resources Tech. Report (in progress).

Hole, F.D., Schmude, K.O. 1959. Soil survey of Oneida County, Wisconsin, Wis. Geol. Survey Bull. 82.

Stephens, J.C., Speir, W.H. 1969. Subsidence of organic soils in the U.S.A. in Tokyo symposium on land subsidence, 2 pp. 523-534.

Terzaghi, K. 1943. Theoretical soil mechanics. Wiley, New York.

Thomas, F.H. 1965. Subsidence of peat and muck soils in Florida and other parts of the United States - a review. Proc. Soil and Crop Sci.Soc. of Fla. pp.153-160.

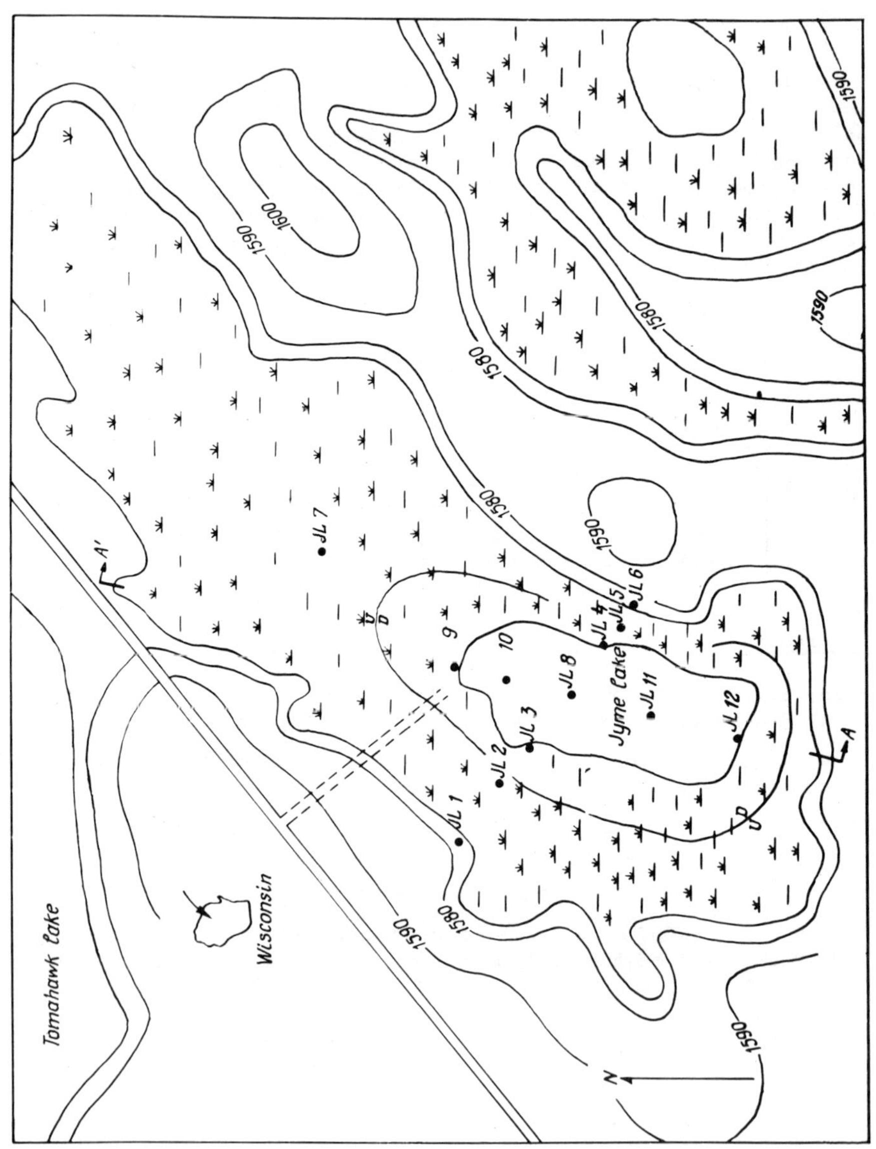

Fig. 112. Map of the Jyme Lake area.

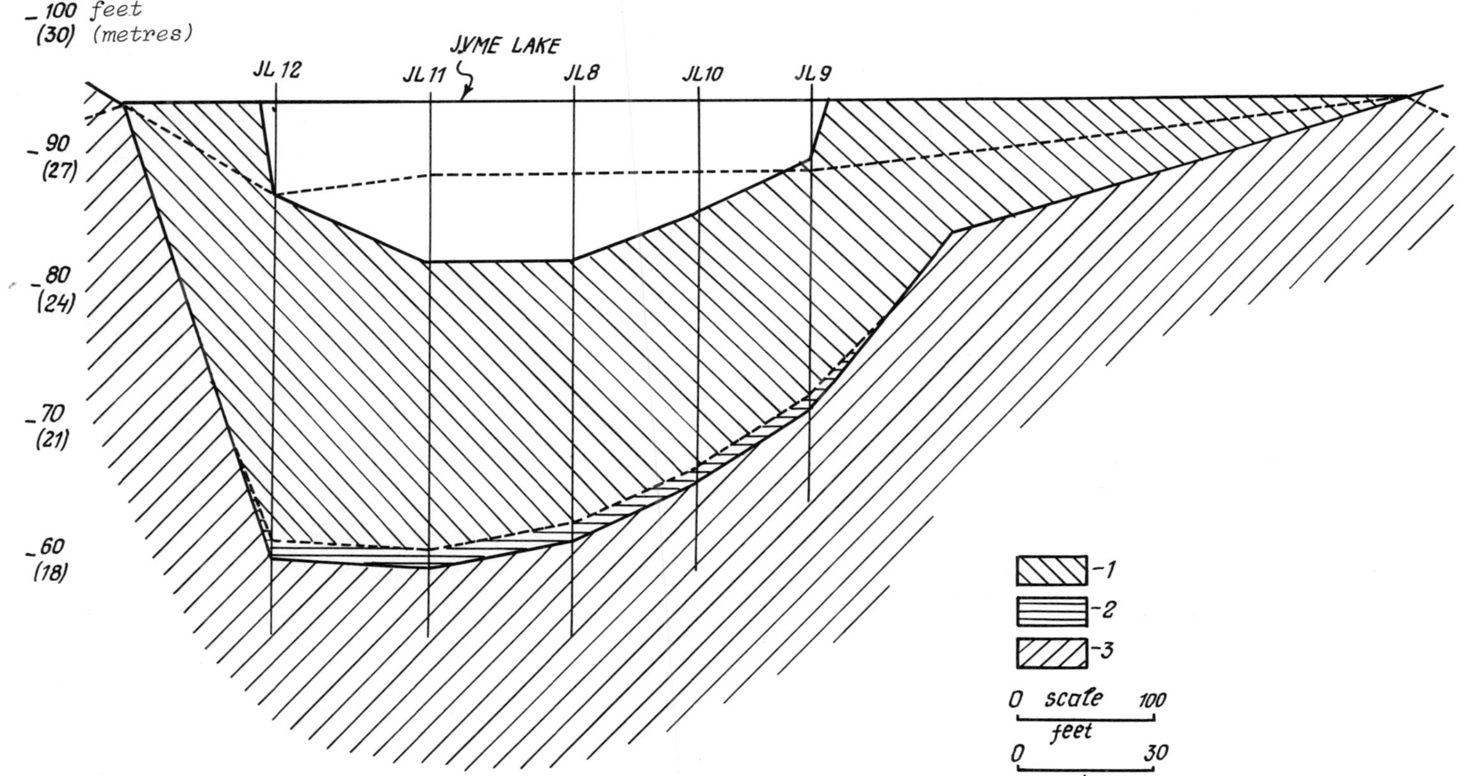

Fig 113. Cross-section A-A' of Jyme Lake.

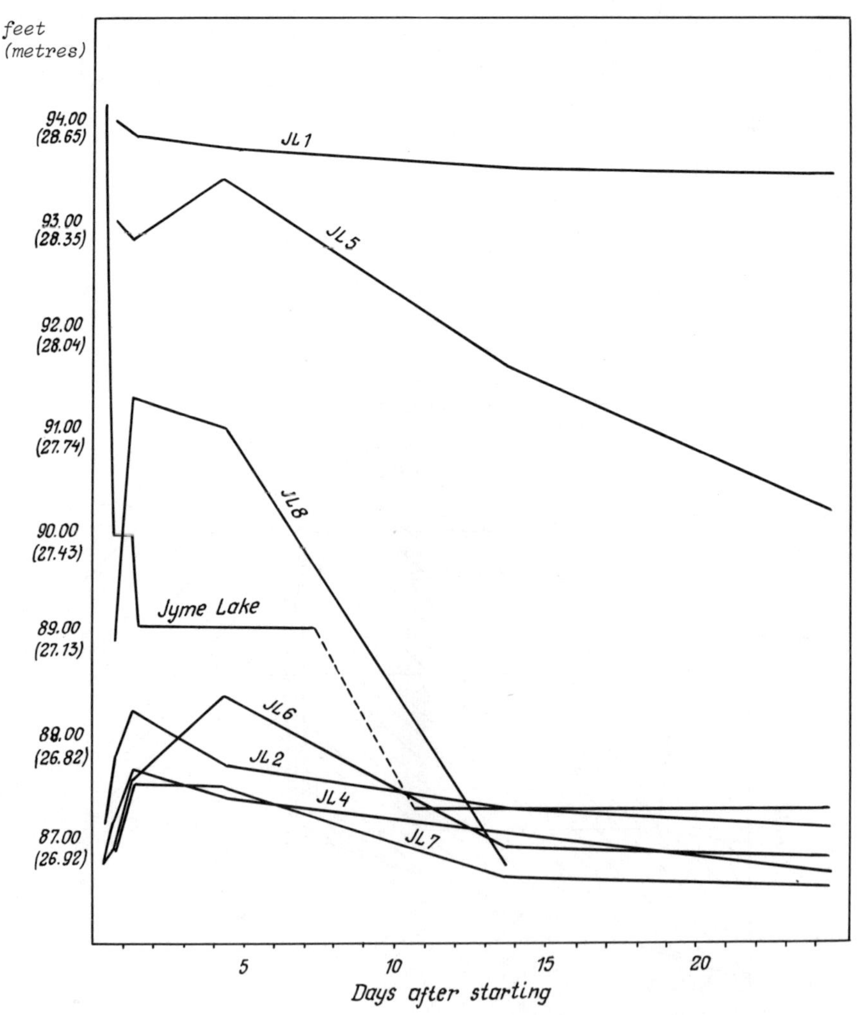

Fig 114. Well and lake hydrographs during pumping.

Fig 115. Bog faulting at the generator on the access road on the third day after pumping started.

THE EFFECT OF RECLAMATION ON RIVERS AND LAKES IN LATVIA

L. I. Glazacheva

Latvian State University, Riga, USSR.

SUMMARY: The effects of reclamation on the water level of lakes and their discharge are analyzed using long-term observations.

From provisional observations of village wells in a selected region, the decrease in the height of the water level is related to the topographical position and the distance from the drainage ditches or regulated rivers.

EFFETS DES MESURES D'ASSECHEMENT SUR LE REGIME HYDRAULIQUE DES RIVIERES ET DES LACS DE LA LETTONIE

RESUME : A partir des données obtenues dans des stations d'observations sur plusieurs années, on a analysé l'influence des mesures d'assèchement sur le niveau d'eau des lacs et sur leur écoulement.

En s'aidant des observations préalables faites dans des puits ruraux d'une région déterminée on a analysé les variations répétées du niveau d'eau en fonction de la position et de la distance des canaux d'écoulement ou des rivières déjà aménagées.

Water-logged and marshy soils occupy about 56% of the whole area of the Latvian SSR. Reclaimable land covers 36000 km^2, of which 22000 km^2 is used for agriculture and 13900 km^2 for forests. By 1971, 38% of the land available for agriculture and about a quarter of the forests had been drained. This means that over 80% of river basins have been regulated and that during the 20 years leading up to the 1960's a great number of lakes, mostly small, had been drained partially or completely.

At the beginning of 1972, the overall area of water bodies, neglecting the newly-constructed large water storages of the Plavinyas and Kegums water power stations, had diminished by more than 100 km^2 (over 10%) since the 1940s. Of the 15 largest lakes having areas of over 10 km^2, 6 are now partially lowered and in 2 the water level is controlled by sluices.

The results of drainage on the hydrological regime of water bodies were examined in rivers and lakes with the help of long-term observations. Against a background of a general decrease of moisture over the area of Latvia from the middle of the thirties and also because of drainage works, both the water level in the lakes and river discharges have tended to decrease.

For natural reasons the water level in the lakes has dropped by 0.5 m on the average. In lakes where the natural outfall has been artificially lowered by the deepening of effluent rivers, the decline of the water level has sometimes exceeded 2.5 to 3 m.

Diagrams in Fig 116 show the fluctuation of the mean annual levels of lakes Burtnieki and Lubanas. In the former, the level dropped by about 1 m, and in the latter by approximately 2.5 m. In the same figure, synthetic diagrams are presented of the variations in level which would have taken place had there been no drainage works. The reconstructed levels are obtained from the relations between the annual water levels in these lakes and the discharges at the outfall. An example of such a plot for the Burtnieki and the Salatsa effluents is shown in figure 117. It was assumed that the annual river discharge had not undergone major changes due to reclamation. This assumption was supported by analysis of data on annual discharges of a number of Latvian rivers. The change of the annual discharge rate was within ± 10%. In one case, some decrease of the discharge rate was found, in others, the rate seemed to increase on account of groundwater drainage.

The effect of drainage works on discharge redistribution within the year is undesirable as the maximum discharge increases and the minimum decreases.

On the Aiviekste river at a gauging station corresponding to a river basin of about 7000 km^2, no appreciable redistribution within the year was found, but the natural regulation coefficient of the discharge in the period 1943 to 1967 following large scale drainage works in the river basin was lower than that for the period 1923 to 1942. The decrease of natural regulation was between 7 and 10% in wet and moderately wet years and up to 20 to 30% in dry years.

In the same river from the late 1940's to the early 1950's, average discharges over 30 days and particularly minimum summer discharges decreased by 30 to 40%. The results of research on the effect of drainage on the regime of the Aiviekste river and Lubanas lake system are considered in the author's monograph entitled; "Effect of reclamation on the regime of the Aiviekste river and the Lubanas lake".

The decrease of the long-term storage of groundwater during drainage sometimes results in an excessive decrease of the groundwater level especially for dry periods.

As a result of the combined effects of natural low moisture content and reclamation by drainage, a marked lowering of the water level in the rural wells has occurred in some regions of the Latvian SSR. From 1968 to 1970, students of the Faculty of Geography at the Latvian University carried out a tentative examination of 687 wells in the Gulbene region. They questioned the local people to find out whether the water level in their wells had decreased after completion of drainage works.

The occurrence of decreased levels in the wells examined was calculated as a function of their location (on a hill, plain, lowland) and the distance from drainage systems (Fig 118).

The most frequent occurrence of decreased levels is found in the wells on the top or upper part of a hill slope; the least frequent occurrence is observed for wells situated in low places.

Without exception, the occurrence of decreased levels is more frequent in cases when the wells are located within 100 to 1000 m from a drainage ditch, drain (regulated river) or partially lowered lake and when the well is not fed by a natural pool or artesian groundwater.

DISCUSSION

V. F. Fashchevsky (Byelorussian SSR) - In what way did you account for cyclic oscillations of climate and river discharge over many years? Have you compared the changes in levels of lakes and discharges of rivers with references systems?

L. I. Glazacheva (USSR) - We have used the data of the runoff and precipitation for one river. Evaporation values are calculated according to Konstantinov. Mass curves of specific discharges, precipitation and evaporation are plotted. In the analysis the normal evaporation and normal precipitation were almost the same as mean annual values.

Under the present conditions it is very difficult to find a reference system, but some attempts have been made. At the present stage of work we have failed to find any interesting data.

I. Riekstс (USSR) - In what way have you managed to estimate the decrease of groundwater accumulated for many years if the lakes were emptied in the twenties and thirties?

L. I. Glazacheva (USSR) - Accurate estimation is the result of analysis and logical deduction. When drainage reclamation is undertaken then groundwater accumulated for many years should decrease irrespective of the time when the works were done, in the twenties or in the sixties.

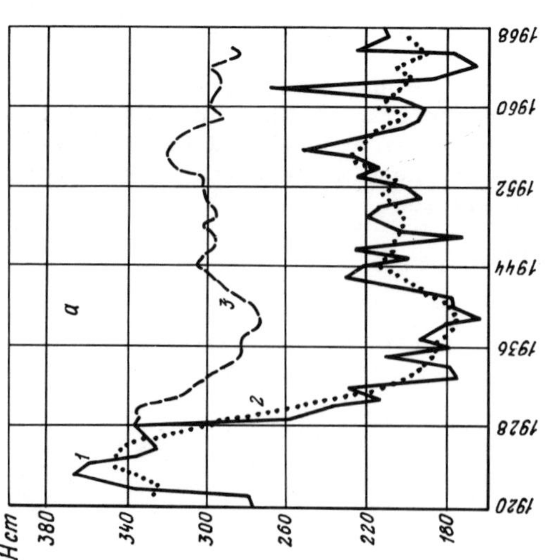

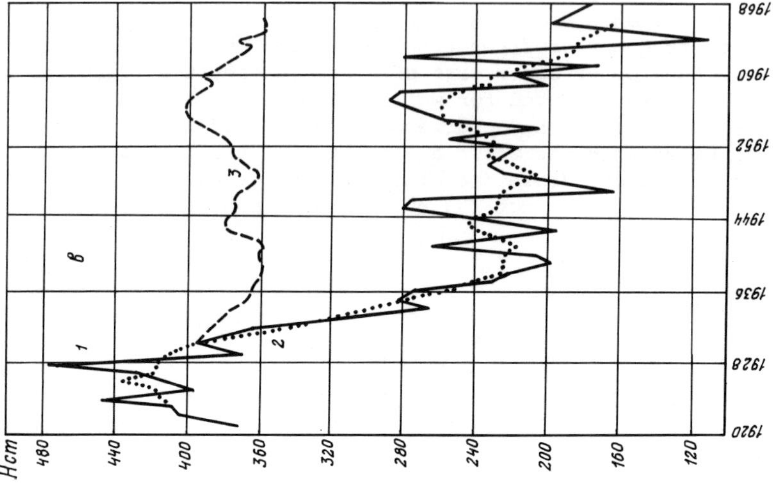

Fig 116. Fluctuations of the mean annual levels of lakes Burtnieki (a) and Lubanas (b).
1. Observed; 2. Five year moving average; 3. reconstructed for conditions of no improvement work.

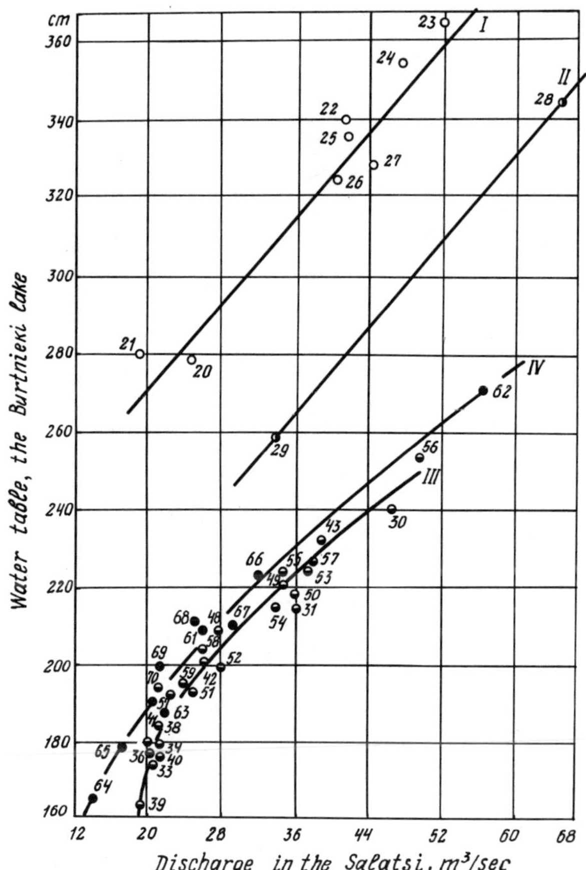

Fig 117.

Plots of mean annual discharges in the Salatsi River versus the level of the Burtnieki Lake.

I. Before improvement works.

II. During improvement works.

III. Stabilized regime after improvement works in the 1920s.

IV. During improvement works in the 1950s and 1960s.

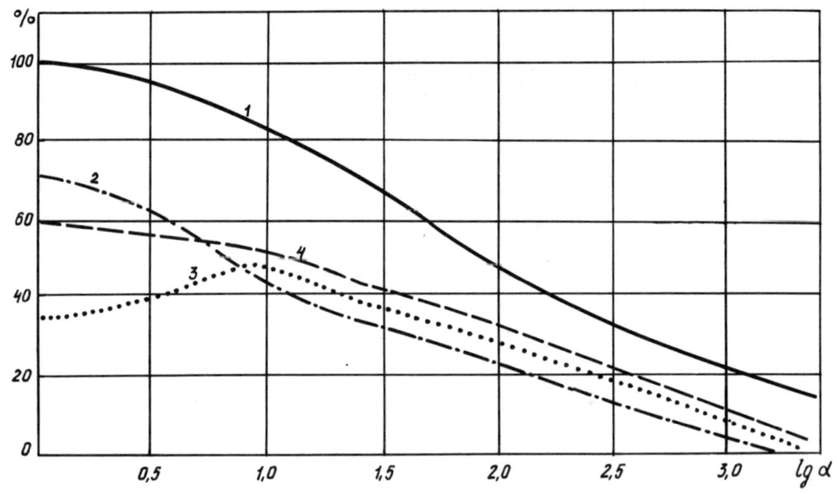

Fig 118. Recurrence of water level decrease in the wells in the Gulbine region as a function of distance from the drainage systems and location in the area:

1. On a hill.
2. In lowland.
3. On a plain.
4. Generalized (with no reference to relief).

INFLUENCE OF FOREST DRAINAGE ON THE HYDROLOGY OF AN OPEN BOG IN FINLAND

S. E. Mustonen and P. Seuna

National Board of Waters,
Helsinki, Finland.

SUMMARY: A drainage experiment based on the control basin method was carried out in south eastern Finland from 1936 to 1969. Two natural basins were calibrated for 22 years. Then 40% of one basin was drained for forestry. Nine years' data on the forest draining period are available. The drained peatland was mostly open bog and swamp with a poor growth of pine. The soil surface settled about 20 cm during the experiment. In addition, the groundwater level dropped about 30 cm. A drastic change in the soil moisture regime largely destroyed the wet peatland vegetation. Transpiration decreased markedly as well as soil surface evaporation. An increase of 43% in annual runoff was observed, relatively evenly distributed throughout the year. Both the spring and summer maximum runoff increased, on average by 31 and 131% respectively, caused mainly by the accelerating effect of the drainage network. Winter and summer minimum runoffs increased many times. Before draining no channels existed and the draining made flow possible even under drought conditions.

It is probable that annual runoff from the peatland will decrease as the new forest becomes established. Also, the increase in runoff from drained forested peatlands may not be as great as in the case of drained open bogs. Conclusions presented here do not concern peatlands drained for agriculture where cultivation changes the effect of drainage considerably.

INFLUENCE DU DRAINAGE FORESTIER SUR L'HYDROLOGIE D'UN MARAIS OUVERT, EN FINLANDE

RESUME : Cette étude donne les résultats des expériences de drainage réalisées aux cours des années 1936-1969 dans le sud-est de la Finlande. L'expérience était basée sur la méthode du bassin contrôlé. Deux bassins naturels ont été comparés pendant 22 ans. Le drainage forestier a été réalisé sur un secteur expérimental et l'on dispose des résultats des observations sur une durée de 9 ans. Cette opération était effectuée sur 40 % de la surface du bassin expérimental avec un pauvre peuplement de pins. Le niveau du sol s'affaissa de 20 cm durant l'expérience. En outre, le niveau des eaux souterraines s'est abaissé de 30 cm. Le changement radical intervenu dans l'humidité du sol a largement anéanti la végétation du terrain tourbeux. La transpiration aussi bien que l'évaporation de la surface du sol ont considérablement diminué. On a observé une augmentation de l'écoulement annuel de 43 %. Cette augmentation s'est répartie d'une façon uniforme au cours de l'année. L'écoulement maximal en été et au printemps a augmenté de 131 % et de 31 % respectivement en moyenne. Cela était dû surtout, évidemment, à l'effet accéléré des travaux de drainage. L'écoulement minimal en hiver et en été a augmenté à plusieurs reprises. Avant les travaux d'aménagement il n'y avait pas de fossés et le drainage a rendu possible l'écoulement pendant la période des basses eaux.

Il est probable que l'écoulement annuel sera en régression quand une nouvelle végétation forestière couvrira les terrains tourbeux. En outre lorsque le drainage est effectué dans des régions tourbières forestières l'écoulement n'est pas aussi considérable. Les conclusions auxquelles on aboutit dans cette étude ne concernent pas les terrains drainés pour l'agriculture. Dans de tels terrains la pratique agricole modifie considérablement les effets du drainage.

Introduction

The total area of peatlands on the earth has been estimated at about 100 million hectares, much of it in regions where intensive forestry is practised. Draining in peatlands for forestry has been used in various countries to increase timber production.

In Finland, there are 10 million hectares of peatland, 3.8 million hectares had been drained by the end of 1971, and more than 3 million hectares will be drained during this decade. Thus, by 1980 about one quarter of the land area of Finland will have been treated in a manner which has considerable hydrological significance.

Research basins and experimental periods

This paper is concerned with the hydrological effects of forest drainage, based on the findings of a drainage experiment carried out in south-eastern Finland (Fig 119). The control basin method was used. In 1935, various hydrological observations were started on two adjacent natural drainage basins, (Mustonen and Seuna, 1971) which were similar in certain respects (Table 65). The control basin (Latosuonoja) comprised in part pine and spruce swamps (15%) and a large cultivated field (19%) long ago reclaimed from peatland. The experimental basin (Huhtisuonoja) had no cultivated fields and open bogs and poorly grown swamps with a poor growth of pine comprised about 45% of the basin. Before drainage, the peat layer was about 1.5 m thick and the mineral soil below the peat was mostly sand and gravel. In 1958 main ditches, 130 cm deep, were dug in the experimental basin (Fig 119) and the peat settled 8 cm on average by 1960; then small forest ditches were dug, 60 cm deep, and by 1969 the settlement amounted to a further 12 cm. The drained area comprised about 40% of the basin and a density of 80 m of main ditch and 225 m of forest ditch per drained hectare. The porosity of the peat increased from 1961-1969 by 50 mm.

The years 1936 to 1957 form the calibration period. The data for 1958 to 1960, when there were only main ditches, cover too short a period to be studied separately. Therefore, only the nine year period from 1961 to 1969 after draining is considered as the treatment period.

The control basin method envisages that both basins are kept completely unchanged throughout the research period except for the treatment of the experiment area. In this case it was not possible. Changes in the basins before 1956 were not measured but they are believed to be insignificant. Forest drainaing and clear-cutting in 1956 affected 7% and 9% respectively of the Latosuonoja basin and 4% and 12% respectively of the Huhtisuonoja basin. The volume of growing stock increased in the Latosuonoja basin and decreased in the Huhtisuonoja basin because of differences in silviculture treatment during the period 1958 to 1969, (Table 65). These changes did not, however, seriously hinder the interpretation of experimental results, which were based on comparisons. The effects caused by changes in growing stock since 1961 can be eleiminated.

Table 66.

Data on the Research Basins of Huhisuonoja and Latosuonoja.

Basin	Drainage area km^2	Peat-land %	Cultivated land %	Mean slope	Tree stand	
					1958 m^3/ha	1970 m^3/ha
Huhtisuonoja	5.03	44	0	5.0	58	39
Latosuonoja	5.34	15	19	8.2	58	74

Annual runoff

The effect of forest draining on annual runoff is shown in Figure 120. The probable undrained runoff (y) in each year after drainage can be estimated by solving the calibration period regression equation (y = 0.910 x - 1.53) for the control basin runoff (x) of the particular year. The effect of forest draining on annual runoff is estimated as the difference between the actual measured Huhtisuonoja runoff and the calculated 'undrained' runoff.

Figure 120 shows that drainage increased annual runoff considerably. The average increase in mean annual runoff for the 9 years was 43%, 3.02 l/s km^2 or 95 mm/year; this increase was highly significant statistically, (Kovner and Evans, 1954). The increase in runoff is largely due to the decrease in evaporation caused by draining. Also the drained peat layer subsided and the water storage of the whole basin decreased by 68 mm over 9 years; this decrease in water storage contributed on average 8 mm/year to runoff.

The volume of growing stock decreased during the 9 year draining period on the experimental basin but increased on the control basin. From measurements of the effect of trees on evaporation (Mustonen, 1965), the annual runoff from the Huhtisuonoja basin was estimated to have increased on average 22 mm/year because of changes in the tree stands.

The calculated increase in runoff due to the reduction in evapotranspiration caused by draining 40% of the basin was thus 95-8-22 = 65 mm/year or about 30% of the mean 'undrained' annual runoff during the period.

No well level observations were available on the firm land around the drained bog. Therefore how much the drainage of the peatland affected the transpiration of the surrounding firm-land forest cannot be estimated. However, the reduction in evaporation was probably largely restricted to the drained peatland. On this assumption the decrease in peatland evapotranspiration was $\frac{100}{40}$, 65 = 163 mm/year.

The effect of draining on evaporation can be explained easily. The peatland on the Huhtisuonoja basin was mainly open bog, partly subject to floods in the spring. Draining lowered the water table, previously close to the surface throughout the year, to 30 cm on the average. This changed the soil moisture conditions especially in mid summer when potential evaporation is high and consequently largely destroyed the open bog vegeation and thus reduced transpiration markedly. As the soil surface dried, evaporation also dropped during rainless periods.

Most of the drained peatlands in Finland are forested. On wooded peatland transpiration continues after draining and the rate may even increase as the soil moisture conditions in the peat become more favourable to tree growth. Furthermore interception by the canopy enhances evaporation, and draining does not, of course, influence interception. The increase in runoff from forested peatland will, therefore, be smaller than from an open bog.

The increase in runoff caused by draining apparently drops after the new growing stand begins more efficient transpiration. In this study, however, pine seedlings, introduced after draining by planting and fertilization, had not yet affected runoff to any demonstrable extent.

Monthly runoff

The increase in runoff for various months is shown in Figure 121. Runoff increased in all months. An interesting point is the relatively great increase during low flow months in winter and in summer; the increase in spring flood time is by comparison, small (in April 23%).

The monthly increase in runoff is statistically highly significant ($P < 0.1\%$) in February, March, June, July and August, significant ($P < 1\%$) in January, May, September, November and December, almost significant ($P < 5\%$) in April and October.

The lowering of the water table and changes in the tree stand also contributed to monthly and annual runoff but the monthly distribution cannot be estimated.

Maximum runoff

Both spring (snowmelt) and summer maximum runoff increased as a result of draining (Figures 122 and 123). The increase was 31% on the average in the spring and 131% in summer. The increase was statistically significant in both cases.

The increase in total runoff described above also affected the maximum runoffs.

It is suggested that the increase in maximum runoff was largely due to the accelerating effect of the ditches on the flow. The flood lakes that normally formed on the natural peatlands also disappeared and their smoothing effect on runoff was eliminated. Drying of the surface layer of the peatland caused a negligible increase in infiltration; during heavy summer rains in particular the moisture deficit of the peat was soon filled and the runoff peak was not reduced by increased infiltration. Thus the runoff caused by heavy rains increased most. After drainage spring runoff cameon average 1.5 days earlier than without drainage. This may be partially due to clear-cutting. When a drained area is situated in the upper part of a river basin it tends to increase flood peaks.

Minimum runoff

The minimum runoff for both winter and summer increased markedly as is shown in Figure 124; both increases were statistically highly significant. Draining made the minimum runoff of the Huhtisuonoja basin rather similar to the control basin; the ditches made flow possible at all seasons of the year, whereas previously there was only a short shallow natural channel on the lower part of the Huhtisuonoja basin. In addition, the main ditches penetrated the pervious mineral soil which acts as an underground drainage channel and augments the effect of drains.

Conclusions

The drainage of a bog for forestry purposes changed the hydrological conditions significantly, as follows:

1. A decrease in evapotranspiration caused by the drop of the groundwater table and the drying of the upper part of the peatland led to an increase in total runoff.
2. An acceleration of flow caused by the drainage network led to an increase in maximum runoff.
3. A possibility of flow throughout the year, during cold winters and dry summers as a result of the drainage network led to an increase in minimum runoff.

Evapotranspiration may increase somewhat later as the new vegetation grows, so that the first effect will decrease. On the other hand, the second and third effects will continue assuming that the drainage network is maintained.

REFERENCES

Kovner, J. L. and Evans, T. C. 1954. A method for determining the minimum duration of watershed experiments. Transactions, American Geophysical Union No. 4, Washington, D.C.

Mustonen, S. E. 1965. Effects of meteorological and basin characteristics on runoff. English abstract. Soil and Hydrotechnical Investigations 12. Helsinki.

Mustonen, S. E. and Seuna, P. 1971. Influence of forest draining on the hydrology of peatlands. Publications of the Water Research Institute 2. Helsinki.

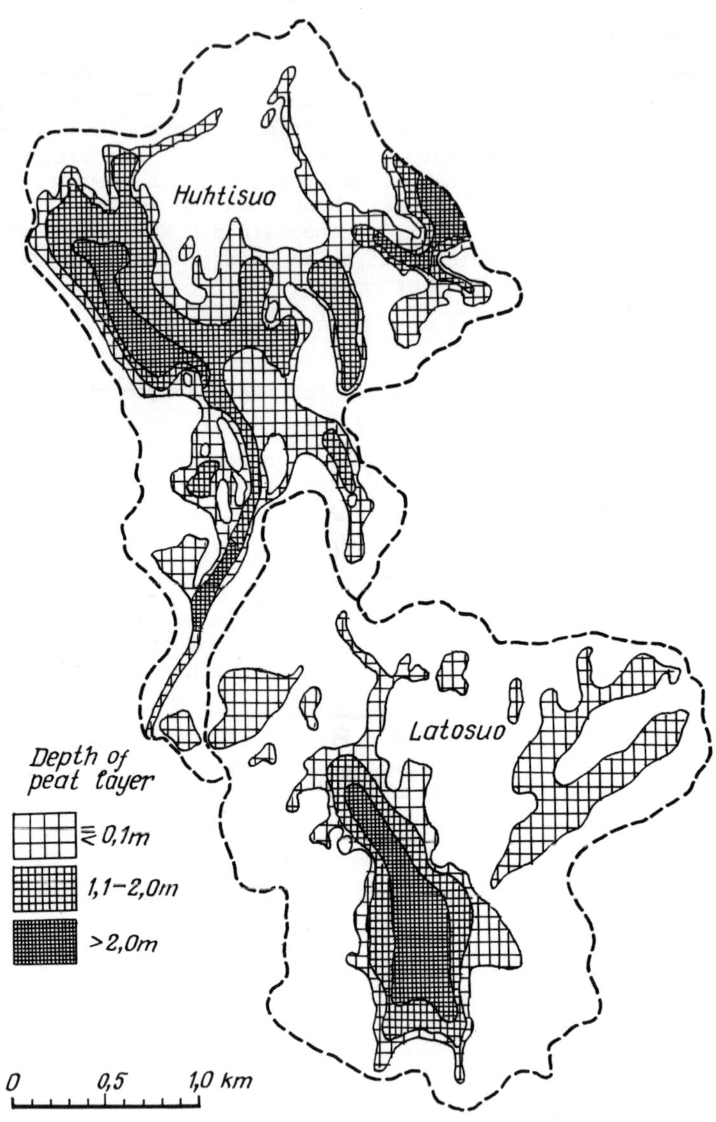

Fig 119. Map showing the control basin, Latosuonoja, and the experimental basin, Huhtisuonoja, after draining in 1970.

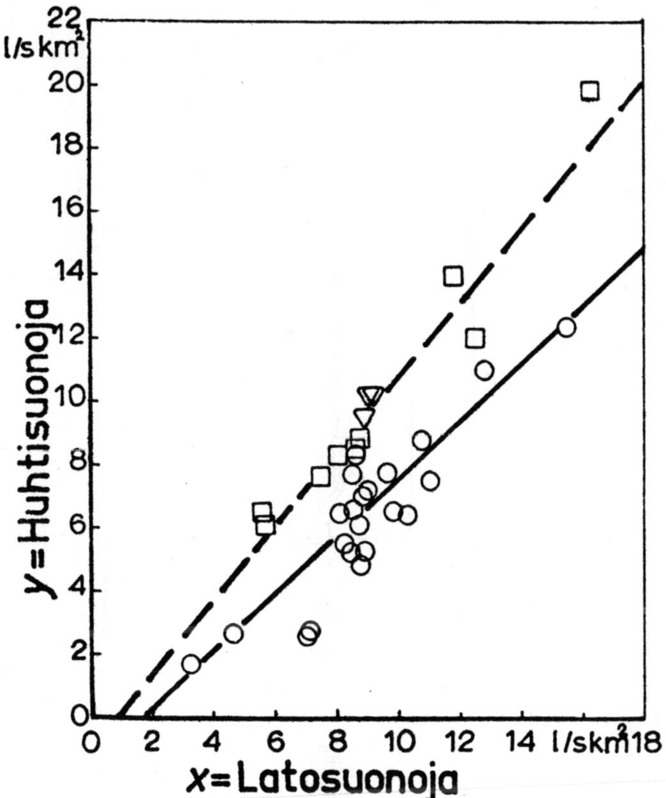

Fig 120. Correlation of the mean annual runoff (l/s km^2) of the basins.

1. 1936 to 1957 (circles and full line y = 0.91x-1.53, r = 0.94).
2. 1961 to 1969 (squares and broken line y = 1.163x-0.88, r = 0.98).
3. Triangles indicate years 1958 to 1960.

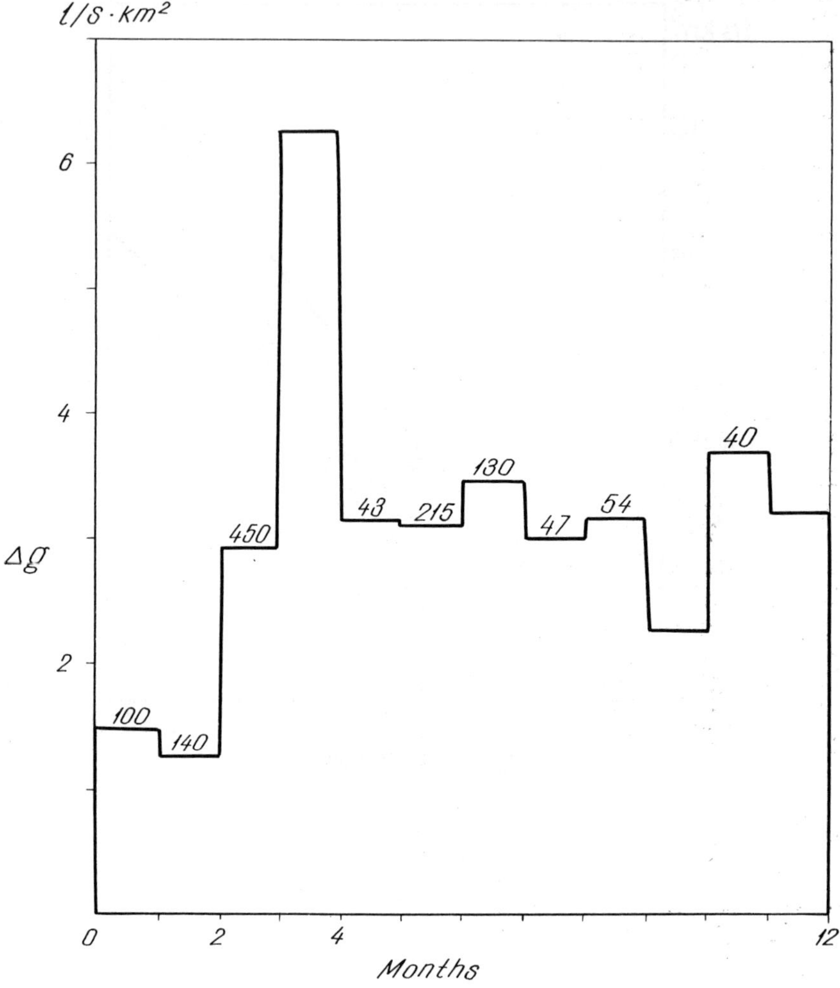

Fig 121. Average increase in mean monthly runoff (l/s km^2) in 1961 to 1969 caused by draining. Figures above the columns indicate the percentage increment.

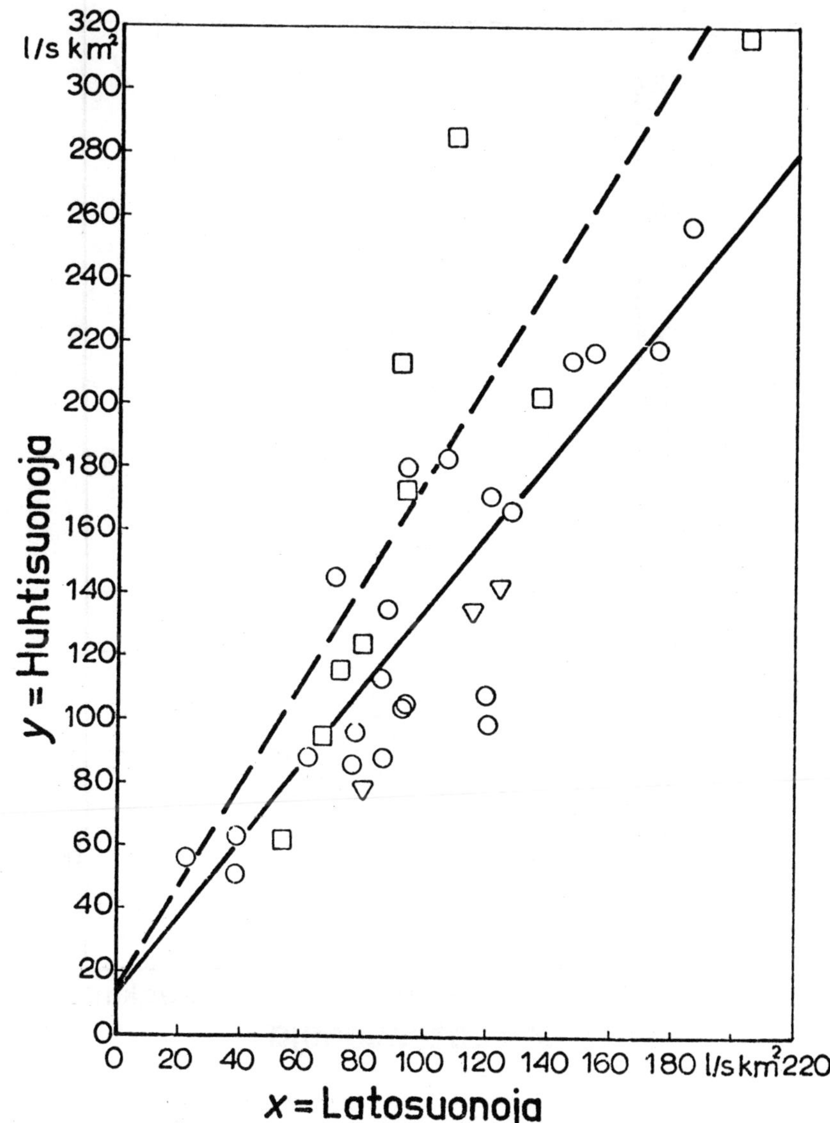

Fig 122. Correlation of the spring maximum runoff of the basins.
1. 1936 to 1957 (circles and full line y = 1.216x + 12.6, r = 0.87).
2. 1961 to 1969 (squares and broken line y = 1.61x + 14.8, r = 0.85).
3. Triangles indicate years 1958 to 1960.

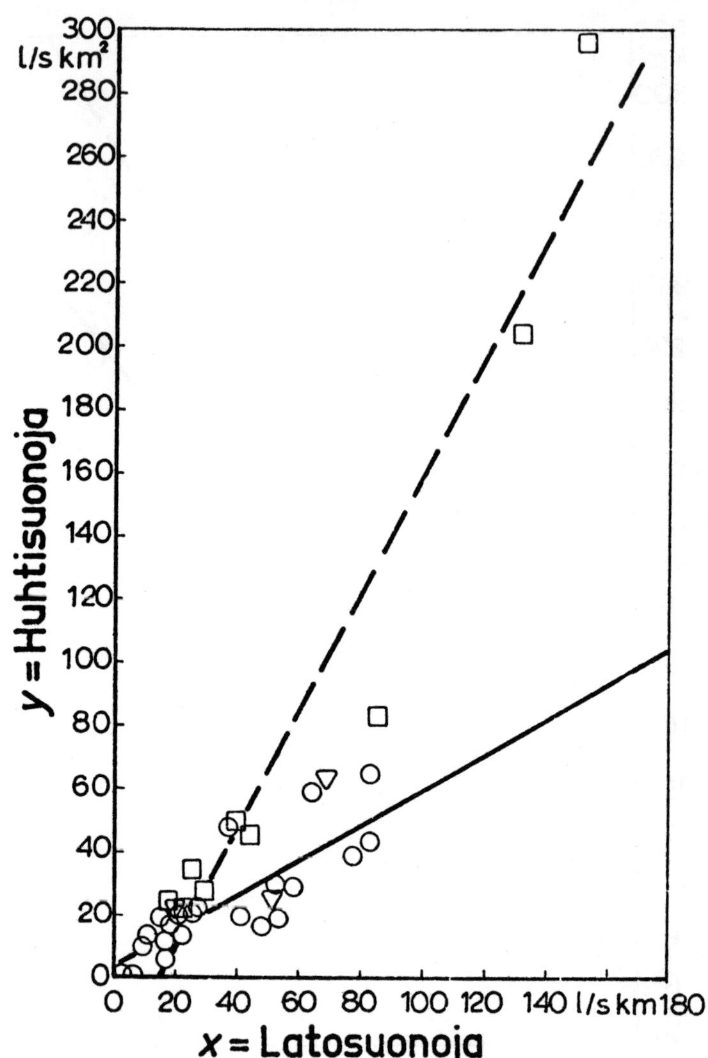

Fig 123. Correlation of the summer maximum runoff of the basins.
1. 1936 to 1957 (circles and full line $y = 0.56x+3.1$, $r = 0.81$).
2. 1961 to 1969 (squares and broken line $y = 1.868x+27.8$, $r = 0.97$).
3. Triangles indicate years 1958 to 1960.

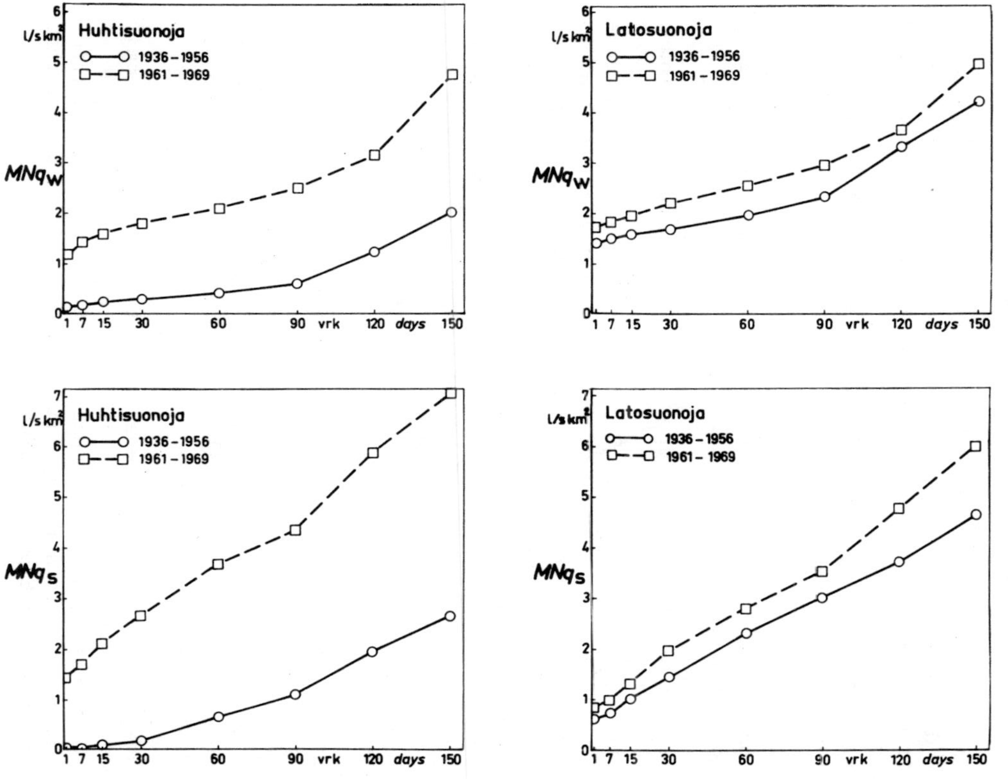

Fig 124. 1-150 days in winter and summe mean minimum runoff MNq_w and MNq_s, respectively, from the basins. 1. 1936 to 1956. 2. 1961 to 1969.

DISCUSSION

F. Zuidema (The Netherlands) - The drain arrangement is supposed to be in accordance with the required groundwater level. Mr Heikutainen reported earlier that the required groundwater level is at 60 cm below the soil surface. You have reported the level at 30 cm. There is some disagreement.

Can you see any danger that trees will fall with so high a root layer and groundwater level (both at 30 cm)?

S. E. Mustonen (Finland) - The groundwater level fell to about 30 cm after drainage. This is an emergency level. But within the summer it may reach 35 to 40 cm. Before the drainage, the groundwater level was at the surface.

Yes, there is some danger if roots are small, but here trees suit the existing conditions.

L. Wartena (The Netherlands) - In your paper you have suggested that evapotranspiration will increase due to water retention by trees. Have you made any measurements?

S. E. Mustonen (Finland) - No direct measurements have been made.

R. Eggelsman (FRG) - Your results concerning increase in the run-off with peat subsidence and decreasing micropores are similar to those of earlier work on high bogs in West Germany. Soil samples from a pasture and from a conifer wood (32 years of age) sited on a drained bog indicated higher moisture contents in the pasture than in the wood. The work was done in autumn 1961 under similar climatic conditions. The lowest water content is found in a soil layer of 1 m that results from retention of rainfall moisture by tree roots. Peat soils under pine woods are always drier than those under pastures or left uncultivated.

S. E. Mustonen (Finland) - Thank you. I agree with you.

V. B. Fashchevsky (BSSR) - What is the peat thickness on the basin? What is the vegetation cover of the basin? Please give some details about the mechanical composition of the soils and the dominating formations under the peat.

S. E. Mustonen (Finland) - The average peat thickness is 130 cm. The vegetation cover consists of different kinds of grass. The mechanical composition is different. Sometimes very low humification is found. Peat is mainly underlaid by sand and gravel that favours water penetration to drains.

GENERAL DISCUSSION

HYDROLOGICAL CLASSIFICATION OF LANDS WITH EXCESSIVE SOIL MOISTURE

D. N. Kats

All-Union Research Institute for Water Engineering and Reclamation, Moscow, USSR.

Prof. Ivitsky showed that the optimal depths of underground water for crops lie in a very narrow range. However, in drained systems the groundwater depths at various growing periods may deviate from drainage norms. This may adversely affect crop yield. To ensure that soil moisture conditions are optimum for the crop, requires detailed seasonal, annual and long-term hydrological forecasts, which unfortunately are not made. I would like to suggest that an accumulation of comparable reference data be made under the IHD Programme, for classification of waterlogged areas is highly desirable world-wide as a basis for compiling a general map of hydrological regions of lands requiring drainage for subsequent agricultural use.

THE EFFECT OF MARSH-RIDDEN AREAS AND DRAINAGE OF THE WATER BALANCE AND HYDROLOGICAL REGIME OF RIVERS

S. M. Perekhrest

Ukrainian Academy of Sciences,
Kiev, Ukrainian SSR.

The topics covered by the Symposium are of theoretical and practical importance, particularly for more effective use of marsh-ridden areas, for the forecasting of possible changes in the water balance and environment resulting from drainage and development of marsh-ridden areas whilst providing measures for protecting the environment from the harmful effect of this human activity.

Hydrological and comprehensive studies of marsh-ridden areas in all countries are not being given sufficient attention. It is therefore impossible to make a scientifically based forecast of all the changes that will occur in the natural elements due to drainage reclamation over vast areas. There is a need to initiate and develop comprehensive investigations of marsh-ridden basins in every climatic zone to estimate the effect of mires and marsh-ridden areas and of reclamation works on the water balance and river regimes.

One of the important elements of the environment considerably affected by mires, as well as by drainage works and intensive agricultural use of the drained lands, is the water balance and water resources of the basin under drainage because water is necessary not only for crop yields but also for the development of all branches of the national economy.

The Board of Investigation of Productive Forces of the Ukr. SSR, Ukrainian Academy of Sciences, has been investigating the whole combination of natural conditions (climate, hydrology, soils, vegetation, geological structure, etc.) which affect the water balance and hydrological regime of rivers of marsh-ridden areas as well as various methods used for the drainage and development of bogs and their effects on the water balance and environment.

The research work was carried out in the upper part of the Dnieper (Polessie) on undrained areas, on drained areas without regulation of the water-air regime of soils and drained areas with regulation of the water-air regime by irrigation which are intensively used for agriculture.

To save time, I will not describe in detail the procedure but will only present the results obtained.

The studies have revealed that formation of runoff and the water balance of marsh-ridden areas are greatly affected by a combination of natural factors such as climate, soils, vegetation, hydrology, relief, etc. After drainage, additional effects appear related to the methods of drainage and the use of the drained lands. Far reaching comprehensive investigations are therefore necessary to study the hydrological regime of marsh-ridden areas before and after drainage.

On a marsh-ridden part of Polessie in the Dnieper basin there are areas where the runoff is 1.5 to 2 times smaller than that from neighbouring basins because of different hydrological conditions. In these basins a considerable part of the streamflow is transferred to other basins in the form of underground flow.

In the upper part of the Dnieper the annual and summer-autumn flows increase considerably with the size of marsh-ridden areas in the basin.

During and in the 3 years after construction of reclamation systems, the streamflow increased by 30 to 50% due to discharge of groundwater accumulated over the ages in the drained area and in the catchment.

Later, when the drained areas are used for agricultural crops with soil moisture control, the annual flow decreased by 20 to 50% and more depending on the dimensions of the drained area and the catchment area. The decrease in flow occurs because considerable quantities of water are spent on irrigating the soil in dry periods to provide high crop yields.

When marsh-ridden areas are drained without soil moisture control, the streamflow changes very little, but such drainage methods cannot provide optimal moisture conditions in dry periods and result in lower crop yields.

Drainage of fen bogs results in the groundwater levels being decreased not only in the drained area but also at a distance of 1 to 5 km away in the catchment area. This reduces crop yields on sand and loamy-sand soils adjacent to the bogs.

As an example, we may take the Kirov Collective Farm near the village of Zavorichi, Brovarsky District, Kiev Region where crop yields on the lands adjacent to the Trubezh flood plain were decreased after construction of drainage ditches without regulating sluices. The data are summarized in the following table

Crop	Crop yield, t/ha			
	Before drainage	After flood plain drainage without regulating sluices		
	1957	1958	1959	1960
Potato	10.5	9.1	5.7	-
Fruits	13.9	7.2	3.6	6.5

After regulating sluices had been built on the drainage canals (1960-1961) with a resulting increase of the groundwater levels by 1.5 to 2 m in the flood plain in summer, crop yields on adjacent lands returned to the original values.

On loamy and podzolic soils, a lowering of the groundwater level has a smaller effect on crop yields.

In the upper part of the Dnieper basin, there are about 6 500 000 ha of bogs and marsh-ridden areas which will be drained in the next 15 to 20 years. As a result of drainage and intensive use of the bog area with regulation of the water-air regime in soils, large water quantities will be required for irrigating in dry years. To meet this demand, construction of water storage is envisaged for the regulation of spring flow, especially in high-water years.

As a result the volume of water in the Dnieper at Kiev will decrease in a medium-water year (P=75%) by about 15% and in a high-water year (P=95%) by 40% in the years 1990 to 2000. The mean annual runoff of the Dnieper at Kiev will decrease by about 25%.

In conclusion, the hydrological regime of marsh-ridden areas depends on a whole combination of natural effects and after reclamation it is also dependant on the methods of drainage and on the use of the drained lands. The runoff relations found for some basins cannot therefore be extended to different areas without comprehensive studies.

EFFECT OF DRAINAGE RECLAMATION ON RIVER FLOW

G. P. Kubyshkin

Institute of Water Engineering and Reclamation, Kiev, Ukrainian SSR.

The effect of drainage reclamation on streamflow is of great importance in the design and operation of drainage systems. To investigate this effect, streamflow data from a network of gauging stations of the Hydrometeorological Service of the USSR have been analysed.

In the northern regions of the Ukraine, insufficient data are available for the comprehensive analysis of streamflow to determine the effect of drainage, particularly since drainage usually extends over only a part of the catchment areas. The data from observation stations allow a comparison of the flow for the periods before and after drainage but, as the observation periods are not always sufficiently long, the conclusions made cannot be considered valid for many of the drained catchments in the Ukrainian SSR.

The following procedure has been adopted in this study. A river with a known flow is selected in a basin in which there are marsh-ridden areas. For each drained basin considered, an adjacent control swamp catchment is chosen with similar topographic, climatic, edaphic and other properties, the control basin either having never been drained or containing negligible areas of drained land. The value of the change in the drained basin is found by comparison of the characteristics of both basins. If comparable data are available, instead of one control basin, two, or more are selected to obtain more reliable data.

Let us consider applications of the above procedure to practical determination of streamflow properties such as mean annual discharge, maximum snowmelt and spring runoff depth.

Analysis has revealed that mean annual discharges of many rivers, for example, the Irpen, decrease after drainage. In this case, the drained area covers only about 3% of the whole catchment area but, as the swamp is located in the lower river basin and has a length of about 100 km, and since the runoff from the whole higher basin is measured at the village of Mostishchi, the effect of drainage on the streamflow is detectable.

Some cases are encountered where drainage does not affect the annual discharge. For example, the drained Nedra basin at the settlement of Berezan shows no alteration in discharge attributable to drainage when its annual discharge records are compared to those of the Udaya at Priluki (a control basin).

Decrease in the flow after swamp drainage implies that a part, sometimes a fairly large part, of the annual discharge is lost because of additional evaporation from drained lands or because a part of the surface flow is lost to groundwater.

As drained swamps are developed for agricultural use by both drainage and irrigation, evaporation and transpiration may somewhat exceed that from undrained swamps. These additional losses by evaporation add to the total surface flow losses.

Double mass curves of annual runoff from neighbouring river basins indicate that the runoff from catchment areas decreases generally but not invariably after drainage.

Increase in the annual flow after reclamation should be expected where river beds and lateral drainage canals become considerably deeper as this will result in additional groundwater flow and also where drainage ditches discharge water from individual basins which were formerly closed.

In the same way, as the data of annual flow, double mass curves of spring flow are plotted for two river sections. As before, a decrease in the spring flow is usually found after drainage, as in the case of the Grezlya, although in certain cases either no change or some increase of the flow is observed, as, for example, in the Tnya basin.

Decrease of spring flow in some years may be attributed to the fact that a part of the volume is absorbed by a dry peat soil above the water table and by mineral soils of the adjacent area. But often, autumn rains and winter snowmelts add a considerable amount of moisture to peat soils and no regulating capacity exists. Frozen peat soil absorbs very little moisture. Also in years with a low spring flood, the whole flow passes along canals discharging water to the flood plains and regulating capacity of peat soils only partially operates.

The maximum discharge is known to depend on the spring flow and the water yield in the rising phase. Consequently, in years with a low spring flood when water is discharged along the canals and there is no overbank flow, maximum discharges from rivers with drained flood plains should not decrease but rather increase in view of the higher flow velocity. In years with a high spring flood, decreased maximum discharges should be expected because of great variation of the velocity distribution across the channel and overbank flow because of regulation of spring runoff depths by the absorbing capacity of peat soils, by the canal networks with sluices and reservoirs. Drainage may not always affect maximum discharge, although for some rivers maximum discharges decrease after drainage, whilst others show increases. Stability of maximum discharge is understandable in the cases where there has been little change in the runoff conditions after drainage of the flood plain.

These studies have shown that drainage works may either increase or decrease streamflow. Further study is required.

In addition to the interrelationships discussed above, the type and magnitude of the changes caused by drainage depend on the dimensions and location of the swamp in the basin and the drainage procedure. With regulation of the soil moisture regime by drainage alone simple networks of discharge canals are constructed. This drainage method is gradually being replaced by regulation of the soil moisture regime both by drainage and irrigation to eliminate the dependence of crop yields on weather conditions. In the latter case the quantity of water evaporated is certainly increased and the regime in water courses on drained areas during the growing period will be different from that of systems of drainage only.

Recently, on drained swamps, drains have been made for rapid discharge of excess water in spring and for supplying water to plants in growing periods, receiving insufficient precipitation. In this case, the natural regime is changed by drainage to the greatest extent and flow losses will be the largest.

To give an answer to all the problems is impossible as yet. More time is required for accumulation of natural data which together with adequate theories of flow formation will provide the solution of many problems.

THE EFFECT OF BOG DRAINAGE ON STREAMFLOW

P. F. Vishnevsky

Ukrainian Research Hydrometeorological Institute, Kiev, USSR.

In my short speech I should like to dwell upon a very important problem discussed at the Symposium, namely, the effect of bog drainage on the streamflow. This problem is at present of great practical importance since on it depends the fate of the water undertakings, which are being designed and realized on a large scale in marsh-ridden areas. Many specialists, especially design engineers, are anxious to know what changes will result from drainage. However, it is impossible to find a single answer to this question.

Data which have been obtained by some workers present at the Symposium on changes of streamflow in particular watersheds cannot be extended to different basins whilst disregarding the mire type, hydrological conditions in the basin and conditions of water supply to the marsh-ridden areas.

In connection with the development of a project for complex use and protection of water and land resources in the West Bug basin, the possible effects of bog drainage on streamflow from this catchment were to be estimated. In the West Bug basin, land reclamation has been carried out for many years, but, during World War I and especially World War II, the existing drainage systems were destroyed. In recent years many systems have been restored or built anew.

Correct estimation of the effect of drainage on streamflow may be made on the basis of water balance data and from evaluation of the effect of drainage on the main components of the water balance, such as precipitation and evaporation. It may be supposed that precipitation is not changed by drainage and will not change in future since it depends on the general conditions of moisture supply to the region which, in turn, depend on the distribution of atmospheric circulation over large areas of the planet. Evaporation is one of the main elements of the water budget. As concluded by Prof. A. I. Ivitsky, evaporation from drained bogs used in agriculture is actually almost the same as that from a natural undrained bog. Thus, if evaporation is not affected by drainage an increase in the streamflow should not be expected from watersheds where marshes will be used in arable agriculture, rather reclamation produces a redistribution of the streamflow.

To evaluate the changes in the streamflow from the West Bug watershed attributable to drainage, double mass curves were plotted of annual flows for rivers with drained areas, and related to control basins where no drainage works were carried out or where they were inconsiderable. The various graphs revealed no decrease in the flow due to drainage.

The graphs of the annual rainfall-runoff relationships which should reflect all effects of human activity, revealed no definite trends in flow. The flow conformed with the precipitation.

Thus, the changes, if any, are inconsiderable and within the accuracy of hydrological studies.

It should be mentioned that swamps in the West Bug basin are mainly supplied by artesian groundwater from the Cretaceous strata widespread in the region, a fact which was presented in the paper by V. E. Alekseevsky and K. P. Tereshchenko. In such a situation, drainage ditches serve to discharge artesian groundwater to rivers. Under these conditions in the West Bug basin a streamflow decrease after drainage cannot be expected. Moreover, a slight increase in the flow may be expected although we have no evidence of the fact as yet. It may be concluded that the effect of drainage on streamflow has not been completely studied.

EXAMPLES OF PEATLAND DISTRIBUTION IN SOUTH EASTERN NORWAY

G. Goffeng

Agricultural University of Norway Asas, Norway I

In Norway great variations in both the types and areal distribution of peatland are common within short distances, depending on sudden changes in geological, topographical and meteorological and other conditions.

In south-eastern Norway, below the level of the late glacial marine transgression, which is at an altitude of about 200 m, the areal peatland frequency is less than 5% concentrated on lowlands floored by Precambrian rocks. Most reclamation of peatland for arable use is found in the higher levels of this zone.

Above this marine limit, the peatland frequency increases suddenly to about 10% and when an average altitude of 500 m is reached, peatland covers 20% of the area. This percentage area is maintained up to the tree-line at about 900 - 1000 m above sea level. The highest peatland frequency is found in those areas with little relative relief and a considerable depth of morainic deposits. Bogs are common, especially above 400 m altitude. Drainage for forest generation take place below about 400 - 500 m above sea level.

Above the tree-line, the percentage area covered by peatland falls rapidly as a result both of less production of organic matter and of topographical conditions.

The highest concentration of peatland in south-eastern Norway, above 50% areal frequency, is situated on parts of the Eocambrian quartz-sandstone plateau where geological, topographical and meteorological conditions favour accumulation of organic matter. The trend of the scarp of this plateau lies across the path of the main, rain-bearing wind, and the terrain level rises from 300 m on the Precambrian areas in the south to an altitude of about 600 m on the plateau. Considerable orographic precipitation is caused by this, and the evaporation is low at this high-level.

This plateau has low relief and is covered to a great extent by deep moraine. In parts the surface has been fluted by glacial erosion which has affected the detailed distribution of the peatlands. The area of peatland is also obviously related to the natural drainage. Younger glaciofluvial deposits in eskers of irregular disposition are also found in the same area. Hardpan formation, which is common in these stratified and originally permeable deposits, cause the natural drainage courses to be lengthened. These peatland areas represent a valuable reservoir for natural runoff regulation of particular importance at times of snowmelt. Frozen ground is rare because of early and deep snowcover.

Water balance studies have started in these areas, and detailed description of the different hydroecological systems are expected to be of importance when interpreting the collected data in the future.

DRAINAGE RECLAMATION FOR FORESTS

S. E. Vompersky

Soil Laboratory, USSR Academy of Sciences
Moscow, USSR

In the last decade, the hydrology of marsh-ridden areas has been attracting the ever-increasing attention of many investigators in connection with reclamation of the areas and improving conditions for forest growth. Apart from their scientific interest, these studies are important for the selection of the best methods of forest drainage and for estimation of the effect of forest drainage on the environment and water balance, in view of the large-scale of the drainage works in swamped forest. Interesting papers presented to the Symposium on this subject are by Dr V. K. Podzharov (USSR), S. E. Mustonen and P. Seuna of Finland.

Nevertheless, science in all aspects of forest drainage and afforesting of bogs is lagging behind the techniques of reclamation for agriculture. In forestry, there are very few examples of experimental drainage and theoretical analysis, an empirical approach dominates and often a method useful for one region is extended to other areas where the expected results are not obtained. Therefore, on the same problem the views and prediction of different specialists often differ as has been revealed in particular at the present Symposium.

Specialists in various countries consider that they know the necessary drainage rates but the situation is much worse as regards criteria of moisture content of the rooting zone and especially for estimates of the effect of forest drainage on the water balance. The latter is known to be of primary importance for the creation of a scientific basis for optimal forest drainage methods.

In view of the above, I should like to present some results of 7 years of observation of the water-balance components in forest drainage catchment experiment in the north-western regions of the RSFSR.

According to 7 years of observation, the normal annual discharge, during the period from the middle of April to the 1st of November was 114 to 115 mm both for undergrowth and medium-aged forest on mesotrophic peat soil with drain spacings of 110 m.

On an oligotrophic bog, without forest and drained by a similar method, the normal annual discharge was 142 mm. Afforested basins have a long spring flood compared to basins free of forest; the contribution of the spring flood to the annual discharge being usually from 68 to 89 per cent. The differences in summer and autumn discharges between afforested bogs and mires free of forests, are considerable when drained at the same rate; the discharges on areas without forests are found to be 2 or 3 times larger.

Smaller spacings between ditches (from 2 to 4 times smaller) with the same depth, only affect the runoff during floods. Specific discharges in low-water periods without rain differ slightly if at all in spite of different drainage rates. Varying effects of ditch depths and spacings were revealed to some extent. It was found that irrespective of the age, with equal ditch depths, pine woods have equal volumes of low-water flow in summer (from 2 July to 17 August), namely, 16 to 17 mm and 6 to 7 mm with depths of ditches of 1.0 to 1.1 m and 0.5 to 0.6 m, respectively. A two-fold decrease of the ditch depth results in a more than two-fold decrease of the summer discharge. The spacing of ditches of equal depth (1.0 to 1.1 m) has much effect on the summer flow, even a four-fold increase (from 55 m to 220 m) decreases the summer flow only from 34 mm to 21 mm (38% decrease). On a bog free of forests, increase in drain spacing from 55 to 108 m and from 108 to 220 m results in a 14% or 13% decrease of the total discharge.

With the same drainage rate, considerably higher groundwater levels are found for oligotrophic peat land covered by low pine than for mesotrophic peat land with a pine wood 70 to 80 years old. Moreover, on a high bog without forest, a decrease of the ditch spacing to 55 m does not produce the drainage rate which is found in a drained forest with a spacing of 101 to 118 m, in spite of a considerably smaller discharge in the latter case.

Evapotranspiration was estimated as the residue of the water-balance equation. On an average over 7 years, it was found to be 45 mm larger for a drained forest than for a drained bog without forest, the drainage rate being the same. Direct calculations made for particular years have revealed some variations in the value, but the variations are often small. For example in 1970, evapotranspiration in pine undergrowth was 315 mm between 4 May and 8 September, in a middle-growth forest it was 295 mm, whereas evapotranspiration from an oligotrophic bog which was drained at the same rate, was 235 mm.

The data presented show that justification of methods and means of drainage of forests and swamps requires a knowledge of the water-balance mechanism in drained forests which must be the object of very careful and comprehensive investigation.

EFFECTS OF SOIL FREEZING ON THE HYDROLOGICAL REGIME OF FORESTED MARSH

Dr. N. M. Fyedorova

Soil Institute,
Moscow, USSR.

I refer to the effect of soil freezing on the hydrological regime with special reference to the swamped forest zone of West Siberia, since freezing and melting effects are fairly widespread in humic swampy regions. Large swamp areas in the taiga zone of West Siberia may be mainly attributed to such factors as high precipitation, low evaporation, a very small regional slope, heavy soils and high groundwater level. In geocryological respects, the area is almost unknown. The area, particularly the Middle Taiga Subzone, is a region where spots of perennially frozen soils, which are melted during warm seasons, are widely distributed among melted soils which are frozen to a depth of 2 m in cold seasons of 8 or 9 months in duration. The temperature of the soils ranges from -5°C to +15°C. Within this range the thermal behaviour of soils is quite different for different types of relief.

First, for most of the year soil temperatures range from +5°C to -5°C; in particular, in the period of negative temperatures, the temperature range is mainly from 0°C to -2°C. These temperature ranges correspond to those of the main phase transformations and of the migration of unfrozen moisture in a frozen layer because of temperature gradients; at such temperatures, soil is almost cemented with ice and becomes actually water impervious. Besides, phase transitions of moisture in frozen soils are accompanied by high rate physical and chemical conversions and transformation of the internal structure of the soil bed, which are not completely reversible when the soil is melted; this can affect the physical, chemical, hydrological and mechanical properties of soils.

The second feature characteristic of the thermal behaviour of soils of the Middle Taiga of West Siberia is instability of the temperature dynamics, particularly of the frost regime over many years. The interrelationship between the temperature and water regimes should be considered in treating variations in surface and groundwater flow, moisture circulation in the whole area and the dynamics of swamps, water logging and loss of forests which are common in West Siberia.

Freezing of soils has a great effect on the whole combination of processes occurring under the conditions referred to as the 'hydrological regime' such as moisture supply to the soil; the type of water retention and migration in the soil beds; moisture discharge and changes of state. These effects are most pronounced in hydromorphic soils in autonomic locations seasonably frozen over many years. General downward moisture migration and internal discharge are essentially limited; freezing has a positive effect increasing the excess moisture accumulated in soils by the formation of temporary perched groundwater. In general, although the soils are highly saturated, for much of the year, including the period of freezing, the moisture content of the soils varies from the minimum moisture capacity to complete saturation. Circulation of moisture in mineral beds occurs at fairly low rates; hydrologically the most active beds are organic mineral layers of small thickness (below 1.5 m).

Freezing of soils causes redistribution of precipitation over the surface area by preventing internal discharge of spring and rainfall water, and by favouring surface flow towards depressions and swamping the latter. So, freezing causes hydrothermal regimes of heteronomous soils (with an additional moisture inflow) to be smaller to those of autonomous soils. Total moisture gain after redistribution of an annual normal precipitation of 500 mm will amount to 500 mm on flat plateaux with impeded drainage; to 340 mm on gentle slopes and to 600 mm in depressions. In the first case small high bogs are formed, in the second case swampy forests occur, in the third case fen bogs are formed which after a very short period become high bogs, sphagnum very often developing over a mineral soil. The boundaries between these zones are variable.

Combinations of temperature and water regimes in loam and sandy loam soils, in particular, very long periods at temperatures from 0°C to -2°C with high moisture content, allow these soils to be considered as a field of intensive cryogenesis. Peculiarities of cryogenic processes in the soils of the region considered are as follows: the gradual moisture redistribution across the frozen bed due to temperature gradients; the formation of temporary perched groundwater tables above and below the frozen layer resulting in moisture redistribution across the layer and phase transitions; the interaction between moisture and the mineral portion of the soil affecting many important properties of the soils, particularly their macro and micromorphology.

As frozen and melted soils in humic regions are widespread both in the northern and southern hemispheres, due attention must be paid to these processes. To aid understanding of the soil genesis in cold humic regions and of their water regime, to identify their properties for economic use the boundary of the cryohypergenesis zone should be traced so as to subdivide the zone into the different cryogenic regions.

HYDROLOGICAL AND HYDROGEOLOGICAL EFFECTS OF RECLAMATION BY DRAINAGE

P. I. Zakrzhevsky

Byelorussian Research Institute of Reclamation and Water Economy, Minsk, BSSR

Over the whole area of the BSSR, large-scale works on reclamation of bogs and marsh-ridden areas are now being carried out. Because of these works, lands which were earlier useless for agriculture, produce high yield crops.

Construction of reclamation systems disturbs the existing dynamic equilibrium of water circulation on the catchment area under drainage and changes moisture regimes in the unsaturated zone, in groundwater levels, in streamflow amounts and temperatures. Under different climate types, favourable conditions for agricultural crops are restricted by various regimes, as may be clearly seen from the papers presented to the Symposium. Therefore, for each region, in addition to general problems of the hydrology of mires, there are particular problems of which the solution is very important for the reclamation of the region.

In Byelorussia such limiting regimes are probably the moisture regime in the unsaturated zone and the streamflow regime. Knowledge of these regimes is of particular interest for the development of the most favourable water regime for agricultural crops and to meet the water requirements of industry with the minimum investment in the construction of water intakes and reclamation systems.

It is therefore very important to forecast the streamflow after the reclamation works have been completed in large areas of Polessie, where all the swamps will be affected. To start the investigations after completion of the works is quite inadmissible. Field observations are required as well as mathematical simulation of the changes in regime.

Mathematical simulation of the hydrological regime should be based on established relations of water migration by surface runoff over the catchment area and underground. Because of very complicated boundary conditions for all hydrological processes over the area simplification of the simulation is only possible with known streamflow behaviour and other data for control catchment areas. Control catchment areas. Control catchment areas are adjacent or nearby basins, one of which contains a drained bog the other a natural bog, located in one geographical region. The dimensions of the paired basins should provide a travelling time of discharge from the most remote point, the source of the stream within 2 to 4 hours. These catchments must be equipped for studying runoff and other hydrological processes such as evaporation and groundwater replenishment.

The results of field experiments on streamflow and its components should allow the difficulties of mathematical simulation of hydrological processes on large catchment areas to be met.

The most difficult period in the operation of intake units is the spring flood; the rate depends on snowmelt. Transformation of snowmelt water into runoff on catchment areas with drained and undrained bogs occurs in different ways. Observations of the streamflow from adjacent watersheds containing drained and undrained bogs have revealed essential differences both in spring flood flows and in low-water periods. On a drained bog, higher discharges because of snowmelt are found 3 to 4 days earlier under a fairly steady temperature regime and these differences may be even larger under an unsteady temperature regime. These differences in the measured flood discharge distribution may be attributed to peculiarities of the runoff formation on drained and undrained bogs.

On drained bogs, snowmelt water levels quickly saturate the ground and result in higher discharge to the drainage network along which the water reaches the closing section within 24 hours.

On an undrained fen bog, snowmelt water flows to the bog and accumulates there causing an elevation of the frozen bog surface (ice and frozen peat dust). Some increase in flow to the water channel appears simultaneously due to filtration in the active layer. Observations in the present study have revealed no free water surface on the bog perhaps because the accumulated snowmelt water lifted frozen peat dust and ice. The latter effect results in a larger void ratio in the lower peat beds, the pores being completely filled and a noticeable increase of the flow from the undrained bog. After all ice particles contained in peat dust are melted, the dust sinks below the water surface and the dust becomes a classical free water area. The flood flow also drops sharply just after the peak.

Drainage network construction changes not only the spring flood discharge, but also the low-water flow. The discharge becomes more sensitive to rain, so that diurnal variations of the discharge are found. At first, the streamflow is sharply increased because of decreased bog water levels, deepened water channels, a more dense drainage network and modification of the groundwater catchment discharging to the new channel network.

The streamflow from two bog basins with no drainage works compares with streamflow from the drained basin as follows.

Years		1965	1960	1967	1968	1969
Precipitation		1.0	1.24	1.29	1.27	1.05
Streamflow from two undrained basins	Rudavka at Rudnya	1.0	0.98	1.33	1.33	0.93
	Yaselda at Bereza	1.0	1.35	1.42	1.63	1.13
Streamflow from drained basin		1.0	1.01	2.56	3.17	1.63

The changes were mainly caused by an increase of the groundwater contribution.

At the time of construction of the drainage system the run-off direct from the bog was about 270 mm and the additional groundwater flow discharged by the drainage system was 358 mm. For the subsequent hydrological year, the values were 23 mm and 197 mm respectively. The additional groundwater flow is caused by an increase of the groundwater catchment shown by measurements of the groundwater levels of the undrained bog; in this area the groundwater level slopes towards the drained bog at a distance of about 2 km from the previous watershed. Piezometers installed in the bog areas have shown vertical pressure gradients where no pressure head was found before drainage.

THE EFFECT OF RECLAMATION BY DRAINAGE ON GROUNDWATER LEVEL

K. F. Yankovsky

Central Research Institute of Complex Use of Water Resources, Minsk, BSSR.

For areas under drainage, lower groundwater levels are designed in advance and, in most cases, are obtained according to the standard drainage rates. On the areas adjacent to the drained lands the dimensions of the zone with lower groundwater levels depends on many factors which are at present insufficiently understood. The available methods for estimation of the effects of reclamation systems on the groundwater level in the adjacent areas and the simulation procedure for the filtration process based on the hydrological parameters of a particular region require verification by natural observations. Correct evaluation of the decrease in the groundwater level from field observations in the zone indirectly affected by reclamation is therefore of particular importance. Lower groundwater levels in the zone indirectly affected by reclamation must also be observed against a background of natural variations of the level, which mainly result from the variable amount of precipitation. Therefore, the method of evaluating the decrease in groundwater level both within drained areas and on adjacent lands involves the use of plotted relationships of the levels versus precipitation.

In the southern part of the Byelorussian SSR a relationship between the spring groundwater levels (March-April) and total precipitation for the period from October to March is demonstrable. There is also a relationship between summer-spring levels and total precipitation from January to August for regions with light soils underlain by sand and from October to August for regions with medium soils underlain by loams and sandy loams moraine. Thus, to get a clear relationship, the period of summation of precipitation should precede the period of averaging groundwater levels.

Analysis of field observations made in the regions considered using the method described above leads to the following conclusions:

1. Plotted relationships of groundwater levels versus precipitation show the amplitude of the natural variations in level for years with different moisture supply, also the decrease in the level due to reclamation by drainage.

2. The effect of reclamation on the groundwater levels of adjacent areas is felt over a distance from 0.3 to 0.5 km to 3 to 5 km from the drained area. The distance depends on the value of the groundwater level decrease in the drained area and on the hydrogeological conditions, such as the thickness of the aquifer and the filtration coefficients and moisture conductivity of the strata.

3. After drainage of surrounding swamps a larger (by 0.5-50cm) decrease of the groundwater levels occurs mainly in late winter and spring in the central area of a watershed (width of affected zone 2 to 10 km) than on the slopes.

LONG-TERM TRENDS IN MARSH RECLAMATION

Professor M. I. Lvovich

Geographical Institute, USSR Academy of Sciences, Moscow, USSR.

A very important aspect of the hydrology of swamps is the formation of long-term forecasts as described in the review report of Dr. Zuidema (the Netherlands) and in other papers.

The basis for long-term forecasts should be a model whose component parts include: firstly, a perfect technological solution such as the double acting drainage systems (ie used for both irrigation and drainage) which were mentioned in a number of papers and on which, I think, we are all in agreement; secondly, high plans for high crop yields meaning crop yields of 4 to 6 tons per hectare as well as high grass productivity (as emphasised in the paper by Professor Ivitsky).

It may be interesting to recall that the well-known Soviet agronomist Williams predicted a maximum crop-yield of 15 tons per ha with optimal conditions of heat, water and plant nutrition. High yields require control of the hydrological regime of the soil in combination with controlled plant nutrition by organic and mineral fertilizers. These conditions can be attained on land that is both irrigated and drained. There are grounds to support the belief that the centre of crop and grass production will shift in future to agricultural areas established on drained land. It is not by accident that some countries have obtained record yields in such areas.

Evidently, on the basis of our past experience, we can expect future hydrological transformations to be far more profound than those that have taken place so far.

Another problem is the role of reclamation activities in the transformation of ecosystems. The study of this aspect of reclamation is an important task. Mankind today is gravely concerned with preservation of fresh-water resources, especially with prevention of their pollution, which is a key problem in marsh-ridden areas where rivers cannot cope efficiently with the removal and regeneration of polluted water under natural conditions.

Lastly, the swamps of the tropical zone need to be mentioned. In the Soviet Union, the area of subtropical swamps is comparatively small and these swamps are being tackled successfully by being turned into a blossoming orchard. However, little research has been done in this field and many developing countries need assistance with their reclamation activities. This particular aspect of the hydrology of swamps undoubtedly deserves attention. It may be necessary to hold a special symposium on the topic.

In conclusion, on behalf of the Commission on Surface Water of the International Association of Hydrological Sciences, I would like to thank the Organizing Committee and to wish all hydrologists specialising in swamps and reclamation further success in their research.

MATHEMATICAL SIMULATION OF GROUNDWATER MOVEMENT

Professor V. F. Shebeko

Byelorussian Research Institute of Reclamation and Water Economy, Minsk, BSSR.

Wide use of mathematical simulation procedures requires knowledge of the hydrophysical properties of soils such as permeability, water yield, saturation deficiency, infiltration of precipitation or irrigation water, in addition to knowledge of the shape of an aquifer, of the location of confining strata and of some other properties. These are of particular significance for the simulation of groundwater movement. Many institutes of water engineering, land reclamation and water economy and offices of hydrological and hydrometeorological services are concerned with investigation and accumulation of such basic data.

B. G. Shtepa, G. V. Bogomolov and Professor D. M. Kats have all stressed the necessity of developing mathematical simulation procedures and of applying them in land reclamation investigations.

I join these gentlemen in their recommendations and encourage the wider use of mathematical simulation procedures in studies of the groundwater regime and of water balance.

SIMULATION OF THE EFFECTS OF DRAINAGE AND IRRIGATION

Professor N. I. Druzhinin

Northern Research Instite of Water Engineering and Reclamation, Leningrad, USSR

Vast areas subject to drainage or irrigation will affect the run-off distribution both in the first years after reclamation and much later.

To study the regime and balance of groundwater in the USSR, large-scale field investigations are carried out and theoretical predictions are made. In recent years, laboratory experiments involving physical and mathematical methods of simulating natural processes using analogue and electronic computers have become of great importance. Mathematical simulation procedures allow selective and differentiated estimation of the following influences on the water tables, discharges and flow velocities:

- non-uniformity of the geological structure both along and across the horizons considered;
- the value and rate of precipitation infiltration and evaporation;
- the significance and effect of each of the geological strata on the streamflow characteristics;
- the value and intensity of pressure flow.

Mathematical simulation procedures produce an illustrative picture of groundwater flow and facilitate the design of reclamation systems which take into account edaphic, geological, hydrological, topographic and climatic conditions. Because of these advantages, it is recommended that investigations be started, to compile a map of water-table contours, isopleths of pore water pressure and depths of groundwater occurrence using the method of electro-hydrodynamic analogy eventually covering all the largest regions of the USSR and starting with the regions where large-scale drainage or irrigation works are planned for the next 15 years.

At present such maps have been compiled for the Dnieper-Don Depression over an area of 437 000 km^2, for the area between the Amu-Daria, Syr-Daria and Zeravshan over an area of 500 000 km^2, for the Karelia Neck over an area of 12 000 km^2 and for the central part of the Kulunda Steppe over an area of 13 200 km^2.

These studies should be extended to the Byelorussian and Ukrainian Polessie, Baraba, Meshchera, Povolzhie, some regions of the north-west non-chernosem zone, West Siberia and Far East.

In water engineering, the method of electrohydro dynamic analogy is used for research work before construction of every large hydro-power station. This method is also used for designing a rational underground contour of constructions such as dams, hydro-power stations, sluices, etc, back filters, for determination of output gradients, filtration discharges and output velocities to drainage devices. Water reclamation systems are very complicated and their effect on the environment (particularly on the water table) may be even more drastic than that of hydropower constructions. To avoid undesirable consequences and harmful processes resulting from reclamation measures, groundwater studies should be made first under natural conditions and then in the process of construction and operation of the reclamation systems. The method of electrohydrodynamic analogy may be of great significance in the study.

Professor V. M. Zubets

President of the Symposium, Byelorussian Research Institute of Reclamation and Water Economy, Minsk, Byelorussian SSR

Now that all the papers included in the Symposium Programme have been delivered, allow me on behalf of the Presidium of the present Symposium to extend our deep gratitude to all the authors who presented informative and interesting papers on the hydrology of marsh-ridden areas.

The Symposium Programme envisaged the presentation and discussion of papers on the following topics covering the general problems of the hydrology of marsh-ridden areas:

- methods and results of investigations of hydrology, climate, geology and hydrogeology, soil and flora of marsh-ridden areas of the temperate zone;

- methods and results of water-budget calculations (analytical procedures, simulation, etc.), their use for design of drainage and irrigation systems of the temperate zone;

- transformation of the water-regime over marsh-ridden areas in the temperate zone by modern reclamation techniques and the forecasting of its effect on hydrometeorological conditions of the environment.

Representatives of thirteen countries have taken part in the discussions. The total number of papers presented is 49.

Allow me briefly to summarize the results of the present Symposium.

The subject of the Symposium is extremely important to-day and as far as I know this is the first Symposium sponsored by UNESCO under the IHD Programme on this branch of hydrological science.

Construction of reclamation systems and transformation of natural conditions over marsh-ridden areas have been developing in the countries of the temperate zone. However, investigations of the hydrology of these areas are not ahead of practical engineering but rather behind it in the solution of many urgent problems.

The aim of the Symposium was to fill this gap, to exchange experience of research work and to outline the aims of further investigations. The Symposium seems to have been a success and to have achieved its objectives.

In the informative papers and discussions particular aspects of climatic, hydrogeological, edaphic, botanic and water-balance investigations of marsh-ridden areas have been treated, as well as the methodology of such investigations and the application of the results to the design, construction and operation of reclamation systems.

The papers on the first topic have been presented by the representatives of the United States of America, Federal Republic of Germany, Polish People's Republic, German Democratic Republic, the Netherlands, Socialist Republic of Roumania, the USSR and the Byelorussian SSR.

The papers presented have shown that comprehensive large-scale investigations on marsh-ridden areas are carried out in the countries of the temperate zone. These investigations have revealed that the main sources of excess moisture are precipitation, ground and artesian water.

For different parts of the temperate zone, maps have been compiled to show the main sources of moisture. Special attention has been paid to the unsaturated zone and to the balance of underground and surface water as a basis for current and long-term forecasts of changes in the regime and balance.

Changes in the hydrophysical and thermal properties of peat and mineral soils resulting from drainage have been investigated. In various countries physical and mathematical simulation techniques have been successfully developed both for the determination of different hydrological parameters and for the estimation of the effects of drainage on the environment. However, hydrometric and hydrometeorological stations which are of vital significance for land reclamation on a local and regional scale have been neglected in a number of regions of the temperate zone.

In some papers, it was stated that more intensive investigations in the field of reclamation hydrology and geological engineering are desirable with a view to the design and operation of drainage systems. Special attention should be paid to the balance and regime of groundwater and to its forecasting on drained lands and adjacent areas.

The papers presented on the second topic deal with the water balance of marsh-ridden areas and application of water-balance calculations to designing drainage-irrigation systems. Water-balance methods have been widely adopted in research, designs and forecasts concerned with undertakings of water economy on water logged lands, in regulation of soil moisture content and in investigations of soil properties and soil-formation processes.

Some papers presented to the Symposium contain valuable information on the values of water-balance elements under regional conditions of the USSR, Ireland, the Hungarian People's Republic, the German Democratic Republic, and the Federal Republic of Germany. Useful methods of investigation and simulation of hydrological processes are described. The discussions have revealed that water-balance methods are highly promising. They are concerned with a wide range of little studied matters under the great variety of natural conditions of marsh-ridden areas and require further development and extension of their use. In view of the above, the further objectives of the hydrology of marsh-ridden areas may be formulated as follows:

- study of moisture exchange between groundwater and the unsaturated zone and between individual soil layers;
- investigation of moisture exchange between the soil and a plant in the root zone under various edaphic, vegetal and

hydrological conditions to envisage optimal conditions for moisture supply to agricultural crops;

- detailed study of particular elements of the runoff from small marsh-ridden areas which have not been adequately investigated as yet;

- development of comprehensive investigations of surface and groundwater in marsh-ridden areas in search of the relationships between them;

- detailed study of relationships of formation and values of water-balance elements in marsh-ridden areas with the aim of improving reclamation methods and other transformations of nature.

Fourteen papers have been presented on the third main topic of the Symposium which is concerned with forecasts of the changes in a water regime of different regions due to reclamation (USSR, the Netherlands, Finland, Federal Republic of Germany and USA). This is the most difficult problem in hydrology since in the majority of cases, the available experimental data refer to small reclamation systems whose area is insignificant in relation to the total area affected by reclamation works.

In the papers presented, interesting results of investigations are given which contain valuable information on changes in the regime of the runoff and water level and suggestions of the reasons for these changes. A large group of the papers is concerned with forecasts for the Polessie Lowland which covers some areas in Byelorussia and the Ukrainian SSR and where mires and marsh-ridden lands occupy more than 30 per cent of the area. The papers are based on the treatment of vast quantities of experimental material and make important forecasts of the changed water regime.

Interesting information has been presented on forecasting the water regime due to forest drainage in Finland and on the groundwater regime and climatic effects on crop yields in the Netherlands.

The papers on the effects of leaching of mineral fertilizers from drained soils on water quality in rivers (Federal Republic of Germany) and of swamps on streamflow distribution in small watersheds (United States of America) are of great interest.

An analysis of the papers presented leads to the conclusion that in temperate countries investigation of the water regime on lands under reclamation and accumulation of experimental data for forecasting of the changes are considered to be of great significance.

The most important requirements in the field are probably an adequate theoretical understanding of the physical processes involved as well as physical and mathematical simulation of all the processes affecting the hydrological cycle which would allow the estimation of changed parameters of the water regime. Establishment of special experimental plots and natural catchment experiments equipped with modern automatic devices for investigating the balance and regime of surface and ground water is extremely urgent.

Exchange of information and experience in the field of scientific and practical research, in maintenance of drainage systems and two-way control of the water regime in water logged lands through National Committees of the IHD is highly desirable. This will assist in the solution of complicated problems of hydrological forecasting of changes in the water regime because of reclamation and other types of human activity.

PARTICIPANTS

Artyushevsky, G.N., BSSR Council of Ministers, Minsk, BSSR.
Balzariavichus, P.I., Dr., Institute of Water Engineering and Reclamation, Kedaynyai, USSR.
Bauer, L., Prof., Institute of Agriculture, Halle, GDR.
Bavina, L.G., Dr., State Hydrological Institute, Leningrad, USSR.
Belotserkovskaya, O.A., Dr., Byelorussian Research Forestry Institute, Gomel, BSSR.
Berdichivets, L.I., Ministry of Reclamation and Water Economy of the BSSR, Minsk, BSSR.
Blyacher, I.S., Byelorussian State Design Institute of Water Economy, Minsk, BSSR.
Bochin, N.A., UNESCO, Paris.
Boelter, Don.H., Forest Service, U.S. Department of Agriculture, Grand Rapids, Minnesota, USA.
Bogomolov, G.V., Academician, Institute of Geochemistry and Geophysics, BSSR Academy of Sciences, Minsk, BSSR.
Bogomolov, Yu.G., Dr., Moscow State University, Moscow, USSR.
Boikova, K.G., Dr., Far East Research Institute of Water Engineering and Reclamation, Vladivostok, USSR.
Bratchikov, I.T., Hydrometeorological Centre, Minsk, BSSR.
Budyka, S.H., Prof., Presidium, BSSR Academy of Sciences, Minsk, BSSR.
Bulavko, A.G., Dr., Central Research Institute on Complex Use of Water Resources, Minsk, BSSR.
Dooge, J., Prof., University College Dublin, Dublin, Ireland.
Druzhinin, N.I., Prof., North Research Institute of Water Engineering and Reclamation, Leningrad, USSR.
Dyachkov, A.B., Dr., State Design Institute for Water Economy, Moscow, USSR.
Eggelsmann, R., Peat Soil Research Institute, Bremen, FRG.
Enea, I., Dr., Meteorological and Hydrological Research Institute, Bucharest, Romania.
Ermolenko, V.V., Water Economic Construction Chief Administration for Polessie, Minsk, BSSR.
Fedorova, N.M., Soil Institute, Moscow, USSR.
Feher, F., Research Institute of Water Resources, Budapest, Hungary.
Fomin, V.R., Byelorussian State Design Institute of Water Economy, Minsk, BSSR.
Gartsman, I.N., Dr., Far East Research Institute of Water Engineering and Reclamation, Vladivostok, USSR.
Gerasimov, V.G., Prof., State University, Gomel, BSSR.
Glazacheva, L.I., Dr., Latvian State University, Riga, USSR.
Glomozda, A.A., Hydrometeorological Service Administration, Minsk, BSSR.
Glybchak, I.P., Ukrainian Institute of Water Economy, Kiev, Ukrainian SSR.
Goffeng, G., Agricultural University of Norway, Aas, Norway.
Goldberg, M.A., Dr., Byelorussian Regional Hydrometeorological Centre, Minsk, BSSR.

Gorbutovich, G.D., Dr., Presidium, BSSR Academy of Science, Minsk, BSSR.
Heikurainen, L., Prof., Academy of Sciences, Tapida, Finland.
Iozopaitis, A.V., Water Engineering and Reclamation, Kedainyai, USSR.
Ivankin, A.G., Water Economic Construction Chief Administration for Polessie, Minsk, BSSR.
Ivanov, K.E., Prof., State University, Leningrad, USSR.
Ivitsky, A.I., Byelorussian Scientific Research Institute for Reclamation and Water Economy, Minsk, BSSR.
Kaluzhny, I.L., State Hydrological Institute, Leningrad, USSR.
Karasev, M.S., Dr., Far East Research Institute for Water Engineering and Reclamation, Vladivostok, USSR.
Kats, D.M., Prof., All-Union Research Institute for Water Engineering and Reclamation, Moscow, USSR.
Khommik, K.T., Dr., Estonian Research Institute of Agriculture and Reclamation, Saku, Estonian SSR.
Kienitz, G., Dr., Research Institute of Water Resources, Budapest, Hungary.
Kiselev, P.A., Prof., All-Union Research Institute of Hydrogeology and Geological Engineering, Moscow, USSR.
Kordukova, N.V., North Research Institute of Water Engineering and Reclamation, Leningrad, USSR.
Kovacs, G., Dr., Research Institute of Water Resources, Budapest, Hungary.
Kubyshkin, G.P., Dr., Institute of Water Engineering and Reclamation, Kiev, Ukrainian SSR.
Lavrov, A.P., Dr., Byelorussian Geological Research Institute, Minsk, BSSR.
Levanovsky, L.B., Dr., Ministry of Reclamation and Water Economy of the RSFSR, Moscow, USSR.
Livshits, I.M., Dr., Byelorussian Polytechnical Institute, Minsk, BSSR.
Lobanok, V.E., Chairman, Organizing Committee, BSSR Council of Ministers, Minsk, BSSR.
Lundin, K.P., Dr., Byelorussian Research Institute for Reclamation and Water Economy, Minsk, BSSR.
Lvovich, M.I., Prof., Geographical Institute, USSR Academy of Sciences, Moscow, USSR.
Madissoon, G.Kh., Estonian Research Institute for Agriculture and Reclamation, Saku, USSR.
Maslov, B.S., Dr., All-Union Research Institute for Water Engineering and Reclamation, Moscow, USSR.
Mikulski, Z., Prof., Warsaw University, Warsaw, Poland.
Murashko, M.G., Dr., Central Research Institute of Complex Use of Water Resources, Minsk, BSSR.
Mustonen, S., Dr., National Board of Waters, Helsinki, Finland.
Narodetskaya, R.Ya., Dr., RSFSR State Design Institute for Water Economy, Moscow, USSR.
Nesterov, E.A., Journal of Water Engineering and Reclamation, Moscow, USSR.
Nesterovich, N.D., Academician, Vice-President of the BSSR Academy of Sciences, Minsk, BSSR.
Nikolsky, Yu.N., Dr., Water Reclamation Institute, Moscow, USSR.

Novikov, S.N., Dr., State Hydrological Institute, Leningrad, USSR.
Orlova, V.P., "Kolos" Publishing House, Moscow, USSR.
Pavluchuk, V.I., Ministry of Reclamation and Water Management, Minsk, BSSR.
Perekhrest, S.M., Prof., Ukrainian Academy of Sciences, Kiev, Ukrainian SSR.
Pkhakadze, P.S., Dr., Georgian Research Institute of Water Engineering and Reclamation, Tbilisi, USSR.
Podzharov, V.K., Byelorussian Research Forestry Institute, Gomel, BSSR.
Rieksts, I.A., Latvian State Design Institute of Water Economy, Riga, USSR.
Romanenko, A.M., Dr., Central Research Institute on Complex Use of Water Resources, Minsk, BSSR.
Rusinov, I.F., Dr., West-Siberian Branch of All-Union Research Institute for Water Engineering and Reclamation, Tyumen, USSR.
Sadikova, V.S., RSFSR State Design Institute for Water Economy, Moscow, USSR.
Schendel, U., Prof., University of Kiel, Kiel, FRG.
Scholtz, A., Dr., Research Institute of Marshes, Potsdam, GDR.
Sedova, V.K., All-Union Research Institute of Water Engineering and Reclamation, Moscow, USSR.
Seleznev, S.M., Chief Administration of Hydrometeorological Service, Moscow, USSR.
Shebeko, V.F., Prof., Byelorussian Research Institute of Reclamation and Water Economy, Minsk, BSSR.
Shtepa, B.S., Dr., Ministry of Reclamation and Water Economy of USSR, Moscow, USSR.
Shimko, K.I., Dr., Byelorussian Polytechnical Institute, Minsk, BSSR.
Shkinkis, C.N., Dr., Latvian Research Institute of Water Engineering and Reclamation, Elgova, USSR.
Sholtes, I., Water Economy State Department, Budapest, Hungary.
Smolsky, N.V., Academician, BSSR Academy of Sciences, Central Botanical Gardens, BSSR Academy of Sciences, Minsk, BSSR.
Sviklis, P.B., Dr., Latvian Research Institute for Water Engineering and Reclamation, Elgava, USSR.
Sych, I., Water Economy State Department, Budapest, Hungary.
Tereshchenko, K.P., Dr., Hydrology and Reclamation Expedition, Lvov, Ukrainian SSR.
Tyulpanov, A.I., Dr., State Planning Committee, Minsk, BSSR.
Van der Molen, W.H., Prof., Agricultural University, Wageningen, The Netherlands.
Velichka, I.I., Ministry of Reclamation and Water Economy, Latvian SSR, Vilnus, USSR.
Vishnevsky, P.F., Dr., Ukrainian Hydrometeorological Research Institute, Kiev, Ukrainian SSR.
Vompersky, S.E., Dr., Soil Laboratory, USSR Academy of Sciences, Moscow Region, USSR.
Vorontsov, A.I., State Committee of the BSSR Council of Ministers on Environment Protection, Minsk, BSSR.
Wartena, L., Dr., N. V. Heidemaatschappy, Research Department Zelhem, the Netherlands.
Yankovsky, K.F., Central Research Institute on Complex Use of Water Resources, Minsk, BSSR.

Zaidelman, F.R., Dr., Moscow State University, Moscow, USSR.
Zaika, V.I., Ukrainian State Design Institute for Water Economy, Kiev, USSR.
Zakrzhevsky, P.I., Dr., Byelorussian Research Institute for Reclamation and Water Economy, Minsk, BSSR.
Zhelobaev, A.A., Dr., State Design Institute for Water Economy, Moscow, USSR.
Zivert, A.A., Dr., Latvian State Design Institute for Water Economy, Riga, USSR.
Zubets, B.M., Prof., Byelorussian Research Institute for Reclamation and Water Economy, Minsk, BSSR.
Zuidema, F., Yesselmeerpolders Development Authorities Science Department, Dronten, the Netherlands.

1. The use of analog and digital computers in hydrology: Proceedings of the Tucson Symposium, June 1966 / L'utilisation des calculatrices analogiques et des ordinateurs en hydrologie: Actes du colloque de Tucson, juin 1966. Vol. 1 & 2. *Co-edition IAHS-Unesco / Coédition AISH-Unesco.*
2. Water in the unsaturated zone: Proceedings of the Wageningen Symposium, June 1967 / L'eau dans la zone non saturée: Actes du symposium de Wageningen, juin 1967. Edited by / Édité par P.E. Rijtema & H. Wassink. Vol. 1 & 2. *Co-edition IAHS-Unesco / Coédition AISH-Unesco.*
3. Floods and their computation: Proceedings of the Leningrad Symposium, August 1967 / Les crues et leur évaluation : Actes du colloque de Leningrad, août 1967. Vol. 1 & 2. *Co-edition IAHS-Unesco-WMO / Coédition AISH-Unesco-OMM.*
4. Representative and experimental basins: An international guide for research and practice. Edited by C. Toebes and V. Ouryvaev. *Published by Unesco.*
4. Les bassins représentatifs et expérimentaux : Guide international des pratiques en matière de recherche. Publié sous la direction de C. Toebes et V. Ouryvaev. *Publié par l'Unesco.*
5. *Discharge of selected rivers of the world / Débit de certain cours d'eau du monde. *Published by Unesco / Publié par l'Unesco.*
   Vol. I: General and régime characteristics of stations selected / Caractéristiques générales et caractéristiques du régime des stations choisies.
   Vol. II: Monthly and annual discharges recorded at various selected stations (from start of observations up to 1964) / Débits mensuels et annuels enregistrés en diverses stations sélectionnées (de l'origine des observations à l'année 1964).
   Vol. III: Mean monthly and extreme discharges (1965-1969) / Débits mensuels moyens et débits extrêmes (1965-1969).
6. List of International Hydrological Decade Stations of the world / Liste des stations de la Décennie hydrologique internationale existant dans le monde. *Published by Unesco / Publié par l'Unesco.*
7. Ground-water studies: An international guide for practice. Edited by R. Brown, J. Ineson, V. Konoplyantsev and V. Kovalevski. (Will also appear in French, Russian and Spanish / Paraîtra également en espagnol, en français et en russe.)
8. Land subsidence: Proceedings of the Tokyo Symposium, September 1969 / Affaisement du sol : Actes du colloque de Tokyo, septembre 1969. Vol. 1 & 2. *Co-edition IAHS-Unesco / Coédition AISH-Unesco.*
9. Hydrology of deltas: Proceedings of the Bucharest Symposium, May 1969 / Hydrologie des deltas : Actes du colloque de Bucarest, mai 1969. Vol. 1 & 2. *Co-edition IAHS-Unesco / Coédition AISH-Unesco.*
10. Status and trends of research in hydrology / Bilan et tendances de la recherche en hydrologie. *Published by Unesco / Publié par l'Unesco.*
11. World water balance: Proceedings of the Reading Symposium, July 1970 / Bilan hydrique mondial : Actes du colloque de Reading, juillet 1970. Vol. 1-3. *Co-edition IAHS-Unesco-WMO / Coédition AISH-Unesco-OMM.*
12. Results of research on representative and experimental basins: Proceedings of the Wellington Symposium, December 1970 / Résultats de recherches sur les bassins représentatifs et expérimentaux : Actes du colloque de Wellington, décembre 1970. Vol. 1 & 2. *Co-edition IAHS-Unesco / Coédition AISH-Unesco.*
13. Hydrometry: Proceedings of the Koblenz Symposium, September 1970 / Hydrométrie: Actes du colloque de Coblence, septembre 1970. *Co-edition IAHS-Unesco-WMO / Coédition AISH Unesco-OMM.*
14. Hydrologic information systems. *Co-edition Unesco-WMO.*
15. Mathematical models in hydrology: Proceedings of the Warsaw Symposium, July 1971 / Les modèles mathématiques en hydrologie : Actes du colloque de Varsovie, juillet 1971. Vol. 1-3. *Co-edition IAHS-Unesco-WMO / Coédition AISH-Unesco-OMM.*
16. Design of water resources projects with inadequate data: Proceedings of the Madrid Symposium, June 1973 / Élaboration des projets d'utilisation des resources en eau sans données

suffisantes : Actes du colloque de Madrid, juin 1973. Vol. 1-3. *Co-edition Unesco-WMO-IAHS / Coédition Unesco-OMM-AISH.*

17. Methods for water balance computations. An international guide for research and practice. *Published by the Unesco Press.*
18. Hydrological effects of urbanization. Report of the Sub-group on the Effects of Urbanization on the Hydrological Environment. *Published by the Unesco Press.*
19. Hydrology of marsh-ridden areas. Proceedings of the Minsk Symposium, June 1972. *Published by the Unesco Press.*

[A.100] SC.74/XX.19/A.